Handbook of Model-Based Systems Engineering

Azad M. Madni • Norman Augustine
Editors-in-Chief

Michael Sievers
Associate Editor

Handbook of Model-Based Systems Engineering

Volume 1

With 645 Figures and 61 Tables

Editors-in-Chief
Azad M. Madni
University of Southern California
Los Angeles, CA, USA

Norman Augustine
Lockheed Martin (United States)
Bethesda, MD, USA

Associate Editor
Michael Sievers
University of Southern California
Los Angeles, CA, USA

ISBN 978-3-030-93581-8   ISBN 978-3-030-93582-5 (eBook)
https://doi.org/10.1007/978-3-030-93582-5

© Springer Nature Switzerland AG 2023

This work is subject to copyright. All rights are reserved by the Publisher, whether the whole or part of the material is concerned, specifically the rights of translation, reprinting, reuse of illustrations, recitation, broadcasting, reproduction on microfilms or in any other physical way, and transmission or information storage and retrieval, electronic adaptation, computer software, or by similar or dissimilar methodology now known or hereafter developed.

The use of general descriptive names, registered names, trademarks, service marks, etc. in this publication does not imply, even in the absence of a specific statement, that such names are exempt from the relevant protective laws and regulations and therefore free for general use.

The publisher, the authors, and the editors are safe to assume that the advice and information in this book are believed to be true and accurate at the date of publication. Neither the publisher nor the authors or the editors give a warranty, expressed or implied, with respect to the material contained herein or for any errors or omissions that may have been made. The publisher remains neutral with regard to jurisdictional claims in published maps and institutional affiliations.

This Springer imprint is published by the registered company Springer Nature Switzerland AG.
The registered company address is: Gewerbestrasse 11, 6330 Cham, Switzerland

Foreword

As an engineering test pilot, astronaut, and especially during my tenure as the 12th NASA Administrator, it became quite apparent to me that a critical missing link in many of our projects and programs was a thorough, practical understanding of systems engineering and its critical importance in design and development of complex systems. Although complex systems have existed for many years – Apollo, the electric grid, the air traffic control system, the Internet – they could generally be designed and operated using well-established engineering practices. However, as systems have become increasingly complex and intertwined, these prior approaches, such as using interface documents to control the connectivity of individually designed components, are no longer sufficient. For example, they may not adequately identify and control the unintended consequences that can suddenly arise when one system element that has incorrectly been believed to be isolated from other system components actually proves to impact the functioning of these other components. Add to this, the unpredictability of human interactions that can take place in modern systems and the possibilities and uncertainties far exceed the ability of humans to process.

Enter Model-Based Systems Engineering (MBSE). Therein a construct is created of an entire system that enables the study of interactions throughout the system and discloses potential outcomes. It permits designers, operators, and maintainers of the system to work in concert from a common representation of the system. Yet, despite its generally recognized benefits, MBSE, when improperly scoped and applied, can, and on occasion has, led to significant inefficiencies and even system failures.

In response to such concerns, the Editors-in-Chief of this handbook have assembled a significant and timely collection of contributions from leading MBSE practitioners across the globe. The stated intent of the handbook is to clarify what MBSE is, where it stands today, and how it can be employed as a cost-effective systems engineering methodology with software tools to aid in modeling, analysis, and decision support. While there are several books and publications that focus on MBSE mechanics and languages, this is the first volume to focus on the methodological and practicable aspects of conducting systems engineering within an MBSE rubric. The handbook contains case studies that reflect the collective wisdom and experience of internationally known experts – and their lessons learned.

As is widely recognized, many of today's systems are far more complex, and their operational environments much more challenging, than ever before. Contrast the spacecraft flown by NASA 10–15 years ago with those flown today, with the latter consisting of multiple interacting computers, instruments, software, and mechanisms that must survive decades in often extremely harsh space environments. Comparing the complexity of the Mariner spacecraft of the 1960s with the James Webb Space Telescope of today is like comparing the computing power and memory capacity of a 1960's IBM 360/91 mainframe computer (once the world's most powerful computer) with a modern desktop computer that operates an order of magnitude faster and can readily accommodate more than twice the main memory. The self-driving vehicle industry has a similar tale to tell. There are as many as 500 million lines of code operating in a self-driving automobile. Even a modern, human-driven vehicle executes as many as 100 million lines of software code. In the same vein, cloud systems can have thousands of clients, servers, and data stores. Streaming services, the Internet of Things, distributed human-cyber physical systems, and fuel-efficient aircraft are continually posing challenges that press the limits of human perception and cognition. Today MBSE is increasingly recognized as essential for meeting modern engineering demands, staying ahead of the competition, and assuring the evolution and survival of enterprises.

This handbook spans topics that range from foundational MBSE concepts to real-world experiences. The sections are independent and can be read in an arbitrary order. Within each section, each individual chapter begins with a domain orientation or tutorial, followed by a discussion of practical "how-to" examples. Readers can peruse the chapters with synergy in content in a sequence of their choosing because the chapters are cross-referenced to related chapters. This feature is especially useful for practitioners and students working on MBSE projects who are also interested in foundational concepts, applications, and lessons learned from prior work. In addition, the handbook indicates how MBSE relates to other disciplines, such as digital engineering and social media. One chapter discusses transdisciplinary systems engineering that has expanded the frontiers of MBSE and provides powerful new approaches to problem solving.

In conclusion, this handbook is a working guide into the many forms, uses, and means of pursuing MBSE projects. It inspires the reader to delve into the exciting and rapidly advancing world of MBSE. Readers will learn to apply MBSE to real-world problems with unique informational and resource constraints, as well as to systems at different scales. It is a must-read for practitioners and students alike.

Major General USMC (Ret) Charles Bolden Jr.
NASA Astronaut
12th NASA Administrator

Preface

We are systems engineers first and foremost. We are also storytellers. And we have been involved with systems engineering for more than five decades. Over that period, we have seen spectacular successes and failures that can be attributed to proper use of systems engineering and its misuse and have learned valuable lessons from them. In its early years, systems engineering was somewhat an orphan, with universities largely guiding and tracking engineers into a single practice, such as electrical engineering, chemical engineering, or mechanical engineering. Yet, most real-world problems that are encountered today include transdisciplinary aspects – that is, "systems engineering." We have seen systems engineering evolve over the past few decades to where we see transformative advances in the field paced by Model-Based Systems Engineering (MBSE), digital twin technology, AI and machine learning, ontologies, and formal methods to prove model correctness.

So how do these different advances come together? Is MBSE the rubric to bring these advances together? How can we capitalize on these advances in the work we are doing in our respective and highly varied organizations? These are but a few of the questions that we aim to address in this handbook, which is a "living document" that allows contributors to add new findings and insights that can advance the practice of systems engineering as it continues to evolve.

We, the Editors-in-Chief, have been working in systems engineering research, practice, and education for decades. Norm Augustine served as chairman and CEO of Lockheed Martin Corp., one of the largest aerospace firms in the world. He led the company to new heights with his innovative blend of systems engineering, systems management, and leadership skills. Previously, he was a faculty member at Princeton University, served as the Acting Secretary of the Army, and chaired the Advisory Committee on the Future of the US Space Program, known as the Augustine Committee. He was the Chairman of the National Academy of Engineering, of the Defense Science Board, and former President of American Institute of Aeronautics and Astronautics. A member of the National Academy of Engineering and National Academy of Science, he led numerous committees such as Rising Above the Gathering Storm, Revisited: Rapidly Approaching Category 5. He has received honorary degrees from 35 universities, is the recipient of the prestigious National Medal of Technology and Innovation, and has been recognized with numerous other international awards and honors. He is the author of *Augustine's Laws*, a highly acclaimed book that sets forth 52 laws that cover engineering and management in an entertaining and informative fashion.

Azad Madni is University Professor and holder of the Northrop Grumman Foundation Fred O'Green Chair in Engineering at the University of Southern California. He is the Executive Director of USC's flagship Systems Architecting and Engineering Program, founding director of the Distributed Autonomy and Intelligent Systems Laboratory, and faculty affiliate of the Ginsburg Institute of Medical and Biomedical Therapeutics. He is the founder and CEO of Intelligent Systems Technology, Inc., a successful hi-tech company specializing in model-based methods for complex systems engineering, education, and training. He pioneered transdisciplinary systems engineering as a next-generation systems engineering discipline and a means to address complex sociotechnical problems that appear intractable when viewed solely through an engineering lens. He also wrote an award-winning book *Transdisciplinary Systems Engineering: Exploiting Convergence in a Hyperconnected World* and is the co-author of *Tradeoff Decisions in System Design*. A member of the National Academy of Engineering and Life Fellow of IEEE, he received the NAE's Gordon Prize for Innovation in Engineering and Technology Education, and the IEEE Simon Ramo Medal in 2023, for exceptional achievement in systems engineering and systems science. Previously, he served as the Executive VP and CTO of a public company and head of the modeling and simulation group on NASA's Space Shuttle Program at Rockwell International, where he pioneered a model-based testing approach which saved substantial sums in physical testing of the Shuttle navigation system. An elected fellow of ten professional societies, he has served as Principal Investigator on 97 R&D projects totaling over $100M.

Contributors to this book are MBSE practitioners selected from experts in the field from several different countries. As MBSE gathered steam in the engineering community worldwide, several questions arose – along with some misconceptions about MBSE. This recognition spurred interest in developing a handbook that would take a broad and deep look at the use of models in doing serious systems engineering. Accordingly, we invited top practitioners to contribute their understanding of MBSE and address areas that cover fundamentals of MBSE as well as how MBSE was successfully applied to real-world problems. At the time, there was a fragmented body of knowledge in MBSE and a pressing need to consolidate MBSE concepts, methods, theories, and practices with real-world examples. This handbook attempts to consolidate many of these contributions by mixing and matching various areas and emphasizing what it takes to derive real value from MBSE. The audience for this book is practitioners, students, and educators seeking a coherent presentation of MBSE concepts, methodologies, assumptions, and illustrative examples, so as to add to the successes of the field and eliminate the failures that have sometimes come from a lack of attention to it.

Los Angeles, USA
Bethesda, USA
July 2023

Azad M. Madni
Norman Augustine
Editors-in-Chief

Acknowledgment

We would like to acknowledge Shatad Purohit, who is pursuing his doctorate under Professor Madni's guidance, for his responsible management of the handbook project and timely interactions with the contributors.

Contents

Volume 1

Part I Introduction 1

1 **Introduction to the Handbook** 3
   Azad M. Madni and Norman Augustine

Part II MBSE Foundations 13

2 **Semantics, Metamodels, and Ontologies** 15
   Michael Sievers

3 **MBSE Methodologies** 47
   Jeff A. Estefan and Tim Weilkiens

4 **SysML State of the Art** 87
   B. Bagdatli, S. Cimtalay, T. Fields, E. Garcia, and R. Peak

5 **Role of Decision Analysis in MBSE** 119
   Gregory S. Parnell, Nicholas J. Shallcross, Eric A. Specking,
   Edward A. Pohl, and Matt Phillips

6 **Pattern-Based Methods and MBSE** 151
   William D. Schindel

7 **Overarching Process for Systems Engineering and Design** 195
   A. Terry Bahill and Azad M. Madni

8 **Problem Framing: Identifying the Right Models for the Job** . 257
   James N. Martin

Part III Technical and Management Aspects of MBSE 287

9 **Model-Based System Architecting and Decision-Making** 289
   Yaroslav Menshenin, Yaniv Mordecai, Edward F. Crawley, and
   Bruce G. Cameron

10	Adoption of MBSE in an Organization	331
	Tim Weilkiens	
11	Model-Based Requirements	349
	Alejandro Salado	
12	Modeling Hardware and Software Integration by an Advanced Digital Twin for Cyber-physical Systems: Applied to the Automotive Domain	379
	S. Kriebel, M. Markthaler, C. Granrath, J. Richenhagen, and B. Rumpe	
13	Integrating Heterogenous Models	417
	Michael J. Pennock	
14	Improving System Architecture Decisions by Integrating Human System Integration Extensions into Model-Based Systems Engineering	441
	D. W. Orellana	
15	Model-Based Human Systems Integration	471
	Guy André Boy	
16	Model-Based Hardware-Software Integration	501
	Joe Cesena	

Part IV Quality Attributes Tradeoffs in MBSE 525

17	Exploiting Digital Twins in MBSE to Enhance System Modeling and Life Cycle Coverage	527
	Azad M. Madni, S. Purohit, and C. C. Madni	
18	Model-Based Mission Assurance/Model-Based Reliability, Availability, Maintainability, and Safety (RAMS)	549
	Luca Boggero, Marco Fioriti, Giuseppa Donelli, and Pier Davide Ciampa	
19	MBSE in Architecture Design Space Exploration	589
	J. H. Bussemaker and Pier Davide Ciampa	

Part V Digital Engineering and MBSE 631

20	Digital Twin: Key Enabler and Complement to Model-Based Systems Engineering	633
	Azad M. Madni and C. C. Madni	
21	Developing Industry 4 Systems with OPM ISO 19450 Augmented with MAXIM	655
	D. Dori	

| 22 | MBSE Testbed for Unmanned Vehicles | 675 |

A. M. Madni and D. Erwin

| 23 | Transitioning from Observation to Patterns: A Real-World Example | 705 |

S. Russell, B. Kruse, R. Cloutier, and D. Verma

Volume 2

Part VI   MBSE for System Acquisition and Management **723**

| 24 | MBSE for Acquisition | 725 |

R. A. Noguchi and R. J. Minnichelli

| 25 | Managing Model-Based Systems Engineering Efforts | 753 |

Mark L. McKelvin

| 26 | MBSE Methods for Inheritance and Design Reuse | 783 |

A. E. Trujillo and A. M. Madni

| 27 | Model Interoperability | 815 |

Tim Weilkiens

| 28 | A Reuse Framework for Mode-Based Systems Engineering | 833 |

Gan Wang

| 29 | MBSE Mission Assurance | 861 |

J. S. Fant and R. G. Pettit

| 30 | Conceptual Design Support by MBSE: Established Best Practices | 895 |

S. Shoshany-Tavory, E. Peleg, and A. Zonnenshain

Part VII   Case Studies **923**

| 31 | Ontological Metamodeling and Analysis Using openCAESAR | 925 |

D. A. Wagner, M. Chodas, M. Elaasar, J. S. Jenkins, and N. Rouquette

| 32 | MBSE Validation and Verification | 955 |

Karen Gundy-Burlet

| 33 | MBSE for System-of-Systems | 987 |

Daniel DeLaurentis, Ali Raz, and Cesare Guariniello

| 34 | NSOSA: A Case Study in Early Phase Architecting | 1017 |

M. W. Maier

| 35 | Cybersecurity Systems Modeling: An Automotive System Case Study 1045
Mark L. McKelvin |

| 36 | Assistive Technologies for Disabled and Older Adults 1079
William B. Rouse and Dennis K. McBride |

| 37 | Multi-model-Based Decision Support in Pandemic Management 1105
A. M. Madni, Norman Augustine, C. C. Madni, and Michael Sievers |

| 38 | Semantic Modeling for Power Management Using CAESAR 1135
D. A. Wagner, M. Chodas, M. Elaasar, J. S. Jenkins, and N. Rouquette |

| 39 | Modeling Trust and Reputation in Multiagent Systems 1153
Michael Sievers |

| 40 | Modeling and Simulation Through the Metamodeling Perspective: The Case of the Discrete Event System Specification 1189
María J. Blas and Silvio Gonnet |

Part VIII   Future Outlook **1229**

| 41 | Exploiting Transdisciplinarity in MBSE to Enhance Stakeholder Participation and Increase System Life Cycle Coverage 1231
Azad M. Madni |

| 42 | Toward an Engineering 3.0 1253
Norman Augustine |

| 43 | Category Theory 1259
S. Breiner, E. Subrahmanian, and R. D. Sriram |

| 44 | Perspectives on SE, MBSE, and Digital Engineering: Road to a Digital Enterprise 1301
H. Stoewer |

Index 1339

About the Editors-in-Chief

Azad M. Madni is a University Professor of Astronautics, Aerospace, and Mechanical Engineering in the University of Southern California's Viterbi School of Engineering. The University Professor designation honors the university's most accomplished, multidisciplinary faculty, who have significant achievements in multiple technical disciplines. He is the holder of the Northrop Grumman Fred O'Green Chair in Engineering and the Executive Director of USC's Systems Architecting and Engineering Program. He is the Founding Director of USC's Distributed Autonomy and Intelligent Systems Laboratory. He is also a Professor in USC's Keck School of Medicine and Rossier School of Education. He is a faculty affiliate of the Keck School's Ginsberg Institute for Biomedical Therapeutics. He is the founder and CEO of Intelligent Systems Technology, Inc., a hi-tech company specializing in transdisciplinary model-based approaches for tackling complex sociotechnical systems problems. He is a member of the Phi Kappa Phi honor society and Omega Alpha Association, an international systems engineering honor society.

Prof. Madni is a member of the National Academy of Engineering and the recipient of the prestigious 2023 NAE Gordon Prize for Innovation in Engineering and Technology Education. A Life Fellow of IEEE, he is also the recipient of the 2023 IEEE Simo Ramo Medal. He is an Honorary Member of ASME and a Life Fellow/Fellow of AAAS, AIAA, INCOSE, IISE, AIMBE, IETE, AAIA, SDPS, and the Washington Academy of Sciences. He is the recipient of approximately 80 prestigious international and national awards from 11 different professional societies. These include the 2019 IEEE AESS Pioneer Award and the 2011 INCOSE Pioneer Award. He has 400+ publications comprising authored and edited books, book chapters, journal articles, peer-reviewed conference publications, and research reports. He has given more than 75 keynotes and invited talks in international conferences and workshops. He is a member of two NAE Lifetime Giving Societies: the Marie Curie Society and the Albert Einstein Society.

He pioneered the field of transdisciplinary systems engineering and wrote an award-winning book, *Transdisciplinary Systems Engineering: Exploiting Convergence in a Hyperconnected World* (Springer, 2018), which presented the founding principles of transdisciplinary systems engineering. He is also the creator of TRASEE™, a new engineering education paradigm based on transdisciplinary systems engineering principles. He is also the co-author of *Tradeoff Decisions in System Design* (Springer, 2016), Co-Editor-in-Chief of the Springer series Systems Engineering Research, and co-author of *3 Ds of Deep Learning – Design, Development, and Deployment* (Springer, 2023).

He transformed USC's Systems Architecting and Engineering Program based on TRASEE and provided a blueprint for other graduate engineering programs to follow. Under his leadership, the program has graduated 3200+ students and is recognized as a top graduate engineering program in the country. He has served as Principal Investigator on 97 R&D contracts and grants totaling more than $100M.

Previously, he was the Executive Vice President for R&D and the Chief Technology Officer of Perceptronics Inc., a simulation-based training and AI company that went public in 1982. Prior to that, as a lead simulation engineer at Rockwell International on NASA's Space Shuttle Program, he led the development of a model-based testing approach that generated substantial savings in navigation system performance testing for the Shuttle Program.

He received his Ph.D., M.S., and B.S. in Engineering from UCLA with a major in Engineering Systems and minors in Computer Methodology and AI. He is also a graduate of AEA/Stanford Executive Institute.

About the Editors-in-Chief

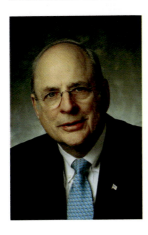

Norman Augustine attended Princeton University where he graduated with a BSE in Aeronautical Engineering, magna cum laude, and an MSE. He was elected to Phi Beta Kappa, Tau Beta Pi, and Sigma Xi.

After graduating he joined Douglas Aircraft where he worked as a Research Engineer, Program Manager, and Chief Engineer, after which he served in the Office of the Secretary of Defense as Assistant Director of Defense Research and Engineering. He then joined LTV Missiles and Space Company, serving as Vice President, Advanced Programs and Marketing. He returned to the government as Assistant Secretary of the Army for R&D and then Under Secretary of the Army and Acting Secretary of the Army. Joining Martin Marietta as Vice President of Technical Operations, he was later elected as CEO and chairman, having previously been President and COO. He served as President of Lockheed Martin Corporation upon the formation of that company and became CEO later the same year. He retired as Chairman and CEO of Lockheed Martin and became a Lecturer with the Rank of Professor on the faculty of Princeton University.

Mr. Augustine was Chairman of the Council of the National Academy of Engineering, the Aerospace Industries Association, the Defense Science Board, and former President of the American Institute of Aeronautics and Astronautics. He is a former member of the Board of Directors of ConocoPhillips, Black & Decker, Proctor & Gamble, and Lockheed Martin and served as a Regent of the University System of Maryland (12 institutions), is a Trustee Emeritus of Johns Hopkins, and a former member of the Board of Trustees of Princeton University and MIT. He has been a member of advisory boards to the Departments of Homeland Security, Energy, Defense, Commerce, Transportation, and Health and Human Services, as well as NASA, Congress, and the White House. He served for 16 years on the President's Council of Advisors on Science and Technology under both Republican and Democratic presidents. He is a member of the American Philosophical Society, the National Academy of Sciences, and the Council on Foreign Relations and is a Fellow of the National Academy of Arts and Sciences and the Explorers Club.

Mr. Augustine has been presented the National Medal of Technology by the President of the United States and received the Joint Chiefs of Staff Distinguished Public Service Award. He has five times received the Department of Defense's highest civilian decoration, the Distinguished Service Medal. He authored *The Defense Revolution*, *Shakespeare in Charge*, *Augustine's Laws*, *Augustine's Travels*, and *The Way I See It*. He holds honorary degrees from 35 universities, is a Distinguished Scholar of the University of Maryland Baltimore, and was selected by Who's Who in America and the Library of Congress as one of "Fifty Great Americans" on the occasion of Who's Who's fiftieth anniversary. He has delivered over 1,500 speeches and lectures and since retiring has served on 59 pro bono committees and commissions of which he chaired or co-chaired 43. He has traveled to 130 countries and stood on both the North and South Poles of the earth.

About the Associate Editor

Michael Sievers
Jet Propulsion Laboratory
Pasadena, CA, USA

University of Southern California
Los Angeles, CA, USA

Michael Sievers is a Senior System Engineer at the California Institute of Technology, Jet Propulsion Laboratory (JPL), and an Adjunct Lecturer in the systems architecting and engineering department at the University of Southern California (USC). He earned a bachelor's degree in electrical engineering and masters and Ph.D. degrees in computer science, all from the University of California, Los Angeles.

Dr. Sievers's graduate studies investigated Very Large Scale Integrated Circuit failure mechanisms, defect testing, and fault-tolerance. In this work, he developed a programmable logic structure and one of the earliest design automation tools used for building a self-checking Hamming Code generator and checker.

He was also a part-time academic at JPL while doing his graduate studies to develop means for automating the control of JPL's large radio antennas that were part of its Deep Space Network. He also designed a self-checking computer module that was part of a larger fault-tolerant computing project.

After completing his academic work, Dr. Sievers focused on developing fault-tolerant spacecraft command and data-handling subsystems for several Earth science experiments and US government applications. He later joined a young company developing special-purpose, high-performance computers for computational biology applications.

On returning to JPL, Dr. Sievers performs research and development in high-performance computing, adaptive optics control, system fault-tolerance and resilience, model-based systems engineering, system reputation and trust, and ground enterprise architectures. His work has earned him numerous NASA and JPL awards. At USC, he teaches classes in system architecture, model-based systems engineering, and system resilience.

Dr. Sievers is an INCOSE Fellow, AIAA Associate Fellow, and IEEE Senior Member. He is the Associate Editor of the *IEEE Open Journal of Systems Engineering* and has written more than 70 publications in refereed journals and conferences.

Section Editors

Norman Augustine
Lockheed Martin Corporation
Bethesda, MD, USA

Barry Boehm
University of Southern California
Los Angeles, USA

Joseph D'Ambrosio
General Motors
Detroit, USA

D. Erwin
University of Southern California
Los Angeles, CA, USA

Jeff A. Estefan
NASA/Jet Propulsion Laboratory
Pasadena, CA, USA

Azad M. Madni
University of Southern California
Los Angeles, CA, USA

R. J. Minnichelli
The Aerospace Corporation
CA, USA

Section Editors

R. A. Noguchi
The Aerospace Corporation
CA, USA

Alejandro Salado
The University of Arizona
Tucson, AZ, USA

M. Sievers
University of Southern California
Los Angeles, CA, USA

Contributors

Norman Augustine Lockheed Martin Corporation, Bethesda, MD, USA
Advisory Services, New York, NY, USA

B. Bagdatli Georgia Institute of Technology, School of Aerospace Engineering, Atlanta, GA, USA

A. Terry Bahill Systems and Industrial Engineering, University of Arizona, Tucson, AZ, USA

María J. Blas Instituto de Desarrollo y Diseño INGAR (UTN-CONICET), Santa Fe, Argentina

Luca Boggero German Aerospace Center (DLR), Hamburg, Germany

Guy André Boy CentraleSupélec, Paris Saclay University, Gif-sur-Yvette, France
ESTIA Institute of Technology, Bidart, France

S. Breiner Information Technology Lab, National Institute of Standards and Technology, Gaithersburg, MD, USA

J. H. Bussemaker Institute of System Architectures in Aeronautics, MDO group, German Aerospace Center (DLR), Hamburg, Germany

Bruce G. Cameron Massachusetts Institute of Technology, Cambridge, MA, USA

Joe Cesena Lockheed Martin, Sunnyvale, CA, USA

M. Chodas Jet Propulsion Laboratory, California Institute of Technology, Pasadena, CA, USA

Pier Davide Ciampa Institute of System Architectures in Aeronautics, MDO Group, German Aerospace Center (DLR), Hamburg, Germany

S. Cimtalay Georgia Institute of Technology, School of Aerospace Engineering, Atlanta, GA, USA

Norman Augustine is Retired.

R. Cloutier University of South Alabama, Mobile, AL, USA

Edward F. Crawley Massachusetts Institute of Technology, Cambridge, MA, USA

Daniel DeLaurentis Purdue University, West Lafayette, IN, USA

Giuseppa Donelli German Aerospace Center (DLR), Hamburg, Germany

D. Dori Technion, Israel Institute of Technology, Haifa, Israel

M. Elaasar Jet Propulsion Laboratory, California Institute of Technology, Pasadena, CA, USA

D. Erwin University of Southern California, Los Angeles, CA, USA

Jeff A. Estefan NASA/Jet Propulsion Laboratory, Pasadena, CA, USA

J. S. Fant The Aerospace Corporation, Chantilly, VA, USA

T. Fields Georgia Institute of Technology, School of Aerospace Engineering, Atlanta, GA, USA

Marco Fioriti Politecnico di Torino, Turin, Italy

E. Garcia Georgia Institute of Technology, School of Aerospace Engineering, Atlanta, GA, USA

Silvio Gonnet Instituto de Desarrollo y Diseño INGAR (UTN-CONICET), Santa Fe, Argentina

C. Granrath FEV.io GmbH, Aachen, Germany

Mechatronics in Mobile Propulsion, RWTH Aachen University, Aachen, Germany

Cesare Guariniello Purdue University, West Lafayette, IN, USA

Karen Gundy-Burlet Crown Consulting Inc., NASA-Ames Research Center, Moffett Field, CA, USA

J. S. Jenkins Jet Propulsion Laboratory, California Institute of Technology, Pasadena, CA, USA

S. Kriebel FEV.io GmbH, Aachen, Germany

BMW Group, Munich, Germany

B. Kruse e:fs TechHub GmbH, Gaimersheim, Germany

Azad M. Madni Systems Architecting and Engineering, Astronautical Engineering Department, University of Southern California, Los Angeles, CA, USA

Intelligent Systems Technology, Inc., Los Angeles, CA, USA

C. C. Madni Intelligent Systems Technology, Inc., Los Angeles, CA, USA

M. W. Maier The Aerospace Corporation, Hill AFB, Ogden, UT, USA

M. Markthaler BMW Group, Munich, Germany

Software Engineering, RWTH Aachen University, Aachen, Germany

James N. Martin The Aerospace Corporation, Chantilly, VA, USA

Dennis K. McBride Intelligent Systems Division, Hume Center for National Security and Technology, Virginia Tech National Security Institute, Blacksburg, VA, USA

Mark L. McKelvin University of Southern California, Los Angeles, CA, USA

Yaroslav Menshenin Skolkovo Institute of Science and Technology, Moscow, Russia

R. J. Minnichelli The Aerospace Corporation, El Segundo, CA, USA

Yaniv Mordecai Massachusetts Institute of Technology, Cambridge, MA, USA

R. A. Noguchi The Aerospace Corporation, El Segundo, CA, USA

D. W. Orellana ManTech International Corporation, Los Angeles, CA, USA

Gregory S. Parnell Department of Industrial Engineering, University of Arkansas, Fayetteville, AR, USA

R. Peak Georgia Institute of Technology, School of Aerospace Engineering, Atlanta, GA, USA

E. Peleg Metaphor Vision Ltd, Kefar-Saba, Israel

Michael J. Pennock The MITRE Corporation, McLean, VA, USA

R. G. Pettit George Mason University, Fairfax, VA, USA

Matt Phillips System Design and Analytics Laboratory, Department of Industrial Engineering, University of Arkansas, Fayetteville, AR, USA

Edward A. Pohl Department of Industrial Engineering, University of Arkansas, Fayetteville, AR, USA

S. Purohit University of Southern California, Los Angeles, CA, USA

Ali Raz Systems Engineering and Operations Research, George Mason University, Fairfax, VA, USA

J. Richenhagen FEV.io GmbH, Aachen, Germany

N. Rouquette Jet Propulsion Laboratory, California Institute of Technology, Pasadena, CA, USA

William B. Rouse McCourt School of Public Policy, Georgetown University, Washington, DC, USA

B. Rumpe Software Engineering, RWTH Aachen University, Aachen, Germany

S. Russell Johnson Space Center, NASA, Houston, TX, USA

Alejandro Salado The University of Arizona, Tucson, AZ, USA

William D. Schindel ICTT System Sciences, Terre Haute, IN, USA

Nicholas J. Shallcross System Design and Analytics Laboratory, Department of Industrial Engineering, University of Arkansas, Fayetteville, AR, USA

S. Shoshany-Tavory Technion – Israel Institute of Technology, Haifa, Israel

Michael Sievers NASA/Jet Propulsion Laboratory, California Institute of Technology, Pasadena, CA, USA

University of Southern California, Los Angeles, CA, USA

Eric A. Specking System Design and Analytics Laboratory, Department of Industrial Engineering, University of Arkansas, Fayetteville, AR, USA

R. D. Sriram Information Technology Lab, National Institute of Standards and Technology, Gaithersburg, MD, USA

H. Stoewer TU Delft, Delft, The Netherlands

E. Subrahmanian Information Technology Lab, National Institute of Standards and Technology, Gaithersburg, MD, USA

Engineering Research Accelerator/Engineering and Public Policy, Carnegie Mellon University, Pittsburgh, PA, USA

A. E. Trujillo Massachusetts Institute of Technology, Cambridge, MA, USA

D. Verma Stevens Institute of Technology/SERC, Hoboken, NJ, USA

D. A. Wagner Jet Propulsion Laboratory, California Institute of Technology, Pasadena, CA, USA

Gan Wang Dassault Systèmes, Herndon, VA, USA

Tim Weilkiens oose Innovative Informatik eG, Hamburg, Germany

A. Zonnenshain The Gordon Center for Systems Engineering, Technion – Israel Institute of Technology, Haifa, Israel

Part I

Introduction

Introduction to the Handbook

Azad M. Madni and Norman Augustine

Contents

Introduction	3
What New Developments in Systems Engineering Are Important to Meeting Twenty-First-Century Challenges?	4
What Is Unique About MBSE?	5
What Kinds of Models Can Be Created Using MBSE?	5
How Can MBSE Be Applied to AI-Driven Systems?	6
How Does MBSE Deal with Complexity?	6
How Should Humans Be Modeled in Complex Systems?	7
How Should Nonlinearities, Including Discontinuities, Be Addressed When Modeling Complex Systems?	8
How Can Complex Systems Be Protected Against Active Human-Initiated Interference?	8
How Can Digital Twin and Digital Thread from Digital Engineering Be Leveraged in Model-Based Systems Engineering?	9
How Can Simulation Benefit Systems Engineering?	9
How Can MBSE Be Used to Protect Against Unintended Consequence Produced by Changes?	10
What Is MBSE's Contribution to Risk Analysis?	10
Is There a Return on Investment from MBSE?	10

A. M. Madni (✉)
Systems Architecting and Engineering, Astronautical Engineering Department, University of Southern California, Los Angeles, CA, USA
e-mail: azad.madni@usc.edu

N. Augustine
Lockheed Martin Corporation, Bethesda, MD, USA

© Springer Nature Switzerland AG 2023
A. M. Madni et al. (eds.), *Handbook of Model-Based Systems Engineering*,
https://doi.org/10.1007/978-3-030-93582-5_3

Introduction

We live in an era of hyperconnectivity and exponential technological trends. The former continues to increase the complexity of systems, while the latter continues to rapidly produce new technology-enabled capabilities. Model-based systems engineering (MBSE) and digital engineering (DE) are two significant examples of such advances.

These advances have on occasion led to confusion and misconception about MBSE and DE even within the engineering community. For example, since virtually all engineering employs models, what distinguishes model-based systems engineering from traditional engineering using models? How does digital engineering differ from MBSE in terms of emphasis, scope, and methodologies? How do models created with MBSE integrate with third-party simulations? How can digital engineering concepts, such as "digital twin" and "digital thread," complement and enhance MBSE? How can digital twin technology be used to facilitate system verification, validation, and testing? Such questions, among others, provided the impetus for this handbook on Model-Based Systems Engineering.

The International Council on Systems Engineering (INCOSE) defines MBSE as "the formalized application of modeling to support system requirements, design, analysis, verification and validation activities beginning in the conceptual design phase and continuing throughout development and later life-cycle phases." This handbook addresses all these various uses of MBSE.

The handbook is for both educators and practitioners responsible for making decisions about what a system is intended to do, and, as importantly, what it is not supposed to do. Putatively, no such book could have been created by a single individual or even a single organization; hence, this volume draws upon a number of authorities from around the world who are experts in the different aspects and uses of MBSE.

Some of the key questions addressed in this handbook are presented in the text which follows.

What New Developments in Systems Engineering Are Important to Meeting Twenty-First-Century Challenges?

Systems engineering today can be thought of as combining two thrusts. The first of these encompasses traditional methods that are often characterized as "old school" and that work well for relatively mature and not overly complicated systems. The second comprises "new methods" that can be characterized as systems engineering for a new era and that are motivated by the complexity and breadth of today's sociotechnical systems, as well as by ongoing advances in computation, engineering, and materials. For the latter, a new type of handbook is required, one which addresses developments such as model-based systems engineering, digital engineering, and other relevant technology-enabled methodological advances. For example, system requirements, as in the past, need to be rigorously specified, but now in a fashion such that they can be addressed by modern methods and algorithms to prove correctness. Similarly, Artificial Intelligence and Deep Learning, often under human direction, can be exploited to enhance and accelerate detailed design efforts.

What Is Unique About MBSE?

While systems engineers have routinely worked with models for several decades, as noted above, MBSE offers a unique value proposition rooted in the way models are created, viewed, and used. Specifically, MBSE is based on a few key guiding tenets: It encompasses a unified model that cuts across disciplines and spans the system life cycle and is used as the authoritative source of truth for all systems engineering activities; documents originate in, or are derived from, models, so that configuration management/control is not a separate activity that suffers from well-known maintenance and escapement problems; and, since MBSE is methodology-neutral, a variety of methodologies and modeling constructs can be employed based on information availability and domain complexity. MBSE is concerned with the creation and exploitation of domain models (not documents) as the primary means of information exchange. The use of formal representation in MBSE can support reasoning and facilitate interoperability among heterogeneous models. The growing emphasis on the use of ontologies to capture domain knowledge reflected in scenarios and use-cases helps circumscribe the regime of applicability (i.e., scope) within which questions can be posed and answered and reasoning performed. Models created with MBSE bridge disciplines, facilitate collaboration among stakeholders, and encourage questioning of assumptions and constraints. Importantly, models in MBSE can be verified in simulation environments (with the aid of virtual prototyping and digital twin development) as part of verification testing, thereby reducing the cost and effort of physical testing.

What Kinds of Models Can Be Created Using MBSE?

Models created using MBSE can range from abstractions to detailed representations. They can be deterministic or stochastic, low fidelity or high fidelity, and approximate or precise. They can reflect a single perspective or multiple perspectives. A particularly challenging problem when constructing models is determining what should be included and what can reasonably be omitted without compromising model validity and usefulness for the intended purpose(s). While omitting key factors can undermine the accuracy and integrity of models, including irrelevant factors can introduce extraneous complexity. A well-known illustration of the former is that of an electric power hurricane recovery model that was employed in a real-world scenario. This model included the provision of poles, wires, vehicles, and much more. However, it failed to include, yes, day care centers. In the aftermath of the hurricane that actually occurred, many workers (e.g., telecom specialists, field technicians) who were needed to lead the recovery were forced to remain at home when critically needed to care for their young ones because schools were closed in the aftermath of the storm. The problem was solved by calling upon the firm's retirees to establish and staff day care centers. This is but one example of the unintended consequences resulting from a key oversight, in this case the omission of a key factor in a hurricane recovery model.

How Can MBSE Be Applied to AI-Driven Systems?

Today, an increasing number of systems employ AI for cognitive tasks, i.e., tasks that require some combination of reasoning, estimation, prediction, and learning. Examples of such tasks include autonomous control and navigation in self-driving cars, planning and decision-making in advanced fighter aircraft and unmanned aerial vehicles (UAV) and teams thereof, and adaptive energy grids. Modeling, verification, and testing of AI-driven systems remain ongoing challenges because such systems invariably exhibit complex, nondeterministic behavior. Typical MBSE challenges for such systems include system modeling with partial knowledge of initial conditions, system state space, and operational environment. For such systems, probabilistic modeling and related verification testing methods continue to show significant promise.

How Does MBSE Deal with Complexity?

System complexity results from any number of characteristics. The first is a large number of components and interfaces in a system. Such systems are often described as complicated systems. The second is random behavior, typically associated with financial markets and customer behaviors in retail markets. The third is dynamically changing organizational structures, processes, and mathematical algorithms. The fourth is reconfigurable, learning systems and organizations, and innovative methods. This latter form of complexity is generally referred to as self-organization complexity.

Complexity has also been characterized as structural complexity, emergent complexity, and sociopolitical complexity. Structural complexity results from multiple stakeholders, multiple components, multiple interfaces, and multiple workflows. Emergent complexity is the result of unavoidable uncertainties and large, unexpected sensitivities to small variations in initial conditions or constraints. Sociopolitical complexity encompasses shifting political trends, conflicting priorities, institutional constraints, and regulations imposed by governing bodies. Addressing such complexities in MBSE needs to begin very early when defining system architectures and concepts of operations.

Various methods have been used to reduce complexity, including abstraction, aggregation, homogenization (i.e., process of making components and interfaces uniform or similar), decomposition, transformation and mapping, reduction in stakeholder conflicts through consensus-building methods, schemes for allocation of functions to hardware, software, and humans, technology selection to facilitate interoperability and standards-compliance, and selective model fidelity, reduced order models, and approximate algorithms to reduce computational complexity. In addition, several principles for complexity management have been employed including "separation of concerns," stable intermediate forms, structural alignment during decomposition, discovery of hidden interactions through story-based simulations, and the use of patterns during systems integration – including human-systems integration (e.g., supervisory control pattern, augmented intelligence pattern). Structural alignment in decomposition pertains to minimizing the number of interfaces among components in the different layers in the decomposition hierarchy, usually by establishing one-to-one correspondence between components in the layers of the hierarchy. Minimizing the number of mappings between architectural perspectives is also a way to reduce structural complexity. Ensuring stable intermediate forms during system development reduces "debugging" complexity when diagnosing mismatches and faults. Explicit representation of context provides a convenient way to separate relevant from irrelevant factors. Subsequent use of context as a filter during scenario execution can reduce complexity during both simulated as well as real-world operation. Ensuring stable intermediate forms during system development reduces debugging complexity when diagnosing mismatches and faults. Explicit representation of context provides a convenient way to separate relevant from irrelevant factors. Subsequent use of context as a filter during scenario execution is an effective way to reduce complexity during simulated and real-world operation. Finally, the size of the system state space and the interaction density of system components are key features that contribute to system complexity. Friedrich Wiekhorst's equation, aptly called "The Monster," defines the number of possible states that exist for a system having a specified number of elements, each of which interacts with every other element in a binary manner. While a two-element system has only four possible states, the states associated with a ten-element system exceed the number of stars in our galaxy!

How Should Humans Be Modeled in Complex Systems?

Some 80% of aircraft accidents are attributable to human error (pilots, controllers, mechanics, etc.) Correspondingly, 98% of automobile accidents are due to human error. The modeling of humans in complex human-machine systems depends on the roles of humans and the functions performed by them. Depending on their roles, humans can potentially be a source of objectivity in resource allocation decisions as well as creative option generation in the face of unprecedented situations. In performing such functions, humans can also become a source of errors and delays. These factors need to be accounted for in the human-machine system model when attempting to quantify, predict, and maximize joint human-machine system performance. This implies that human-systems integration methods (i.e., methods used to integrate human performance models with system design) should be incorporated within the MBSE framework and invoked at appropriate points during the overall system integration process. Human performance-modeling perspectives and levels of abstraction depend on the roles that humans play and functions they perform in the human-machine system. For example, if the human-machine system is concerned with planning and decision-making, then the anthropometric properties of the human do not have to be represented in the human model. Also, the roles that humans play can vary from that of a supervisor/monitor of machine tasks with override privileges (such as in the case in early autonomous vehicles) to that of an actual participant in shared task performance (such as in robotic surgery). Depending on the types and frequency of human-machine interactions, human performance models vary significantly. Model fidelity and model abstraction level are both key considerations that come into play when modeling humans in joint task performance. For relatively simple human-machine systems, humans can be modeled as a time delay, a state machine, or a simple rule-based system. For complex human-machine systems operating in the face of uncertainty, humans can be modeled using Hidden Markov Models. When planning, decision-making, and control are involved in uncertain operating environments, humans can be modeled using Partially Observable Markov Decision Process (POMDP) methods. Ultimately, systems can be designed to protect against many forms of human error, an example of which would be the "experiments" being unwisely conducted on the nuclear reactor in the Chernobyl power plant.

How Should Nonlinearities, Including Discontinuities, Be Addressed When Modeling Complex Systems?

In complex sociotechnical systems, since nonlinearities can manifest themselves in a variety of forms, there is no standard approach for modeling nonlinearities. In relatively simple cases, linearized models can sometimes suffice. In other special cases, the nonlinearity can be approximately modeled using methods such as quadratic functions, polynomials, exponentials, and logarithmic functions. In general, nonlinear systems are difficult to model and simulate because the characteristics of such system can change abruptly or over time with the slightest variations in system and environmental parameters. For example, several years ago a tree fell on a power line in Ohio interrupting power in much of New England and Southeastern Canada for several days. To model this type of complex system requires addressing such a discontinuity (a type of nonlinearity) using problem domain-specific methods.

How Can Complex Systems Be Protected Against Active Human-Initiated Interference?

System failures attributable to hostile interference by humans, both internal and external, have become increasingly commonplace as systems become more complex, broadly distributed, and interconnected. Many of today's most critical systems, ranging from the Internet to the electric grid, were developed with little or no regard to the possibility that their functioning might be intentionally disrupted. However, today the widespread use of systems ranging from Internet applications and integrated healthcare to automated vehicles and digital banking can be expected to significantly increase demands for safety and security. Consequently, these requirements need to be addressed on day 1 of system development.

How Can Digital Twin and Digital Thread from Digital Engineering Be Leveraged in Model-Based Systems Engineering?

Digital twin, a concept that originated in Product Lifecycle Management, is now a part of Digital Engineering. A digital twin is a dynamic digital representation of a physical system or device that has bidirectional communication with the physical system and can operate in parallel to the physical system operating in the real world. A digital twin can represent the full physical system, a subsystem, or only specific aspects of the system needed by specific applications. A digital twin is dynamically updated in real time/near real time as the development status of the physical system changes during design or as the state of the physical system changes during operation; for example, through wear or environmental forces. Digital twins have recently been used to represent humans. For example, digital twins have been used to represent patients. Digital twins of patients are generated from multimodal patient data, population data, and real-time updates of patient and environment variables. Digital twins are also being created for personalized medicine. Personalized medicine requires the integration and processing of vast amounts of data. Digital twins for personalized medicine are computationally treated with a variety of medical interventions to identify the optimal interventions for a particular patient. By definition, a full-featured digital twin cannot exist in isolation; it requires a physical counterpart (its physical twin) to exist in some state of development or use. Digital thread is the communication infrastructure that offers the means to achieve digital connectivity throughout the system or product life cycle. It provides an integrated view of an asset throughout a product or system's life cycle. A companion concept to digital twin, it assures that data from all stages in a system's lifecycle are fed back into system ideation, conceptualization, and creation. Given these characteristics, the concepts of "digital twin" and "digital thread" are valuable complements to MBSE in all stages of system development. Specifically, they offer a cost-effective means to enhance system verification and testing in a variety of contexts. This is of particular consequence in validating system performance envelopes and failure response protocols.

How Can Simulation Benefit Systems Engineering?

Simulation has important uses in systems engineering including, but not limited to, the following: (a) accelerating the design of the physical system by developing and experimenting with digital twins; (b) exploiting knowledge of state and status of the digital twin to determine how long a system or product can be expected to remain in operational use, and use that information for predictive maintenance; and (c) proactively scheduling and performing system "shutdown" for maintenance. A seemingly benign change in the placement of bolts made during the construction of the Kansas City Hyatt Hotel ultimately led to a structural collapse killing 114 people.

MBSE, and particularly the use of a digital tower, could quickly point out the hazard of the change that was made. The Internet of Things (IoT), which serves as a bridge between the physical and digital worlds, offers a complementary capability to Digital Twin in, for example, preventive maintenance by supporting analytics and AI-based optimization. It informs the Digital Twin with data and insights from real-world system operation. It offers greater flexibility in product mobility, location, and monetization options (e.g., selling a capability versus selling equipment).

How Can MBSE Be Used to Protect Against Unintended Consequence Produced by Changes?

Various failures of systems in the past have as their root cause changes to one element of the system that propagated in unexpected ways. As dependence on systems of systems becomes increasingly common and as systems incorporate expanded spatial reach accompanied by ever greater volumes of software, vulnerability to escapements can be expected to grow as a result of unforeseen interactions among system elements and between the system and the environment. The extreme form of this phenomenon is present in what is called the "butterfly effect." MBSE is well-suited for identifying and eliminating such "sneak" pathways using different types of model-based methods in conjunction with system simulation.

What Is MBSE's Contribution to Risk Analysis?

An essential aspect of complex system design and operation is risk analysis. MSBE permits potential failure modes to be identified, their consequences determined, and protective measures defined and incorporated. Such circumstances often cannot be explored in the operational world due to safety concerns for both humans and hardware. Also, using these models, risk-based validation methods can be incorporated that negate the most significant risks when schedule and budget constraints prevent full-blown validation. An early form of such endeavors was the use of simulators to train pilots to respond to in-flight failures, such as the one in 2009 when an Airbus A320 was severely damaged when it struck a flock of birds, yet landed in the Hudson River without fatalities.

Is There a Return on Investment from MBSE?

MBSE can be effectively used to explore alternative system designs and their performance/cost trade-offs. It can also provide a strong basis for optimizing total lifetime costs associated with various system configurations. When coupled with digital twin technology, MBSE can enhance optimization of life cycle costs by replacing time-based maintenance with predictive and condition-based maintenance. The Return on Investment (ROI) stems from the fact that MBSE provides the necessary structure for complexity management, rapid retrieval and reuse of artifacts and the ability to explore candidate system models and methods in simulation before resorting to hardware testing in the costly and sometimes hazardous real world. Since MBSE can support a wide variety of modeling approaches and algorithms for learning and optimization, part of the ROI will depend on the choice of the most appropriate methods and algorithms. It is important to note that an organization needs to have the proper commitment, preparation, and requisite resources in place to realize the ROI from adoption of MBSE. As important, certain industries that experience infrequent change and operate in stable market dynamics may not derive the same benefit as industries that experience frequent change in policies, regulations, and technology-enabled processes.

The above questions, and others, are addressed in various chapters in this handbook. As increasingly integrated capabilities are sought by users, the role of Model-Based Systems Engineering is poised to play a vital role in the definition of requirements, design, testing, production, and operation of modern systems and systems-of-systems. The authors who have contributed to this volume sincerely wish that the material presented in each chapter will introduce new avenues for providing important capabilities – and do so in a fashion that prevents the problems that in the past have too often plagued complex systems and complex systems development.

Part II

MBSE Foundations

Semantics, Metamodels, and Ontologies

2

Michael Sievers

Contents

Introduction	16
Problem Statement	16
Key Concepts and Terminology	19
State-of-the-Practice	27
Best Practice Approach	31
What Is the Ontology Structure and Ownership?	34
How Much Commonality Is There?	35
What Is the Process for Developing an Ontology?	36
Ontology Development and Analysis Tools	40
Illustrative Example	41
Competency Questions	41
Existing Ontologies	41
Key Terms	42
Concepts, Concept Hierarchy, Slots, and Facets	42
Instances	43
Chapter Summary	43
Cross-References	44
References	44

Abstract

Ontologies and *metamodels* define a controlled domain vocabulary and are foundational to MBSE...

Keywords

Ontologies · Metamodels · Semantic domain · Controlled vocabulary · Shared conceptualization · Model patterns

---

M. Sievers (✉)
NASA/Jet Propulsion Laboratory, California Institute of Technology, Pasadena, CA, USA
e-mail: msievers@jpl.nasa.gov

© Springer Nature Switzerland AG 2023
A. M. Madni et al. (eds.), *Handbook of Model-Based Systems Engineering*,
https://doi.org/10.1007/978-3-030-93582-5_2

Introduction

Domain semantics, metamodels, and ontologies are foundational formalisms that make possible the primary benefits of MBSE: design collaboration among all stakeholders, model sharing, model reuse, terminology consistency, and reduction of ambiguity and unnecessary redundancy. Domain semantics is the meaning assigned to system concepts, i.e., the "physics" behind system components and their interaction. Metamodels consist of generalized object types, object relationships, object attributes, and compositional rules. Metamodels define abstract semantics and syntax for modeling languages for expressing models within a domain of interest. An ontology is an instance of a metamodel for a specific problem or system. All metamodels are ontologies but ontologies need not be metamodels. A modeling language contains the objects and relationships used in creating a model that conforms to a metamodel and represents aspects of a system. Modeling languages may be informal (e.g., natural languages), semiformal (e.g., SysML, DoDAF), or formal (e.g., Promela) [2].

Problem Statement

All engineers are familiar with system models that are useful in communicating concepts and features. However, before the introduction of MBSE, models were typically ad hoc representations that frequently caused confusion, misunderstanding, and even disaster. One need only look at the root cause of the Mars Climate Orbiter (MCO) failure [1] as an example – confusion over units led to the incorrect commanding of the vehicle as it was beginning orbit insertion. Rather than achieving its intended orbit, MCO instead fell into the Mars atmosphere and disintegrated. Problems also occur when the same term is used differently by different subject matter experts. Consider the term "bandwidth" which can mean "bits per second" to a computer person but to a signal processing person is the difference between the upper and lower frequencies in a continuous band of frequencies.

An illustration of communication confusion due to ad hoc models is shown in Fig. 1a. In the figure, we see a shaded map of the United States that models a portion of the Earth. By convention, we can assume that the lines delineate states, but of course, the physical Earth has no such delineation. Figure 1a does not unambiguously help us understand the meaning of the colors, we need the legend in Fig. 1b.

Figure 2 shows an older flu map in which shading and colors indicate flu spread in the United States, and Fig. 3 shows the number of confirmed COVID-19 cases per 100,000 residents. Except for the implied use of lines for delineating states, these figures have different ad hoc vocabularies even though all three show similar content. A stakeholder familiar with one vocabulary might easily misunderstand the content of a model that uses a different vocabulary that reuses the symbols from another vocabulary.

In the last example, consider Fig. 4 which shows a system model as a block diagram. It is unclear what the lines mean that connect boxes. They could indicate

2 Semantics, Metamodels, and Ontologies 17

Fig. 1 (**a**) Map of the United States (**b**) Map legend. (Source: https://www.cdc.gov/flu/weekly/usmap.htm)

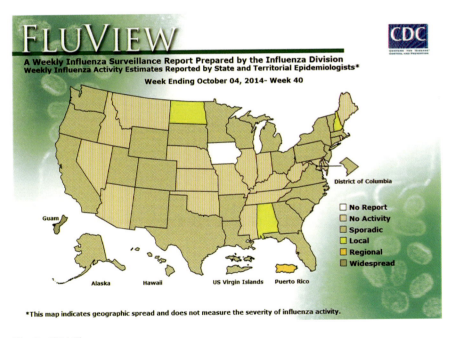

Fig. 2 2014 Flu map

relationships, data flows, hierarchy, or maybe something else. It is not obvious whether the boxes or circles indicate functions, behaviors, and are physical components, or possibly something else. Does the "eye" imply a stakeholder, a camera, or a viewpoint? As with the previous examples, a detailed legend is required and even if present might not result in a complete understanding of the model. Moreover, it is

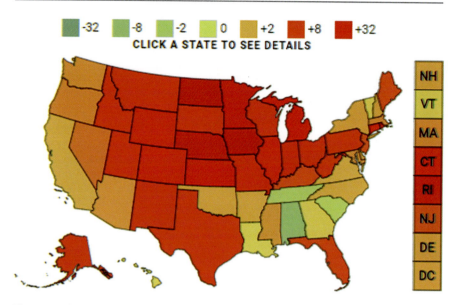

Fig. 3 Confirmed cases of COVID-19 per 100,000 residents. (Source: https://time.com/5800901/coronavirus-map)

Fig. 4 Block diagram system model

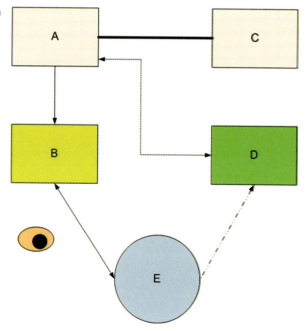

not evident what aspects of a system the model addresses or whether it satisfactorily covers the aspects of interest.

Figures 1, 2, 3, and 4 highlight several important points:

- Models answer a set of specific questions about a domain.
- Interpreting a model requires understanding the model vocabulary.
- An explicitly defined and agreed to vocabulary reduces confusion in model interpretation.
- Model vocabularies based on ad hoc terminology can lead to incorrect assumptions made about the meaning of a model resulting in interpretation errors.

As an amusing but germane anecdote, an interesting cognitive study was done some years ago related to so-called *false memories* [3]. Our brains remember associations and interpret semantics far better than recalling details. A consequence is that we can believe we recall something that is not present. As an example, try to remember the words in this list:

BANANA, SWEET, APPLE, TREE, FIG, CRUNCHY, HEALTHY, PIE, ORANGE, PEEL, STRAWBERRY, TART, AND LEMON

Now try to remember if the words FIG, PIE, and FRUIT were in the list. Most people can remember a few words but also make the connection that the list relates to FRUIT. However, the word FRUIT does not appear in the list even though many people will believe it had. The point is, as, with this list, models may suffer from misinterpretation when false memories resulting from ad hoc associations and semantics lead to incorrect interpretation.

Key Concepts and Terminology

Semantics and Semantic Domain

Semantics concerns the meaning of something. As the examples in Figs. 1, 2, 3, and 4 illustrate, the lines, colors, and shading confer meaning and potentially add misunderstanding to a map or block diagram. The semantics of the reasoning we perform using a model and the results expected determine the model vocabulary. Additionally, semantics helps scope a model not only in terms of what it contains but also enables determining when a model is sufficiently complete for its purposes. These points are important because models are made for defined purposes and are complete when they satisfy those purposes. Moreover, model sharing, collaboration, and reasoning require explicitly defined vocabularies.

A semantic domain defines shared terminology associated with a specific area of interest or problem. For example, consider the sport of soccer (football in most of the world). Terms such as "offside," "touchline," "striker," "referee," throw-in, "pitch," and "kickoff" have specific and well-understood meanings. The use of these terms is both expected and necessary when describing the action in a soccer match, comparing teams, looking at statistics, and so forth. Although known by common usage, many of these terms are codified and documented in the *Laws of the Game* published by the International Football Association Board (IFAB) [4].

The semantic domain for a spacecraft might include terms such as "bus," "instrument," "solar array," "reaction wheel assembly," "mass," and "command and data handling." Unlike soccer, however, these terms are not universally used by the space flight community. For example, some in the community use "flight computer" while others might refer to the same component as the "spacecraft control processor." Also, units of measure are not universal, e.g., a supplier in Europe might provide a subsystem having dimensions measured in centimeters to a prime contractor in the United States that bookkeeps dimensions in inches. Adding to the potential confusion is that different stakeholders might use terms such as "spacecraft," "vehicle," "observatory," or "collection segment" for the entity that is launched into orbit. Even the term "collection segment" could be interpreted only as the vehicle in orbit by some but might also include the ground antennas and receivers by others.

Creating a semantic domain for soccer understood by multiple stakeholders is relatively straightforward because there is an acknowledged controlling authority that determines the terminology. Conversely, as with many engineering endeavors, a universally agreed upon semantic domain may not be possible, and often, attempts at creating one can cause unresolvable contention. Approaches for dealing with heterogeneous semantics are discussed in section "Best Practice Approach."

Metamodels

Metamodels define the *abstract syntax* of a modeling language used for expressing models in a domain of interest. Metamodels comprise generic object types, relationships between object types, attributes of object types, and the rules for combining objects and relationships. Models represent a semantic mapping of a domain ontology to a system while metamodels are the basis of metamodel ontologies used for defining the domain ontology (Fig. 4) [5]. Metamodels may be hierarchically defined, that is, a meta-metamodel can define the terminology of a metamodel, i.e., a metamodel is an *instance-of* a meta-metamodel. An *instance-of* relationship exists between two entities when one of the entities is an *instance* of the other, that is, is a member of or is created from the other entity. A *dog* is an *instance-of pet*, for example, and inherits the properties of *pet*.

The *is-a* association is related to the *instance-of* relationship. An *is-a* relationship requires that an entity be an *instance-of only* one other entity in a system. A few examples:

- Felix *is-a* cat => Felix is an *instance-of* cat and no other entity.
- Cycling *is-a* sport => cycling is uniquely an *instance-of* sport.
- Ground squirrel *is-a* squirrel => ground squirrel is an *instance-of* squirrel.
- But...
- The culprit is a squirrel is not an *is-a* relationship because the culprit is not a unique *instance-of* squirrel.

2 Semantics, Metamodels, and Ontologies

When properly applied, *is-a* relationships lead to conclusions such as the following:

Joe *is-a* man.
    Man needs rest.
    Joe needs rest.

But imprecisions can lead to erroneous conclusions:

Joe *is-a* man
    Man *is-a* species
    Joe *is-a* species

The corrected version is as follows:

Joe is an *instance-of* man.
    Man is an *instance-of* species.
    Joe is an *instance-of* a species instance.

As illustrated in Fig. 5, a *system* is *represented* by a *model* which is an *instance-of* a *metamodel*. The model is *expressed-in* a *modeling language* that uses the *syntax, semantics, constraints, and patterns* defined by a *metamodel*. There is a semantic mapping from the metamodel to a *metamodel ontology* which is the specialization of metamodel concepts needed for *instantiating* a *domain ontology* (more on ontologies in section "Ontologies").

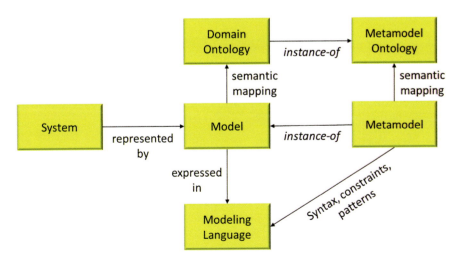

Fig. 5 Model, metamodel, and ontology relationships

- In a *strict metamodel*, if a model element, m_i, is an *instance-of* model m_j, then every element in m_i must be an *instance-of* some element in m_j. In a model comprising L levels, m_0, m_1, m_2... and m_{L-1}, every element of a m_i level model must be a unique *instance-of* a m_{i+1} level model for all i < L-1, and any other relationship other than *instance-of* relationship between two elements m_j and m_k implies that level(j) = level(k) as shown in Fig. 6.

Loose or ambiguous metamodeling does not follow strict rules allowing placement of model elements where needed rather than by *instance-of* relationships (Fig. 7). Loose metamodeling can simplify development, but there are two important side effects:

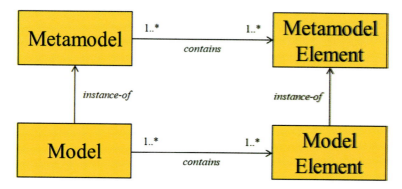

Fig. 6 A strict metamodel

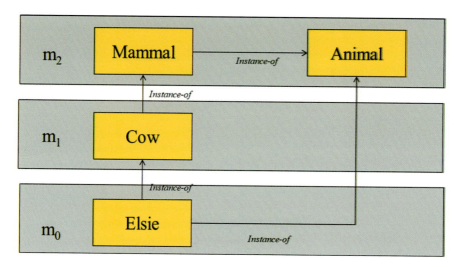

Fig. 7 Example of a loose metamodel

Fig. 8 Metamodel example

- The loose assignment of elements permits inheritance, associations, and other relationships that cross metalevel boundaries which degrade model integrity if model elements are grouped into similar-purpose subgroups.
- More significantly, loose metamodeling erodes well-established object-oriented decompositions in which an instantiation can be at the same level as the metamodel it is an *instance-of* [6].

We look at two clarifying metamodel examples: a map and a programming language.

Figure 8 shows a graphical application of Fig. 5 in which a map is shown as a model of the real world. The metamodel defines state boundaries as gray lines, national parks as green coloring, and cities as dots. The meta-metamodel includes text, lines, and colors that are used in the metamodel.

Programming languages such as Python, C, C++, etc. are built on text-based meta-metamodels. Text is fashioned into a metamodel that comprises text-based entities and combination rules. A modeling language is developed from the metamodel that defines the operators, syntax, and semantics of the programming language.

For example, the metamodel for programming languages might include the following:

- Alphabet: $\Sigma = \{a - z, A - Z, 0 - 9\}$
- Special characters: $\Psi = \{* / + - = [\ ]\ (\ ),\ `\ ;\}$
- All possible strings over Σ: Σ^*
- A string from the set of all possible strings: $s \in \Sigma^*$
- One of the special characters from the set Ψ: $\psi \in \Psi$
- A line of text: $\alpha = s[\psi[\ ]s];:$ (consists of a string, s, and optionally, a special symbol, ψ, followed by an optional white space and another string s followed by a line break (;)
- The set of all lines of text consisting of α: α^*

A modeling language, L, uses Σ^* for expressing constants and variables while Ψ defines the set of operators and delimiters:

- var declares the name of a variable or constant from *s*.
- * multiplication.
- / division.
- \+ addition.
- − subtraction.
- = assignment.
- statement *a*.
- ; statement termination.
- ( ) function arguments and groupings.
- /* [[*s*][*ψ*]]* */ comment (any combination of alphabet and special characters delimited by /* and */).

Rules are needed for using brackets and parenthesis and so forth, but for simplicity, we ignore this detail for this example. Using the modeling language, *L*, we can create models of Newton's Second Law and a resistor voltage divider.

- /* Newton's Second Law */
   var f, m, a
   m = 20; a = 40
   f = m * a
- /* Voltage Divider */
   var v, e, i, R,r1, r2;
   r1 = 100; r2 = 1000; v = 28
   R = r1 + r2
   i = v/R
   e = i * r1; /* equals v * r1/(r1 + r2) ... a voltage divider */

In summary, metamodels:

- Comprise the rules and grammar of a modeling language
- Have little or no semantic content
- Represent the general class of all models that can be expressed by the modeling language, i.e., a model is an instantiation of a metamodel for a given semantic domain
- A system is represented by a model that conforms to a metamodel

Ontologies

Ontologies are the *controlled vocabularies* that comprise agreed-upon sets of explicitly enumerated, unambiguous, and nonredundant terms, relationships, and constraints necessary for building and interpreting models [7]. These vocabularies are established with inputs from domain experts and are configuration managed by a controlling authority. Ontologies are foundational to MBSE and essential in model scoping, creation, and usage.

Ontologies have many uses including the following:

- Sharing a common understanding of how information is structured and named among actors such as stakeholders and software databases
- Supporting domain knowledge reuse by facilitating specializations for multiple domains
- Enforcing the need for explicit domain assumptions which reduces ambiguity and simplifies changes should these become necessary
- Separating domain and operational knowledge which allows creating a system from its components independent of the components and the products it produces
- Enabling domain knowledge reasoning when the ontology is represented by formal, declarative statements

Ontologies capture an agreement on terminology and relationship usage but are neither a dictionary nor taxonomy. Ontologies enable consistent communication within a semantic domain but are not necessarily complete. Although the terms taxonomy and ontology are sometimes used interchangeably, the terms are fundamentally different. Taxonomies are hierarchically ordered and used to name, describe, and classify terms in a domain. Taxonomy ordering is determined by a set of consistent and unambiguous rules. Conversely, ontologies link domain concepts with relationships in ways that support deeper understanding.

For example, Fig. 9 shows part of a vehicle taxonomy [5]. The primary concept *Vehicle* is decomposed into specific types of vehicles: *Air Vehicle*, *Land Vehicle*, and *Water Vehicle*. This taxonomy can answer a question such as "What are the types of *Air Vehicles*?" or "What *Vehicle* type is a *Bicycle*?" However, it cannot answer the question, "How many passengers fit in a *Rowboat*?" without enumerating *Rowboat* types based on seating capacity, for example. Another enumeration based on speed and type is needed if we wanted to know how fast a *Rowboat* can travel. Taxonomies are easily traversed by binary and tree searches, clustering [8], taxonomic reasoning [9, 10].

Ontologies are the concepts and relationships for a specific domain and are more formal and more information-rich than taxonomies. Figure 10 shows an ontology using *is-a* relationships in which the subconcepts *Water Vehicle*, *Land Vehicle*, and *Air Vehicle* acquire the properties of the parent concept, Vehicles [5]. That is, a model in which Air Vehicle is instantiated will have the properties Lift Capacity, Weather Restrictions, and Max Distance in addition to the inherited properties, Propulsion Type, Steering Type, and Usage. Concept properties may be restricted as necessary; Fig. 11 shows constraints on a type of *Vehicle* called *Rowboat Type 1* [5].

As shown in Figs. 10 and 11, common components of ontologies are concepts, relationships, instances, axioms, restrictions, and rules.

Concepts: represent things or sets of *things* within a semantic domain. A *boxer* is a concept in the domain, *dog*. *Primitive concepts* are those having only the necessary condition(s) for belonging to a class, e.g., all members of the class, *Mammal*, must-have hair, three middle ear bones, mammary glands, and a neocortex. An animal lacking any of these properties cannot be a mammal. *Defined concepts* are properties that are both necessary and sufficient for class membership. The

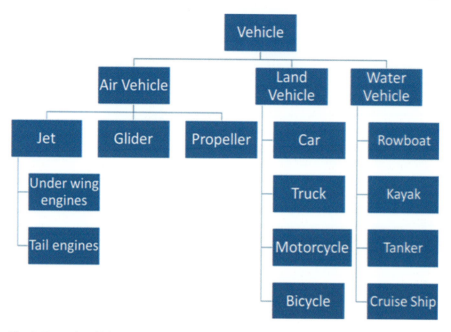

Fig. 9 Example vehicle taxonomy

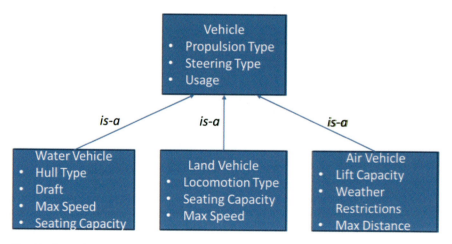

Fig. 10 Example vehicle ontology

condition that an adult person is ≥ 1.5 m in height is a necessary and sufficient condition of the class, *At Least or More Than 1.5 Meters Tall*.

Relationships are the associations made between entities and entity instances. Relationships might define parent/child associations, source to destination flows, composition, allocations, decompositions, and more.

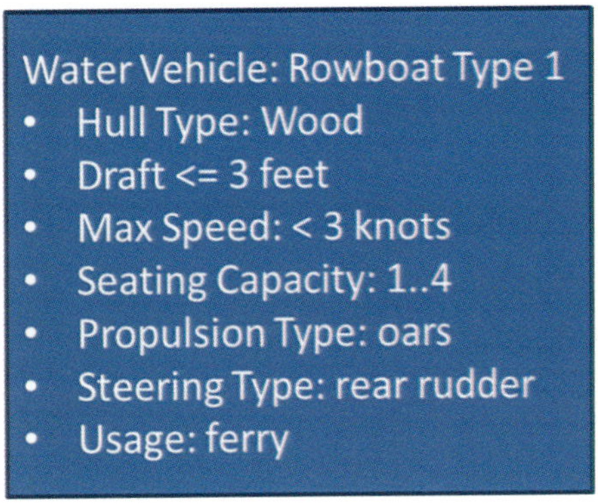

Fig. 11 Constraints on properties of *Rowboat Type 1*

> Water Vehicle: Rowboat Type 1
> - Hull Type: Wood
> - Draft <= 3 feet
> - Max Speed: < 3 knots
> - Seating Capacity: 1..4
> - Propulsion Type: oars
> - Steering Type: rear rudder
> - Usage: ferry

Instances: are concrete or abstract things represented by a concept. In a strict sense, because ontologies are domain conceptualizations, they should not include instances. An ontology and its associated instances are called a *knowledge base*. In some situations, determining whether a thing is a concept of an instance is difficult and depends on the domain and its application. In a spacecraft, a GPS receiver could represent a general class of GPS receivers or it could be an instantiation of a time and location device.

Axioms: are assertions that constrain class and instance values. An example rule might require that a spacecraft have dimensions less than x, y, and z to fit in a particular launch vehicle fairing.

Restrictions: determine what must be true for accepting an assertion as an input. The time of a spacecraft launch must occur within a launch window that avoids space debris.

Rules: are assert-guarantee statements describing logical conclusions made from assertions. If we assert that a spacecraft fits within a particular fairing, then its dimensions are less than x, y, and z.

State-of-the-Practice

Metamodel concepts, metamodeling languages, and tools have been around since the late 1970s [11] and are related to entity-relationship (ER) models [12] which are abstract data models. Metamodels and ER models comprise three primary constructs:

- Entities: sets of "things" having similar properties of interest within a domain. Entities may be decomposed into subtypes as suiting the domain and the intended usage of an entity set.
- Attributes: properties of an entity including identification, quantity, physical or logical characteristics, constraints, and so forth.

- Relationships: associations made between entities and entity instances. Relationships may be *strong* (an entity exists independent of other entities or *weak* (an entity's existence depends on another entity)). Common relationships include generalization/specialization, physical or logical flows from sources to destination, composition, aggregation, dependency, refinement, control, etc.

Over time, additional features and formalities have been added to the early metamodel and ER concepts. The paper by Bommel et al. explores processes for developing metamodels using a rule-based framework [13]. Hofstede et al. look at formal metamodel languages for data manipulation [14]. An examination of several constructs by Vianna et al., looks at metamodeling methods for the design and analysis of computer experiments [15].

More recently, a paper by Schön et al. discusses metamodeling issues related to requirement engineering (RE) in agile software development environments [16]. A common issue with RE in agile development is the lack of common terminology and means for organizing artifacts, discovering, iterating, and reviewing requirements. A metamodel is presented that comprises several metaclasses as shown in Table 1, and examples of metamodel instantiation are discussed for a Kanban-based development process that focuses on continuous improvement.

A large number of metamodel applications show their diversity and provide insight into development and usage. An interesting application assesses battlefield damage using Bayesian network simulation [17]. In this application, a metamodel-based mapping from conventional fault tree diagrams to a Bayesian representation supports an evaluation of the extent of damage that might occur in a hypothesized battlefield explosion. A biotechnology metamodel described in [18] looks at medical device code generation and debugging. That application defines abstractions for basic physical and software components such as controllers and switches and then extends those basics with abstractions for code generators, execution platforms, and run-time systems. Coordinated, multimetamodels used for healthcare systems are described in [19]. There are also many other papers on metamodeling fundamentals and applications [20–24].

Ontologies examples and development processes are well covered in the literature. A very readable and informative paper by Noy and McGuinness discusses developing ontologies through an amusing wine example [25]. Their approach

Table 1 RE metamodel for agile development

Metaclass	Metaclass properties
User	Users of the software product
Stakeholder	Individuals interested in and/or influence the product design
RE problems and patterns	Summary of problems and pattern usage
Domain	Name and domain description
Requirement	Description and value of a requirement
System	System infrastructure, goals, and description
Context	Description of system environment

comprises a multistep process that begins by determining domain scope and proceeds through looking for existing ontologies, enumerating terms, defining entities and an entity hierarchy, defining entity properties, determining property constraints, and ends by creating instances.

The semantic web comprises a set of standards set by the World Wide Web Consortium (W3C) that make data on the Internet machine-readable. Enabling technologies for the semantic web include the Resource Description Framework (RDF) [26] and Web Ontology Language (OWL) [27] which is supported by the Protégé language [28].

Creating a pizza ontology using Protégé [29] illustrates many of this chapter's concepts. The Protégé language consists of three primary components: *concepts* that define an aggregation of things, instances of concepts called *individuals*, and relationships that link concepts and individuals called *properties*. The term *triples* represents an ontological component in the form of subject-verb-object. For example, the phrase: "A pizza has cheese" in triples form might be "pizza hasTopping cheese" which can be represented graphically as in Fig. 12.

Assuming that we already have thought about the pizza domain, the main ontology categories of the Protégé might be the following:

- Pizza
- Pizza_base
- Pizza_topping

The main categories are associated with the following instances:

- Pizza_base ← Thick crust, Thin crust...
- Pizza_topping ← Tomato_topping, Mozzarella_topping, Spicy_beef_topping, and Pepperoni_topping...
- Pizza (primitive) ← Margherita, Hot_and_spicy, and Seafood...
- Pizzas (define) ← Vegetarian, Cheesey...

Basic pizza ontology properties are the following:

- Has_topping
- Has_base

Fig. 12 Graphical representation of a Protégé triple

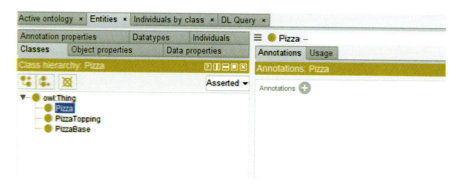

Fig. 13 OWL representation of pizza ontology

Figure 13 shows an OWL representation of a skeleton pizza ontology in the Protégé workbench. Protégé assumes that concept names are unique. Sameness and uniqueness are expressed using relationships *Equivalent-To* and *Disjoint-With* as shown in Fig. 14.

Properties are added that link concepts:

- A pizza has a deep pan base (hasBase).
- A pizza has mozzarella cheese topping (hasCheeseTopping).
- A pizza has tomato and cheese topping (hasTomatoTopping) and (has CheeseTopping).

Inverse properties are also possible. For example, "a pizza has a thick crust" has the inverse property, "A thick crust is a base of a pizza." That is, (*isBaseOf*) is the inverse of (*hasBase*) and (*haseBase*) is the inverse of (*isBaseOf*), similarly for toppings as shown in Fig. 15.

OWL also defines several useful relationship primitives such as functional and inverse functional, transitive, symmetric and antisymmetric, and reflexive and irreflexive. Properties can further link instances from a domain to instances in a range. That is, the domain of pizza can be linked to the range of pizza toppings through the hasTopping relationship. Similarly, the domain of pizza toppings can link to the range of pizzas using the isToppingOf relationship as shown in Fig. 16.

Property restrictions are class definitions that group instances by one or more object properties. An example might be a class of instances that have *at-least-one* "*hasTopping*" relationship to instances that are members of MozzarellaTopping. Existential restrictions describe a subset (*some*) of values from a set of restrictions. For example, a pizza must have a pizza base. Figure 17 illustrates a more complete pizza ontology that may be used with Protégé's reasoner for finding class inconsistencies. Figure 18 shows an error message if Class *Trouble* is replicated.

2 Semantics, Metamodels, and Ontologies

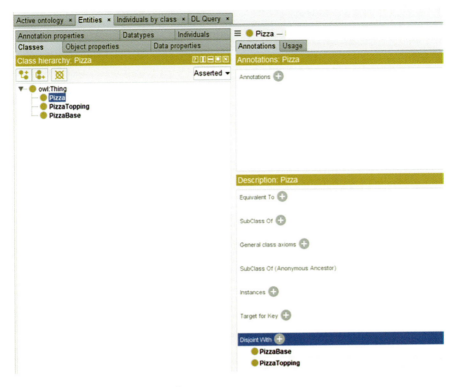

Fig. 14 Example use of *disjoint-with*

Best Practice Approach

Rigor and adherence to metamodels and ontologies are essential for developing unambiguous and well-formed models, and every reasonable attempt should be made in creating, maintaining, and enforcing consistent project-wide semantics. But equally important is remembering that models are made for representing certain aspects of a domain defined by stakeholder concerns and needs. Moreover, engineering projects have cost and schedule constraints that pace work and drive the artifacts produced in the development and deployment lifecycle. As with most systems engineering activities, the work that should be done and the work that can be done within cost and schedule almost always end up by descoping or eliminating work. MBSE offers no magic bullet here but can support a deeper understanding of the risks incurred when planned work must be scaled back. Looking at the ultimate goal of a project which is delivering a satisfactory system, it is reasonable to think about the time and effort necessary for achieving a gold-standard project ontology. Also important is the recognition that it is unlikely that any metamodel or ontology will be perfect. As systems engineers, we want model representations that are "good enough" for getting our job done while not fretting over every nuance. Moreover,

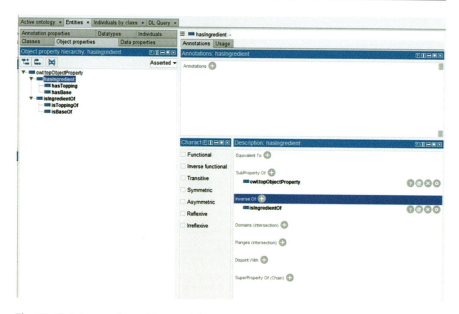

Fig. 15 Object properties and inverse object properties

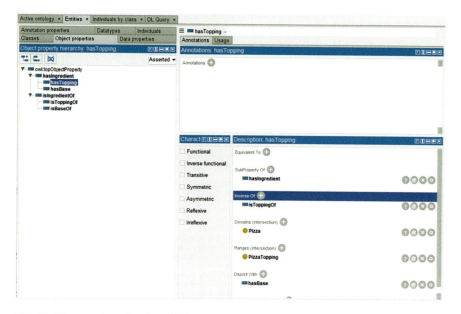

Fig. 16 Pizza ontology showing additional relationship primitives, domain, and range

terminology, focus on the design hierarchy, and systems engineering processes will almost certainly differ in large, multiagency collaborations. We are not advocating abandoning the concept of a unified ontology under certain conditions. However,

2 Semantics, Metamodels, and Ontologies   33

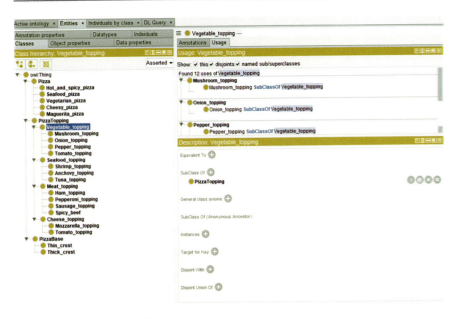

Fig. 17 Expanded pizza ontology

Fig. 18 Reasoner-reported error for duplicated use of a class

from a real-world perspective, we do advocate consideration of practicalities in ontology construction, configuration management, and use.

Consider an example comprising three semiautonomous organizations developing various aspects of a weather analysis and prediction system-of-systems (SoS) that provides a suite of analysis products developed from data captured by a diverse collection of space, aircraft, and ground sensors. One organization might be responsible for defining the logical architecture in terms of high-level functions and interfaces. Another organization might be responsible for creating the infrastructure for managing weather scientist requests, collecting weather data, sending the weather data to the proper analysis applications, and then delivering the processed products to the requestor. At another level are the weather scientists who specify what data they want captured, when it should be collected, what processing they need, and where the results go. While all of these stakeholders are interested in the same system, they have different interests and use different terminology. Returning to the earlier point about limitations in effort and the desire to find "good enough" solutions, this three-tiered system seems fraught with unsolvable challenges.

OWL and Protégé already anticipate ontological conflicts through the use of *Equivalent-To* and *Disjoint-With* relationships. Moreover, as noted in the previous section, the reuse of concept names for different purposes is flagged as an error. This is a very powerful means for both allowing stakeholders to create their own semantics and still have an overall consistent ontology. Of course, a drawback is that either all parties need access to the Protégé file or someone (or designated group) has to own the Protégé file and take inputs from all stakeholders.

The paper by Rabbi et al. [19] describes a collaborative approach based on a linguistic extension of Diagram Predictive Frameworks (DPF) using formalized metamodeling and model transformations. Their approach makes use of category theory and graph transformations with application to healthcare applications. The essence of the approach is the definition of *atomic constraints* defined by predicate signatures (a set of named predicates each having a graphical representation). Predicates impose constraints on some aspect of a metamodel graph and the associated graphical representation (call an *arity*). Layers of metamodel are coordinated by assigning coordination predicates between layers that define how *arities* relate to each other.

Formal, rigorous methods are encouraged when practical; however, the reality is that these methods are likely not sensible in most situations. For the most part, system engineers are not familiar with formal methods. But more importantly, a system engineer's job is developing a system and not in developing mathematically provable system representations that require significant effort in a usually tight schedule. Evaluating the degree of ontology formality and coordination could profit by thinking about the three factors below.

What Is the Ontology Structure and Ownership?

We generally think of an MBSE model housed in a centralized repository that is accessible by all stakeholders. The central source of truth concept operates well when stakeholders all work within the same problem space, have easy access to the central repository (which could be distributed and virtualized), and use identical or compatible tools. However, there are frequent situations in which one or more of these conditions do not hold. Consider the weather analysis and prediction SoS that collects inputs from heterogeneous and geographically distributed sensors and produces multiple data products that are disseminated to geographically distributed users. This SoS could include localized applications for gathering and filtering inputs, distributed applications that fuse and analyze data from numerous sources, computational and communication framework applications, data storage and

2 Semantics, Metamodels, and Ontologies

retrieval applications, and so forth. Given the breadth and depth of this SoS, it is likely that stakeholders will each have their vision of the SoS's purpose, structure, and behaviors and each will insist on using their domain-specific semantics.

There are general modeling types that influence ontology form and content:

- In the canonical multiview, MBSE model stakeholders collaborate through a central source of truth. For these models, semantics are consistent with a *simple* ontology which is in a single, hierarchical structure. For many efforts, a simple model is ideal and there are good tools available for collaboration and analysis. While large, simple ontologies are conceptually possible, avoiding unintended consistency flaws requires a great deal of attention to development and configuration management by the ontology owners and modelers.
- Hierarchical multimodeling comprises hierarchically combined distinct models that take advantage of unique knowledge and semantics. While each model might use its own ontology, the semantics of inter-ontology linkage relationships must agree since inconsistent intramodel semantics can cause global reasoning problems. Federated, multimodel, structures can avoid some of the problems that might result in attempting a single, simple ontology for diverse stakeholders.
- Amorphous heterogeneity comprises multiple distinct models that are arbitrarily combined. An example is a model that allows multiple, simultaneous interaction mechanisms between model components. Typically, amorphous heterogenic models are associated with ontological *semantic heterogeneity* that occurs when domain-wide agreements are either not possible, not practical, or not needed. Amorphous heterogeneity implies loose inter-model semantics which could impact big picture reasoning and collaboration. There may be a need for a mapping that translates semantics from one ontology to another.

How Much Commonality Is There?

Somewhat related to the previous question is determining the degree of commonality in stakeholder viewpoints. Consider the situation in Fig. 19a that shows overlapping

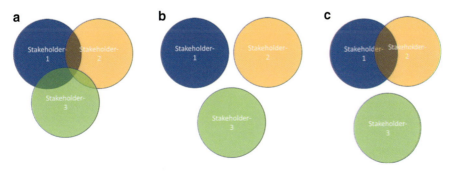

Fig. 19 (a) Overlapping viewpoints (b) Disjoint viewpoints (c) Partially overlapping viewpoints

viewpoints for three stakeholders. Sufficient overlap in viewpoint semantics might exist such that a simple ontology suffices, even if the ontology is large and is implemented by semiautonomous organizations that use incompatible tools. Figure 19b depicts a situation in which there is no viewpoint overlap which might imply semantic heterogeneity if views conforming to those viewpoints have irreconcilable semantic differences. Lastly, Fig. 19c illustrates a gray area in which some viewpoints overlap, but not necessarily all viewpoints. The situation in Fig. 19c could indicate that certain aspects of an ontology need high consistency while other portions can be specialized for disconnected stakeholder viewpoints.

What Is the Process for Developing an Ontology?

As a general rule, a domain ontology should be owned by all stakeholders in the sense that once established, all stakeholders should agree to them and will abide by them. Additions or modifications to the ontology are controlled by a configuration management authority so that individual stakeholders cannot unilaterally make changes. But as noted above, an unresolvable divergence of opinion may occur.

Noy and McGuiness [25] describe a common-sense and very practical ontology development process consistent with the simple-model concept described above. By considering the issues above though, the Noy and McGuiness process is readily adaptable to most situations. Their process comprises seven steps:

Step 1: Determine the Domain and Scope

Four fundamental questions are addressed in this step: (1) What is the domain for the ontology? (2) How will the ontology be used? (3) What are the questions and issues that must be answerable by instantiations of the ontology? (4) Who will use and maintain the ontology? Answers to these questions will not and need not be static but rather reflect an understanding at a particular time in the development of the ontology and the model(s) that are instantiated from it.

The domain question should be answered as narrowly as possible while still including factors of interest to all stakeholders. We would likely not want a domain as big as all spacecraft, for example, but rather want a focus on spacecraft that provides weather information. With time, the domain might extend to weather-collecting aircraft and ground sensors as needed for addressing stakeholder viewpoints.

The second question involves understanding whether the ontology will be used purely for reasoning as might be performed in Protégé, instantiated in one or more MBSE models, or a combination of both. When used for reasoning, the ontology will necessarily require significant rigor so that results are consistent and reliable. When used as the basis for models, rigor may be relaxed as noted above, but reductions in rigor can result in confusion and inconsistencies.

Question 3 may be addressed by creating a collection of competency questions. These are questions that stakeholders want to be answered by the ontology or an instantiation of the ontology. Example questions for a weather spacecraft might be the following:

1. What sensors are needed for measuring water content in the snow?
2. What sensors are needed for measuring wind speed?
3. What sensors are needed for measuring ground temperature?
4. What sensors are needed for measuring barometric pressure?
5. What sensors are needed for measuring atmospheric carbon dioxide?
6. How can the system evaluate climate and carbon dioxide changes?
7. What inclination provides consistent measurements independent of sun position?
8. What is the weather like now in London?
9. What is the 5-day weather outlook for Los Angeles?
10. What orbit covers the greatest percentage of the Earth's surface?
11. How many spacecraft are needed to guarantee measurements of all covered areas at least every hour?
12. How will sensor data get collected and weather analyses get disseminated?

Determining who uses and maintains the ontology relates to the discussion in section "What is the Ontology Structure and Ownership?" and is impacted by the competency questions. Sensor-related questions are important issues for spacecraft designers and climatologists. The list of competency questions implies a diverse knowledge base comprising weather science, spacecraft design, spacecraft orbits, ground systems, application processing, and weather instrumentation. Stakeholders are likely orbitologists, meteorologists, ground systems designers, spacecraft operations technicians, spacecraft designers, instrumentation designers, and the general public.

While overlap exists in the concerns of these stakeholders, Fig. 19c most likely represents their viewpoints and to an extent the influence they have on the content and structure of the ontology. Consider that someone in London or Los Angeles will want to know if rain is likely in the next few days but is likely unconcerned about how the prediction was reached. Climatologists will want a say in sensors, orbits, data storage, data analysis, and result dissemination but will not care about spacecraft design details. Spacecraft designers likely care about the mass, dimensions, thermal characteristics, power consumption, uplink, and downlink needs, attitude control requirements, and data interfaces that support the instrumentation but likely are uninterested in the semantics of the data collected by those instruments.

Ideally, a simple, system-wide ontology is the best option for any domain; given the factors above, a good candidate for the weather system might be a federated ontology. Depending on the best practical ontology structure, the next steps may be completed by a single team, multiple collaborating teams, or multiple independent teams.

Step 2: Look for Existing Ontologies

Ontologies have been created for great many domains that may be identical to or very similar to a domain of interest. In many cases, these ontologies have been well tested and used by a large community and therefore worthy of consideration, even if tailoring is needed for a specific application. Specific ontologies are easily found by web searches, in references [35–37], or by looking at the abbreviated online ontology list below:

- Indiana University: https://info.sice.indiana.edu/~dingying/Teaching/S604/OntologyList.html
- EMBL-EBI: https://www.ebi.ac.uk/ols/ontologies
- University of Michigan: https://guides.lib.umich.edu/ontology/ontologies
- W3C: https://www.w3.org/wiki/Good_Ontologies
- Aber-OWL: http://aber-owl.net/ontology
- Meteorological Ontology:
- Protégé Ontology Library: https://protegewiki.stanford.edu/wiki/Protege_Ontology_Library

Below are a few of a large number of spacecraft-specific ontologies:

- Satellite Parts Ontology https://indico.esa.int/event/310/contributions/4572/attachments/3495/4643/Presentation_-_1400_-_Kobkaew_Opasjumruskit.pdf
- Space Mission Design Ontology [30]
- Space Systems Data Exchange Ontology [31]

Step 3: Enumerate Key Terms

In this step, stakeholders create an informal list of the key terms that are either needed for addressing competency questions or desired discussion topics. The list includes term properties if known and some concept of what is important about the term. For a weather satellite, important terms might be mass, orbit, instrumentation types, instrumentation accuracy, Earth coverage, and so forth.

Steps 4 and 5: Define Concepts, Concept Hierarchy, and Concept Properties

Key terms and competency questions drive the definition of ontology concepts, structure, and concept properties. In Step 4, Noy and McGuinness develop classes and class hierarchy which is equivalent to defining ontology concepts and concept hierarchy. In Step 5, they create class properties for the classes in Step 4. However, it is also noted that because Step 4 and Step 5 are tightly entwined, the recommendation is doing them in parallel.

Concepts and concept hierarchy can be defined in many ways that loosely fit into three categories: top-down, bottom-up, and combination:

- Top-down begins by enumerating the most general domain concepts and then decomposes those into specializations. For example, we might start with the concept, *spacecraft*, and create subconcepts, "*attitude determination and control, telecommunications, electrical power distribution, command, and data handling, thermal control, propulsion,* and *instrumentation*."
- Bottom-up starts with the most specific concepts and then groups these into more general concepts. For example, we might begin with *reaction wheel, star tracker,* and *inertial measurement unit* and generalize these as *attitude determination and control*.
- Combination employs both top-down and bottom-up as suits each concept. In some cases, specific subconcepts may not yet be known in which case top-down is used. In other situations, it may not yet be clear how to group concepts into the most general concepts so a bottom-up approach could allow grouping into intermediate concepts.

No one way is better than another and is best chosen based on stakeholder preferences and viewpoints and the level of detail available in the domain. However, Step 4 by itself will not suffice for fully answering the competency questions. Additional information is needed; in the form of concept, properties are added in Step 5 that describe the internal details of the concepts developed in Step 4.

Many of the unallocated terms from Step 3 are likely properties of the concepts from Step 4. Each unallocated term is examined and a determination made of the concept that best describes it. Some terms may belong to more than one concept, and some concepts may need properties not found in the list of key terms; however, best practice assigns properties to the most general concept consistent with the property. For example, if the most general concept in a domain is *system electronics*, then attach the *mass* property here rather than at subconcepts because subconcepts will inherit the *mass* property from *system electronics* making the ontology less prone to confusion and more readily changed if needed.

Terms allocated to a concept become *slots* included with the concept. There are four general types of properties:

- Intrinsic properties are aspects of a concept that must exist if the concept exists. For example, a physical component will have a mass.
- Extrinsic properties are external properties such as the name of a manufacturer that supplied a physical component.
- Part properties define physical or logical entities that define a concept hierarchy, e.g., a reaction wheel has a controller, interface board, motor, and a rotating mass.
- Relationships explain how a concept interacts with, depends on, is used by, etc., other concepts or subconcepts.

Step 6: Define Slot Facets

Facets are constraints on slots such as cardinality, range of values, and units. Not all slots require facets, and overuse of facets can hamper ontology utility by making it overly restrictive. Consider that if an ontology constrains the *mass* of an electronic component, then any component outside of that constraint cannot exist in an instantiation of the ontology. Typical facet types include the following:

- *Strings*: comprises textual information.
- *Number*: usually specified as floating-point or integer.
- *Boolean*: value must be either true or false.
- *Enumerated*: specifies an acceptable value range.
- *Instance*: indicates that a slot is an instance of a concept; the allowed concepts are its *range,* and the concept that the instance belongs to is its *domain*.

Noy and McGuinness recommend additional considerations for Instance slot range and domain:

- Use the most general concepts or concept when defining range and domain, but overly general selections are problematic. An *Instance* slot should describe all relevant domain concepts, and all instances of range classes should pertain to the slot. However, if the *thing* is the most general concept, then using it for the range of an Instance is too broad.
- Remove subconcepts if the range of an *Instance* includes both a concept and its subconcept(s).
- Add a concept to the range or domain if a slot contains all of its subconcepts but not the concept itself.
- If the list of concepts within the range or domain of a slot contains only a subset of the subconcepts, then the concept should be reconsidered for use as a range definition only.

Step 7: Create Instances

In this final step, instances of concepts are defined by selecting a concept, creating a specific instance of that concept, and then assigning slot values that are consistent with slot facets. Typically, instantiations become features of models but more importantly support answering the competency questions. There is a somewhat fuzzy line between what should be a concept and what should be an instance of a concept. To an extent, the decision can be based on the purpose of the ontology and the needed granularity for assessing competencies. Another consideration is that an overly specific ontology might limit reuse.

Ontology Development and Analysis Tools

We close out this section by listing a few tools available for ontology development and reasoning:

- Protégé [28]
- RDFox: https://www.w3.org/2001/sw/wiki/RDFox
- Oracle Spatial and Graph: https://www.w3.org/2001/sw/wiki/Oracle

2 Semantics, Metamodels, and Ontologies

- Apache Jena: https://www.w3.org/2001/sw/wiki/Apache_Jena
- Mobi: https://www.w3.org/2001/sw/wiki/Mobi
- AllegroGraph: https://www.w3.org/2001/sw/wiki/AllegroGraph
- DogmaModeler: http://www.jarrar.info/Dogmamodeler
- Onto-Animal Tools: http://www.hegroup.org/ontozoo

Illustrative Example

This section demonstrates the application of the steps discussed in section "Ontology Development and Analysis ToolsS9" for a spacecraft that monitors the effects of climate change. It is assumed that this spacecraft performs functions similar to the Soil Moisture Active/Passive (SMAP) spacecraft [32] and the Orbiting Carbon Observatory (OCO) [33].

Competency Questions

What is the orbit?
What instruments are available for space-based moisture observation?
What instruments are available for measuring atmospheric CO_2?
What is the ground pointing accuracy?

Existing Ontologies

In section "What is the Process for Developing an Ontology?", we listed a few ontologies that are suitable candidates for our weather spacecraft. In particular, the Satellite Parts Ontology seems like a good starting point because it includes typical physical components and component properties such as the following:

- Star tracker
- Magnetometer
- Battery
- Magnetic torque rods
- Reaction wheels
- Solar panels
- Battery
- Sun sensors
- Propulsion
- Earth sensors
- Antennas
- Computers

These are all valid concepts for an ontology; however, for this exercise we want more generality. Additionally, some of the components might not make sense for our satellite. For our example, we will use concepts such as command and data handling, electric power, attitude determination and control, telecommunications, and thermal control. The components in the Parts Ontology then are instances of a more general ontology used for this example.

Searches for space-based CO_2 sensor ontologies and soil moisture sensor ontologies were not fruitful; however, NASA studies mention several CO_2 sensors [34], and the sensors used by SMAP are described in [32]. This exercise will generalize these as *Climate Sensors*.

Key Terms

Our ontology will need the following terms: orbit type, orbit_altitude (apogee and perigee), orbit inclination, orbit period, attitude accuracy, climate sensor, sensor accuracy, sensor type, sensor mass, sensor power, ground-pointing accuracy, bus, and spacecraft.

Concepts, Concept Hierarchy, Slots, and Facets

Many options are available for capturing the knowledge needed for the competency questions making use of key terms. For simplicity, we can separate the ontology into two concepts: things related to basic spacecraft operation we allocate to the *bus* concept, and things related to the sensors are allocated to the *climate sensor* concept. The spacecraft concept is a composition of *bus* and *climate sensor*.

The candidate ontology below comprises two primary concepts: *bus* and *climate_sensor*. The main concepts respectively have slots for their related terms. Slot facets are shown in parentheses. Instances of spacecraft have a textual name, there must be at least one *climate_sensor*, orbit_type is restricted to the five options, and units are specified for several slots: kilometers (Km), meters (m), kilograms (Kg), watts, and *orbit_inclination* are specified in degrees as a floating-point number.

spacecraft (name:text) *bus*
spacecraft_type (text)
spacecraft_mass (Kg)
orbit_type (polar, sun-synchronous, elliptic, equatorial, and circular)
orbit_period (minutes:float)
orbit_apogee (Km)
orbit_perigee (Km)
orbit_inclination (degrees:float)
attitude_accuracy (degrees:float)
climate_sensor (1..*), (name:text)
ground_resolution (Km)
sensor_type (text)
sensor_mass (Kg)
sensor_power (watts)

Instances

Although creating instances is not always necessary, instances help confirm that the competency questions are represented in the ontology. Using the SMAP example, an instantiation might be the following:

spacecraft: SMAP *bus*
spacecraft_type (3-axis stabilized)
spacecraft_mass (94)
orbit_type (sun-synchronous)
orbit_period (98.5)
orbit_apogee (680.9)
orbit_perigee (683.5)
orbit_inclination (98.1237)
attitude_accuracy (0.1)
climate_sensor (SAR)
ground_resolution (1)
sensor_type (synthetic aperture radar)
sensor_mass (58.6)
sensor_power (550)
climate_sensor (RAD)
ground_resolution (30)
sensor_type (radiometer)
sensor_mass (30)
sensor_power

Power consumption for the radiometer is left blank indicating that this value is not available. Although we cannot fully determine sensor power, we can confirm that this ontology supports the knowledge base necessary for answering the competency questions. Note too that this example uses a somewhat arbitrary format for the ontology rather than using OWL or Protégé. Languages such as SysML, UML, Promela, and so forth are common languages for ontology representation.

Chapter Summary

This chapter provides an overview of metamodels, semantics, and ontologies which are the foundations of MBSE. At its core, MBSE depends heavily on consistent terminology and relationships. Without that consistency, model views may be unclear, misinterpreted, and possibly conflict with other views.

Through theory and examples, we have provided recommendations for developing and using metamodels and ontologies. While careful development is essential, equally important is using the metamodels and ontologies throughout a domain. Other chapters in this handbook examine specialized ontologies and ontology usage.

Cross-References

▶ Model-Based System Architecting and Decision-Making
▶ Overarching Process for Systems Engineering and Design

References

1. Mars Climate Orbiter Mishap Investigation Board, Phase 1 Report, November 10, 1999, http://sunnyday.mit.edu/accidents/MCO_report.pdf
2. Promela Manual http://spinroot.com/spin/Man/Intro.html
3. (Eugene N. Parker, 2006), H.L., and McDermott, K.B., "Creating False Memories: Remembering Words Not Presented in Lists," *Journal of Experimental Psychology: Learning Memory, and Cognition*, 24, 803–814, 1995
4. *Laws of the Game* published by the International Football Association Board (IFAB) (https://ussoccer.app.box.com/s/xx3byxqgodqtl1h15865/file/683837728699)
5. Madni, A.M, Sievers, M.W (2018) Model-based systems engineering: Motivation, current status, and research opportunities, Systems Engineering, 21:172–190
6. Alvarez, J.A., Evans, A., Summut, P., *MML and the metamodel architecture, Workshop on Transformationsin UML (WTUML'01)*, ETAPS'01, Genova, Italy, 2001.
7. Gruber, T.R., A Translation Approach to Portable Ontology Specification. Knowledge Acquisition 5: 199–220, 1993
8. Grygorash, O., Zhou, Y. and Jorgensen, Z., "Minimum Spanning Tree Based Clustering Algorithms," Proc. 18th IEEE International Conf. on Tools with Artificial Intelligence (ICTAI'06), pp. 73–81, 2006
9. Tenenberg, J., "Taxonomic Reasoning," *IJCAI'85 Proceedings of the 9th international joint conference on Artificial intelligence.* (1): 191–193, 1985
10. Bergamaschi, S. and Satori, C. (1992). "On Taxonomic Reasoning in Conceptual Design," *ACM Transactions on Database Systems* (TODS), 17(3): 385–422, September, 1992
11. Teichroew, D., Macasovic, P., Hershey, E., Yamamoto, Y., Application of the entity relationship approach to information processing systems modeling. In Entity-Relationship Approach to Systems Analysis and Design, P. P. Chen (Ed.), Amsterdam: North-Holland, 1980
12. Chen, P., "The Entity-Relationship Model – Toward a Unified View of Data," ACM Transactions on Database Systems (1) 9–36, March 1976
13. van Bommel, P., Proper, H.A., Hoppenbrouwers, S., and van der Weide, Th. P., "Exploring Modelling Strategies in a Meta-Modelling Context," R. Meersman, Z. Tari, P. Herrero et al. (Eds.): OTM Workshops 2006, LNCS 4278, pp. 1128–1137, 2006
14. Hoftede, A.H.M, Proper, H.A., and van der Weide, Th. P., "Formal Definition of a Conceptual Language for the Description and Manipulation of Information Models," Information Systems 18(7): 489–523, 1993
15. Vianna, F., Simpson, T., Balabanov, V., and Toropov, V., "Metamodeling in Multidisciplinary Design Optimization: How Far Have We Really Come?," AIAA Journal 52(4): 670–690, April, 2014
16. Eva-Maria Schön, E-M, "A Metamodel for Agile Requirements Engineering," https://www.scirp.org/journal/paperinformation.aspx?paperid=90250
17. Li˙ Y., Xing, C., Fang, S.. and Duan˙ Y. "Research on Damage Tree Bayesian Network Simulation Metamodel Based on Fault Tree Analysis," Journal of Physics: Conference Series, Volume 1314, 3rd International Conference on Electrical, Mechanical and Computer Engineering 9–11 August 2019, https://iopscience.iop.org/article/10.1088/1742-6596/1314/1/012089/pdf

18. Tolvanen, J.-P., Djukić, V., and Popov, A., "Metamodeling for Medical Devices: Code Generation, Model-Debugging and Run-Time Synchronization," Procedia Computer Science 63 (2015) 539–544, https://www.sciencedirect.com/science/article/pii/S187705091502517X
19. Rabbi, F., Lamo, Y., and MacCaull, W., "Coordination of Multiple Metamodels, with Application to Healthcare Systems," Procedia Computer Science 37, 473–480, 2014
20. Assar, S. (2015). Meta-Modeling: Concepts, Tools, and Applications. *IEEE International Conference on Research Challenges in Information Science (RCIS)*, 2015, Athens, Greece, May 13–15.
21. Saeki, S., Kaiya, H. (2006). "On Relationships Among Models, Meta Models, and Ontologies," *Proc. 6th OOPSLA Workshop on Domain-Specific Modeling*," 2006.
22. Jenkins, S. (2010). "Ontologies and Model-Based Systems Engineering," Presentation at *INCOSE International Workshop 2010*, Workshop on Model-Based Systems Engineering, Phoenix, AZ.
23. Clark, T., Evans, A., Kent, S. (2002) "Engineering Modeling Languages: A Precise Meta-Modeling Approach," *International Conference on Fundamental Approaches to Software Engineering*, 2002, pp. 159–173
24. Beydoun, G. et al., "FAML: A Generic Metamodel for MAS Development," in *IEEE Transactions on Software Engineering*, 36(6): 841–863, Nov.–Dec. 2009
25. Noy, N., and McGuinness, D., "Ontology 101: A Guide to Creating Your First Ontology," http://www.ksl.stanford.edu/people/dlm/papers/ontology101/ontology101-noy-mcguinness.html
26. W3C, "Resource Description Framework (RDF)" https://www.w3.org/RDF
27. W3C, "Web Ontology Language (OWL)," https://www.w3.org/OWL
28. Protégé, https://protege.stanford.edu
29. Pizzas in 10 Minutes, https://protegewiki.stanford.edu/wiki/Protege4Pizzas10Minutes
30. Berquand, A., Moshfeghi, Y., and Riccardi, A., "Space mission design ontology: extraction of domain-specific entities and concepts similarity analysis.," ResearchGate https://www.researchgate.net/publication/338400758_Space_mission_design_ontology_extraction_of_domain-specific_entities_and_concepts_similarity_analysis
31. Bally, J., Boneh, T., Nicholson, A., and Korb, K., "Developing an Ontology for Meteorological Forecasting Process," Decision Support in an Uncertain and Complex World: The IFIP TC8/WG8.3 International Conference, pp. 70–81, 2004
32. Entekhabi, D. et al. "The Soil Moisture Active Passive (SMAP) Mission." Proceedings of the IEEE 98.5 (2010): 704–716
33. Eldering, A., O'Dell, C. W., Wennberg, P. O., Crisp, D., Gunson, M. R., Viatte, C., Avis, C., Braverman, A., Castano, R., Chang, A., Chapsky, L., Cheng, C., Connor, B., Dang, L., Doran, G., Fisher, B., Frankenberg, C., Fu, D., Granat, R., Hobbs, J., Lee, R. A. M., Mandrake, L., McDuffie, J., Miller, C. E., Myers, V., Natraj, V., O'Brien, D., Osterman, G. B., Oyafuso, F., Payne, V. H., Pollock, H. R., Polonsky, I., Roehl, C. M., Rosenberg, R., Schwandner, F., Smyth, M., Tang, V., Taylor, T. E., To, C., Wunch, D., and Yoshimizu, J.: The Orbiting Carbon Observatory-2: first 18 months of science data products, Atmos. Meas. Tech., 10, 549–563, https://doi.org/10.5194/amt-10-549-2017, 2017
34. NASA Carbon Monitoring System, Phase 1 Report, May 2014, https://carbon.nasa.gov/pdfs/CMS_Phase-1_Report_Final.pdf
35. Hansen, L.J., Lanza, D., and Pasko, S., "Developing an ontology for standardizing space systems data exchange," 2012 IEEE Aerospace Conference, Big Sky, MT, pp. 1–11, 2012
36. Guizzardi, G. and Almeida, João Paulo, "Engineering Ontologies and Ontology Engineering," February, 2020, ResearchGate https://www.researchgate.net/publication/339313928_Engineering_Ontologies_and_Ontologies_for_Engineering
37. De Farias, T.M., Roxin, A., and Nicolle, C., "FOWLA, A Federated Architecture for Ontologies," August 2015, ResearchGate https://www.researchgate.net/publication/280047091_FOWLA_A_Federated_Architecture_for_Ontologies/link/57693fc608ae3bf53d3315c1/download

Michael Sievers is Senior Systems Engineer/Systems at the California Institute of Technology, Jet Propulsion Laboratory (JPL), Pasadena, CA, and an adjunct lecturer in Systems Architecting and Engineering Department, University of Southern California (USC), Los Angeles CA. He is responsible for the design and analysis of spacecraft avionics, software, end-to-end communication, and fault protection as well as ground system modeling and mission concepts of operation. He has conducted research at JPL and USC in the areas of trust and resiliency and performed a DSN study that investigated dynamic trust assessments. At USC he teaches classes in systems and systems of systems architecture, resilience, and model-based systems engineering. Dr. Sievers is an INCOSE Fellow, AIAA Associate Fellow, and IEEE Senior Member.

MBSE Methodologies

3

Jeff A. Estefan and Tim Weilkiens

Contents

Introduction	48
Key Concepts and Definition	49
Best Practice Approach	50
Object-Oriented Systems Engineering Method (OOSEM)	51
Object-Process Methodology (OPM)	53
Integrated Systems Engineering and Pipelines of Processes in Object-Oriented Architectures (ISE&PPOOA)	55
Illustrative Example/Case	57
The Systems Modeling Toolbox (SYSMOD) in Review	58
A Few Examples of SYMOD Core Concepts in Greater Detail	59
Challenges and Gaps	70
Evaluating and Comparing Methodologies	73
Methodology Adoption and Tailoring	74
Expected Advances in the Future	79
Chapter Summary	81
Systems Modeling Toolbox (SYSMOD)	81
Cross-References	82
References	82

Abstract

A Model-Based Systems Engineering (MBSE) methodology can be characterized as the collection of related processes, methods, tools, and environment used to support the discipline of systems engineering in a "model-based" or "model-

J. A. Estefan (✉)
NASA/Jet Propulsion Laboratory, Pasadena, CA, USA
e-mail: Jeffrey.A.Estefan@jpl.nasa.gov

T. Weilkiens
oose Innovative Informatik eG, Hamburg, Germany
e-mail: tim.weilkiens@oose.de

driven" context. In this chapter, a cursory description of some of the leading MBSE methodologies used in industry today is provided. It is expected that during an organization's candidate MBSE methodology assessment process (including impact to native processes and procedures), a tool survey and assessment will occur concurrently or shortly thereafter, followed by selection and piloting of relevant tools. This latter effort requires participation from the organization's systems engineering practitioner community because that is the community that will most heavily be using the tools. Although a tool survey is beyond the scope of this chapter, guidance and a novel technique for MBSE methodology assessment is provided. In addition, future directions and areas of standardization of various MBSE methodologies is also described.

Keywords

Process · Methods · Tools · Environment · Methodology · Methodologies · MBSE methodology · MBSE methodologies · Object-Oriented Systems Engineering Method (OOSEM) · Object-Process Methodology (OPM) · Systems Modeling Toolbox (SYSMOD) · Integrated Systems Engineering and Pipelines of Processes in Object-Oriented Architectures (ISE&PPOOA)

Introduction

A *methodology* is defined as a collection of related processes, methods, and tools [1]. A *MBSE methodology* can thus be characterized as the collection of related processes, methods, and tools used to support the discipline of systems engineering in a "model-based" or "model-driven" context [2]. In this chapter, section "Key Concepts and Definition" characterizes the difference between methodologies and processes, methods, tools, and environment. Section "Current State of the Practice/Current State of the Art" provides a brief survey of a selected set of vendor-neutral MBSE methodologies that are widely published and currently used in the current state of the practice today, and most of which are continually evolving to keep up with the current state of the art. Section "Illustrative Example/Case" provides an illustrative example of selected methods and practices for the Systems Modeling Toolbox (SYSMOD) methodology as an example [3]. In addressing current challenges and gaps, section "Challenges and Gaps" describes both a novel approach for evaluating and comparting candidate MBSE methodologies as well as a recommended set of training and consulting practices to help an organization succeed in MBSE methodology adoption. These are the primary challenges, and not so much a candidate set of "off-the-shelf" MBSE methodologies per se. Finally, section "Expected Advances in the Future" looks at a few key expectations for advancement in the future of MBSE methodology development including industry standardization.

Key Concepts and Definition

The word *methodology* is often erroneously considered synonymous with the word *process*. For purposes of this handbook, the following definitions from [1] are used to distinguish methodology from process, methods, and tools:

- A *process* (*P*) is a logical sequence of tasks performed to achieve a particular objective. A process defines "WHAT" is to be done, without specifying "HOW" each task is performed. The structure of a process provides several levels of aggregation to allow analysis and definition to be done at various levels of detail to support different decision-making needs.
- A *method* (*M*) consists of techniques for performing a task; in other words, it defines the "HOW" of each task. (In this context, the words "method," "technique," "practice," and "procedure" are often used interchangeably.) At any level, process tasks are performed using methods. However, each method is also a process itself, with a sequence of tasks to be performed for that particular method. In other words, the "HOW" at one level of abstraction becomes the "WHAT" at the next lower level.
- A *tool* (*T*) is an instrument that, when applied to a particular method, can enhance the efficiency of the task, provided it is applied properly and by somebody with proper skills and training. The purpose of a tool should be to facilitate the accomplishment of the "HOWs." In a broader sense, a tool enhances the "WHAT" and the "HOW." Most tools used to support systems engineering are computer- or software-based, which is also known as Computer-Aided Engineering (CAE) tools.

Based on these definitions, a *methodology* can be defined as a collection of related processes, methods, and tools. A methodology is essentially a "recipe" and can be thought of as the application of related processes, methods, and tools to a class of problems that all have something in common [1]. Extending this definition a bit further, an *MBSE methodology* can thus be characterized as the collection of related processes, methods, and tools used to support the discipline of systems engineering in a "model-based" or "model-driven" context [2].

Associated with the above definitions for process, methods (and methodology), and tools is environment. An environment (E) consists of the surroundings, external objects, conditions, or factors that influence the actions of an object, individual person, or group [1]. These conditions can be social, cultural, personal, physical, organizational, or functional. The purpose of a project environment should be to integrate and support the use of the tools and methods used on that project. An environment thus enables (or disables) the "WHAT" and the "HOW."

A visual graphic that depicts the relationship between the so-called "PMTE" elements (Process, Methods, Tools, and Environment) is illustrated in Fig. 1 along with the effects of technology and people on the PMTE elements.

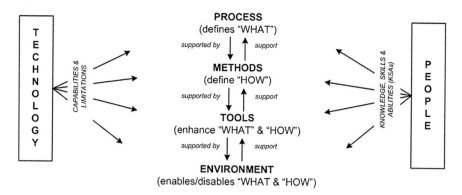

Fig. 1 The PMTE elements and effects of technology and people [1]. (Copyright © 1996 CRC Press, Inc. Used by permission)

As stated by Martin [1], the capabilities and limitations of technology must be considered when developing a systems engineering development environment. This argument extends, of course, to an MBSE environment. Technology should not be used "just for the sake of technology." Technology can either help or hinder systems engineering efforts. Similarly, when choosing the right mix of PMTE elements, one must consider the knowledge, skills, and abilities (KSA) of the people involved [1]. When new PMTE elements are used, often the KSAs of the people must be enhanced through special training and special assignments.

Best Practice Approach

In support of the International Council on Systems Engineering (INCOSE) MBSE Focus Group, an initial comprehensive survey of MBSE methodologies was conducted in 2007 and published as Revision A. A Revision B update to the survey paper was released the following year in the early half of 2008 under the auspices of the INCOSE MBSE Initiative [2]. In an attempt to keep the survey current, for subsequent years, methodologist was asked to publish a short overview of each respective methodology on an open MBSE Wiki forum provided by the Object Management Group™ (OMG™) in order to capture any methodology gaps since the 2008 Rev. B survey paper. This is part of a now archived yet still accessible challenge/activity team known as "Methodology and Metrics" (see http://www.omgwiki.org/MBSE/doku.php?id=mbse:methodology).

For purposes of this handbook, a select subset of the MBSE methodologies captured on the Methodologies and Metrics Wiki page is included in the subsections below. The selected methodologies described herein are for purposes of illustration only and were chosen because they are (a) independent of any product vendor or vendor offering, (b) have had or continue to have visibility within the INCOSE community and/or international standards committees, and (c) have sufficient literature to support cross-reference and use by the reader of this handbook. Further, the subset described herein does not imply one methodology is superior over another. In section "Evaluating and Comparing Methodologies," an approach for evaluating and comparing methodologies is described to assist individuals and organizations in methodology selection. Finally, it should be noted that newer methodologies continue to evolve, and the reader is encouraged to survey the literature as they are published given the fluid and emerging nature of the MBSE discipline in the past few years.

Object-Oriented Systems Engineering Method (OOSEM)

Overview

- Integrates top-down (functional decomposition) approach with model-based approach (see Figs. 2 and 3)
- Leverages object-oriented concepts
  - Uses OMG™ SysML™ (formerly UML®) to support specification, analysis, design, and verification of systems
- Intended to ease integration w/object-oriented S/W development, H/W development, and test
- Includes the following activities:
  - Analyze stakeholder needs
  - Define system requirements
  - Define logical architecture

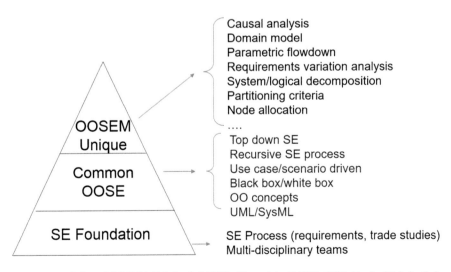

Fig. 2 Foundation of OOSEM (Friedenthal 2020). (Copyright © 2012–2020. Sanford Friedenthal. Used by permission)

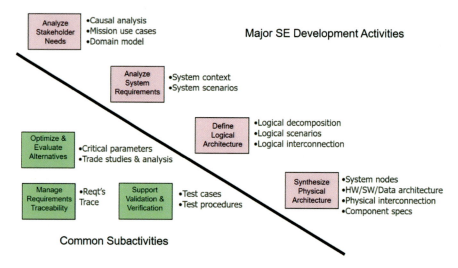

Fig. 3 OOSEM activities and modeling artifacts (Friedenthal 2020). (Copyright © 2012–2020. Sanford Friedenthal. Used by permission)

- Synthesize candidate-allocated architectures
- Optimize and evaluate alternatives
- Validate and verify system
• Is tool- and vendor-neutral

Tool Support
- OOSEM was created as tool- and vendor-neutral, model-based methodology. A dedicated process framework tool for OOSEM does not exist; however, an Eclipse Process Framework (EPF) version of OOSEM developed by J. D. Baker exists (see http://www.omg.org/cgi-bin/doc?syseng/2009-12-09).
- Tool support for OOSEM can be provided by COTS-based OMG™ SysML™ tools and associated requirements management tools. Other tools required to support the full system life cycle should be integrated with the SysML and requirements management tools, such as configuration management, performance modeling, and verification tools. Note, however, this is true for any methodology.
- A more complete set of OOSEM tool requirements is provided in the referenced OOSEM tutorial (see the 9.1.2 References section of this chapter).

Offering/Availability
- The OOSEM tutorial and training materials are available on the INCOSE Connect site under the INCOSE OOSEM WG page (see http://www.incose.org/login/?ReturnURL=https://connect.incose.org/Pages/Home.aspx). **Note:** Access requires INCOSE membership.
- Unlike other industry-provided MBSE methodologies, OOSEM is not a formal offering that can be purchased from any specific vendor, including professional

3 MBSE Methodologies

services. Support services may be available by contacting representatives of the INCOSE OOSEM Working Group.

Selected Resources
- See 9.1.2 References section under OOSEM.

Object-Process Methodology (OPM)

- Formal paradigm to systems development, life cycle support, and evolution (see Figs. 4 and 5)
  - Combines simple OPDs (diagrams) with OPL (constrained natural language)
  - Basic building blocks:
    - **Object** – Thing that exists or has the potential of existence physically or mentally
    - **Process** – Pattern of transformation that the object undergoes
    - **State** – Situation object can be at
- Ratified as ISO/PAS 19450:2015, a publicly available specification of the International Organization for Standardization (ISO) in December 2015
- [Reflective] methodology refers to system life cycle as *system evolution*
  - In OPM, "process" is a reserved word; therefore, "system process" is used.
  - System Developing process (SD1) contains three main stages, each can be "zoomed" of which can be further zoomed.
  - Requirement specifying
  - Analyzing and developing
  - Implementing
  - [Also, using and implementing stage]
- Visual models tool- and vendor-specific (see **Tool Support** below)

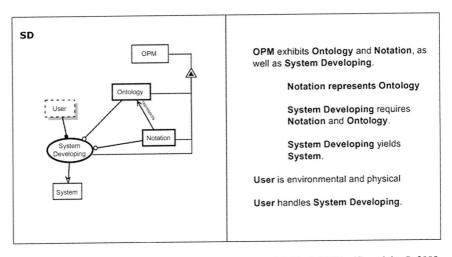

Fig. 4 The top level specification of the OPM metamodel (Dori 2002). (Copyright © 2002. Springer. Used by permission)

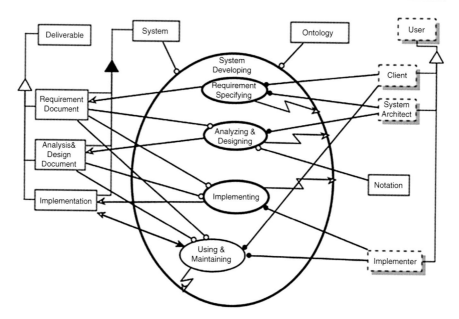

Fig. 5 Zooming into System Developing (Dori 2002). (Copyright © 2002. Springer. Used by permission)

Tool Support
- Tool support for OPM is provided via OPCAT software available under the OPCAT installation page of Enterprise Systems Modeling Laboratory (ESML) of Technion Israeli Institute of Technology website (see esml.iem.technion.ac.il/opcat-installation/). This product suite supports the concepts related to the OPM metamodel for the system development process, including modeling support of the System Diagram (SD).
- The OPCAT tool suite has a configurable template for a number of types of document artifacts, including but not limited to System Overview, the Current State, Future Goals, Business or Program Constraints, and Hardware and Software Requirements. In addition, OPCAT has facilities for animated simulation, requirements management, and other advanced features.

Offering/Availability
OPCAT software for OPM systems modeling, systems engineering and life cycle support, as well as support services and education and training material can be obtained via the ESML of Technion Israeli Institute of Technology (see esml.iem.technion.ac.il/).

Selected Resources
- See 9.1.2 References section under OPM.

Integrated Systems Engineering and Pipelines of Processes in Object-Oriented Architectures (ISE&PPOOA)

Overview
- ISE&PPOOA (Integrated Systems Engineering and PPOOA) provides an integrated process, methods, and a tool for systems engineering of software-intensive mechatronic systems (see Figs. 6 and 7).
  - The ISE part of the process includes the first steps of a systems engineering process applicable to any kind of system, not only the software intensive ones. The ISE subprocess integrates traditional systems engineering best practices and MBSE.
  - The PPOOA part of the process emphasizes the modeling of concurrency as earlier as possible in the software engineering part of the integrated process.
- ISE&PPOOA provides a collection of guidelines or heuristics to help the engineers in the architecting of a system.
- One of the project deliverables is the functional architecture representing the functional hierarchy using the SysML block definition diagram. This diagram is complemented with activity diagrams for the main system functional flows. The N2 diagram is used as an interface diagram where the main functional interfaces are identified. A textual description of the system functions is also provided as part of the deliverable.
- Other of the deliverables is the physical architecture, representing the system decomposition into subsystems and parts using the SysML block definition diagram. This diagram is complemented with SysML internal block diagrams for each subsystem and activity and state diagrams as needed. A textual description of the system blocks is also provided. The heuristics used for the particular architecture solution are identified and documented.
- ISE&PPOOA makes easy the architecture evaluation from time responsiveness characteristics.

Tool Support
- ISE&PPOOA was created as tool- and vendor-neutral.
- Currently, tool support for PPOOA is provided via a free add-on for Microsoft Visio 2003 that can be requested at http://www.ppooa.com.es/ppooa_visio.htm.

Offering/Availability
- A 3-day tutorial dealing with the architecting and evaluation of real-time systems is offered by ISE&PPOOA experts (e.g., http://www.ppooa.com.es/seminar.pdf).
- White papers and articles are available on the PPOOA website: http://www.ppooa.com.es/publications.htm.

Selected Resources
- See 9.1.2 References section under ISE&PPOOA.

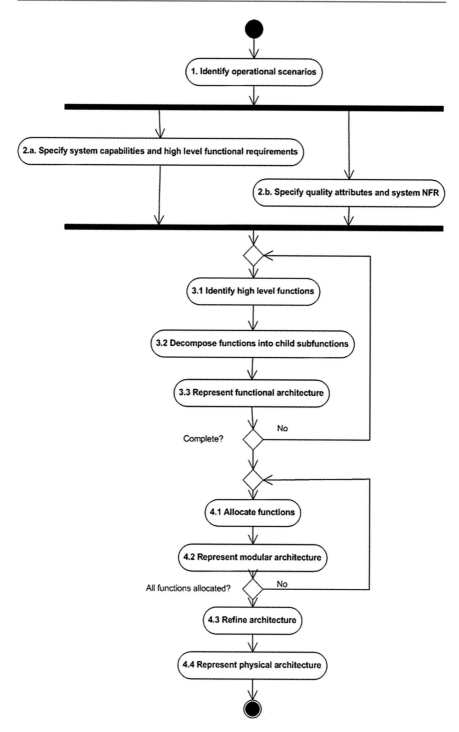

Fig. 6 Integrated Systems Engineering (ISE) subprocesses (Fernandez and Hernandez 2019). (Copyright © 2019. Artech House. Used by permission)

3 MBSE Methodologies

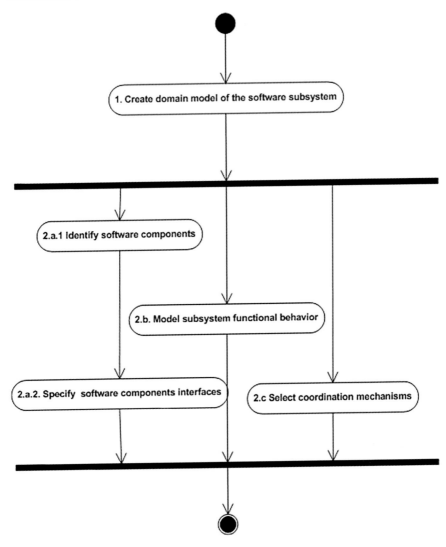

Fig. 7 Process Pipelines in OO Architectures (PPOOA) subprocesses (Fernandez and Hernandez 2019). (Copyright © 2019. Artech House. Used by permission)

Illustrative Example/Case

In this section, a greater level of specificity is provided for the SYSMOD methodology to be used for elaboration of a select set of methods of this MBSE methodology as well as a few selected examples of the work products produced as a result of these methods.

The section begins by providing a brief summary of the SYSMOD methodology that includes the full set of core concepts and the relationships as well as a table of artifacts documenting all of the SYSMOD methods at the time of this writing.

This is followed by elaboration of a set of four selected SYSMOD methods, namely, (1) tailor the MBSE methodology, (2) analyze the problem, (3) identify system use cases, and (4) model the logical architecture. With the exception of tailor the MBSE methodology, the primary SYSMOD output products from each SYSMOD method described is also included. Note that all SYSMOD product examples cited are about the same system; a fictional example is a forest fire detection system (FFDS). The FFDS is a multifaceted system, sophisticated enough for a valuable demonstration, and easy to understand without specific domain knowledge [3].

The Systems Modeling Toolbox (SYSMOD) in Review

Recall from section "Integrated Systems Engineering and Pipelines of Processes in Object-Oriented Architectures (ISE&PPOOA)" that SYSMOD is a general-purpose methodology to create requirements and architecture specifications, mainly used to specify updated and new products or parts [3, 4]. It provides a set of widespread, well-known methods such use case analysis that can be adapted for specific purposes.

The SYSMOD methodology is comprised of the following three core "artifact" kinds:

- *SYSMOD products*: Crucial artifacts for system development like requirements or architecture descriptions
- *SYSMOD methods*: Best practices for creating an SYSMOD product
- *SYSMOD roles*: Work descriptions of a person; responsible for SYSMD products and a primary/additional performer of SYSMOD methods.

Figure 8 illustrates the relationships between SYSMOD methods, products, and roles. A role is responsible for one or more (1..*) methods and supports zero or more (0..*) methods as a coworker. A method has exactly one role that is responsible for the method and some roles as additional performers. Each method requires zero or 0..* products as inputs and produces 1..* products as outputs. Exactly one role is responsible for a product.

As illustrated in Fig. 8, a process is part of 0..* methodologies and performs 1..* methods. A tool is also part of 0..* methodologies and facilitates 1..* methods. (**Note**: The reader is referred back to section "Key Concepts and Definition" for a definition of the general terms of Process, Methods, Tools, and Environment (PMTE) as well as Methodology and more specifically MBSE methodology.)

A complete list of the most current SYSMOD methods along with their associated input and output products and their primary performing roles is summarized in Table 1.

3 MBSE Methodologies

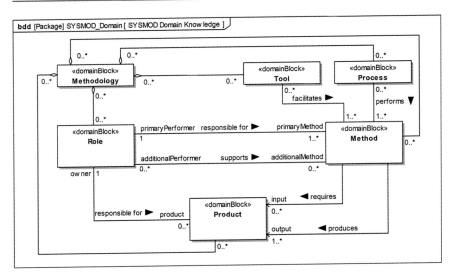

Fig. 8 Overview and relationship of SYSMOD artifact kinds

Although SYSMOD is a toolbox and not a process per se, some default processes are provided to demonstrate a typical logical order of execution of the SYSMOD methods. In practice, a project typically uses a customized set of methods in a different order, including recursions, iteration, and loops [4]. One suggested logical grouping and order of execution of the set of detailed SYSMOD processes to support performance of SYSMOD methods, along with the primary SYSMOD roles that perform those methods, is depicted in the following use case diagram (see Fig. 9).

Each of these high-level default SYSMOD processes is further decomposed into more detailed subprocesses and articulated in the latest edition of the SYSMOD booklet series (see Ref. [3]). It should also be noted that Appendix A of this publication provides a mapping of the ISO/IEC 15288:2015 standard processes to the SYSMOD methods without distinction between full and partial coverage. This mapping is included in Table 1 shown previously.

A Few Examples of SYMOD Core Concepts in Greater Detail

Copyright Notices: *(1) The material used in this section was adapted from the latest edition of the MBSE4U booklet series entitled "SYSMOD, Systems Modeling Toolkit: Pragmatic MBSE with SysML," 3rd edition and used by permission [3]. Copyright (c) 2020. MBE4U. (2) The SYSMOD examples shown herein and adapted from [3] were created with the use of the CATIA/No Magic Cameo Systems Modeler from Dassault Systèmes. Copyright © 2020. No Magic, Inc.*

Table 1 Summary of SYSMOD methods

SYSMOD method	SYSMOD role (primary performer)	SYSMOD [input] product	SYSMOD [output] product	ISO 15288 process
Tailor the MBSE methodology	MBSE methodologist	None	MBSE methodology	Infrastructure management
Setup and maintain the SME	SME administrator	MBSE methodology	System modeling environment	Infrastructure management
Deploy the MBSE methodology	MBSE methodologist	MBSE methodology	None	Infrastructure management
Provide MBSE training and coaching	MBSE methodologist	MBSE methodology	None	Human resource management
Analyze the problem	Project manager	None	Problem statement	Business mission analysis
Describe the system idea and the system objectives	Project manager	Problem statement	System idea and system objectives	Business mission analysis
Describe the base architecture	System architect	System idea and system objectives	Base architecture	Business mission analysis
Identify stakeholders	Requirements engineer	Base architecture, system idea, and system objectives	Stakeholders	Business mission analysis, stakeholder needs, and requirements definition
Model risks	Requirements engineer	System idea and system objectives	Risks	Risk management
Model requirements	Requirements engineer	Stakeholders, system idea, system objectives, and base architecture	Requirements	Stakeholder needs and requirements definition and system requirements definition
Identify the system context	Requirements engineer	Requirements	System context	Business mission analysis, stakeholder needs and requirements definition, and architecture definition
Identify system use cases	Requirements engineer	System context and requirements	System use cases	Stakeholder needs and requirements definition and system requirements definition
Identify system processes	Requirements engineer	System use cases	System processes	Stakeholder needs and requirements definition and system requirements definition

(continued)

Table 1 (continued)

SYSMOD method	SYSMOD role (primary performer)	SYSMOD [input] product	SYSMOD [output] product	ISO 15288 process
Model use case activities	Requirements engineer	System use cases and requirements	Use case activities	Stakeholder needs and requirements definition and system requirements definition
Model the domain knowledge	Requirements engineer	Use case activities	Domain knowledge	Stakeholder needs and requirements definition and system requirements definition
Specify test cases	System tester	Requirements and use case activities	Test Cases	Verification and validation
Model the functional architecture	System architect	Use case activities	Functional architecture	Architecture definition
Model the logical architecture	System architect	System use cases, system context, requirements, base architecture, and functional architecture	Logical architecture	Architecture definition
Model the product architecture	System architect	Logical architecture	Product architecture	Design definition
Revise an architecture with scenarios	System architect	Physical architecture	Scenarios	Architecture definition and design definition
Define system states	System architect	Physical architecture	System states	Stakeholder needs and requirements definition, system requirements definition, architecture definition, and design definition
Model the test architecture	System tester	Physical architecture	Test architecture	Verification and validation

SYSMOD Method: Tailor the MBSE Methodology

Tailor a given MBSE methodology to the specific needs of an organization or project.

SYSMOD Process
- SYSMOD infrastructure process

SYSMOD Role
- MBSE methodologist

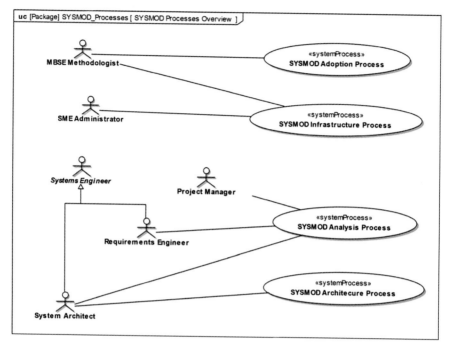

Fig. 9 Logical grouping of SYSMOD processes

SYSMOD [Input] Product
- None

SYSMOD [Output] Product
- MBSE methodology

ISO 15288 Process
- Infrastructure management

Purpose

Adapting the MBSE methodology to the specific needs of the engineering project avoids useless overhead on engineering tasks that have no real value to the project and stakeholders.

Description

An MBSE methodology is a collection of processes, methods, products, roles, and tools. A predefined MBSE methodology like SYSMOD does not fit one-to-one to an organization or project. Some steps need to be more emphasized; others are superfluous and can be skipped, and some engineering artifacts are not even covered by the methodology and need to be added. That is not a lack of methodology. The tailoring is a standard task before you apply the methodology to your projects.

You can use SYSMOD itself to elaborate your needs for a methodology and to derive the methods. In this case, the MBSE methodology itself is the system of interest.

SYSMOD Method: Analyze the Problem

SYSMOD Process
- SYSMOD adoption process
- SYSMOD analysis process

SYSMOD Role
- Project manager

SYSMOD [Input] Product
- None

SYSMOD [Output] Product
- Problem statement

ISO 15288 Process
- Business mission analysis

Purpose

Find out the real problem that should be resolved by the system. Sometimes, it is not the initial problem statement, and the real problem must be identified by a profound analysis.

Description

An illustrative example of the importance of finding the real problem is the quote by Henry Ford: "If I had asked my customers what they wanted, they would have said a faster horse." The quote hits the point. Even if it is doubted that Henry Ford had really said this, it illustrates very well that the real problem to be solved and the real needs to be satisfied require a profound analysis.

Design Thinking is an approach to solve problems creatively with a particular focus on problem analysis. The origin of Design Thinking lies in the 1960s (Archer 1965). An essential recent publication about Design Thinking is Change by Design: How Design Thinking Transforms Organizations and Inspires Innovation from Tim Brown (2009), and the Design Thinking Playbook (Lewrick et al. 2018).

Design Thinking is well known in the business and engineering domain and used as a standard method for around 15 years. The main drivers were the Stanford University, the company IDEO, and the Hasso Plattner Institute.

Part of the Design Thinking process is the redefinition of the initial problem statement to find the real challenge that satisfies the needs of the stakeholders. Design Thinking does not ask what the users want but what the users need.

The problem statement is codeveloped with a prototype in an iterative process. A first problem statement leads to the first solution attempts that again clarify the problem statement and so on. The user is always in the center of the Design Thinking process. Besides the visible problem, Design Thinking addresses also the invisible emotional needs of the user.

Reinhard Haberfellner et al. describe another approach with the problem-solving process (Fig. 10) (Haberfellner et al. 2019).

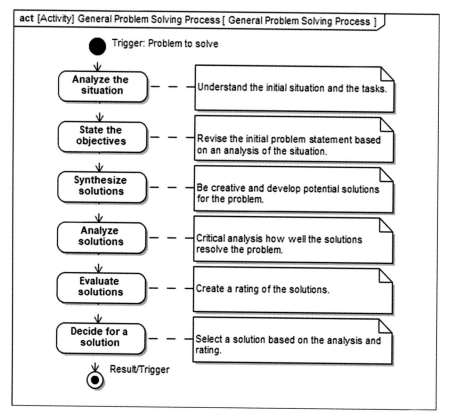

Fig. 10 Problem-solving process

The problem-solving process is based on the five steps of problem-solving by John Dewey, published in 1910 (Dewey 1910). The process starts with the analysis of the problem before it proceeds with the development of solutions.

The process in Fig. 10 looks sequential, but it depicts only the logical sequence and not the execution order, which includes lots of feedback cycles.

Example of SYSMOD Product: FFDS Problem Statement

Document
Figure 11 shows an extract of a document that covers the problem statement.

Model
Figure 12 depicts the abstract system element with the problem statement shown in a separate compartment.

3 MBSE Methodologies

System	Forest Fire Detection System (FFDS)
Problem Statement	
How can we provide a forest fire detection system for forest authorities that can be scaled to any forest size, is affordable, highly reliable and accurate in detecting forest fires.	

Fig. 11 Problem statement in a text document

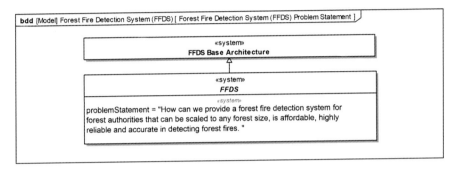

Fig. 12 Problem statement in the model

The system element is located in the package *FFDS_Core* as depicted in the header of the diagram. Here, it is tightly coupled with the base architecture by using the generalization relationship. If the base architecture also has a problem statement, it is not inherited by the system element. The *problemStatement* is a property of the stereotype «*system*», and the generalization relationship does not cover stereotypes and stereotype properties.

SYSMOD Method: Identify System Use Cases
Identify all services of the system provided for the actors and stakeholders of the system.

SYSMOD Process
- SYSMOD adoption process

SYSMOD Role
- Requirements engineer

SYSMOD [Input] Product
- System context requirements

SYSMOD [Output] Product
- System use cases

ISO 15288 Process
- Stakeholder needs and requirements definition
- System requirements definition

Purpose

The system use cases are the services and the essential ends of a system. The system is developed and operated to achieve these services.

Description

The system use cases provide an outside-in view on the system functions from the perspective of the system actors. That supports the system development to build a system that satisfies the real needs of the system actors. An inside-out perspective on the functional requirements of a system can also lead to a system that satisfies all of them but with the risk that they do not consider the usability needs of the users.

An illustrative example, in most cases, is remote controls for a projector (Fig. 13). They provide all required functions but not from the perspective of the users. The remote control offers lots of buttons, and the primary use cases are hard to find.

The system use case description should at least include:

- The associated system actors
- The trigger that starts the use case
- The result of the use case
- A brief textual description (2–5 sentences)
- Traceable paths to the nonfunctional requirements that are relevant for the use case
- Traceable paths to the functional requirements that are covered by the use case

The set of use cases is a flat list without a hierarchy of super- and sub-use cases. The use case activities specify the functional decomposition of the system use cases.

Fig. 13 Remote control

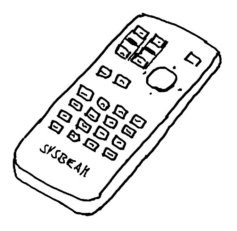

Example of SYSMOD Product: FFDS System Use Cases

Document
Figure 14 depicts a system use case description in a document.

Model
Figure 15 depicts system use cases of the actor operator in an OMG SysML™ use case diagram.

Figure 16 highlights the operator use case detect and report fire. The comment symbol shows the properties of the use case. The diagram also shows the refine relationships from the use case to some functional requirements. The use case activity owned by the use case is also a functional requirement derived from these high-level functional requirements. Since the information represented by the refine relationships between the use case and the requirements is redundant, the relationships could be skipped. They are only shortcuts and illustrate the relationships on the use case level.

Figure 17 depicts some system use cases of the actor maintenance, who is a specialization of the actor operator. That means the actor can perform the same use cases. However, it does not necessarily mean that the actor maintenance has the same access rights. This is a different aspect of the system.

Figure 18 depicts the system use cases in the system model package structure.

Detect and report fire		continuous
Description	Continuously observe the region of interest and check for indications of a forest fire. Report the potential fire to the operator and to the responsible fire departments.	
Actorsr	Operator, Fire, Fire Department	
Trigger	Operator activates observation mode.	
Continuous Result	Potential forest fires are reported.	
Precondition	FFDS is in ready mode.	
Postcondition	None	

Fig. 14 A FFDS system use case in a document

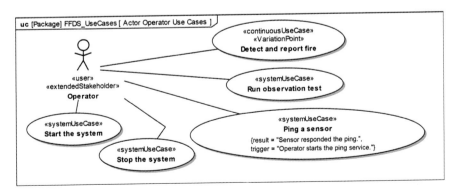

Fig. 15 Actor operator use cases

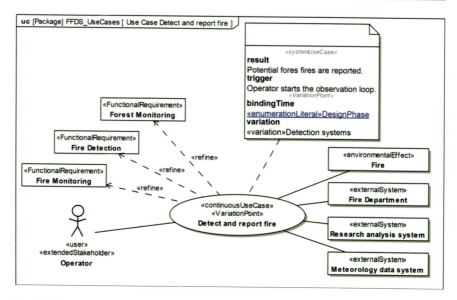

Fig. 16 Use case detect and report fire

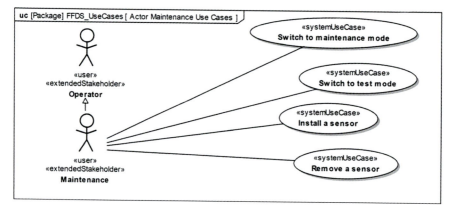

Fig. 17 Actor maintenance use cases

SYSMOD Method: Model the Logical Architecture
Model a physical architecture on a high abstraction level that satisfies the given requirements.

SYSMOD Process
- SYSMOD architecture process

SYSMOD Role
- System architect

SYSMOD [Input] Product
- System use cases
- System context requirements

3 MBSE Methodologies

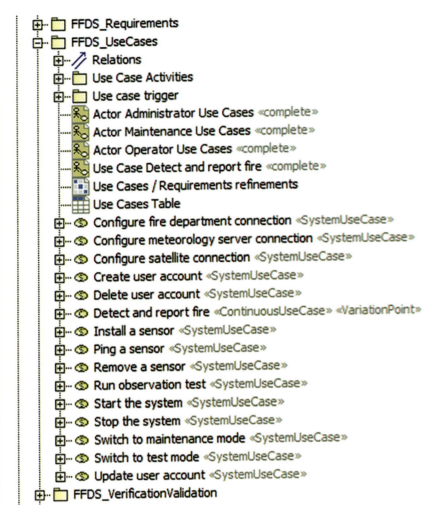

Fig. 18 FFDS system use cases in the model package structure

- Base architecture
- Functional architecture

SYSMOD [Output] Product
- Logical architecture

ISO 15288 Process
- Architecture definition

Purpose

The logical architecture describes the technical concepts and principles of the system.

Description

The logical architecture is an abstract physical architecture. Thinking top-down in the development process, the logical architecture is the first version of a physical architecture. It covers architectural and technical principles and concepts, for example, an electric motor, a valve, or a display as technical concepts. Further down in the intended top-down approach, the product architecture is then a concrete and more detailed specification of the concepts.

The logical architecture must conform to the base architecture. If strongly coupled, the logical architecture is a specialization of the base architecture. If loosely coupled, there are only allocate relationships between both architectures.

If you do not need a separate logical architecture, you can merge this step and the architecture with the product architecture. Finally, you have a physical architecture with logical and concrete product aspects.

Example of SYSMOD Product: FFDS Logical Architecture

Document

Figure 19 depicts a description of the logical architecture in a text document.

Model

Figure 20 depicts the definition of the logical architecture elements in a SysML block definition diagram (system breakdown, product tree). Check carefully if the value of the diagram view exceeds the effort in creating the visualization. Alternatively, you can display the block definitions in a table.

Figure 21 depicts the information aspect of the structure of the logical architecture in a SysML™ internal block diagram. This is the main view of an architecture. Typically, it is not valuable to show all different aspects like information, mechanical, electrical, and others in a single diagram. Instead, each aspect is shown in a separate internal block diagram. The diagram name indicates the aspect. However, finally, the stakeholders decide about the view and the format and content of the diagrams.

Figure 22 depicts the logical architecture in the model package structure.

Challenges and Gaps

In terms of gaps, it was stated earlier in section "Current State of the Practice/Current State of the Art" that the candidate list of MBSE methodologies described herein is not a comprehensive set of methodologies but rather methodologies that have substantial presence in the literature and that are currently in active use, being advanced, and/or actively supported at the time of this publication. The earlier cited MBSE methodology survey was published a dozen years ago [2]; however,

3 MBSE Methodologies

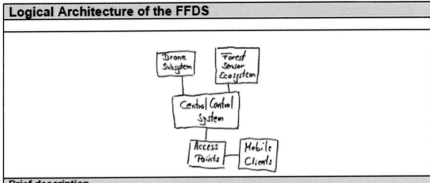

Fig. 19 FFDS logical architecture in the document

following the release of that report, methodologists were encouraged to provide a cursory description of their methodologies on the publicly available MBSE Wiki supported by the Object Management Group™ (OMG™). This was intended to facilitate more rapid visibility into gaps in methodologies for the benefit of the broad international systems engineering community that were either not included in the 2008 survey or created since that survey. Brief descriptions of these methodologies and primary points of contact can be found on that OMG MBSE Methodology and Metrics Wiki page (see http://www.omgwiki.org/MBSE/doku.php?id=mbse:methodology).

With respect to challenges, the primary challenges associated with MBSE methodologies are not necessarily complexity, availability, or even supportability as one might expect but rather the following: (a) evaluating and comparing methodologies and (b) methodology adoption and tailoring within an organization. Each of these challenges is further described in the following subsections of this chapter.

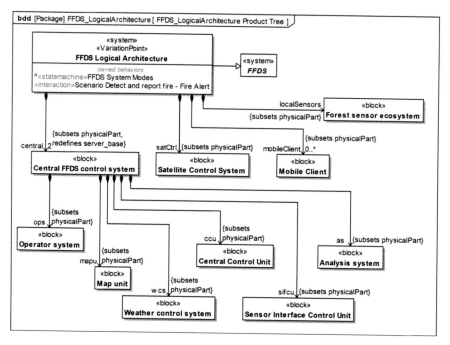

Fig. 20 FFDS logical architecture product tree

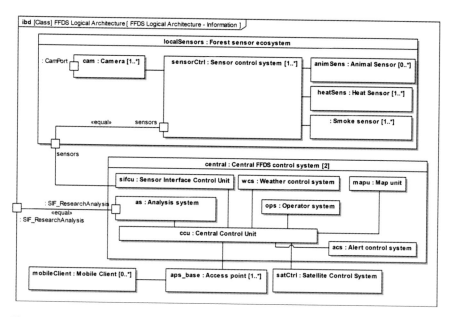

Fig. 21 FFDS logical architecture

3 MBSE Methodologies

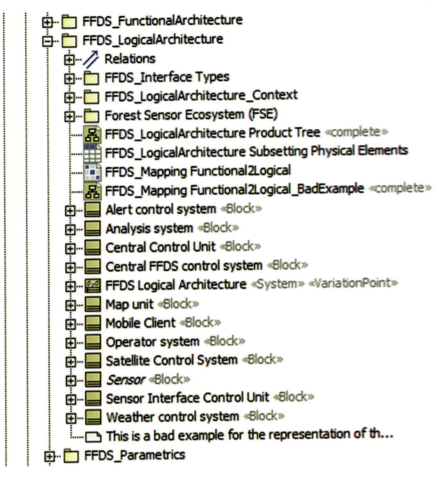

Fig. 22 FFDS logical architecture in the model package structure

Evaluating and Comparing Methodologies

In addition to the modeling language and the tool, the methodology is an essential building block in the implementation of MBSE. As this handbook shows, there are a variety of methodologies. The Framework for the Evaluation of MBSE Methodologies for Practitioners (FEMMP) [4] enables the assessment of the methodologies by providing an overview without the ambition to rank the methodologies.

The evaluation process for a single methodology consists of five steps:

1. Compile a summary of the methodology with further references.
2. Highlight the unique and most important features of the methodology.
3. Apply the methodology on a simple case study to illustrate it. For example, a steam engine case study is used in Weilkiens et al. [4]. The case study should fit the context in which the methodology is to be applied.

4. Evaluate the methodology using the framework criteria with answers, explanations, and comments.
5. Discuss and conclude the evaluation (findings, remarkable things, discoveries).

The criteria from step 4 are grouped into five aspects:

1. Essentials: Criteria to be met for the methodology to be considered as MBSE methodology, for example, coverage of ISO 15288 process steps (ISO/IEC 15288:2002 2013)
2. Practicality: Criteria for the practical application of the methodology, for example, the language, engineering scope, or significant features of the methodology
3. Efficiency: Criteria that make the methodology efficient, for example, background validation or (semi)automatic generation of reports
4. Usability: Criteria for the user experience of the available tools used to apply the methodology
5. Support: Criteria how well the methodology is supported by its owners, the community, and tool vendors, for example, by providing documentation or training

Each criterion defines its scope, such as process, language, or tool. People and technology would be other useful scopes. FEMMP is not limited to the predefined scopes.

A weighting controls the focus during the evaluation. Each criterion has a weighting between 1 and 3, with 3 being the ones to focus on first. The weight does not necessarily imply importance.

The evaluation of each criterion can be specified in one of the following metrics:

- Yes/no plus a free text justification
- List of items plus a free text explanation
- Qualitative assessment scale: A, fully compliant; B, acceptable performance; C, limited applicability; G, generalization; and X, not addressed

Table 2 depicts the table with all criteria that were used for the first applications of FEMMP.

FEMMP is a novel framework for the evaluation of MBSE methodologies. Experiences with the framework will lead to further adaptations and calibrations of the framework.

Methodology Adoption and Tailoring

Once an MBSE methodology has been determined to be a good fit for an organization or specific department within an organization, it is time to put the methodology into practice; this does not happen automatically. There needs to be a concerted effort to tailor the selected methodology and provide training to the practitioner

Table 2 Framework for the Evaluation of MBSE Methodologies for Practitioners [4]. (Copyright ©2016. IEEE. Used by permission)

ID	Area	Category	Title	Description	Type	Wt
A-00-P	Essentials	Process	ISO Standard	What process steps of ISO 15288 are covered?	List	3
A-01-P	Essentials	Process	Framework	What views from the reference framework are used?(MODAF, DODAF …)List		3
A-02-L	Essentials	Language	Philosophy	Are Model Elements clearly distinguished from Diagram Elements? (separation of content from representation)	Y/N	3
A-03-T	Essentials	Tool	Precision	How precise does the tool implement process semantics and sequence? (Is the process well enforced, can "wildcard" elements be used, e.g., an "association relationship," are constrained clearly communicated and controlled, are "work arounds" allowed that reduce the model quality)	Scale	3
B-00-L	Practicality	Language	Language	What Modelling Language is used? (If NOT SysML: How well does it define the real-world semantics of the engineering, are elements strictly typed, is their meaning unambiguous, do they have a defined purpose etc.)	List	3
B-01-M	Practicality	Methodology	Scalability	How well does the model scale? (suitable for large projects, "grows" with time without becoming cumbersome, does it require partitionaing, e.g., in a tree)	Scale	3
B-02-M	Practicality	Methodology	Scope	For what engineering purpose is the methodology suited(innovation, improved products, refactoring, reverse engineering,...)?	List	2

(continued)

Table 2 (continued)

ID	Area	Category	Title	Description	Type	Wt
B-03-M	Practicality	Methodology	Tailoring	How easy is it to tailor the methodology? (add, delete or change processes or process steps, object definitions or toggle tool features on and off)	Scale	3
B-04-P	Practicality	Process	Consistency	Is the process self-contained? (are in/outputs to all steps connected)	Y/N	3
B-05-M	Practicality	Methodology	Variants	How well does the methodology support the variant management?	Scale	3
B-06-M	Practicality	Methodology	Complexity	How often is the methodology "interrupted"? (by external processes and/or non-integrated tools)	Scale	2
B-07-T	Practicality	Tool	Connectivity	How easily can the information be exchanged with other tools? (What standard API are provided by the tool, what API can be added, Is import and export based on open protocols, is it guided, e.g., by a wizard can it be rolled-back, what the quality control mechanism etc.)	Scale	3
B-08-L	Practicality	Language	Integration	How well can the model be integrated with specialty engineering models? (CAD, PNID, Project Management, Document Mangement)	Scale	1
B-09-M	Practicality	Methodology	Simulation	How well does the methodology provide for an integrated simulation?	Scale	2
B-10-M	Practicality	Methodology	Redundancy	How well does the methodology prevent duplication? (of work, model elements, artefacts, communications and reports)	Scale	2
C-00-T	Efficiency	Tool	Perspectives	To what level is the creation of experts' perspectives automated? (can views be defined on the model or do they require manual re-work)	Scale	1

(continued)

Table 2 (continued)

ID	Area	Category	Title	Description	Type	Wt
C-01-T	Efficiency	Tool	Checking	Does the tool support consistency checking of the model? (Automated detection of wrong content and/or formats, flagging of, "loose ends" etc.)	Y/N	2
C-02-T	Efficiency	Tool	Reporting	How quickly are standard/custom reports, is design documentation created? (select templates or views, filter reports, re-use of settings, define aggregation, required level of experience, potential level of automation)	Scale	1
C-03-T	Efficiency	Tool	Admin	How well does the tool help to minimise work that isn't creating any value? (low admin, auto versioning and back-up)	Scale	1
C-04-T	Practicality	Tool	Reuseablity	Does the tool allow to reuse any type of Modelling Element across projects? (sharing the same object with the same lifecylce in any project)	Y/N	2
D-00-T	Experience	Tool	Navigation	How easy is it to find the correct model element? (are elements links, users "guided" in the process, information well aggregated, need to "jump" screens)	Scale	2
D-01-T	Experience	Tool	Intuitition	How intuitive is the tool to work with? (compliance with UX conventions, standard tool reactions, e.g., tool tips, double/right click, drag&drop, delete, Keyboard shortcuts, spell check, familiar operations, e.g., as MS-Office)	Scale	2
D-02-T	Experience	Tool	View	How easy is it to configure the UX dynamically? (define a matrix with sorting & filtering of columns and rows, store customised view, annotation, comment)	Scale	1

(continued)

Table 2 (continued)

ID	Area	Category	Title	Description	Type	Wt
D-03-T	Experience	Tool	UI	How readable is the UI? (Good use of screen estate and colour, zoom, can fonts and sizes be changed, is information well presented...)	Scale	1
E-00-M	Help	Methodology	Documentation	How well is the methodology supported? (books, manuals, case studies, online help, community, websites, interactive support, user feedback etc.)	Scale	3
E-01-M	Help	Methodology	Training	How well is training supported? (availability, consultants, coaches, e-training, background knowledge required)	Scale	1
E-02-T	Help	Tool	Support	How well is the tool supported? (vendor response times, 24/7 helpline etc.)	Scale	1

community. In addition, various commercial and open-source process modeling tools exist in which a process engineer can create a Web-based framework of a particular methodology. A couple of examples include the IBM Engineering Lifecycle Optimizer – Method Composer (formerly IBM Rational Method Composer or RMC) (see http://www.ibm.com/us-en/marketplace/method-composer) and the EPF from the Eclipse Foundation, respectively (see projects.eclipse.org/projects/technology.epf).

A key element of the training plan is to hold several so-called Methodology Adoption Workshops or "MAWs" in which process engineers are trained in the selected methodology and have developed training material and perhaps even developed a *tailoring* of the methodology using one of the aforementioned process/method modeling tools. These MAWs could be held as weeklong sessions or split into smaller training segments of a few days at a time over the course of several weeks. Early MBSE adopters within the organization should be targeted first, followed by senior staff and management. It is also highly recommended that external training and MBSE industry leaders be brought in for consultation and training if such training is provided. Consulting services could also be secured to help tailor the selected methodology for the organization. The important point to note here is that continuous training while staying current in the state-of-the-practice is paramount to successful MBSE methodology adoption.

Expected Advances in the Future

In addition to being described in a number of MBSE textbooks with application to introductory systems engineering problem sets and in the context of industry-leading modeling languages, the MBSE methodologies surveyed in this chapter continue to evolve and have been applied in a variety of real-world industry applications, for example, medical robotics (Dori 2014), aircraft design and landing constraints (Li et al. 2019), planet sustainability [3], and space exploration (Spangelo et al. 2012; Fernandez et al. 2020), to name a few.

With respect to future advances of the Object-Oriented Systems Engineering Method (OOSEM), there is a dedicated INCOSE OOSEM Working Group currently led by WG chair Howard Lykins (INCOSE OOSEM WG 2020). In addition, there are plans to update the methodology based on OMG™ SysML™ v2. According to S. Friedenthal (S. Friedenthal, private communication, Sep. 28, 2020), that will result in new opportunities to simplify the methodology and to apply it in new ways. For example, there are specific constructs in SysML™ v2 that address modeling of variability. This will make it much more straight forward to specify variant configurations. Also, there is a standardized Application Programming Interface (API) with more sophisticated query capability. The methodology will likely evolve to leverage this capability in ways that were not practical previously, such as for change impact assessment.

The Object-Process Methodology (OPM) made great strides in moving toward industry standardization such as the release of the ISO Publicly Available Specification (PAS) ISO/PAS 19450:2015 entitled *Automation systems and integration – Object-Process Methodology*. This ISO/PAS was produced under the auspices of the ISO/TC 184/SC 5 Technical Committee on interoperability, integration, and architectures for enterprise systems and automation applications.

OPM has since moved beyond the ISO 19450:2015 specification to fusing conceptual-qualitative modeling with computational modeling (Dori 2020). This enables expanding the span of the life cycle stages that MBSE covers to later (downstream) life cycle stages, including detailed design, model-based testing, simulation, and execution. The OPM model is currently being prepared to serve not just as a blueprint of the system to be but also the actual engine that drives the operational system. OPM methodologist refer to this as "Systems Operating System (SysOS)"; just as a computer has an OS, so does any system that combines humans, hardware, and software (Soskin et al. 2020). The OPM SysOS can already connect to Internet of Things (IoT) sensors and devices and drive them based on data received in real time from the environment. Recent OPM-based developments include IoT security modeling and deep learning to convert free text to OPM models (D. Dori, private communication, Sep. 30, 2020).

The methodology of SYSMOD will use the new possibilities of OMG™ SysML™ v2. The same applies to the related method Variant Modeling with SysML (VAMOS). Although the methodology is independent of the modeling languages, new features enable new possibilities.

Further developments of SYSMOD will include new technologies and concepts such as the Digital Twin and AI. Special attention will be paid to the user, the systems engineer, and methods for collaborative engineering, and more views on the methodology, for example, a documentation based on the Essence framework, will be included.

And advances are being planned with the Integrated Systems Engineering and Pipelines of Processes in Object-Oriented Architectures (ISE&PPOOA) methodology as well, including the following (J. Fernandez, private communication, Sep. 28, 2020):

(a) Creation of ISE&PPOOA profiles for diverse commercial MBSE tools to facilitate the simulation and execution of the models
(b) Extending the ISE&PPOOA conceptual model with new concepts related to particular heuristics and patterns to deal with nonfunctional requirements and quality models
(c) Complementing the ISE&PPOOA MBSE methodology with best practices related to other engineering domains
(d) Using other software architectural frameworks instead of PPOOA architectural framework for implementing the software architecture

As stated in section "Best Practice Approach," the list of MBSE methodologies described in this chapter is reflective of a limited set of leading methodologies that are current, have a high level of maturity, and have been widely described in supporting MBSE textbooks and industry literature. That said, newer and emerging MBSE methodologies should be considered in an organization's overall assessment of potential MBSE methodology adoption. An example of one such methodology is the Arcadia method developed by Jean-Luc Voirin, a Systems Technical Director at Thales [6, 7]. Arcadia is also supported by a dedicated open-source MBSE tool known as Capella (see http://www.eclipse.org/capella/). This and other MBSE methodologies will be reported in future revisions of this handbook as they begin to mature and gain widespread industry adoption.

In closing, exciting new industry trends such as "Digital Transformation (DT or DX)," Data Science, and Artificial Intelligence and Machine Learning as well as the ever-expanding role of the discipline of systems engineering into domains beyond the traditional aerospace and defense sector emerge, and it will be exciting to watch how these trends and disruptive technologies will influence how the MBSE methodologies surveyed herein evolve and respond. To learn more, the interested reader is encouraged to get involved in various industry forums such as the Workshop on MBSE held annually during the INCOSE International Workshops and Symposiums (see http://www.incose.org/IW2021 and http://www.incose.org/symp2021) and the annual Conference on Systems Engineering Research (CSER) (see cser.info). **Note:** Past INCOSE International Workshop MBSE Workshop materials are posted on the

OMG™ MBSE Wiki site at the following URL: http://www.omgwiki.org/MBSE/doku.php#mbse_events_and_related_meetings.

Chapter Summary

In this chapter, a select set of industry-leading MBSE methodologies were introduced along with novel techniques for how to evaluate and compare methodologies as part of an organization's methodology selection process. An illustrative example of one particular methodology, i.e., the Systems Modeling Tool (SYSMOD), was described. In addition, a set of recommended practices for methodology adoption were described to address MBSE methodology challenges, including a novel approach for methodology assessment as well as continuous training by means of Methodology Adoption Workshops (MAWs), use of commercial and open-source process/method modeling tools, and engagement of professional MBSE consulting services to assist with methodology tailoring, training, and adoption. And finally, new insights into future advances in the MBSE methodology field were profiled.

Systems Modeling Toolbox (SYSMOD)

Overview

- User-oriented approach for requirements engineering and system architectures (see Fig. 23)

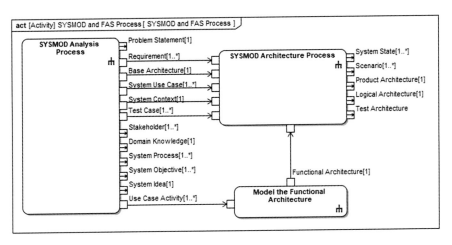

Fig. 23 SYSMOD and FAS process [3]. (Copyright © 2020. MBSE4U. Used by permission)

- Includes processes for the adoption of MBSE in the organization and to set up an MBSE infrastructure
- Guidelines and examples provided for each task
- Includes tasks to specify problem statement, system idea and objectives, requirements, stakeholders, base architecture, systems context, use cases, use case activities, domain model, test cases, logical architecture, physical architecture, scenarios, and test architecture
- Provides a mapping to the ISO/IEC/IEEE 15288:2015 standard [5]
- Provides interfaces to other methods like Functional Architectures for Systems (FAS) or Variant modeling with SysML (VAMOS)
- Tool vendor-independent methodology

Tool Support
- Free SYSMOD plugins for use with CATIA/No Magic Cameo Systems Modeler and Sparx Systems Enterprise Architect that provides SYSMOD stereotypes, icons, and diagrams, available at mbse4u.com/download/
- Tool support for SYSMOD can be provided by COTS-based OMG™ SysML™ tools.

Offering/Availability
- Official source and latest version published in the book *SYSMOD: The Systems Modeling Toolbox*: mbse4u.com/sysmod/
- SYSMOD training and coaching provided by consultancy http://www.oose.de/

Selected Resources
- See 9.1.2 References section under SYSMOD.

Cross-References

▶ Model-Based System Architecting and Decision-Making
▶ Overarching Process for Systems Engineering and Design

References

General Refences

1. J. N. Martin, Systems Engineering Guidebook: A Process for Developing Systems and Products, CRC Press, Inc.: Boca Raton, FL, 1996.
2. J. A. Estefan, "Survey of Model-Based Systems Engineering (MBSE) Methodologies," INCOSE Technical Document, Publ. No: INCOSE-TD-2007-003-01, Ver/Rev. B, INCOSE MBSE Initiative, International Council on Systems Engineering, Seattle, Washington, Jun. 10, 2008.
3. T. Weilkiens, SYSMOD – The Systems Modeling Toolbox, 3rd edition, Pragmatic MBSE with SysML, MBSE4U, 2020.

4. T. Weilkiens, A. Scheithauer, M. Di Maio and N. Klusmann, "Evaluating and Comparing MBSE Methodologies for Practitioners," 2016 IEEE International Symposium on Systems Engineering (ISSE), Institute of Electrical and Electronics Engineers, Oct. 3–5, 2016.
5. ISO/IEC 15288:2005, *Systems Engineering – System Life Cycle Processes*, International Organization for Standardization/International Electrotechnical Commission, Jun. 8, 2005.
6. J-L. Voirin, *Model-based System and Architecture Engineering with the Arcadia Method, 1st Edition*, ISTE Press – Elsevier: Kidlington, Oxford, UK, 2017.
7. S. P. Alaia, "Evaluating ARCADIA/Capella vs. OOSEM/SysML for System Architecture Development," Masters Thesis, Purdue University, West Lafayette, IN, Aug. 2019.

MBSE Methodology References

Object-Oriented Systems Engineering Method (OOSEM)

S. Friedenthal, "An Introduction to MBSE with SysML: Part 4 – MBSE Method," Training Vugraphs, Sandford Friedenthal, Feb. 10–14, 2020.

S. Friedenthal, A. Moore, and R. Steiner, *A Practical Guide to SysML: The Systems Modeling Language*, Third Edition (Refer to Chapter 17), Morgan Kaufmann OMG Press, 2015.

P. Pearce and M. Hause, "ISO-15288, OOSEM and Model-Based Submarine Design," Systems Engineering, Test and Evaluation Conference of The Systems Engineering Society of Australia (SESA) and The Southern Cross Chapter of The International Test and Evaluation Association (ITEA) and INCOSE Region VI Incorporating the 6th Asia Pacific Conference on Systems Engineering (APCOSE 2012), Brisbane, Australia, Apr. 30, 2012.

"Object-Oriented Systems Engineering Method (OOSEM) Tutorial," Ver. 02.42.00, Lockheed Martin Corporation and INCOSE OOSEM Working Group, Apr. 2006.

H. Lykins, S. Friedenthal, and A. Meilich, "Adapting UML for an Object-Oriented Systems Engineering Method (OOSEM)," *Proceedings of the INCOSE 2000 International Symposium*, Minneapolis, MN, Jul. 2000.

S. Friedenthal, "Object Oriented Systems Engineering," *Process Integration for 2000 and Beyond: Systems Engineering and Software Symposium*, New Orleans, LA, Lockheed Martin Corporation, 1998.

S. Spangelo, D. Kaslow, C. Delp, B. Cole, L. Anderson, E. Fosse, B. Gilbert, L. Hartman, T. Kahn, and J. Cutler, "Applying Model Based Systems Engineering (MBSE) to a Standard CubeSat," in *Proceedings of IEEE Aerospace Conference*, Institute of Electrical and Electronics Engineers, Big Sky, MT, Mar. 2012.

INCOSE OOSEM Working Group URL: http://www.incose.org/incose-member-resources/working-groups/transformational/object-oriented-se-method, International Council on Systems Engineering, San Diego, CA, 2020.

Object-Process Methodology (OPM)

D. Dori, *Model-Based Systems Engineering with OPM and SysML*, Springer: New York, New York, 2016.

ISO/PAS 19450:2015, *Automation systems and integration – Object-Process Methodology*, International Organization for Standardization/Publically Available Specification, Dec. 2015.

Y. Grobshtein and D. Dori, "Generating SysML views from an OPM model: Design and evaluation", *Systems Engineering,* **14** (3): 327–340, Feb. 2011.

I. Reinhartz-Berger and D. Dori, "A Reflective Metamodel of Object-Process Methodology: The System Modeling Building Blocks," *Business Systems Analysis with Ontologies*, P. Green and M. Rosemann (Eds.), Idea Group: Hershey, PA, USA, pp. 130–173, 2005.

D. Dori, I. Reinhartz-Berger, and A. Sturm, "OPCAT – A Bimodal Case Tool for Object-Process Based System Development," *5th International Conference on Enterprise Information Systems (ICEIS 2003)*, pp. 286–291, 2003.

D. Dori, *Object-Process Methodology: A Holistic Systems Paradigm*, Springer-Verlag: Berlin Heidelberg, Germany, 2002.

D. Dori, "Medical Robotics and Miscommunication Scenarios. An Object-Process Methodology Conceptual Model.," *Artificial Intelligence in Medicine*, **62**(3) pp. 153–163, 2014.

L. Li, N. L. Soskin, A. Jbara, M. Karpel, and D. Dori, *Model-Based Systems Engineering for Aircraft Design with Dynamic Landing Constraints Using Object-Process Methodology*, IEEE Access, Institute of Electrical and Electronics Engineers, pp. 61494–61511, 2019.

ISO/PAS 19450:2015, *Automation systems and integration – Object-Process Methodology*, International Organization for Standardization, Publicly Available Standard, Dec. 2015.

N. L. Soskin, A. Jbara, and D. Dori, "The Model Fidelity Hierarchy: From Text to Conceptual, Computational, and Executable Model," *IEEE Systems Journal*, Intstitute of Electrical and Electronics Engineers, 2020.

Systems Modeling Toolbox (SYSMOD)

T. Weilkiens, J. G. Lamm, S. Roth, and M. Walker, *Model-Based System Architectures*, 2nd edition, Wiley, 2022. https://www.wiley.com/en-us/Model+Based+System+Architecture%2C+2nd+Edition-p-9781119746652

T. Weilkiens, *VAMOS – Variant Modeling with SysML*, MBSE4U Publishing, 2016.

T. Weilkiens, *SYSMOD – The Systems Modeling Toolbox*, 3rd Edition, *Pragmatic MBSE with SysML*, MBSE4U Publishing, 2020.

T. Weilkiens, *Systems Engineering with SysML/UML: Modeling, Analysis, Design*, Morgan Kaufmann OMG Press, 2008.

ISO/IEC 15288:2015, *Systems Engineering – System Life Cycle Processes*, International Organization for Standardization/International Electrotechnical Commission, May 2015.

2003.INCOSE MBSE Challenge Team SE^2, MBSE Cookbook, mbse.gfse.de

B. L. Archer, Systematic Method for Designers, Council of Industrial Design, H.M.S.O., 1965.

T. Brown, *Change by Design: How Design Thinking Transforms Organizations and Inspires Innovation*, Harper-Business, 2009.

M. Lewrick, P. Link, and L. Leifer, *The Design Thinking Playbook: Mindful Digital Transformation of Teams, Products, Services, Businesses and Ecosystems*, Wiley, 2018.

R Haberfellner, O. de Weck, E. Fricke, and S. Vössner. *Systems Engineering – Fundamentals and Applications*, Birkhäuser Basel, 2019.

J. Dewey, *How we think*, D.C. Heath & Co., 1910.

Integrated Systems Engineering and Process Pipelines in OO Architectures (ISE&PPOOA)

J. L. Fernandez, and C. Hernandez, *Practical Model-Based Systems Engineering*, Artech House, Boston, MA, 2019.

C. Hernandez and J. L. Fernandez-Sanchez, "Model-Based Systems Engineering to Design Collaborative Robotics Applications," 2017 IEEE International Symposium on Systems Engineering (ISSE) Proceedings, Institute of Electrical and Electronics Engineers, Vienna, Austria, Oct. 11–13, 2017.

J. L Fernandez, "An Integrated Systems and Software Engineering Process". Tutorial. Aula Artigas. Industrial Engineering School. ETSII-UPM. Madrid (Spain), May 18, 2012.

J. L. Fernandez and G. Marmol, "An Effective Collaboration of a Modeling Tool and a Simulation and Evaluation Framework," 18th Annual International Symposium, International Council on Systems Engineering, Systems Engineering for the Planet, Utrecht, The Netherlands. Jun. 2008.

J. L. Fernandez and W. Mason, "A Process for Architecting Real-Time Systems", 15th International Conference on Software and Systems Engineering (ICSSEA), Paris, France, Dec. 2002.

J. L. Fernandez, C. Hernandez, J. A. Martinez, E. Diez and I. Valiente, "NASA MBSE Challenge Application of the ISE&PPOOA Methodology to the In-Space Habitat," Report on NASA MBSE Challenge, National Aeronautics and Space Administration, Jan. 6, 2020.

Jeff A. Estefan is a Principal Engineer and Technical Consultant for the Multimission Ground Systems and Services (MGSS) program office at NASA's Jet Propulsion Laboratory (JPL) in Pasadena, California. Jeff currently provides consulting expertise to program managers and decision-makers in matters related to long-term strategic planning, program level architectures, technology portfolio management, and systems engineering insight and oversight of the MGSS program office's primary product line known as NASA's Advanced Multi-Mission Operations System (or AMMOS), which is used my multiple NASA space science missions across NASA field and partner centers (see ammos.nasa.gov).

Jeff earned his M.S. in Applied Mathematics from the University of Washington and his B.S. in Mathematics from Washington State University and served as an Aeronautical Engineering Duty office for the US Naval Air Reserve from 1988 to 1998. He received an honorable discharge from military service with the grade of Lieutenant Commander.

Tim Weilkiens is a member of the executive board of the German consulting company oose, a consultant and trainer, a lecturer of master courses, a publisher, a book author, and an active member of the OMG and INCOSE community. Tim has written sections of the initial SysML specification and is still active in the ongoing work on SysML v1 and the next-generation SysML v2.

He is involved in many MBSE activities, and you can meet him on several conferences about MBSE and related topics.

As a consultant, he has advised a lot of companies from different domains. The insights into their challenges are one source of his experience that he shares in his books and presentations.

Tim has written many books about modeling including Systems Engineering with SysML (Morgan Kaufmann, 2008) and Model-Based System Architecture (Wiley, 2022). He is the editor of the pragmatic and independent MBSE methodology SYSMOD – the Systems Modeling Toolbox.

SysML State of the Art

B. Bagdatli, S. Cimtalay, T. Fields, E. Garcia, and R. Peak

Contents

Introduction (Problem Statement, Key Concepts, Terms, and Definitions)	88
Decision-Making, Surrogate Modeling, and Parametric Requirements	88
An Example of Multifidelity Multiphase Tool Integration Through SysML	100
Design, Operate, and Manage Digital Ecosystems Using SysML and MBSE	103
Additional Project Examples	113
Chapter Summary	114
Cross-References	114
References	115

Abstract

This chapter focuses on state-of-the-art applications of SysML. A few example applications are selected to illustrate the power of the SysML language and the breadth of possible problems that can be tackled with its systematic application. The first example describes several approaches to integrating detailed design analyses with SysML authoring tools to enable informed decisions. The second example explores the challenges posed by the different levels of analyses of fidelity encountered throughout a product development cycle. The third example discusses how MBSE practices help manage digital ecosystems. Finally, a few additional case studies are discussed that illustrate the wide variety of projects employing SysML and provide a complete picture of the current state of the art.

Keywords

SysML · MBSE · Digital ecosystem · Decision-making

B. Bagdatli · S. Cimtalay · T. Fields · E. Garcia (✉) · R. Peak
Georgia Institute of Technology, School of Aerospace Engineering, Atlanta, GA, USA
e-mail: burak.bagdatli@ae.gatech.edu; cimtalay@gatech.edu; tfields6@gatech.edu; elena.garcia@ae.gatech.edu; Russell.Peak@gatech.edu

© Springer Nature Switzerland AG 2023
A. M. Madni et al. (eds.), *Handbook of Model-Based Systems Engineering*,
https://doi.org/10.1007/978-3-030-93582-5_13

Introduction (Problem Statement, Key Concepts, Terms, and Definitions)

This chapter focuses on state-of-the-art applications of the Systems Modeling Language (SysML). A few example applications are selected that illustrate the power of the SysML language and the breadth of possible problems that can be tackled with its systematic application. These examples are only a small subset of the increasingly widespread use of SysML and Model Based Systems Engineering (MBSE) methods, but they represent a variety of applications and problem spaces. The first example is related to the incorporation of decision-making methods within an MBSE model either directly or via plugins communicating with external software.

Decision-Making, Surrogate Modeling, and Parametric Requirements

Decision-making, even in the engineering domain, is a broad discipline. One perspective for the top-down design decision support process contains the following main steps: establish the need, define the problem, establish the values, generate feasible alternatives, evaluate alternatives, and make decisions [1].

This section goes through an example that demonstrates some of the generalized decision-making concepts, entities, and their relationships in the SysML model and an application to a representative engineering decision-making problem. The scoped engineering decision-making problem and solution comprise a methodical selection of suitable models and tools that implement them in the aerodynamics analysis discipline for criteria such as time-to-run analysis, model fidelity, and geometry. While other methods exist and may be more suitable for different types of requirements, throughout this section, the focus will be on using computer analyses to perform requirement checking. Demonstrating methodical decision-making and keeping records in SysML for reasoning, consistency, and continuity are among the modeling benefits. During the design process, aerodynamics discipline engineers evaluate aircraft components with various alternative tools based on the design phase. The collaboration between system engineers and decision-makers is implemented in a decision-making environment. The SysML authoring tool enables capturing and guiding how aerodynamics analysis engineers can select suitable analysis tools in a methodical decision-making process in conjunction with decision-making techniques implemented in an interactive Python-based application. System engineers can manage the decision data by establishing the decision engineering problem-conceptual and domain-specific levels, exporting decision data, and importing decision results in the SysML authoring tool. Decision-makers decide by setting preferences and methods in a Python-based application, and ranked results are sent to the SysML model. The following paragraphs elaborate on the methodology starting with a description of decision-making techniques.

The methodical decision-making techniques comprise Analytical Hierarchy Process (AHP) and Technique for Order Preference by Similarity to Ideal Solution (TOPSIS). AHP and TOPSIS are Multi-Criteria and Multi-Attribute Decision-Making (MADM) techniques [2, 3]. They both rely on identifying and selecting an analysis option for aerodynamics during the design of an aircraft of a defined phase. The recommended approach applies AHP and TOPSIS to a problem of interest rather than selecting decision-making methodologies. Additional background in systematic means for selecting decision-making methodologies is available [4]. MADM techniques select the best-compromise solution from a few alternatives based on prioritized attributes of alternatives. Figure 1 shows the decision tree for MADM.

Factors needed before solving a MADM problem include well-defined and measurable criteria, preference information on the criteria, alternatives, and a decision-making technique. The criteria are measures of performance for an alternative. Preference information on the criteria can be relative weights or pairwise criteria comparisons. Alternatives are the candidates among which the best solution is selected. The decision-making technique contains decision rules and algorithms to formulate the decision problem and provide guidance to the decision-maker.

AHP utilizes preference information for obtaining a scale of preferences by pairwise comparisons among the criteria and the alternatives. AHP breaks down the problem into its constituent elements, such as criteria, subcriteria, and alternatives. AHP facilitates evaluating MADM problems with the hierarchical structure of attributes.

TOPSIS is based on the concept that the chosen alternative should be the closest to the positive ideal solution and the farthest from the negative ideal solution. It uses

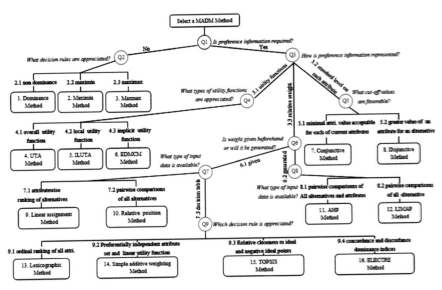

Fig. 1 Decision tree for MADM technique selection [3]

relative weights as the representation of preference information. The methodical decision-making techniques, TOPSIS and AHP, are of algorithmic and mathematical nature, and they are implemented in the Python language [5] as an application outside the SysML tool. The SysML tool, with its scripting capability with its API and simulation plug-in, enables communications with these applications. It sends the generalized or engineering decision data, executes the applications, and retrieves the ranking results. Specifically, custom-developed activities with opaque actions parse the SysML elements, write Java Script Object Notation (JSON) [6] files containing decision data, invoke the Python application, and read the JSON files containing ranking data. Opaque actions in Unified Modeling Language (UML) and SysML are utilized to execute implementation-specific instructions and include scripting languages [7]. SysML tool and its simulation plugins enable the execution of activities with opaque actions.

The SysML model is treated as the source of truth. Balancing content and modeling between the SysML and surrounding applications is an art that requires expertise. It can also be subjective.

These decision-making techniques are modeled in SysML by applying suitable classification and patterns. Specifically, the SysML model contains the packages with their elements and relationships. The main packages are the Decision-Making Procedure, Generic Entities, Methodologies, and Environments.

Generic decision concepts such as decision data, process, and decision result are modeled as blocks and behavior in SysML. They are the parts of the Decision Context block as seen in Fig. 2. Decision Data and Result are also a kind of Decision Artifact. Those relationships are shown as generalizations. SysML diagrams and relationships matrices help model and view various aspects of the system. System engineers generate model relationships manually in SysML tool, just as they would if spreadsheets were used. One advantage of SysML tools is that they contain all the SysML elements, relationships, diagrams, tables, and relationships matrices in one place. This capability enables modeling and viewing diagrams, tables, and matrices of various aspects of the same system. For instance, the generalization relationships between the diagram and matrix in Fig. 2 are interrelated. Furthermore, one can also

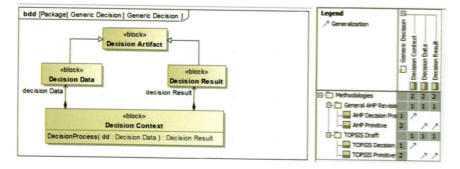

Fig. 2 Generic decision concepts in SysML

4 SysML State of the Art

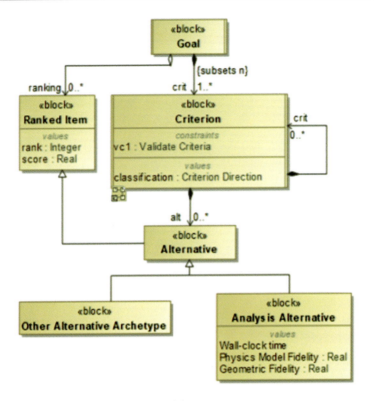

Fig. 3 AHP conceptualization in SysML model

show the composition relationships in the matrix to see the other aspect of the system.

AHP and TOPSIS problems specialized from the Decision Context block have the same parts, as well as their own conceptual SysML elements. For instance, AHP has concepts such as goal, criteria, alternatives, and ranks modeled in SysML. AHP and TOPSIS modeled in this stage can be applied to any decision-making problem in that category. One approach can be seen in Fig. 3.

Further specialization was applied to the scoped engineering decision-making problem for the selection of several alternative aerodynamics tools. The criteria are minimum run time, maximum model, and geometry fidelity. Run time is quantitatively measured in seconds whereas aerodynamics model and geometry fidelity are subjective-quantitative (i.e., Low: 1 and High: 5). Alternatives are analyses with various physical fidelities. Preferences were defined based on the decision-making techniques. AHP has pairwise comparisons, whereas TOPSIS uses criterion weighting.

Furthermore, specific external interfaces and views are desired so that decision-makers need not browse the SysML model directly. Applying decision-making techniques alongside the architecture models can enable and formalize the practice of decision-making and project review.

After deciding on what analyses to use for the system's design, the focus shifts to performing the analyses to check whether the design meets its requirements and to derive subsystem requirements. Derived requirements are then allocated to subsystems decomposed from the main system block. How requirements are derived and allocated to subsystems requires decisions as part of the system development. The decisions are recorded in the SysML model for traceability.

Following the decisions regarding the requirements, the system and its subsystems must be physically defined and sized to satisfy the complete requirement set. This step involves further decisions that can be accomplished with the design team communicating and collaborating to find an optimum design that satisfies the requirements. To enable such activities, two environments are needed:

1. A set of physics- or regression-based analysis engines outside of SysML tools that provide assessments of a baseline design to determine whether requirements are met.
2. A model-based engineering platform to bring design teams and their assessments together to make design iterations collaboratively and document assumptions.

As a systems engineering language, SysML and SysML authoring tools can capture the logic and the rationale behind the requirement, analysis, and design decisions. To perform the analyses and make sound decisions, additional tools and software are needed. However, these tools need to be represented and documented by the systems engineers and have stable interfaces with the SysML model to cut down on the manual entry of information that is error-prone.

Three steps to the overall solution are discussed next. The first involves setting up the analysis environment for disciplinary analyses involved in system design. Second, once the engineering analyses are executed many times in a design space of interest, regressions to polynomial equations capturing the performance trends as a function of design variables can be created that can be imported into SysML using constraint blocks in parametric diagrams. The third solution creates useful visualizations that specify whether requirements are satisfied and help find feasible alternatives if the initial baseline is not compliant. The focus is on making decisions using visualizations consistent with the system model. Advantages, disadvantages, and software requirements for each solution are developed after their descriptions.

SysML authoring tools are well-suited for determining validity of a system's design via requirement checks. For clarity, this check does not validate the system requirements but checks the current design against its requirements to ascertain feasibility. Some checks can even be (semi-)automated for consistent and rigorous checks for the system under development if possible [8]. However, after performing the checks, if the current system design is not meeting the requirements, usually it is not easy to know what parameter to change in a system's design to make it compliant. This difficulty is different to the traceability that systems engineering methods and tools provide. It is easier to trace a requirement failure to multiple systems; however, it is not as easy to decide what design change is necessary to meet the requirement. An example can help: Imagine an aircraft's range is not sufficient to meet its design range requirement. The problem can be traced fairly easily to inefficient aerodynamics, propulsion, or structure design. However, it is not clear whether a change in aerodynamics design will be sufficient or the most cost-efficient design activity to meet the requirement. The solution may be found by a single design discipline or a combination of them. Traceability is not suited to answer such questions. The binary nature of the assessment requires several iterations between the systems engineering and system design groups.

Making matters worse, the systems engineer cannot see whether the requirements are even technically achievable. Such aggressive but common requirements necessitate new technology development and use within the system's design, or implementations may take too long or cost too much to fit within the program schedule and budget. However, the design teams can discover the possibility of infeasibility with the assumed technology levels with the power of their engineering analysis models that can be used to predict performance (predictive models). It is important to distinguish responsibilities for systems engineering and engineering design functions as SysML, and its tools are not competing with detailed analyses such as finite element analysis. The two sides need to work together, not hierarchically. If systems engineers can help designers understanding the context of requirements and functions in general, designers will be better equipped to find effective solutions. In return, if design engineers can help systems engineers see the effect of allocated or derived requirements on the totality of the vehicle or give them the ability to trade off requirements against each other, costly iterations can be removed from the design time line.

Systems engineering teams cannot necessarily use analysis-based discoveries to determine which requirement is too constraining because such infeasibility may be caused by conflicting sets of requirements usually. Relaxing one or the other requirement may work, but it may not be easy to decide which requirement(s) to modify. Additionally, the rich outputs of a computer analysis run cannot be easily compared to SysML requirements due to the difficulty in capturing time-history behaviors, interacting component-timing relationships, and multiple complicated conditions.

Whether the initial design is infeasible or infeasible, decisions are needed to solve the problem of closing the design. If feasible solutions can be found by modifying the design variables, the process is called *satisficing*. However, if no feasible solutions exist, the design problem with its requirements and assumptions must be modified until a satisfactory solution arises. These modifications are not to be left to automated tools, and human decision-makers are needed. The modifications can be grouped into three different categories:

1. Adding technologies to the system that make it perform better than traditional designs simulated in the engineering analyses.
2. Investigating a different solution concept with a different system architecture.
3. Trading off requirements against each other and relaxing them if possible (some requirements cannot be modified, e.g., certification requirements).

While the solution process for the satisficing problem is straightforward, it still requires coordination between the design and system engineering teams. Additionally, each of the proposed solutions will have undesirable side-effects and is not purely beneficial. Each technology, advanced concept, and requirement change may have negative impacts on other subsystems, and the pros and cons must be carefully considered.

The systems engineering work focuses on capturing and tracing requirements to design analyses. SysML tools are currently limited to specifying analyses that include algebraic constraints that ensure the system meets requirements; however, more complex requirements cannot be easily captured in the language. Examples include time-based requirements and specification of detailed condition-based rules for system components. While requirements can be captured in text, they are difficult to execute and verify in SysML tools. However, information in a SysML model can guide analyses that can be much more complex and thorough. The link must be set up manually and collaboratively between systems engineers and design engineers.

Keeping variables in design requirements, parameters, and performance consistent between the teams that use different analysis software should be an automated process to reduce human errors and task monotony. Additionally, not every variable must be exchanged between a SysML model and an executable physics-based model. For example, all potential failure conditions in a system's structure must be listed in SysML such that the analysis knows what to check for; however, the structural margins for each elements must not be recorded in the SysML model—a collective *all margins pass requirements* is enough. It also allows teams to be up to date with each other's modifications on the design being worked on. In this work, the focus is on the way the design and its requirements can be modified in the SysML model based on analysis results.

One difficulty that will be faced with teams using detailed geometries or other physical system information is the necessity to keep their models geometric if the SysML model uses higher-level, more abstract variables. For example, in SysML, a wheel's diameter may be exposed, but on the physics-based analysis side the wheel and tire need more descriptions detailed. If the wheel diameter is increased, a parametric model must update the tire groove pattern consistently and ideally automatically with an optional engineer review. Otherwise, the systems engineering group and engineering design group's understanding of the system will diverge over time. Potential consequences may be the loss of traceability of design changes impacting other subsystems, miscalculating of requirement parameters, mismatching parts, and inconsistent parametrization between design teams.

Adding technologies and changing concepts may be similar to the satisficing problem in that the overall system's performance is improved. Capturing the impacts of technologies to the physics of the system's, design engineers can predict the possible improvements using their *predictive models.* However, new concepts and technologies may alter the system decomposition and even the parameter space, meaning that the systems engineer must work with the technology and design teams to understand the architectural changes in the system.

The third scenario involves modifying the system requirements themselves to achieve feasibility. There are two possibilities for such a change. The first is accepting a lower level of performance compared to initial system expectations. From a management perspective, developing an inferior but more dependable or affordable system may make more sense if new concepts and technologies are risky or expensive to develop. However, the primary possibility for this scenario arises from the need to modify previously derived requirements.

Consider a scenario where the total system mass meets high-level system requirements. For example, an aircraft's range performance can be estimated initially using a simple equation called Breguet's Equation, the details of which are not necessary for the discussion. In this equation, to meet a specific aircraft range, the ratio of mission start and mission end weights may be determined and set as a total weight requirement for the complete system. Subsystems—such as wing, fuselage, engine, and landing gear—will now need budgets to be developed within subsystem design groups. In other words, the systems engineer must allocate some mass budget to each system using derived requirements. However, these requirements are not as rigid as the original, highest-level system requirements, because reduction of mass elsewhere could allow one subsystem to be heavier, or if multiple subsystems are overweight the system with the easiest path to lightening could be focused on. Additionally, some mission requirements may be deemed less critical and relaxed if they drive up mass and energy budgets. Therefore, it is important to treat some requirements as flexible and allow trade-offs between them.

SysML currently does not facilitate easy trade-offs between the requirements. The requirements are stored as pure text and require human interpretation; therefore, requirement values—whether they are numerical values or strings from a list or even equation representations—are difficult to modify automatically. The values are even more difficult to propagate in derived, copied, and flow-down requirements automatically. Therefore, if requirements are updated as a result of a trade-off, careful systems engineering work is required to update the SysML requirements that are traced, derived, and otherwise connected to other parts of the model. In the upcoming version of SysML v2 [9], formal requirements are introduced where textual requirements are represented as machine-readable statements and requirement variables that can be automatically checked or modified as needed. Decomposing the requirement text into pieces such as values and comparators enables the propagation of updates from requirements to design parameters and vice versa. In addition to a text requirement statement, SysML v2 will have the capability to specify a requirement definition with a formal requirement statement that has value properties and constraints. For instance, a mass requirement has *allowed* and *calculated* masses that are value properties whereas a *calculated mass* less than the *allowed mass* is a constraint. The new capability enhances the methods described in this section.

The decision to change the design or the derived subsystem requirements does not rest on a single person within development teams. While SysML and analysis tools support tracking changes, the bridge between them must have similar change tracking and review capabilities. A collaboration between design teams is needed to execute such changes. It is also important to note that trade-off analyses or design improvements need to be backed by quantitative methods such as disciplinary design analyses based on a physics analysis of varying complexity whenever possible. Systems engineers may not have access to such tools or the expertise to use them, especially in a multidiscipline system design. Conversely, disciplinary engineers may not have access to systems engineering tools or the expertise to use them. Imagine a group of Computational Fluid Dynamics specialists accessing the system model. They will not need access to the full system decomposition or requirements derivation. Nor will they need to update anything other than a limited number of parameters that describe the aerodynamics performance of the system being developed. It is not greatly valuable to train such specialists and provide them with expensive licenses for systems engineering software. It is better to coordinate with a systems engineering team who set up the necessary *hooks* in the system model for disciplinarians to update/fill based on analysis results.

There are currently two possible approaches to solve the communication problems between the systems engineering and systems design groups. Both require executing a set of analysis cases within a *design space* ahead of time and creating surrogate models of the analysis results that are mathematical equations if appropriate such as when dealing with continuous results and when the physics relationships are not trivial enough for SysML tools to handle on their own such as a mass rollup. These surrogate models can be executed with generic mathematical computation libraries that can be found within SysML-authoring software; however, creating surrogate models requires machine learning or statistical regression methods and tools. Many open-source, free, or commercial software are available for these tasks, and they do not need to be connected with the SysML tools. The idea is to replace computationally expensive analyses with more accessible equations with minimal error.

Briefly, the creation of a surrogate model is described. This step happens outside of the use of SysML and regular systems engineering functions and is used as a bridge between analysis and SysML. To create surrogate models, original physics-based models need to be exercised in a multidimensional space (design space) spanned by the influential design variables for the system. Statistical methods such as *designs of experiments* [10] reduce the number of times the original analyses need to be exercised while extracting enough information for the regression methods and allow for creation of simpler mathematical *curve fits* to the data. If the original engineering analyses are non-trivial and expensive to execute, regression equations allow for the exploration of the design space thoroughly by reducing the computational burden. Such regressions can also be used within optimization loops where the analyses are executed multiple times until the system's metrics are optimized using a pre-determined cost function.

Surrogate models may be embedded into SysML constraint blocks and used as a constraint property on parametric diagrams. Parametric diagrams allow for tight connection of system parameters using equations. For example, if width, depth, and height were variables of a cube system, its volume can be captured in the SysML model using a constraint block. Parametric diagrams constructed with constraint blocks can catch mistakes in which a system's variables are modified inconsistently

4 SysML State of the Art

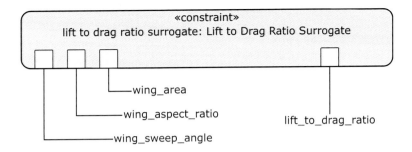

Fig. 4 Polynomial regressions such as response surface equations can be relatively easily embedded within SysML models with the variables of the regression exposed as parameters (squares in the diagram) and the equation itself embedded into the constraint block's constraint definition. Here the constraint equation is omitted for brevity as regression equations can be quite long complicated

in other uses. Parametrics are commonly featured as part of the four pillars of SysML [11]. One such example is shown in Fig. 4. The underlying equation is of polynomial form with regression coefficients before every linear, and quadratic, term. The equation's form is not shown in the constraint block representation – only the variables are shown. In this way, a second-order polynomial equation will appear no different than any other form as long as the input and output variables are the same. This fact is important in the act of replacing complex physics-based computations with the same input and output relationships with the surrogate model. In SysML, their representation will appear the same; however, actual analyses will require an external executable call, but surrogate models can be executed in a SysML authoring software such as MagicDraw directly. The equation is simple enough for interpreting directly by SysML authoring software such as MagicDraw's math module for constraint blocks [12, 13]. The contents of the equation are not important for the discussion and are left out for brevity.

Alternative computation capabilities may be needed if the equations are more complex (e.g., vector equations with matrices, nonlinear functions, or equations with significant length). External computations can be delegated to custom scripts that run on common engines such as Python, Matlab, or R to perform design performance calculations. Using files to communicate with external scripting engines is a common and robust means for using external tools. The file(s) will be used as an input to the computation, for example, a geometry and flight condition can be used as inputs to perform aerodynamics calculations. The SysML-authoring software only needs to write a file to disk and read in the results from an external analysis tool. Not all outputs will be needed in the SysML model, only the ones useful in describing the system and checking requirements. At the end of this operation, as many requirements or constraints as possible are evaluated and checked for compliance within SysML. For requirements that SysML cannot compare to a simple value, the analyses are expected to return feasible designs or return a list of error messages. If the system is not compliant, the systems engineer can try other options until all requirements are satisfied properly or some other agreed upon solution is found. The remedies will include interacting with the subsystem design teams and other stakeholders.

Embedding or calling surrogate models of predictive models provides the systems engineer with a quick check of whether the system meets its requirements regardless of how complicated the underlying process of the measurement or calculation. Conversely, the capability allows for modifying the derived requirement values (constraints) as needed to achieve a valid overall design. It is important to note that the surrogate model must have sufficient accuracy and not violate any assumptions used in the model building using several standard visualizations and model checks [14]. When dealing with low-accuracy physics-based models or surrogate models, the recommended practice is to exercise the higher-fidelity analyses at the end of decision cycles to ensure compliance. Surrogate models are best used for exploration and exploitation, not for final design baseline decisions. Therefore, the surrogate models can provide strong confidence in designs meeting requirements; however, they cannot provide absolute certainty. For higher degrees of confidence, high-fidelity design analyses, scale tests, and prototype tests can be used in addition to surrogate models.

If embedding equations or running scripts locally are not desired, a systems engineering collaboration software can be used as a platform to keep the SysML model up to date with design activities via a common database where system parameters can be saved. The OpenMBEE project can be used for such a setup [15]. In OpenMBEE, the systems engineer sets up parameters and performance indicators for designs and system requirements and *publishes* them to a shared database. By sharing the necessary addresses and IDs of variables that need attention with the system designers, a systems engineer can determine whether other analyses are needed.

Once the database is established, design teams use it for maintaining aspects of the system within their scope. Model-Based Engineering (MBE) platforms, such as OpenMBEE, offer application-programming interfaces (API) implemented as web services that facilitate connecting to external applications. OpenMBEE's API is documented on its website [15]. Applications need only comply with API interfaces rather than the MBE platform. On the MBE side, the main burden is to maintain the server.

Through APIs, any analysis environment within the web is available to execute specialized analyses and update the data in the database. The API calls may need to happen outside of the pure analysis software; however, many common scripting languages are equipped to perform Hypertext Transfer Protocol (HTTP) Requests. Examples include Python, R, and MATLAB. The reader is referred to the specific details of implementation in each scripting language [16, 17, 35]. Quality of life improvement libraries exist for certain platforms that simplify API use via freely available software. OpenMBEE offers several such libraries for MATLAB, Python, Jupyter, Mathematica, and more [18].

The OpenMBEE setup allows disciplinarians receiving the design parameter and performance updates back into their platform, e.g., a structure analysis software. Additionally, Model Management System (MMS) plug-ins can be installed into

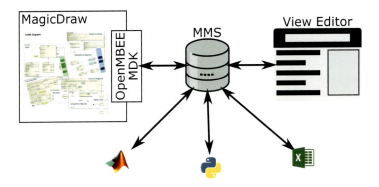

Fig. 5 Analysis tools can connect to the systems engineering environment via OpenMBEE APIs

MagicDraw that achieve the same capability for the MBSE-practitioner. MMS is a collection of software tools that provide the database and web services for an OpenMBEE implementation. Figure 5 depicts the central database approach for accessing analysis results from within MagicDraw. The information contained in the MagicDraw environment can be synced to a database and accessed from external software. Additionally, the ViewEditor web service can display documentation and data contained in the database via an easy to access and edit webpage interface. An example of View Editor page is given in Fig. 6.

Each of the three methods comes with pros and cons. For example, embedding entire surrogate models into constraint blocks is the most SysML-native solution; however, it is tiresome if there are many such models and when they need to be updated as more analysis executions are performed. Connecting directly to analyses is extremely fast and expandable but requires writing tool-specific, custom scripts or macros and is prone to errors or failure as analysis software and SysML authoring software are updated. Organizations that rely on their own automations will be required to keep them up to date as the software they use are updated. The reluctance to update to avoid breakdowns can keep an organization behind the state-of-the-art capabilities. Finally, OpenMBEE allows many teams to work on the same system within their own computers and disciplinary tools and environments; however, it requires nontrivial setup, maintenance, and management of a network server. It also must be noted that when tools that connect to MMS get updated and break backward compatibility, OpenMBEE also stops working correctly. Therefore, it is important that software vendors and plug-in authors work in responsible ways.

The best practice recommendation is to use embedded surrogates for limited-scope and short-term work, connected analyses for individual projects or small engineering design groups, and OpenMBEE for collaboration between multiple design teams. In this section, an example use for OpenMBEE is given; however, readers are encouraged to think in other ways as collaboration with OpenMBEE can be useful for their practical problems and projects. It is a powerful environment that can truly unlock the potential for MBSE-driven engineering activities. Options other

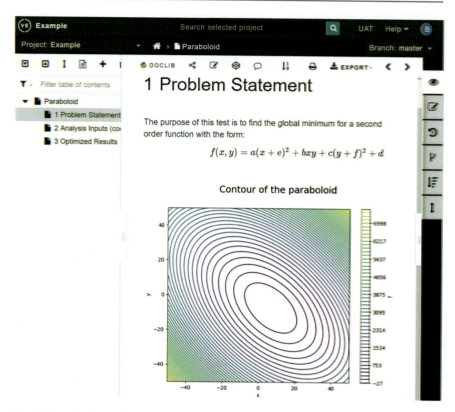

Fig. 6 Example view editor page that can be edited in a web browser

than OpenMBEE are available if SysML is not a necessity. For example, the 3DEXPERIENCE platform made available by Dassault Systèmes [19] enables the exchange of information between analyses, 3D modeling, and requirements applications; however, it is a closed environment, and external collaboration options may be limited.

An Example of Multifidelity Multiphase Tool Integration Through SysML

This section summarizes work carried out by NASA in collaboration with the Jet Propulsion Laboratory and the California Institute of Technology. This work is considered a preeminent example of the capability provided by SysML when integrating analyses across different development and fidelity levels. The application is summarized in a presentation by Erich Lee entitled: "End-to-End Integrated Resource Analysis on Europa Clipper" [20].

4 SysML State of the Art

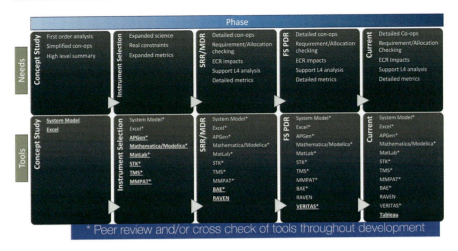

Fig. 7 Europa Clipper tool evolution

NASA's Europa Clipper mission provided an excellent case study for SysML application. This project encompassed a comprehensive suite of analysis tools integrated into a single-source-of-truth SysML model. The SysML model was used throughout development as a basis for consistent mass and power allocations. The tools (Fig. 7) ranged from Microsoft Excel spreadsheets to time-dependent behavioral models in Modelica. Tool integration was further complicated by the need to use tools at different levels of fidelity during each development phase. Whereas early development phases may require only high-level performance metrics, later phases necessitate detailed metrics and constraints with more detailed behavioral analyses; see Fig. 7. Both high and low levels of analyses, despite their profound differences, interacted with a decision support dashboard that providing data to decision makers.

The Europa Clipper team approached the problem through standardization. The analysis tools were modified and wrappers created as needed to produce data products in formats such as JSON or comma-separated variable (CSV) files. Standard metadata produced by tools at each level of fidelity enabled the transfer of information between the tools and the data repository. Defined workflows for data-processing and a robust comparison mechanism enabled the display of data in the decision-making dashboard regardless of the analysis tool that generated the data. This platform is shown in Fig. 8, with a metadata repository that enables handoff across tools (MPS server) and a Tableau server for visualization and cross-checking of analysis results [21].

Configuration management and full Engineering Change Request (ECR) control of all analysis tools in the integrated model is also a key element of success for this type of endeavor. Figure 9 shows the ECR workflow employed in this project. Note that verification of results by ECR initiator is a critical element of this workflow for ensuring consistency. The ECR process included the tool operator and other Europa

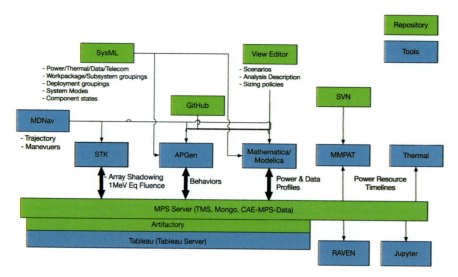

Fig. 8 Europa Clipper single-source-of-truth model architecture overview

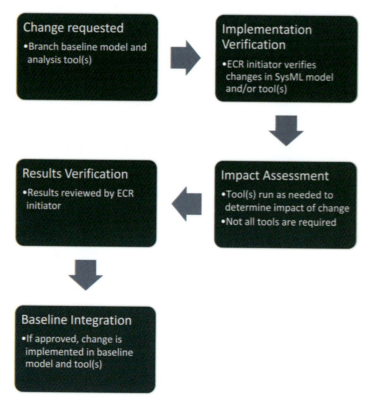

Fig. 9 Engineering change request workflow

Clipper personnel as needed. For example, the engineering change control board (CCB) reviewed documentation based on the model, and CCB decisions were subsequently recorded manually in the model by the modeling systems engineer. The CCB did not interact directly with the model in this project; however, that is a future goal to further streamline changes.

While the Europa Clipper project was successful in its creation of a single-source-of-truth model using SysML, there were several challenges that had to be overcome. First and foremost, this type of effort constitutes a large upfront investment balanced against the repeatability, consistency, and flexibility gained through the environment. To ensure success, future projects wishing to follow this approach must guarantee traceability through all design phases and determine the best means for providing continuous integration that ensures the latest knowledge captured. Further, this type of SE environment generates a significant amount of data that must be managed and validated. Finally, stakeholders and users must be fully engaged as the single-source-of-truth model is not a substitute for the technical experts, but rather supports and harmonizes their efforts. In this case, the project involved inputs from technical stakeholders involved in the tracking of mass, power, and electrical resources, as well as mission designers considering the overall mission margins. This reflects the limited modeling scope achieved for this project but could be expanded to encompass additional stakeholder perspectives and their tools [22].

Design, Operate, and Manage Digital Ecosystems Using SysML and MBSE

It is well-known that digital transformation has swept the world by bringing digital capability to tools, processes, systems, and products through high-functioning computing systems, programs, and cloud-based technologies that automate and increase process efficiency. The digital transformation has expanded and proven its value in several areas of industry such as business management, supply chain operations, and systems engineering [23].

As the systems engineering paradigm is shifting to model-based systems engineering (MBSE), research is being conducted to understand how MBSE can be utilized for digital transformation of systems engineering practices. MBSE aims to use models articulated in a modeling language (e.g., SysML) to represent system knowledge and information to accomplish systems engineering tasks. Managerial and technical disciplinary aspects of a project can use the model representation of the system as the "authoritative source of truth" throughout the system's lifecycle to observe the most current system information as well as traceability throughout the system. Also, many engineering tools for a project may be integrated with SysML models, whether directly or indirectly, for better communication and collaboration for systems engineering activities involving requirements, design, analysis, and validation and verification tasks [24].

Mechanical design tools have been digitally transformed into Computer-Aided Design (CAD) tools. Other engineering discipline's design tools have been digitally

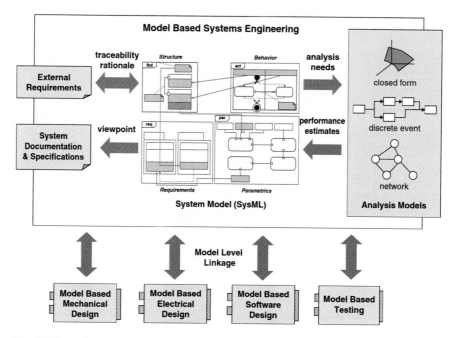

Fig. 10 Example of model-based systems engineering DECO [26]

transformed such as electrical, structural, and thermal design tools. These transformations were to increase the effectiveness and accuracy of the design process. In the same way, the systems engineering discipline has been in the process of digitally transforming. Digitally transforming tools, processes, and products individually has its advantages; however, combining each digital transformation into an ecosystem for design and/or management is of great interest. This is called a Digital Ecosystem (DECO), and an illustration in the context of MBSE can be seen in Fig. 10. For more clarification, a DECO is an integrated system of processes, models, tools, people, repositories, interconnections, and workflows to bring together information and knowledge so that the decision-maker can make the most informed decision about a program, project, and/or system. The purpose is to improve executing the system architecting process, manage the complex dependencies of the system model, better enable integration of external analyses and design space exploration, and support more informed decision-making.

There is a distinction between a DECO and a pure MBSE platform. An MBSE platform has the intention of accomplishing systems engineering tasks through a model-based or computer-aided methodology and is oftentimes referred to as an authoritative source of truth. A DECO aims to integrate external analyses, processes, etc., with SysML models to further enable this idea of the MBSE model being the authoritative source of truth. There is an overlap between a DECO and an MBSE platform such as the example given in the previous section. The Europa Clipper infrastructure example above could be considered a Digital Ecosystem that uses the integrated information with the SysML models to accomplish systems engineering tasks. Digital Ecosystems are very complex and, thus, difficult to design and manage. Companies are exploring how MBSE and SysML can be exploited for successfully implementing MBSE at the local system level all the way up to the enterprise-wide level [25]. The following examples show how model-based systems engineering principles and SysML are used for designing, operating, and managing DECOs.

One of the earliest-known published examples of a production DECO is discussed by Karban et al. in [27]. Karban uses the Executable Systems Engineering Method (ESEM) for integrating requirements with executable behavior and performance models for system level analysis. ESEM is an extension of INCOSE's Object-Oriented Systems Engineering Method (OOSEM) [28]. Karban's work demonstrates how model executability is essential for requirements analysis and verification by simulation. The use case for the project is the Thirty Meter Telescope (TMT) system under development by the TMT Observatory Corporation (TOC) [29]. The TMT analyses provide power and mass roll-ups for various operational scenarios and demonstrate requirements satisfaction. Specifically, Karban's project focused on modeling and analyzing the Alignment and Phasing System (APS) of the TMT which consists of a set of analysis patterns using SysML structural, behavioral, and parametric diagrams to produce an executable model. The simulation capability of the model is driven by Cameo Simulation Toolkit (CST) plug-in. The executable nature of the ESEM method enables automation of system analysis using the system model.

Karban uses a set of integrated software tools available for use in other projects. For example, these tools include Dynamic Object Oriented Requirements System (DOORS) for requirements management, MagicDraw for SysML modeling, CST for simulation capability, and OpenMBEE for version and access control and generating engineering products (e.g., Interface Control Document). Karban's work and the need for multiple tool coordination across a community led to JPL's Open Computer-Aided Engineering (OpenCAE) platform. Figure 11 shows a current illustration of OpenCAE. OpenMBEE, the open-source portion of OpenCAE, acts as a multitool and multirepository integration platform with modeling and web-based capabilities. OpenMBEE provides the DECO with collaboration expanding across engineering disciplines and management by supplying a framework for version and access control, workflow, API access, and document generation through a web-based interface. OpenCAE, due to the complex ecosystem, requires design, management, and support to ensure complete functionality for all users.

The OpenCAE DECO is modeled using MBSE methodologies to aid in its design, operation, and management as shown in Fig. 12. Figure 12 illustrates an early example of how SysML models may be used to "engineer" a production DECO. The figure shows SysML architecting of JPL's OpenCAE including OpenMBEE, and its integration with MagicDraw and CAE tools. The red arrow in the figure points to the OpenMBEE portion of the modeled DECO. The call-out boxes in the image label each portion of the ecosystem represented in SysML. Having the model-based engineering ecosystem represented in an MBSE model

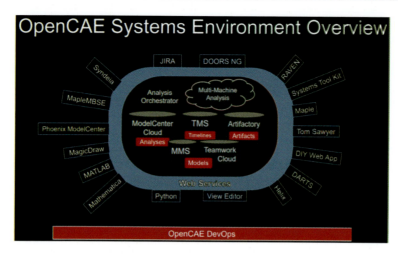

Fig. 11 OpenCAE systems environment overview [36]

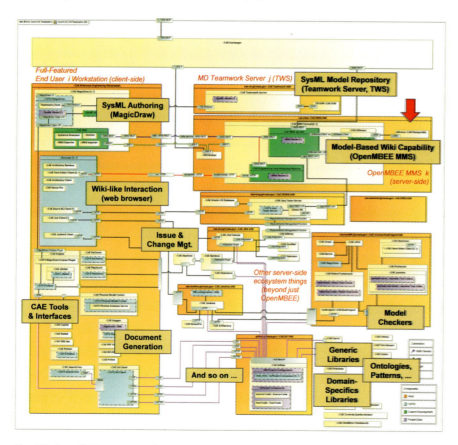

Fig. 12 OpenCAE ecosystem SysML architecting of a digital ecosystem involving OpenMBEE [27]

helps to manage the digital ecosystem by enhancing the communication between tools, including ensuring protocols are compatible, firewall ports are on, etc. Another advantage of this digital ecosystem is managing what is located where and on which machine in the network.

The following illustrates using SysML in the context of MBSE for system engineering a DECO. Ryan Noguchi et al. of The Aerospace Corporation has applied MBSE methodologies to architect, implement, and operate an MBSE system [30]. The Aerospace Corp team developed an approach for improving the process of transitioning to MBSE and the practice of establishing an MBSE capability within an organization. Using system architecting principles with MBSE methodologies and tools, one can advance the architecting, implementation, and operation of an MBSE capability.

Many harmonious pieces are needed when putting MBSE into action for a project. These pieces include databases; models; and descriptive, design, and analytical tools. Tools developed from scratch do not have to be produced or utilized in a segregated fashion, and if integration is beneficial to the overall workflow, then these tools can be included in a DECO. In addition, commercial off-the-shelf Application/Product Lifecycle Management (ALM and PLM, respectively) and Requirements Management (RM) tools, and others, do not have to be utilized in a segregated fashion. Rather, these pieces can be architected and/or utilized in a consistent and comprehensible MBSE system where the necessary information can be captured and represented as elements in SysML models alongside system elements which can be beneficial for traceability throughout the system's lifecycle.

This is a different type of DECO than Karban's example. This project uses MBSE capabilities that apply an MBSE methodology to an MBSE system of a real system. Figure 13 shows what this means. Karban's work focuses on the integration of software tools into a DECO whereas Noguchi's work focuses on managing a DECO represented by an MBSE system (personnel, roles, responsibilities, model compatibilities, etc.). Hence, the name of the approach is MBSE^2 because it is MBSE of an MBSE system of a real system.

The value of this approach is that MBSE applied to an MBSE system strengthens the practice of the architecting process and helps manage the intricate web of connections of the real system. This approach enables the ability to model the

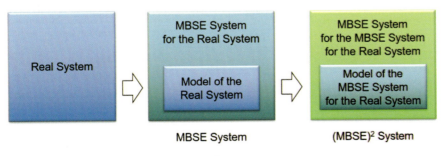

Fig. 13 Depiction of what is meant by MBSE^2 [30]

MBSE system and MBSE^2 system in the same environment, allowing for the elements of each to be connected, and provides customers and stakeholders with beneficial information about modeling processes, tools, procedures, and operations before official procurement. This is useful when conceiving, eliciting, and communicating requirements to support the acquisition. The approach arises from the need of organizations to share information and models about the system in development. A program may have distributed partners working together, and these organizations will require each of the individual models (requirements models, system models, databases, etc.) from a particular organization to be compatible with models from other organizations to provide knowledge and understanding of a complete system for systems engineering and program management purposes across each institution. The Aerospace Corp team explains, "a parent organization and its subordinate—e.g., a prime contractor and its subcontractor, or an enterprise and its constituent programs—may need to share models to support their respective systems engineering and decision processes." This idea becomes very advantageous when organizations are looking to utilize MBSE to enable systems engineering through a DECO across the enterprise.

The MBSE^2 approach demonstrates its capabilities by applying MBSE to descriptive models of an MBSE system. In order to effectively exhibit the approach, the team utilizes a generalized systems-architecting process and then begins illustrating how to apply MBSE to each step of the systems-architecting process thus revealing how to construct an MBSE system of an MBSE system. The team states that the MBSE^2 process involves modeling the Purpose Analysis, Problem Space Exploration and Refinement, Solution Space Exploration and Refinement, and Harmonization & Analysis of the MBSE system. Each step is briefly defined as follows [30]:

- Purpose Analysis identifies objectives, concerns, and decisions and traces them to stakeholders.
- Problem Space Exploration and Refinement defines questions, expressed as functional requirements, that the models need to answer for the stakeholders. The questions are decomposed and traced to the decisions identified in the Purpose Analysis step. This decomposition and traceability supply prioritization criteria for the questions.
- Solution Space Exploration and Refinement identifies and partitions model components that are needed to satisfy its requirements (answer questions). As this step is better developed, the trade-offs are more easily identifiable and comprehendable, and the model can be updated based on the revealed details. As these details are implemented, it "serves as documentation of the recommended modeling patterns to enable the distributed modeling effort to be executed with greater internal consistency."
- Harmonization and Analysis identifies views needed to answer questions. This view links the question and the related model components. These views include queries, templates, scripts, and others.

4 SysML State of the Art

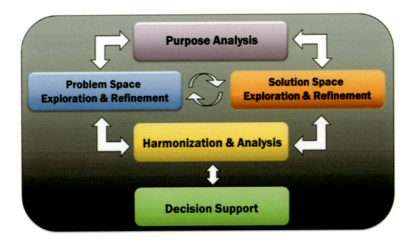

Fig. 14 Architecting process [30]

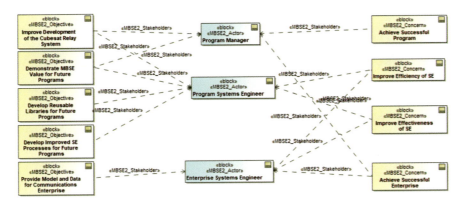

Fig. 15 MBSE^2 modeling of Purpose Analysis [30]

Figure 14 illustrates the MBSE^2-architecting process. Figure 15 shows an excerpt from the MBSE^2 model which displays the modeling of the purposes of the MBSE system. This figure is meant to depict the modeling concept and technique for the "Purpose Analysis for descriptive models" step of the systems-architecting process, and it shows some of the modeling logic needed for implementing MBSE for MBSE systems across an enterprise by portraying traceability relationships to key MBSE system drivers and stakeholders. More detail about each of the steps can be found in reference [30].

The Aerospace team found the Agile development process to be helpful in architecting and implementing the MBSE system and proved useful for rapid learning and feedback. The MBSE^2 approach was then used to inform and manage this development process by monitoring progress (impact analysis, coverage gap

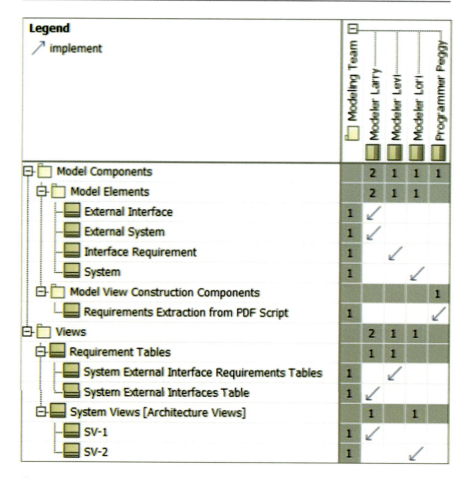

Fig. 16 Example of assigning model elements to model developers [30]

analysis, etc.). Feedback from the agile development method is used to adjust the MBSE^2 model to more accurately capture the modeling plan.

The MBSE^2 model captures the elements and system of connections described in the architecting process. SysML was viewed as very practical when handling the network of dependencies created with this type of model.

The plan for modeling the project was to split the modeling effort into four sprints. These sprints were explicitly assigned to modeling elements. Once the modeling elements were created, more relations and views helped to keep track of what questions would be answered by which sprints. Also, modeling views enabled the observation of what model components are needed. Next, the model elements and views are assigned to model developers as seen in Fig. 16. Modeling the effort in this way aids in planning and managing the modeling effort when executing the MBSE^2 approach. The MBSE^2 System is used to supervise the advancement of the modeling effort within a sprint. While the modelers report their progress, the MBSE system within the MBSE² System can be iteratively updated. This means the completed model elements and views can be identified. More questions answered by the model means more progress has been made. Also, as the approach is executed for a sprint, feedback and lessons-learned can be noted and used in the following sprint.

To demonstrate the operation of the MBSE² system, MBSE reviews were conducted. For example, a System Requirements Review (SRR) use case showed the effectiveness of MBSE² to revolutionize performing systems-engineering reviews. However, this is not the only application of this process. Other use case examples, as identified in Noguchi's paper, include source selection activities, enterprise portfolio management and other applications of enterprise systems engineering, mission, and product assurance activities.

The Purpose Analysis step in SRR use case example identifies the program manager, program systems engineer, enterprise systems engineer, and subject matter experts, a few of the stakeholders involved in an SRR. It also aims to define the objectives for the SRR such as what information is to be evaluated for maturity at a particular milestone. Another objective would be to compare the information against the success criteria. The stakeholders would be assigned to the objective for which they are responsible. For example, the subject matter expert may be assigned to the objective of ensuring the information has met the success criteria for their discipline. The Problem Space Exploration and Refinement step in this example would determine what questions (related to review success criteria, e.g., standards) need answering and what decisions need informing such as relevant trade-offs identification and understanding for better informing for decisions pertaining to alternatives. Solution Space Exploration and Refinement step serves to identify the model components that can be composed to answer the questions from the previous step. The Harmonization and Analysis step identifies the model views needed to answer questions. Model views would be extremely useful in an SRR.

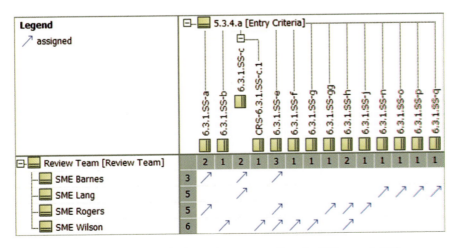

Fig. 17 Example model views assigning reviewers to questions for SRR example [30]

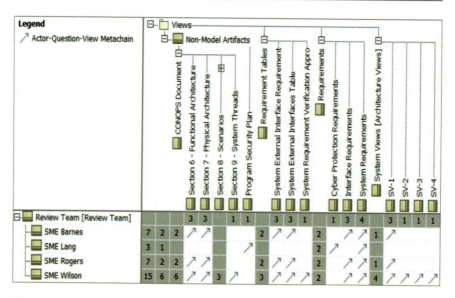

Fig. 18 Example model views assigning views to SMEs [30]

Figure 17 is an example of a Dependency Matrix view in SysML. The Dependency Matrix permits the portrayal of any kind of relations through element properties. These tables use the model information to quickly visualize dependency criteria. The dependency tables compactly illustrate relations of a system that cannot be represented by a single-page diagram on account of the large size of the system. Unimportant information can be filtered from these tables enabling the user to scope the relations to be observed. This SysML semantic allows for deeper model analysis and is certainly advantageous when studying scoped relations for a type of element.

Views like a Dependency Matrix can be utilized in an SRR. Figure 17 uses the Dependency Matrix to assign reviewers (systems engineers, SMEs, etc.) to questions. Figure 18 then shows using the model information to create a matrix assigning particular views to certain SMEs for review. The model can be updated to reflect whether the reviewers have finished evaluating views and solving questions. Dynamic views are used to identify which areas of the model need reevaluation through metachain analysis. If elements in the Real System within the MBSE System have been changed, then the element in the MBSE System reflects that elements can be (through metachain analysis) flagged as complete or reassess. This enables users to rapidly observe which areas require reevaluation. This type of process allows for a continuous (or daily) SRR as compared to a static one time SRR.

While this approach to a DECO is dissimilar to Karban's example, it represents the flexible nature of the idea of DECOs. Karban's example showed the integration side of a DECO while Noguchi's example shows the management side of a DECO. It can be seen that Noguchi's approach can utilize SysML and MBSE to manage and operate a DECO spread across an enterprise. The MBSE^2 approach is a compelling demonstration of how MBSE principles and SysML can be used to "systems engineer a digital ecosystem." To architect, manage, and function the DECO, MBSE[2] is useful for management of implementation across organizations and operating an MBSE system that is a DECO.

Additional Project Examples

In addition to the examples mentioned thus far, SysML related research projects from various domains such as aircraft design process, manufacturing, cost, and acquisitions, as well as supporting newer SysML versions in the committees, have been conducted with our team. The concept applied in SysML ranges from requirements engineering, decomposition, interfaces, analysis, behaviors, to complexity management. Some of these projects have been listed below.

- Development of System Model and Documentation for Aircraft Design [8, 34].

One of the main goals of this project was the development on an MBSE model for coordinating aircraft design activities. Product and analysis representations in SysML provide traceability from system and analysis requirements to physical components and analysis codes. The system model is interfaced with various engineering environments and tools for interactive documentation, multi-disciplinary analysis, design space exploration, and visualization.

- Production System Reference Modeling for queuing analysis of workstations and query of resources, processes, and parts [31].

This project investigated applying MBSE to composite parts manufacturing and assembly. The project develops a discrete event logistics system-based production system model in SysML and demonstrates its potential for analysis integration and abstractions for product, process, and resource.

- Model Health & Complexity Management Project on finding complexity measures of large SysML Models and how to manage the complexity [32].

The project's high-level objectives were to monitor the quality and complexity of a model, determine factors affecting tool performance, and develop guidelines and tools to facilitate the mentioned objectives. Part of the approach was to leverage SysML graph structure and to develop measures of model complexity and health. Additional goals were to investigate graph analysis techniques to detect model issues and model partitioning. Factors affecting tool performance were investigated by conducting the design of experiments on various model sizes and computing configurations.

- MCE (Model Centric Engineering) for UAV System for managing SysML/Simulation and OpenMBEE for web-based documentation syncing with database and UAV system model [33].

One of the goals of this multiphase, multipartner project was to develop model-based acquisition under NAVAIR SET (System Engineering Transformation) framework in the context of surrogate pilot experiments. The pilot is developing an experimental surrogate UAV system with a research and rescue mission. Based on the interaction of various stakeholders, several system models with the needed engineering environment were developed to the represent specific scope and partition of the broad spectrum of the acquisition domain.

The system models were integrated via project usages. Some of those models were the surrogate mission model supporting different scenarios, surrogate UAV system with its subsystems and key parameters as if created by government agency, and surrogate contractor System RFP model for UAV System. Surrogate contractor assessed, refined, and extended UAV system model. It also traces back to Government surrogate UAV system and mission models. It is supported by design models that address multiphysics analysis and design aspects and traces back to government UAV system and mission models. Also, the surrogate acquisition model, which includes models for Statement of Work and Technical Evaluation Criteria supported source selection.

A few MBSE capabilities were developed to help the system model's communication with interactive documentation and various engineering software. Those are collaboration environments for the Authoritative Source of Truth, View & Viewpoints for OpenMBEE and other libraries to generate the specifications from the models based on stakeholder views.

Chapter Summary

This chapter focused on several examples illustrative of the current state of the art in SysML applications. The examples consider multiple approaches for the integration of analyses in a SysML environment and discuss the challenges associated with incorporating different levels of analysis fidelity. Several different applications are covered in the chapter to demonstrate the wide applicability of these methods across industries and products. The state of the art in any given field is always evolving, especially in a field as active as SysML at this time. Therefore, future chapters covering this topic may be necessary to capture advances in the field.

Cross-References

▶ Semantics, Metamodels, and Ontologies

References

1. D. N. Mavris, *Methodology, Technology Identification, Evaluation, and Selection. IPPD,* Atlanta: Lecture slides, AE, Georgia Institute of Technology, 2003.
2. G. A. Hazelrigg, System Engineering: An Approach to Information-Based Design, New Jersey: Prentice Hall, Inc., 1996.
3. Y. Li, "An Intelligent Knowledge-based Multiple Criteria Decision Making Advisor for Systems Design," Atlanta, 2007.
4. Y. Li, N. Weston and D. N. Mavris, "An approach for multi-criteria decision making method selection and development," in *29th international congress of the aeronautical sciences*, USA, 2008.
5. Python Software Foundation, "Python Language Documentation," 2022.
6. ECMA International, *The JSON Data Interchange Syntax,* Geneva, 2017.
7. T. Weilkiens and B. Oestereich, UML 2 Certification Guide: Fundamental and Intermediate Exams, Oxford, UK: Elsevier Science, 2010.
8. K. A. Reilley, S. Cimtalay and D. N. Mavris, "Analysis-Centric Template Enabling Requirements Validation Through Engineering Models," in *Canadian Aeronautics and Space Institute Aero*, Laval, Quebec, 2019.
9. OMG, "Systems Modeling Language v2 Request for Proposal," 12 2017. [Online]. Available: https://www.omgsysml.org/SysML-2.htm.
10. National Institute of Standards and Technology, "What is design of experiments (DOE)," 2012. [Online]. Available: https://www.itl.nist.gov/div898/handbook/pmd/section3/pmd31.htm. [Accessed 26 04 2021].
11. Object Management Group, "What is SysML?," 2021. [Online]. Available: https://www.omgsysml.org/what-is-sysml.htm. [Accessed 26 04 2021].
12. No Magic, "Constraint Blocks in Magic Draw," 2021. [Online]. Available: https://docs.nomagic.com/display/SYSMLP2021x/Constraint+Block. [Accessed 26 04 2021].
13. No Magic, "SysML Parametric Diagram in Magic Draw," 2021. [Online]. Available: https://docs.nomagic.com/display/SYSMLP2021x/SysML+Parametric+Diagram. [Accessed 26 04 2021].
14. R. H. Myers and D. C. Montgomery, Response surface methodology: process and product optimization using designed experiments, Wiley-Interscience, 202.
15. OpenMBEE, "Open Model Based Engineering Environment," [Online]. Available: https://www.openmbee.org/. [Accessed 19 03 2021].
16. R Foundation, "Tools for Working with URLs and HTTP," [Online]. Available: https://cran.r-project.org/web/packages/httr/index.html. [Accessed 26 04 2021].
17. Mathworks, "Send HTTP request message and receive response," [Online]. Available: https://www.mathworks.com/help/matlab/ref/matlab.net.http.requestmessage.send.html. [Accessed 26 04 2021].
18. OpenMBEE, "OpenMBEE Github Organization," 2021. [Online]. Available: https://github.com/Open-MBEE. [Accessed 26 04 2021].
19. Dassault Systèmes, "3DEXPERIENCE Platform," 2021. [Online]. Available: https://www.3ds.com/3dexperience. [Accessed 26 04 2021].
20. E. Lee, "End-to-End Integrated Resource Analysis on Europa Clipper," 2019. [Online]. Available: OpenMBEE.org.
21. T. Bayer, "Is MBSE Helping? Measuring Value on Europa Clipper," in *IEEE Aerospace Conference*, 2018.
22. T. J. Bayer, Interviewee, *Electronic mail communication.* [Interview]. 25 August 2021.
23. A. Ustundag, Industry 4.0: Managing The Digital Transformation, Springer, 2018.
24. H. Kim, D. Fried, P. Menegay, G. Soremekun and C. Oster, "Application of Integrated Modeling and Analysis to Development of Complex Systems," *Procedia Computer Science*, vol. 16, pp. 98-107, 2013.

25. B. L. Papke, G. Wang, R. Kratzke and C. Schreiber, "Implementing MBSE – An Enterprise Approach to an Enterprise Problem," *INCOSE International Symposium, 30: 1550-1567.*
26. S. Friedenthal, A. Moore and R. Steiner, "Chapter 18 - Integrating SysML into a Systems Development Environment," in *A Practical Guide to SysML (Third Edition)*, The MK/OMG Press, 2015, pp. 507-541.
27. R. Karban, F. G. Dekens, S. Herzig, M. Elaasar and N. Jankevicius, "Creating systems engineering products with executable models in a model-based engineering environment," *Modeling, Systems Engineering, andProject Management for Astronomy VII, 99110B,* 2016.
28. S. Friedenthal, A. Moore and R. Steiner, A Practical Guide to SysML: The Systems Modeling Language 3rd Ed., Morgan Kaufmann OMG Press, 2014.
29. "TMT, "Thirty Meter Telescope"," [Online]. Available: http://www.tmt.org.
30. R. A. Noguchi, J. Martin and M. J. Wheaton, "(MBSE)2: Using MBSE to Architect, Implement, and Operate the MBSE System," 2020.
31. L. McGinnis, "Bringing MBSE to the Design of Aircraft Production Systems," in *Global Product Data Interoperability Summit,* USA, 2019.
32. R. S. Peak, S. Cimtalay, M. Ballard, M. Mayokonda, I. Ogev and A. Alarcon, "System Model Complexity and Health Managemen," 2021. [Online]. Available: https://www.researchgate.net/project/MCHM-System-Model-Complexity-and-Health-Management.
33. M. Blackburn, R. Peak, S. Cimtalay, T. Fields, et al, "Transforming Systems Engineering through Model-Centric Engineering," SERC project report, 2021. http://sercuarc.org.
34. B. Bagdatli, F. Karagoz, . K. A. Reilley and D. Mavris, "MBSE-enabled Interactive Environment for Aircraft Conceptual Sizing and Synthesis," in *AIAA Scitech Forum,* 2019.
35. Python Software Foundation, "Extensible library for opening URLs," 2021. [Online]. Available: https://docs.python.org/3/library/urllib.request.html#module-urllib.request. [Accessed 26 04 2021].
36. E. W. Brower, C. Delp, R. Karban, M. Piette, I. Gomes, E. van Wyk, *OpenCAE Case Study: Europa Lander Concept: Model-Based Systems Engineering Products in the OpenCAE Model-Based Engineering Environment with Europa Lander as a Case Study.* Torrance, California: 2019 Annual INCOSE International Workshop, 2019. https://trs.jpl.nasa.gov/handle/2014/50351.

Burak Bagdatli, Research Engineer II, Georgia Institute of Technology. Burak Bagdatli received his BS, MS, and PhD degrees in Aerospace Engineering from Georgia Institute of Technology. Before starting graduate school, he worked as a design engineer on a medium altitude high-endurance UAV at Turkish Aerospace Industries. He is currently a research faculty member at Georgia Institute of Technology working within the Aerospace Systems Design Laboratory. His primary research areas are system of systems architectures, stochastic process simulation, and interactive visualizations. Burak currently works on model-based systems engineering for exploring design spaces for commercial airline designs, large-scale manufacturing data analysis and machine learning, and long-term goal setting for low-carbon aviation. He is teaching two graduate-level classes on topics relating to system of systems architecting, modeling, simulation, design, and interpretation. He also teaches guest lectures in advanced design methods classes.

Selçuk Cimtalay, Senior Research Engineer, Georgia Institute of Technology. Selçuk received his doctoral degree in 2000 in Mechanical Engineering from Georgia Tech. He has several years of teaching, research, and industry experience. Dr. Cimtalay has conducted research on various projects on model-based systems engineering (MBSE) with industrial projects. Before joining the research faculty at Georgia Tech, he worked for 7 years for the software start-up company ClickFox LLC as a technical analyst/quality engineer, and 3 years for a consulting company, Hybrid Solutions Inc., as a lead technical analyst. Selçuk mentors ASDL undergrads and graduate students in their SysML-related learning and research. He serves as an instruction associate in SysML short courses for professional engineers. He has also served on the instruction team for the SysML-related course

(ASE 6005) in the Professional Masters in Applied Systems Engineering (PMASE). He is a certified systems modeling professional and member of INCOSE.

Taylor Fields, Research Engineer, Georgia Institute of Technology. Taylor Fields received his BS in Aerospace Engineering from Georgia Tech in May of 2018. Shortly thereafter, he received his MS in Aerospace Engineering from Georgia Tech in December of 2019. During graduate school, working in Aerospace Systems Design Laboratory (ASDL), MBSE, and designing space systems were the focus of research. In 2020, Mr. Fields began his career by joining the research faculty at Georgia Tech. He is a Research Engineer in the Advanced Methods Division of ASDL within the School of Aerospace Engineering where MBSE is the focus of his research.

Elena Garcia, Senior Research Engineer, Georgia Institute of Technology. Elena Garcia was born and raised in Madrid, Spain. She graduated in 1996 from the University of Virginia with a BS in Aerospace Engineering. In December of 1997, she received an MS degree in Aerospace Engineering and joined the PhD program under the supervision of Dr. Dimitri Mavris. Her graduate research initially focused in the cost estimation area as it relates to aircraft affordability. Her PhD dissertation (2002) then considered a variety of technologies that could be applied to alleviate capacity problems taking a system-of-systems perspective of the National Airspace System. Ms. Garcia is currently a Senior Research Engineer and Advanced Methods Division Chief at the Aerospace Systems Design Laboratory (ASDL) within the School of Aerospace Engineering. Dr. Garcia's role is key in furthering the state of the art in systems engineering methods, ensuring application of the latest methods across the ASDL research portfolio, and transitioning methods into the classroom instruction.

Russell Peak, Senior Research Engineer, Georgia Institute of Technology. Russell Peak, PhD, is a Senior Researcher at Georgia Tech in the Aerospace Systems Design Lab (ASDL - www.asdl.gatech.edu) where he is MBSE Branch Chief. Russell specializes in knowledge-based methods for modeling and simulation, standards-based product lifecycle management (PLM) frameworks, and knowledge representations that enable complex systems interoperability. He originated the multi-representation architecture (MRA) – a collection of patterns for CAD-CAE interoperability – and composable objects (COBs) – a noncausal object-oriented knowledge representation. This work provided a conceptual foundation for executable SysML parametrics. After 6 years in industry at Bell Labs and Hitachi, he joined the research faculty at Georgia Tech. Since 1996, he has been principal investigator on numerous projects with sponsors including BAE Systems, Boeing, IBM, JPL, Lockheed, NASA, Rockwell Collins, Sandia, Shinko (Japan), TRW Automotive, US DoC (NIST), and DoD. He has authored 135+ publications (including several Best Paper awards), holds several patents, and is an active member in AIAA and INCOSE. He currently leads the Digital Ecosystems Challenge Team (DECO) in the INCOSE MBSE Initiative. Russell represents Georgia Tech on the OMG SysML task force, is a Content Developer for the OMG-Certified Systems Modeling Professional (OCSMP) program, and holds the highest OCSMP certification, Model Builder – Advanced (MBA). Since August 2008, he has led a SysML/MBE/MBSE training program that has conducted ~388 short courses for ~7440 professionals (as of Dec 2020) in the USA and internationally.

Role of Decision Analysis in MBSE

5

Gregory S. Parnell, Nicholas J. Shallcross, Eric A. Specking,
Edward A. Pohl, and Matt Phillips

Contents

Introduction	120
Key Concepts and Definitions	121
Current State of Practice/Current State of the Art	122
MBSE Capabilities to Enable Trade-Off Analyses	123
Illustrative Example/Case Study	131
Case Study Background	131
Case Study Motivation	132
Modeling Requirements for Enabling Decision-Making in System Design	132
Implementing the Integrated System Model	135
Best Practice Approach: Creating a Custom MODA Model for Use in Integrated Trade-Off Analysis	137
Using the Model-Based Integrated Decision Support Tool to Inform Design Decisions	140
Discussion of Case Study Observations	143
Challenges and Gaps	144
Chapter Contribution	145
Expected Advances in the Future	146
Summary	146
Cross-References	147
Appendix: List of Acronyms and Their Meaning	147
References	147

Abstract

This chapter examines the use of model-based systems engineering (MBSE) tools to perform trade-off analysis of alternative systems decisions throughout the

---

G. S. Parnell (✉) · E. A. Pohl
Department of Industrial Engineering, University of Arkansas, Fayetteville, AR, USA
e-mail: gparnell@uark.edu; epohl@uark.edu

N. J. Shallcross · E. A. Specking · M. Phillips
System Design and Analytics Laboratory, Department of Industrial Engineering, University of Arkansas, Fayetteville, AR, USA
e-mail: njshallc@uark.edu; especki@uark.edu; mmphilli@uark.edu

© Springer Nature Switzerland AG 2023
A. M. Madni et al. (eds.), *Handbook of Model-Based Systems Engineering*,
https://doi.org/10.1007/978-3-030-93582-5_14

system life cycle. Specially, we seek integrated models that automate the simultaneous evaluation of the performance, effectiveness, stakeholder value, and cost of multiple alternative system designs. We used Web of Science to perform a structured literature search to identify papers that describe the use of MBSE tools to support automated analysis of alternatives and trade-off analyses. While we found no papers that use the terms decision science and MBSE. We did find papers that used decision analysis and MBSE. We also found very few papers that claimed to use MBSE to provide analysis of design alternatives or tradespace exploration. Based on the literature search insights, we identified and described the required and desired capabilities to perform automated trade-off analyses of performance, effectiveness, stakeholder value, and cost for multiple system design alternatives using integrated models. We provide an illustrative case study of an unmanned aerial vehicle (UAV) design trade-off analysis that uses a model-based engineering tool, ModelCenter® Integrate, to develop an integrated modeling tool in order to simultaneously evaluate the performance, effectiveness, and stakeholder value of UAV designs. We performed eight iterations of tool development with increasing fidelity models. Iteration 1 identifies 13 Pareto optimal alternatives. However, by the end of iteration 8, only four of these designs remain feasible, though not all remain as Pareto optimal. By delaying major design decisions and using higher-fidelity models, we are able to prevent the selection of a suboptimal and potentially infeasible design. By integrating decision analysis and life cycle cost models with physics models and simulations, we are able to take advantage of the benefits of model-based engineering practices to support system decision-making.

Keywords

MBSE · Decision science · Decision analysis · Multiple objective decision analysis · Trade-off analysis · Decision-making · Tradespace · Set-based design · Integrated models

Introduction

Engineering systems are becoming increasingly complex due to demanding customer needs, advances in technology, interconnectedness, increasing software capabilities, and adversary actions. Model-based systems engineering (MBSE) has demonstrated the ability to define, describe, and manage a baseline system configuration in several phases of a systems life cycle. For example, "The software and electronics of modern automobiles are becoming increasingly complex. Ford Motor Company has been applying MBSE to manage design complexity including architecture, requirements, interfaces, behavior, and test vectors. Ford has established digital design traceability across their onboard electrical and software systems by applying multiple integrated modeling technologies..." [1].

While project managers and systems engineers need to have a system baseline to assess cost, schedule, and performance, they routinely need to identify and evaluate alternative system designs throughout the system life cycle. Systems decisions made in every life cycle stage require evaluation of alternatives and trade-off analyses. Therefore, the creation and evaluation of the decision tradespace is a critical system engineering activity throughout the systems life cycle [2]. As noted in the reference, the creation and evaluation of the tradespace are especially important in conceptual and preliminary designs since the cost of redesigns is much less early in the life cycle. The purpose of this chapter is to assess the use of MBSE to support the creation and automated analysis of alterative systems decisions.

This chapter is organized as follows. In the first section, we present our literature search to identify the use of MBSE to support system decision-making. In the second section, using insights from our literature survey, we identify and describe the required and desired capabilities to perform trade-off analyses for alternative decisions in performance, effectiveness, stakeholder value, and cost. In the third section, we provide an illustrative example of integrated modeling using ModelCenter. Finally, we provide a summary of the paper.

Key Concepts and Definitions

- Decision analysis: A philosophy and a social-technical process to create value for decision-makers and stakeholders facing difficult decisions involving multiple stakeholders, multiple (possibly conflicting) objectives, complex alternatives, important uncertainties, and significant consequences [3]
- Decision: The commitment of resources to implement a choice between alternatives
- Design decision: The selection of a design alternative or set of design alternatives
- Digital engineering: "The creation of computer readable models to represent all aspects of the system and to support all the activities for the design, development, manufacture, and operation of the system throughout its lifecycle" [4]
- Integrated model-based engineering: The use of integrated models to automate the simultaneous evaluation of the performance, effectiveness, stakeholder value, and cost of multiple alternative system designs
- Life cycle cost (LCC) model: A model that calculates the potential cost over the system life cycle given a set of design decisions and assumptions about the operations and maintenance concepts in the operational environments
- Measure of performance (MOP): An engineering performance measure that provides design requirements that are necessary to satisfy a measure of effectiveness [4]
- Measure of effectiveness (MOE): The metrics by which an acquirer will measure satisfaction with products produced by the technical effort [4]
- Model-based engineering (MBE): "An engineering approach that uses models as an integral part of the technical baseline that includes the requirements, analysis, design, implementation, and verification of a capability, system, and/or product throughout the acquisition life cycle" [5]

- MBSE: "The formalized application of modeling to support system requirements, design, analysis, verification and validation activities beginning in the conceptual design phase and continuing throughout development and later life cycle phases" [4]
- Stakeholder value: Value is defined by meeting the objectives and requirements of the stakeholders [2].
- Set-based design (SBD): It is a complex system design method that enables robust system design by (1) considering a large number of alternatives, (2) establishing feasibility before making decisions, and (3) using experts who design from their own perspectives and use the intersection between their individual sets to optimize a design [6].
- System effectiveness model: A model that calculates the potential stakeholder MOE given a set of design decisions and assumptions about the environment
- System performance model: A model that calculates the potential system measure of performance given a set of design decisions and assumptions about the environment
- Trade-off: "Decision making actions that select from various requirements and alternative solutions on the basis of net benefit to the stakeholders" [7]
- Trade-off study: An engineering term for an analysis that provides insights to support system decision-making in a decision management process [2]
- Tradespace: A multidimensional space that defines the context for the decision, bounds the region of interest, and enables Pareto optimal solutions for complex, multiple stakeholder decisions [2]
- Value model: A single- or multiple objective model that assesses the value of the system design to the stakeholders based on measures of achievement of their objectives.

Current State of Practice/Current State of the Art

In this section, we describe our literature survey methodology to determine the documented use of MBSE in peer-reviewed journals to analyze systems design alternatives and enable trade-off analysis.

Methodology. We began the literature survey by selecting the Web of Science (WoS) as our primary search database since it is both extensive and commonly used, which increases the likelihood of finding relevant articles. WoS allows researchers to search titles, abstracts, author keywords, and Keywords Plus, which uses words that appear in the titles of the authors' references, to find relevant articles. Next, we defined phrases to use as search criteria. These search phrases focused on MBSE or Model-Centric Engineering and how they related to trade-off analyses. Since our focus is on system decisions, we also used terms involving decisions (e.g., decision science and decision analysis). We did seven distinct searches as described in Table 1. We found that decision science is not a term used in the literature in conjunction with MBSE. These searches identified 13 unique articles.

Next, we reviewed these 13 papers for the following information: types of models in the paper, modeling software, decision analyzed, decision model, and

5 Role of Decision Analysis in MBSE

Table 1 Web of Science (WoS) search results

Screening number	Keyword	# of papers	Unique papers
1	"MBSE" AND "decision science" OR "decision sciences"	0	0
2	"MBSE" AND "decision analysis"	5	5
3	"MBSE" AND "trade-off" OR "trade-off" OR "trade study"	8	6
4	"Model-Centric Engineering" AND "decision science" OR "decision sciences"	0	0
5	"Model-Centric Engineering" AND "decision analysis"	0	0
6	"Model-Centric Engineering" AND "trade-off" OR "trade-off" OR "trade study"	0	0
7	"Model-Centric Engineering"	2	2
	Total used	15	13

visualization of the tradespace. Table 2 provides answers to these questions. We used acronyms to save space in all tables and provide the complete name for each acronym in the Appendix.

Screening Results. After compiling the information in Table 2 on the 13 journal papers, we identified the most relevant articles. A relevant paper requires the use of the MBSE software or significant system modeling, alternative identification, decision models to evaluate the alternatives, and tradespace visualization. Three papers passed this screening.

Detail Review Results. All three articles were published in journals dealing with systems or systems engineering between 2017 and 2019. An SBD literature survey published in 2020 discussed many of the topics relevant to MBSE and decision analysis but did not provide a new methodology specific to either [21]. The most cited article received seven citations after 2 years of publication. All the articles used multiple objective models of stakeholder value and visualized the tradespace. Two of the three articles used the models to evaluate system design alternatives. One of the articles made extensive use of integrated models to simultaneously evaluate the decision tradespace but did not use the MBSE software. Table 3 shows the full review of the articles.

MBSE Capabilities to Enable Trade-Off Analyses

In this section, we describe the MBSE capabilities required to perform automated analysis of the performance, effectiveness, stakeholder value, and cost of multiple system decisions and fully evaluate the feasible decision tradespace. We know that some MBSE tools have a few of these capabilities. However, we have not been able to identify an MBSE tool that provides all or many of these capabilities. Table 4 provides a summary of our required and desired MBSE capabilities and describes their implementation in the case study.

Table 2 Results of the first screening

Title	Types of models	Modeling software	Decision	Decision model	Tradespace visualized?	Pass screening?
Early Design Space Exploration with Model-Based System Engineering and SBD [8]	Physics based, Life cycle cost (LCC), and value model	Excel	Creating better alternatives for the design of a UAV	Multiobjective decision analysis (MODA)	Yes	Yes
Trade-off analysis for SysML models using decision points and Constraint Satisfaction Problems (CSPs) [9]	Optimization, Block Definition Diagrams (BDD), and Parametric Diagram	Papyrus SysML, and Object-Oriented System Engineering Method (OOSEM)	Hardware decisions regarding sensors, processors, and interface networks on a UAV	Multiobjective Optimization Problem with Constraints (CSMOP)	Yes	Yes
Integrating model-based system engineering with set-based concurrent engineering principles for reliability and manufacturability analysis of mechatronic products [10]	BDD and Requirements Diagrams	SysML, Gephi, and Python	Design of electric compressors in car air-conditioning systems	MODA	No	Yes
Informing System Design Using Human Performance Modeling [11]	BDD, requirements, simulation, and Parametric Diagram	IMPRINT and SysML	Design of an algorithm for Vigil Sprint Monitoring System	Multiattribute Model	Yes	Yes
Rapid model-based interdisciplinary design of a CubeSat mission [12]	Simulation, LCC, and physics	MATLAB	Two studies: (1) Optimal satellite placement for earth study. (2) Satellite captures data at a set rate over its lifetime, maximize this.	Multiobjective Genetic Algorithm	No	Yes

SBD: The state-of-practice and research opportunities [13]	N/A	N/A	N/A	No	No	No
An Analysis of Theories Supporting Agile Scrum and the Use of Scrum in Systems Engineering [14]	N/A	NA	N/A	No	No	No
Applying Composable Architectures to the Design and Development of a Product Line of Complex Systems [15]	Capability Model, Variation Model, and Parametric Diagram	Excel and MATLAB	Design of a communications spacecraft by Lockheed Martin	Architect Variability Model	No	No
Leveraging Variability Modeling Techniques for Architecture Trade Studies and Analysis [16]	Simulation, BDD, requirements, cost, and inventory	Excel, Model Center, and SysML	Design of a grounded radar system	No	No	No
A Design Task-Oriented Model Assignment Method in MBSE [17]	N/A	SysML	Which model should be assigned to which task in a radar project?	MODA	No	No

(continued)

Table 2 (continued)

Title	Types of models	Modeling software	Decision	Decision model	Tradespace visualized?	Pass screening?
An MBSE Approach to Tradespace Exploration of Implanted Wireless Biotelemetry Communication Systems [18]	BDD, Parametric Diagrams, Requirements Diagrams, Integrated analytical models, and safety models	IBM Rhapsody	Design of an implanted wireless biotelemetry system	No	No	No
Transforming systems engineering through digital engineering [19]	Information Model, 3D Models, simulation, performance model, and Assessment Model	Vitech Core	Assessing the technical feasibility of transferring US Navy Systems Engineering techniques and acquisition practices to MBSE	No	No	No
Model-centric engineering with the evolution and validation environment [20]	N/A	N/A	N/A	No	No	No

5 Role of Decision Analysis in MBSE

Table 3 Detailed review results

	Early Design Space Exploration with Model-Based System Engineering and SBD [8]	Trade-off analysis for SysML models using decision points and CSPs [9]	Informing System Design Using Human Performance Modeling [11]
Year of publication	2018	2019	2017
Journal	Systems, 6(4)	Software and Systems Modeling, 18(6)	Systems Engineering, 20(2)
Authors	Specking et al.	Leserf et al.	Watson et al.
Number of citations	Google Scholar: 11, Web of Science (WoS): 5	Google Scholar: 5, WoS: 2	Google Scholar: 5, WoS: 3
Models	Physics, Life Cycle Cost, and Value	Optimization, Block Definition Diagram (BDD), and Parametric Diagram	BDD, requirements, simulation, and Parametric Diagram
Use of MBSE	Yes	Yes	Yes
Software	Excel	Papyrus SML and OOSEM	IMPRINT and SysML
Decision	UAV discrete (e.g., sensor) and continuous (e.g., wingspan) design alternatives	Hardware decisions regarding sensors, processors, and interface networks on an UAV	Design of an algorithm for the Vigil Sprint Monitoring System
How many alternatives did they analyze?	Used SBD to evaluate 100,000 alternatives and 2576 feasible	36	6
What type of decision model?	MODA	Multiobjective Optimization Problem with Constraints (CSMOP)	Multiattribute Model
What are the axes on the tradespace visualization?	X: LCC Y: Value	X: Failure Rate Y: Cost	X: Workload Y: Alternatives

Stakeholder Requirements. Stakeholder requirements document user needs. MBSE tools include stakeholder requirements. These are obviously essential for evaluating the feasibility of system alternatives and performing trade-off analysis.

Systems Requirements. System engineers develop system requirements to guide system designers. MBSE tools include system requirements. They are obviously essential for evaluating the feasibility of system designs and performing trade-off analysis.

Define and Describe Systems Alternatives. Many MBSE tools only define the baseline system design. To improve the design or solve a design problem (e.g., a

Table 4 MBSE capabilities to enable trade-off analyses

Capability	Required	Desired	Illustrated in case study
Stakeholder requirements	In current MBSE tools		In MODA model
System requirements	In current MBSE tools		In performance and effectiveness models
Define and describe systems alternatives	Discrete alternatives	Continuous design parameters	Generate designs using Monte Carlo simulation
Modeling environment enabling custom modules	Incorporate user-defined modules for a variety of analysis uses		Use ModelCenter with custom Java and Python modules
Integrated models	Use integrated models to automate the evaluation of system alternatives and exploration of the decision tradespace		Integrate performance, effectiveness, value, and cost models
System performance models	Determine if alternatives meet the system requirements (MOE)	Varying fidelity and multiresolution modeling	Model UAV system and a sensor performance in various operating environments
Ility models	Perform ility calculations for the alternative system designs		Not implicitly modeled and use parameter inputs
System effectiveness models	Determine if alternatives meet the stakeholder requirements (MOE)	Include ilities data using the mission chain analysis of system alternatives	Use integrated multiobjective value model to measure system effectiveness
LCC model	Calculate the LCC of each alternative	Include ilities data in the LCC model	Use an integrated cost model
MODA value model	Calculate the stakeholder value using multiple effectiveness measures of each alternative		Use an integrated multiobjective value model to measure system effectiveness
Engineering economic analysis models	Perform economic analysis of design alternatives		Not done
Quantify uncertainty	Put distributions on inputs and perform Monte Carlo simulation on outputs		Use Monte Carlo simulation
Tradespace visualization	Provide performance, effectiveness, stakeholder value, and cost data that can be used to visualize the	Provide Graphical User Interface for tradespace visualization	Produce outputs allowing for trade-off analysis of

(continued)

5 Role of Decision Analysis in MBSE

Table 4 (continued)

Capability	Required	Desired	Illustrated in case study
	tradespace in another software		stakeholder value and LCC
Pareto optimal solutions	Incorporate algorithms to identify Pareto optimal designs	Evaluate design sets using SBD	Pareto optimal solutions identified during tradespace analysis

baseline design does not meet a requirement), system analysts need to identify and evaluate design alternatives. It is required that we have discrete alternatives (e.g., multiple alternative concepts, alternative subsystems, alternative components, etc.). To fully identify the tradespace and perform SBD, it is desirable to have continuous design parameters (e.g., the wingspan of a UAV).

Modeling Environment Enabling Custom Software Modules. Custom software modules are a required capability that would be necessary for many uses, such as evaluation of design alternatives and trade-off analyses. This capability enables users to add appropriate models not available in the MBSE software. For example, custom modules are often essential for performance, effectiveness, stakeholder value, and cost modules.

Integrated Models. One of the most important capabilities for evaluating many alternatives and performing trade-off analyses that fully explore the decision tradespace is integrated model. An integrated model can simultaneously evaluate alternative designs using performance, effectiveness, stakeholder value, and cost models without subjective evaluation by experts or stakeholders. The system design decisions and scenario assumptions provide all the data needed to calculate the ilities. The models, design decisions, and ilities provide the data for calculating the performance, effectiveness, stakeholder value, and cost.

System Performance Models. Systems analysts use system performance models to evaluate alternatives versus systems requirements. It is desired that MBSE tools have the capability to include varying fidelity models and multiresolution models. The models are essential to evaluate the feasibility and performance of system designs prior to physical testing.

Ility Models. The ilities are critical to system success. Systems analysts need to evaluate the ilities of design alternatives. This requires Reliability, Availability, and Maintainability models and other models for other ilities, including survivability, resilience, etc. The ilities are required to calculate system effectiveness.

System Effectiveness Models. System effectiveness models are required to calculate the system effectiveness of the alternatives versus the stakeholder requirements. It is desired that these models include ilities data used in the mission chain analysis of system alternatives.

LCC Model. System designers and analysts need to understand the system cost for the entire life cycle and use cost estimation best practices to estimate these costs. This includes using parametric modeling methods, such as the Constructive Cost Model (COCOMO) for software [22] and Constructive Systems Engineering Cost Model (COSYSMO) for system engineering estimation [23]. LCC models have a critical capability of comparing system alternatives. It is desired that the LCC models use the ilities in their calculations.

MODA Value Model. For complex systems, there are typically many stakeholder objectives and measures of effectiveness. MODA models are commonly used to evaluate the stakeholder value of systems alternatives and perform trade-off analyses [4]. The multiple objective value can be calculated by using the system effectiveness measures for the systems alternatives and a MODA value model, typically the additive value model. MODA is a decision theory approach. See Chap. 8, Generating and Evaluation Alternatives [2], for a discussion on and comparison of seven additional alternative evaluation techniques, namely, the Pugh Method, Axiomatic Approach to Design, TRIZ, Design of Experiments, Taguchi Approach, Quality Function Deployment, and Analytic Hierarchy Process.

Engineering Economic Analysis Models. Economic analysis models are required to evaluate the economic value of system design alternatives over the system life cycle. These models often use the LCC model.

Quantify Uncertainty. System design alternatives are not deterministic. The major uncertainties in system design include scenarios, mission, environment, technology, interfaces with other systems, and adversary actions. MBSE capabilities should include the ability to quantify uncertainty in the ilities, system performance, system effectiveness, stakeholder value, and LCC as a function of those uncertainties. The capability to put distributions on inputs and perform Monte Carlo simulation on outputs is required to visualize the impact of uncertainty on the decision tradespace.

Tradespace Visualization. Data to visualize the performance, effectiveness, stakeholder value, and cost data are required. This visualization can be performed using another software package or with a Graphical User Interface (GUI) in the MBSE tool.

Pareto Optimal Solutions. For large tradespaces, trade-off analyses require the capability to incorporate algorithms to identify Pareto optimal design alternatives or sets. To support SBD, the tools must be able to identify and evaluate design sets.

Our literature search found very few papers that describe the use of MBSE tools to perform the analysis of design alternatives or to identify and explore the decision tradespace. A finding supported by Shallcross et al. [21], system developers identify and evaluate system alternatives in every life cycle stage. The important decisions will determine the success of the system development, deployment, and operation. Currently, MBSE tools focus primarily on describing the system baseline. We identified and described MBSE capabilities and integrated models needed to provide automated evaluation of system alternatives and decision tradespace exploration throughout the system life cycle.

Illustrative Example/Case Study

In the previous section, we identified the MBSE requirements to support trade-off analyses. In this section, we provide an in-depth discussion of integrating decision analysis tools and methods in a specific MBSE design application. An unmanned aerial vehicle (UAV) design case study, implemented in the ModelCenter process workflow environment, demonstrates key ideas regarding the use of integrated models to inform system design decisions. We begin by providing an overview of the case study followed by a discussion regarding motivation and objectives. Given these objectives, we discuss modeling requirements to facilitate decision-making under uncertainty and then explain how we implemented these requirements in ModelCenter. We then provide a model demonstration and conclude with a discussion of observations.

Case Study Background

To demonstrate the fundamentals of MBSE-enabled decision analysis, we use a modified UAV design case study [2]. The original case study provided a plausible system design example to explore engineering and analytical methods enabling the design of resilient systems. The case study seeks to design a small UAV for surveillance missions. The UAV must completely satisfy 11 functional performance and design requirements, given the seven primary design decisions seen in Table 5. These decisions include five discrete options for the UAV engine and sensor suite, and two continuous options, regarding the wingspan and operating altitude, which we bin into discrete categories for graphical displays. The various combinations of all available design options produce 204,120 unique design sets, assuming five wingspan bins and seven altitude bins. The decision-maker wants to select the best design alternative from these sets. However, this decision requires effective design space exploration and analysis methods to handle the design space complexity and objectively assess the various UAV designs.

Table 5 UAV case study design decisions

Design decision	Decision type	Available design options
Engine type	Discrete choice	Piston (P) and electric (E)
Electro-optical (EO) sensor resolution (pixels)	Discrete choice	200 × 200, 400 × 400…, 1800 × 1800
EO sensor field of view (degrees)	Discrete choice	15, 30, …, 90
Infrared (IR) sensor resolution (pixels)	Discrete choice	200 × 200, 400 × 400…, 1800 × 1800
IR sensor field of view (degrees)	Discrete choice	15, 30, …, 90
Wingspan (ft.)	Continuous choice	2–12
Operating altitude (m)	Continuous choice	300–1000

Recent SBD studies have used the UAV case study to develop quantitative methods to perform efficient analysis of complex and multidimensional design spaces [21]. SBD is a concurrent engineering methodology that develops and analyzes a large number of unique design options, organized as sets within the design space. In this context, a set contains a number of unique system designs sharing at least one common design attribute [24]. As a design methodology, SBD is ideal for applications with multiple design decisions, each with several potential options, such as those of the UAV case study. To address the complexities of the UAV design problem, Small et al. incorporate concepts from MBE, multiple objective decision analysis (MODA), and trade-off analysis to develop an integrated model and tradespace analytics tool [25]. Their methodology used Monte Carlo methods to produce a large number of system alternatives by generating random combinations of the available design options. They evaluated each alternative by calculating performance using low-resolution parametric models, whose outputs feed the integrated MODA and LCC model. This allowed them to generate the system tradespace and assess each design alternative in terms of stakeholder value and total life cycle cost. Subsequent work by Specking et al. validated the methodology and demonstrated an improved tradespace exploration performance over other methods such as the use of genetic algorithms [8].

Case Study Motivation

The case studies of [8, 22, 27] demonstrated the usefulness of integrating decision analysis and LCC models within a model-based engineering framework to inform stakeholder decisions. However, an SBD state-of-practice survey identified knowledge and methodology gaps for complex system design management [21]. The survey specifically identified a lack of quantitative methodologies informing program management decision-making, just as striking was the limited number of SBD methodologies using techniques such as MBSE and multiresolution modeling in system design. Our case study question is, "How can we combine and adapt the best practices and methods first described by [27] with other techniques like multiresolution modeling and information theory to enable program management decisions and ultimately deliver higher value to the system stakeholders?" While the original tradespace analytics tool provided an excellent example of the functionality required in model-based engineering applications, it lacked certain features and capabilities required for our case study. Additionally, it was evident that our new methodology would also require the development of a new set of decision models tailored to the program manager, in addition to those required for the system stakeholders.

Modeling Requirements for Enabling Decision-Making in System Design

The original tradespace analytics tool developed by [27] is an Excel-based tool containing nine modules for the parametric, MODA, and life cycle models, a random

5 Role of Decision Analysis in MBSE

number generator, and a user interface known as the control panel. The tool enables analysis of multiple alternatives by propagating many design solutions through the integrated parametric physics, MODA, and LCC models to create the system tradespace. The tool provides several advantages in terms of portability, an integrated Monte Carlo simulation engine, and the ability to enable near-real-time analysis of design and requirement changes. However, the tool has several drawbacks as well. The first issue is the tool's inability to incorporate high-resolution models and simulations. While the tool can use higher-resolution parametric models and even some simple Monte Carlo simulations, it is unable to directly integrate high-resolution models such as flight simulations and discrete event simulations required for effective uncertainty resolution and risk reduction. Furthermore, modifying the existing parametric models is a difficult process due to the complex linkages and interdependencies between the different modules.

The second major issue was the use of an Excel workbook as our modeling environment. Excel provides a high level of model portability and user familiarity but also has limitations regarding (1) its compatibility with Linux operating systems typically used in high-performance computing (HPC) and (2) the upper limit of the maximum system tradespace size. The first issue limits the tool's ability to take full advantage of advances in computing power and higher-resolution models. The second issue is the result of the number of available rows present within an Excel spreadsheet, limiting the tradespace size to approximately 1 million design points (The total number of rows in an Excel 2019 worksheet is 1,048,576 [35].). This ultimately limits the tool's usefulness to components and simple system design applications, not requiring the generation of an extremely large number of alternatives to explore the tradespace. We provide an example of this issue in Table 6. Recall that the seven UAV design decisions and available options produce 204,120 individual design sets. These design sets are continuous design spaces, due to the presence of the two continuous decision variables. Thus, it would take a minimum of 204,120 design points to explore the UAV tradespace with possibility of producing a representative design for each individual design set. However, the randomness of Monte Carlo design generation may not produce a representative design in each set, requiring the generation of more design points. As shown in Table 6, generating 204,120 points only enumerated 127,556 of the possible design sets, leaving 38% of the design sets unrepresented in the study. Increasing the number of design points to

Table 6 UAV case study tradespace exploration results

Number of generated design points	Number of represented design sets	Number of unrepresented design sets	Percent unrepresented
204,120	127,556	76,564	38%
400,000	175,193	28,927	14%
600,000	193,353	10,767	5%
800,000	200,165	3955	2%
1000,000	202,564	1556	1%

600,000 increases the percentage of unrepresented sets to 5%. However, even with generating 1 million design points, we fail to fully explore the case study's design space. At this point, we had reached the upper limits of Excel's capabilities requiring a reevaluation of our modeling methods.

We based our program management decision-making on the SBD tenets of delayed decision-making and uncertainty resolution [21]. To facilitate our case study, we needed a new modeling environment capable of fully exploring a system's design space and using higher-resolution models and simulations. We viewed both aspects as crucial elements to understanding and ultimately reducing sources of epistemic design uncertainty while building resilience toward sources of aleatory uncertainty. After analyzing the problem, we determined that the modeling platform needed to satisfy the following criteria:

(i) Contains or can integrate MODA and LCC models
(ii) Is modular and easily modifiable
(iii) Has multiresolution modeling capable
(iv) Has ability to efficiently generate over 1 million design points
(v) Is HPC compatible while retaining model portability

To enable complex system design and analysis, the new modeling platform required the ability to generate tradespaces containing at least 1 million design points. By initially generating an extremely large number of unique designs, we increase the likelihood of producing representative designs in each set, enabling better understanding of potential design feasibility and performance. We create the initial tradespace using low-resolution models, which produce highly conceptual designs. As a result, a significant amount of design and performance uncertainty exists requiring increased resolution during the design process. Multiresolution modeling offers a method to reduce this uncertainty [26]. Thus, our new platform requires the ability to easily integrate discrete event and scenario simulation in addition to higher-resolution physics and cost models. This ultimately requires a modular design allowing the analyst to easily add and remove modules throughout the design process. This modularity also applies to the requirement that the modeling environment contains integrated MODA and LCC models. These models can either be preexisting within the modeling environment's architecture, or we need the ability to create and integrate custom MODA and LCC models. Finally, we have the dual requirements of retaining model portability while also enabling Linux, and by extension HPC, compatibility. We define model portability as the characteristic enabling distributed and easy access to the modeling environment and modules. For example, Excel is highly portable as most personal computers run Microsoft Office software, allowing easy access and use. To achieve the final requirement, we chose general programming languages, such as Java and Python for our models and simulations, as they are both commonly used and compatible with most operating systems including Linux. Now that we had defined our modeling requirements, it was time to implement them in a new modeling platform.

Implementing the Integrated System Model

The purpose of this case study and model-based decision support tool was to better inform decision-making in complex system design. This overarching purpose directly influenced the functions, architecture, and l design of the decision support tool. The integrated trade-off analytics framework seen in Fig. 1 provides the logic for the tool's design and the linkages between the individual modules. We show the framework as an influence diagram, providing the relationships between design and modeling decisions, key uncertainties affecting system performance, and measures of value and affordability [27].

For our case study, we choose to implement the new UAV model in ModelCenter version 11, a model integration software package, developed by Phoenix Integration (https://www.phoenix-int.com/, accessed 8 Dec 20). We chose ModelCenter due to its ability to integrate many popular modeling and analysis tools, conduct trade and optimization studies, and enable distributed collaboration between multiple design teams [28]. ModelCenter also provides the ability to create custom models and script packages using various programming languages. The ModelCenter interface requires a Windows operating system and is not compatible with Linux operating systems. However, the ModelCenter software package also includes Analysis Server, which enables the creation, distribution, and execution of model components. Analysis Server is Linux compatible, providing the ability to construct and execute the model on a Linux-based server. Thus, ModelCenter provided the requisite capabilities to enable our case study.

Model development was an iterative process creating the five major model versions, shown in Table 7, which we refer to as the Model-Based Integrated SBD Decision Support Tool. Model version 1 was a proof of concept, demonstrating that we could recreate the original tradespace analysis tool in ModelCenter. This version

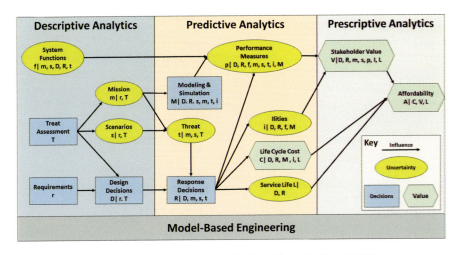

Fig. 1 Integrated trade-off analytics framework. (Adapted from Small et al. [27])

Table 7 Iterative development of the Model-Based Integrated SBD Decision Support Tool

Model version	Module type/ modeling language	Workflow	No. of parallel processes	Linux compatible?	Runtime: 100 K design points
1	Excel workbook modules	Process	1	No	34 h
2	Excel script wrapper modules	Process	4	No	20 h
3	Excel script wrapper modules	Data	4	No	14 h
4	VB script, Java, and Python modules	Data	4	No	1.5 h
5	Java and Python modules	Data	4	Yes	<1.5 h

used individual Excel spreadsheet modules, linked together with a process workflow. Using a laptop computer, with a 6th generation Intel Core i7 processor running at 2.80 GHz, this version was able to generate 100,000 design points in approximately 34 h. While this version successfully demonstrated a ModelCenter implementation of the tradespace analysis tool, its excessive runtime inhibited the creation of sufficiently large tradespaces and was not suitable for our objectives. In the second version, we replaced the Excel workbook modules with Excel script wrappers. The use of script wrappers enabled parallel workflow processing, reducing the runtime to 20 h. However, this implementation still required the use of Excel worksheets for each module, which ModelCenter called using the script wrapper files, resulting in continued unsatisfactory runtimes and compatibility. In the third version, we changed the workflow from a process-oriented workflow to a data workflow, which provided improved computational efficiency over the previous versions. In model version 4, we eliminated the use of Excel script wrappers in all modules, replacing them with modules written in Visual Basic (VB), Java, and Python. Using the same Dell Latitude computer, this model version reduced the 100,000-point runtime to approximately 1.5 h, which we viewed as sufficient for our purposes. However, the use of VB script was still incompatible with Linux operating systems, requiring the development of the fifth and final version of the tool. Version 5 used nine Java and two Python-based modules, enabling Linux compatibility and resulting in runtimes under 1.5 h with four parallel processors.

The final model version contained the 11 modules shown in Fig. 2 and described in Table 8. The order in the table corresponds to the modules seen in Fig. 2. We selected Java as our primary modeling language to take advantage of accessible Java simulations used later in higher-resolution UAV simulations. However, we also included two Python modules to demonstrate ModelCenter's ability to integrate different types of models. The first 10 modules were adapted from the original model built by [27]. These include a control panel providing the primary means for controlling and executing the model, a random number generator required for SBD tradespace creation and exploration, five UAV design and performance

5 Role of Decision Analysis in MBSE

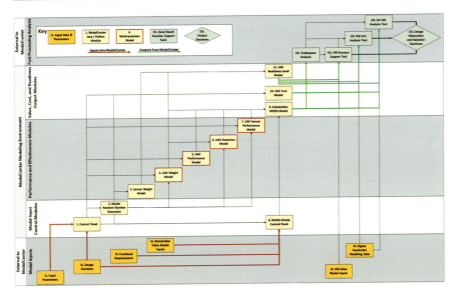

Fig. 2 The model-based integrated decision support tool flowchart

modules, as well as integrated MODA and LCC models. The new tool also contains an 11th module assessing a system's technology, integration, and manufacturing readiness levels, based on the combination of design options. We added this last module to enable the development and testing of our program management decision methodology. Our modeling approach allowed us to implement the physical and functional requirements mandated by the integrated trade-off analytics framework. In addition to the tool itself, Fig. 2 provides a comprehensive view of the complete methodology, taking inputs and data guiding system development and assessment (bottom row) to ultimately inform complex design maturation and a selection decision (top row).

Best Practice Approach: Creating a Custom MODA Model for Use in Integrated Trade-Off Analysis

One of the primary requirements for the new modeling environment was the ability to use existing or creating custom MODA models. ModelCenter does not contain a native decision analysis package, but it does provide the user the ability to create custom decision analysis modules. The development of the case study MODA model requires additional discussion as it pertains to overall model development. For our model, we developed a multiple objective value model adapted from the original tradespace analysis tool. We base our MODA model on the functional value hierarchy seen in Fig. 3. A value hierarchy enables the identification of system objectives and value measures to enable the identification of key aspects of the decision problem and facilitate the quantitative evaluation of alternatives [3]. This

Table 8 Model-Based Integrated SBD Decision Support Tool module descriptions

Module	Name	Purpose	Modeling language
1	Control panel	Controls model inputs, tradespace generation, and model resolution	Java
2	Model Random Number Generator	Generates discrete and continuous uniform random numbers representative of the design decisions and available options	Python
3	Sensor Weight Model	Models UAV sensor weight based on design decision parameters	Java
4	UAV Weight Model	Models total UAV weight based	Java
5	UAV performance model	Models UAV performance given design decision parameters	Java
6	UAV Detection Model	Models an adversary's ability to observe the UAV in flight given a specific size and operating altitude	Python
7	UAV Sensor Performance Model	Models the sensor's ability to locate and identify ground targets	Java
8	MODA control panel	Controls inputs required by the primary MODA model; contains the MODA model swing weight matrix	Java
9	Stakeholder MODA model	Models stakeholder value given UAV design and performance data	Java
10	UAV Cost Model	Models the total UAV life cycle cost	Java
11	UAV Readiness Level Model	Assigns each design a technology, integration, and manufacturing readiness level given primary design parameters	Java

hierarchy's purpose is the selection of the best system capable of performing surveillance missions. We achieve this purpose by assessing systems based on four primary functions regarding transportability, maneuverability and endurance, survivability, and sensor performance. The hierarchy provides a qualitative description of what is important and what we should measure in the case study; however, it requires an implementation strategy linking our primary design decisions to our performance measures [29] (Fig. 4).

To implement the value hierarchy and integrate the MODA model into the case study, we use the UAV assessment flow diagram developed by Cilli [30]. The assessment flow diagram allowed us to quantitatively assess each potential alternative by mapping the seven primary design decisions to the performance objectives and life cycle costs. Additionally, it identifies the primary calculations required to assess the performance measures. For example, we calculate the performance measure *Detect Human in Daylight*, in Module 7, using an electro-optical (*EO*) *Probability of Detection* model taking the EO Resolution, EO FOV, and operating altitude design decisions as inputs. The model then sends the performance measure outputs

5 Role of Decision Analysis in MBSE

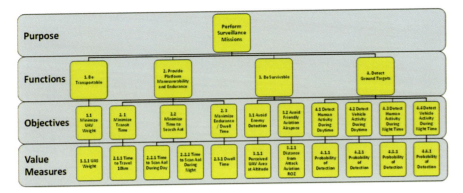

Fig. 3 UAV case study functional value hierarchy. (Adapted from [27])

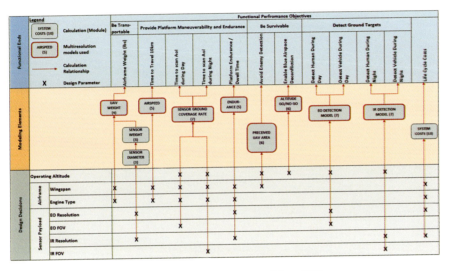

Fig. 4 UAV case study assessment flow diagram. (Adapted from [28])

to the stakeholder MODA model, in module 9, which assesses design feasibility and assigns each alternative a value score for use in subsequent trade-off analysis.

We calculate total stakeholder value $v(x)$ using the additive multiobjective value model given in Eqs. 1–3 [31]. In this model, $v(x)$ is the sum product of n performance measure values $v_i(x_i)$ and their normalized swing weights w_i. A normalized swing weight assesses a particular value measure's importance in relation to the other value measures. We calculate w_i using a swing weight matrix, which enables a decision-maker to explicitly describe each value measure's swing weight using an unnormalized weight f_i, where $f_i \in [0, 100]$.

$$v(x) = \sum_{i=1}^{n} w_i v_i(x_i) \qquad (1)$$

$$w_i = f_i \Big/ \sum_{i=1}^{n} f_i \qquad (2)$$

$$\sum_{i=1}^{n} w_i = 1 \qquad (3)$$

We calculate a performance measure's value $v_i(x_i)$ using value functions such as those seen in Fig. 5. A value function assesses a system's a priori potential value using a given system's performance as an input. A value score of *0* equates to achieve some minimal acceptable performance requirement, while a value score of *100* equates to meet or exceed some ideal performance requirement [3]. This case study uses 11 value functions and normalized swing weights to calculate an alternative's total stakeholder value.

In the original tradespace analytics tool, [27] used a macro-enabled additive value model, originally created by Kirkwood for use in spreadsheet modeling [32]. In order to integrate the MODA model into the new integrated decision support tool, we needed to convert the model and swing weight matrix from Excel to Java. This process resulted in the development of two separate modules, 8 and 9, containing the MODA control panel and multiple objective additive value models. The MODA control panel contains the swing weight matrix and allows the user to control and adjust inputs to the MODA model. We separated the MODA model into two separate modules to take advantage of efficiencies offered by ModelCenter's data workflow structure and to enable easier modifications to either the swing weight matrix or the additive value model. These decision models, in conjunction with the LCC model, provide the data enabling tradespace exploration and trade-off analysis activities to inform design decisions.

Using the Model-Based Integrated Decision Support Tool to Inform Design Decisions

Following model development and verification, the case study used the model-based integrated decision support tool to develop and mature the UAV designs over the

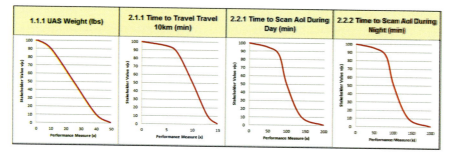

Fig. 5 Example of UAV case study value curves

5 Role of Decision Analysis in MBSE

Table 9 Case study model resolution level by design iteration

Design iteration	Model resolution level				Total unique design sets	No. of generated design points	No. of feasible designs
	Module 4	Module 5	Module 6	Module 7			
1	Base	Base	Base	Base	204,120	1,000,000	10,624
2	Base	Base	Base	1	45,360	600,000	29,344
3	Base	Base	Base	2	17,500	600,000	32,425
4	1	1	Base	2	12,600	600,000	34,712
5	2	2	1	3	6300	400,000	37,107
6	2	3	1	4	648	200,000	54,895
7	2	4	2	4	96	200,000	116,118
8	2	4	3	4	16	100,000	92,934

course of eight modeling iterations shown in Table 9. These iterations eventually reduced the number of unique design sets, under consideration, from the original 204,120 to 16 highly feasible and resilient design sets. Following each iteration, tradespace and uncertainty analysis informed a combination of multiresolution modeling and design selection decisions, prioritizing development of promising and high-value designs, resulting in eventual design convergence. We primarily used higher-resolution models in Modules 4–7, to resolve uncertainty associated with airframe design and performance, survivability, and sensor performance. This case study used five different levels of model resolution. The base level consisted of the original deterministic and parametric models developed by [25]. Level 1 models were deterministic moderate-resolution physics and sensor glimpse models, while Level 2 models used higher-resolution physics and glimpse models with probabilistic input variables [33]. Level 3 models were Monte Carlo simulations, producing 1000 replications per design point. Finally, Level 4 models used a combination of Monte Carlo and discrete event simulations to assess airframe and sensor performance in different operational scenarios. We ran all design iterations on a Linux server, using 16 Intel Xeon E5-2680 processors running at 2.70 GHz. Computational costs increased with each design iteration, due to the use of higher-resolution models. It initially took the tool approximately 4 h to generate the iteration 1 tradespace, using 16 parallel processes. However, it would have taken approximately 72 h to generate a similar-sized tradespace in iteration 8. Thus, we balanced the number of generated design points with the number of remaining design sets and model resolution, allowing us to explore the design space with less total design points.

The output of each iteration was a comprehensive data set describing the multidimensional system design space. This data-enabled design set and uncertainty analysis used a program management MODA model, informing our design selection modeling decisions. The data set also produced the system's feasible tradespace, such as that seen in Fig. 6. This example shows the initial feasible UAV tradespace, containing 10,624 feasible designs. The tradespace organizes the design points into

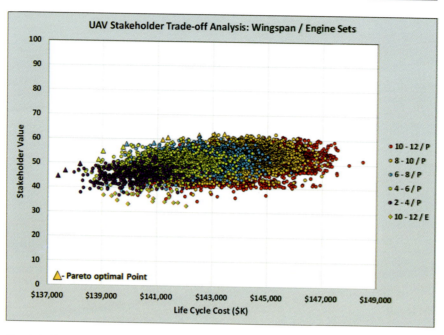

Fig. 6 Initial stakeholder tradespace

sets using wingspan bin size and engine type. Trade-off analysis considering stakeholder value and life cycle costs allowed us to identify 13 Pareto optimal design points identified with triangles in the figure. However, design performance and feasibility were subject to significant uncertainty due to the initial use of low-resolution models. Thus, we delayed selecting an alternative during this early phase of the design process to resolve uncertainty and reduce risk.

Using the model-based integrated decision support tool and our program management decision methodology, we were able to mature the design space to eventually converge on the final 16 candidate design sets [13]. By this time, we had finalized several design decisions, selecting an engine (*P*), wingspan (*10 ft.*), and operating altitude (*400 m*). This left only the EO and IR resolution and the EO and IR FOV decisions, each with two options apiece, open for deliberation. In the eighth and final iteration, we generated 100,000 design points, resulting in the tradespace seen in Fig. 7a. Unlike the initial tradespace, which we generated using low-resolution deterministic models, this tradespace was created using high-resolution models and probabilistic simulations and contains 92,934 design points. Thus, Fig. 7a provides the range of expected performance for each of the final UAV design candidates, given different scenarios and operating conditions. In Fig. 7b, we describe the representative designs in terms of the average stakeholder value and life cycle cost and compare each design's performance over the range of values. At this point, design performance and feasibility were so similar across the set of alternatives that no one design achieved stochastic dominance over any other design.

5 Role of Decision Analysis in MBSE

Fig. 7 (**a**) Final stakeholder tradespace analyzing design sets and (**b**) stakeholder tradespace analyzing final representative design solutions

However, certain options provided advantages in terms of highest average stakeholder value (*Design 12*), minimum stakeholder value variance (*Design 10*), lowest average life cycle cost (*Design 9*), and minimum cost variance (*Design 3*). Ultimately, our use of a set of integrated physics, cost, and decision models and our ability to increase model resolution throughout design development resulted in the development of high-value and highly resilient designs for final consideration.

Discussion of Case Study Observations

Throughout the case study, we viewed design development as a series of sequential design and modeling decisions, whose objective was to mature and converge the design space and reduce uncertainty. A key observation was that design selection and modeling decisions required coordination to resolve uncertainty. Simply selecting or eliminating design options, without increasing model resolution, failed to significantly reduce performance or feasibility uncertainty between iterations. Thus, the modular implementation of the integrated decision tool allowed us to efficiently revise and update modules between iterations, enabling effective uncertainty resolution. Additionally, the tool enabled us to prioritize development of promising high-value designs, while retaining a large pool of potential, though lower-value designs, to test and assess throughout the design process. This allowed us to deliberately delay major decisions, retain design flexibility throughout design development, and prevent the selection of infeasible or suboptimal designs.

For example, during design development, there were concerns regarding airframe feasibility, specifically its ability to fly and carry a payload. Prior to design iteration 4, we replaced the original parametric UAV performance models, located in Module 5, with a higher-resolution physics models requiring us to specify an airfoil shape and a chord length for the wings. For this model, we selected a National Advisory Committee for Aeronautics (NACA) 2412 airfoil, with a variable chord length dependent on wingspan. Additionally, we assumed a constant atmospheric temperature (10 °C), atmospheric pressure (101 kPa), and ground elevation (0-m) in this model. This higher-resolution model showed that designs using wingspan less than 6 ft were infeasible, which included several of the original Pareto optimal designs identified in the initial tradespace. For iteration 5, we added uncertainty to the physics models, about temperature which we varied from $-10\,°C$ to $50\,°C$, altitude which ranged from 0 to 4000 m, and atmospheric pressure which was dependent on both temperature and altitude, to assess design performance in various operational conditions. The results of this iteration demonstrated a significant lack of design resilience, regarding payload and operational environment, for alternatives using wingspans less than 8 ft, ultimately eliminating these designs from further consideration. We observed similar effects when we increased model resolution in other modules, highlighting the efficacy of using multiresolution modeling to enable design decision-making.

The purpose of our SBD methodology and integrated decision tool is to identify and select a highly resilient and Pareto optimal design delivering value and affordability to the stakeholders. Following design iteration 1, we identified 13 Pareto optimal alternatives. However, by the end of iteration 8, only four of these designs remained feasible, though not necessarily optimal. Again, by delaying major design decisions, we were able to prevent the selection of a suboptimal and potentially infeasible design. By integrating decision analysis and LCC models with physics models and simulations, we were able to take advantage of the benefits on model-based engineering practices to enable decision-making in complex system design.

Challenges and Gaps

In the 1970s, the famous British statistician George E. P. Box stated, "All models are wrong, but some are useful." This statement is still true today, especially as organizations move toward MBSE. MBSE requires making assumptions and gathering data. Analysts use these assumptions, the data, and variable relationships to develop the computer-based models vital to the implementation of MBSE. As models are an abstraction of the physical world, model accuracy is a function of both the model resolution and input data. Given it is likely that development programs will use higher-resolution models during the design process, input data quality becomes of paramount importance in correctly modeling the system.

Therefore, organizations must strive to develop repositories of data and models to enable MBSE and MBE. Organizations should be careful when developing these repositories, ensuring the curation of verified and validated data and models. Additionally, organizations should ensure proper data and model documentation. It is important to maintain a history of the source of the data or model, all the necessary information to document the context for its original use, and the individuals who created or updated the models.

Enterprise adoption of MBSE must overcome both technical and social obstacles. The technical obstacles come from the lack or compartmentalization of the data or models within organizations. Compartmentalization may inhibit the understanding and usefulness of existing data and models, limiting their use. Additionally, compartmentalization increases the bureaucracy and cost associated with maintaining and using data and models, leading to the second social obstacle. Organizations must change their culture to encourage cross-functional sharing and collaboration. This is particularly important for model integration. Teams should share input and output formats, needs, and requirements. Additionally, engineers and designers must move away from providing only designs to also providing models, enabling system analysis and maturation. This requires a change in organizational mindset by helping them see the value of model-based engineering practices.

All teams across the organization must see this value. They need to see how the additional time required to document and create the data and models, in addition to providing them, help the organization make better decisions. If everyone sees how they contribute to the value in the process and the key decisions, they are more likely to follow leadership guidance and eventually become advocates to help with the organization transformation toward MBSE.

This takes time since the focus shifts from designs to models and from data/models to decisions. This focus on decisions and enabling decision-making requires the need for strong, traceable decision-making processes. This is why we recommend using decision analysis best practices since they use the axioms of probability, the utility theory, the philosophy of systems analysis, and the five decision analysis axioms as their foundation [34]. Decision analysis best practices enable many of the required and desired MBSE tool capabilities discussed above, such as system performance, system effectiveness, and MODA value models. In addition, decision analysis best practices enable analysts to create tradespaces, quantify uncertainty, and find Pareto optimal solutions.

Current MBSE tools will need to evolve to support enterprise MBSE adoption. They will need to integrate with various databases and varying model fidelity. Several current tools have some of these capabilities, but none of them have all the required and desired capabilities to enable trade-off analyses and a traceable decision-making process.

Chapter Contribution

This chapter provides several contributions. We base the work on the current state of decision analysis and MBSE literature. We use the result of the literature review to recommend 14 MBSE tool capabilities that enable trade-off analyses. These MBSE tool capabilities include (1) stakeholder requirements, (2) system requirements, (3) ability to define and describe systems alternatives, (4) modeling environment enabling custom software modules, (5) integrated models, (6) system performance models, (7) ility models, (8) system effectiveness models, (9) LCC models, (10) a MODA value model, (11) engineering economic analysis models, (12) ability to quantify uncertainty, (13) tradespace visualization, and (14) ability to identify Pareto optimal solutions. We use a UAV illustrative case study to demonstrate the integration of decision analysis techniques with MBSE, which uses most of our recommended MBSE tool capabilities. These contributions demonstrate how MBSE can support systems engineering and project management decision-making.

Expected Advances in the Future

The adoption of MBSE is an emerging practice in industry and will continue to grow with the recent push for the use of digital engineering and the establishment of Digital Twins for systems. Many professional organizations, such as International Council on Systems Engineering (INCOSE), are developing and sharing the benefits, best practices, and gaps in applying digital engineering and MBSE (https://www.incose.org/products-and-publications/webinars). We believe that SE leaders will adopt MBSE tools and practices to improve effectiveness and efficiency to remain competitive. We expect continued development and integration of new MBSE and other digital transformation techniques. These techniques will lead to new standards and practices that will cross all organization functions. Overtime, this will lead to required culture changes. Throughout these evolutions, MBSE tools will improve. We believe that future MBSE tools will have an integrated platform that support performance modeling at different resolutions, value models, life cycle cost, and schedule models that show tradespace feasibility, uncertainty, and risk. These tools will also provide affordability analysis, economic comparison analyses, and other trade-off analyses throughout the enterprise.

Summary

This chapter began with a literature review of the papers that use MBSE to perform trade-off analyses. We found no papers that use the terms decision science and trade-off analysis. Expanding our search, we found 13 possible papers that had MBSE and decision analysis. Only three of these papers developed a tradespace and described a trade-off analysis. Using these papers and material from [2], we identified the required and desired MBSE capabilities needed for trade-off analyses (Table 4).

Next, we described a UAV design case study, implemented in the ModelCenter process workflow environment, that demonstrated the required capabilities for the use of integrated models to inform system design decisions. The last column of Table 4 summarizes the implementation of required capabilities in the case study.

Finally, we conclude with three sections. The first section describes the challenges and gaps. We believe that the biggest challenge will be overcoming the organization barriers to cooperation and sharing of models and data. The second section summarizes the chapter contributions: the literature search, the MBSE requirements for trade-off analysis, and the illustrative example that illustrated the trade-off analysis using ModelCenter. The third section summarizes the expected future advances in MBSE that will enable the tools to perform trade-off analyses and help systems engineers identify, design, develop, and deploy improved system designs that meet stakeholder requirements.

Cross-References

▶ MBSE Methodologies
▶ Overarching Process for Systems Engineering and Design
▶ Problem Framing: Identifying the Right Models for the Job

Appendix: List of Acronyms and Their Meaning

AC: Air-conditioning
BDD: Block Definition Diagram
CSMOP: Constraint Satisfaction Multicriteria Optimization Problem
GUI: Graphical User Interface
HPC: High-performance computing
ID: Influence diagram
IMPRINT: Improved Performance Research Integration Tool
LCC: Life cycle cost
MATLAB: Matrix Laboratory
MBSE: Model-based systems engineering
MODA: Multiple objective decision analysis
OOSEM: Object-Oriented System Engineering Method
RAM: Reliability, Availability, and Maintainability
SBD: Set-based design
SML: System Modeling Language
SysML: Systems Modeling Language
UAV: Unmanned aerial vehicle

References

1. A. INCOSE, "A world in motion: systems engineering vision 2025," *International Council on Systems Engineering, San Diego, CA, USA*, 2014.
2. G. S. Parnell, *Trade-off analytics: creating and exploring the system tradespace*. John Wiley & Sons, 2016.
3. G. S. Parnell, M. Terry Bresnick, S. N. Tani, and E. R. Johnson, *Handbook of decision analysis*, vol. 6. Hoboken, New Jersey: John Wiley & Sons, 2013.
4. "Decision Management. Retrieved from: Systems Engineering Body of Knowledge (SEBoK)." [Online]. Available: https://sebokwiki.org/wiki/Guide_to_the_Systems_Engineering_Body_of_Knowledge_(SEBoK). [Accessed: 21-Nov-2020].
5. J. Bergenthal, "Final Report Model Based Engineering (MBE) Subcommittee," *NDIA Systems Engineering Division-M\&S Committee*, 2011.
6. D. J. Singer, N. Doerry, and M. E. Buckley, "What Is Set-Based Design?" *Naval Engineers Journal*, vol. 121, no. 4, pp. 31–43, 2009.
7. "Systems and software engineering – System life cycle processes," Geneva, Switzerland, 2015.
8. E. Specking, G. Parnell, E. Pohl, and R. Buchanan, "Early Design Space Exploration with Model-Based System Engineering and Set-Based Design," *Systems*, vol. 6, no. 4, p. 45, 2018.

9. P. Leserf, P. de Saqui-Sannes, and J. Hugues, "Trade-off analysis for SysML models using decision points and CSPs," *Software and Systems Modeling*, vol. 18, no. 6, pp. 3265–3281, 2019.
10. M. F. Borchani, M. Hammadi, N. Ben Yahia, and J.-Y. Choley, "Integrating model-based system engineering with set-based concurrent engineering principles for reliability and manufacturability analysis of mechatronic products," *Concurrent Engineering*, vol. 27, no. 1, pp. 80–94, 2019.
11. M. Watson, C. Rusnock, M. Miller, and J. Colombi, "Informing system design using human performance modeling," *Systems Engineering*, vol. 20, no. 2, pp. 173–187, 2017.
12. C. Lowe and M. Macdonald, "Rapid model-based inter-disciplinary design of a CubeSat mission," *Acta Astronautica*, vol. 105, no. 1, pp. 321–332, 2014.
13. N. Shallcross, G. S. Parnell, E. Pohl, and E. Specking, "Informing Program Management Decisions Using Quantitative Set-Based Design," https://doi.org/10.1109/TEM.2021.3078387, pp. 1–15, 2021.
14. M. Bott and B. Mesmer, "An Analysis of Theories Supporting Agile Scrum and the Use of Scrum in Systems Engineering," *Engineering Management Journal*, vol. 32, no. 2, pp. 76–85, 2020.
15. C. Oster, M. Kaiser, J. Kruse, J. Wade, and R. Cloutier, "Applying composable architectures to the design and development of a product line of complex systems," *Systems Engineering*, vol. 19, no. 6, pp. 522–534, 2016.
16. J. Ryan, S. Sarkani, and T. Mazzuchi, "Leveraging variability modeling techniques for architecture trade studies and analysis," *Systems Engineering*, vol. 17, no. 1, pp. 10–25, 2014.
17. X. Wang, W. Liao, Y. Guo, D. Liu, and W. Qian, "A Design-Task-Oriented Model Assignment Method in Model-Based System Engineering," *Mathematical Problems in Engineering*, vol. 2020, 2020.
18. B. Hull, L. Kuza, and J. Moore, "A model-based systems approach to radar design utilizing multi-attribute decision analysis techniques," in *2018 Systems and Information Engineering Design Symposium (SIEDS)*, 2018, pp. 197–202.
19. M. A. Bone, M. R. Blackburn, D. H. Rhodes, D. N. Cohen, and J. A. Guerrero, "Transforming systems engineering through digital engineering," *The Journal of Defense Modeling and Simulation*, vol. 16, no. 4, pp. 339–355, 2019.
20. J. G. Süß, A. Leicher, H. Weber, and R.-D. Kutsche, "Model-centric engineering with the evolution and validation environment," in *International Conference on the Unified Modeling Language*, 2003, pp. 31–43.
21. N. Shallcross, G. S. Parnell, E. A. Pohl, and E. Specking, "Set-Based Design: The State-of-Practice and Research Opportunities," *Systems Engineering*, vol. 23, no. 5, pp. 557–578, 2020.
22. J. Baik, "COCOMO II, model definition manual, version 2.1," *Center for Software Engineering at the University of Southern California*, vol. 18, pp. 45–49, 2000.
23. R. Valerdi, "Academic COSYSMO User Manual-A Practical Guide for Industry and Government," 2006.
24. Z. Wade, G. S. Parnell, S. Goerger, E. Pohl, and E. Specking, "Convergent set-based design for complex resilient systems," *Environment Systems and Decisions*, vol. 39, no. 2, pp. 118–127, 2019.
25. C. Small, R. Buchanan, E. Pohl, G. S. Parnell, M. Cilli, S. Goerger, and Z. Wade, "A UAV Case Study with Set-based Design," in *INCOSE International Symposium*, 2018, vol. 28, no. 1, pp. 1578–1591.
26. N. Shallcross, G. Parnell, and E. Pohl, "Enabling Design Decisions in Set-Based Design with Multiresolution Modeling," in *Proceedings of the International Annual Conference of the American Society for Engineering Management*, 2020, pp. 1–8.
27. C. Small, G. Parnell, E. Pohl, S. Goerger, B. Cottam, E. Specking, and Z. Wade, "Engineering Resilience for Complex Systems," in *15th Annual Conference on Systems Engineering Research*, 2017.

28. M. Bigley, C. Nelson, P. Ryan, and W. Mason, "Tutorials and examples of software integration techniques for aircraft design using model center," *Blacksburg, USA: Virginia Polytechnic Institute and State University*, 1999.
29. C. Small, "Demonstrating set-based design techniques-a UAV case study," 2018.
30. M. Cilli, "Decision Framework Approach Using the Integrated Systems Engineering Decision Management (ISEDM) Process," in *Model Center Engineering Workshop, Systems Engineering Research Center (SERC)*, 2017.
31. R. L. Keeney, H. Raiffa, and others, *Decisions with multiple objectives: preferences and value trade-offs*. Cambridge university press, 1993.
32. C. W. Kirkwood, "Strategic decision making," *Duxbury Press*, vol. 149, 1997.
33. L. D. Stone, J. O. Royset, A. R. Washburn, and others, *Optimal search for moving targets*. Springer, 2016.
34. R. Howard, "The foundations of Decision Analysis revisited. In 'Advances in Decision Analysis: from Foundations to Applications,'" W. Edwards, RF Miles Jr and D. von Winterfeldt., Ed. Cambridge University Press: Cambridge, UK, 2007, pp. 32–56.
35. "Excel specifications and limits." [Online]. Available: https://support.microsoft.com/en-us/office/excel-specifications-and-limits-1672b34d-7043-467e-8e27-269d656771c3. [Accessed: 23-Aug-2020].

Dr. Gregory S. Parnell is a Professor of Practice in Department of Industrial Engineering, University of Arkansas, and is Director of the MS in Operations Management and the MS in Engineering Management programs. His research focuses on decision and risk analysis. He is editor of Trade-off Analytics: Creating and Exploring the System Tradespace (2017), lead editor of Decision-Making for Systems Engineering and Management (2nd Ed, 2011), and lead author of the Handbook of Decision Analysis (2013). He is a fellow of the International Committee for Systems Engineering, the Institute for Operations Research/Management Science, and the Military Operations Research Society. He has a PhD from Stanford University, is a Certified Systems Engineering Professional, and is a retired Air Force Colonel.

LTC Nicholas J. Shallcross is an active-duty Army Officer with over 17 years of operational and staff experience at multiple echelons. LTC Shallcross holds a BS in Mechanical Engineering from the Virginia Military Institute and an MS in Operations Research from the Air Force Institute of Technology. He is currently pursuing a PhD in Industrial Engineering from the University of Arkansas. His research interests include decision analysis, set-based design, risk analysis, simulation, stochastic processes, and operations assessments. He is a member of ASEM, INCOSE, INFORMS, and MORS.

Dr. Eric A. Specking serves as the Assistant Dean for Enrollment Management and Retention for the College of Engineering at the University of Arkansas. Specking received a BS in Computer Engineering, an MS in Industrial Engineering, and a PhD in Engineering from the University of Arkansas. His research interest includes decision quality, resilient design, set-based design, engineering and project management, and engineering education. During his time at the University of Arkansas, Eric has served as Principal Investigator, Coprincipal Investigator, or Senior Personnel on over 40 research projects totaling over $6.6 million, which produced over 50 publications (journal articles, book chapters, conference proceedings, newsletters, and technical reports). He is an active member of the American Society for Engineering Education (ASEE) and International Council on Systems Engineering (INCOSE) where he has served in various leadership positions.

Dr. Edward A. Pohl is a Professor and Head of the Industrial Engineering Department and holder of the Twenty-First-Century Professorship at the University of Arkansas. He has participated and led reliability-, risk-, and supply chain-related research efforts at the University of Arkansas. Before coming to Arkansas, Ed spent 21 years in the United States Air Force where he served in a variety of engineering, operations analysis, and academic positions during his career. Ed received his PhD in Systems and Industrial Engineering from the University of Arizona. He holds an MS in Systems Engineering from the Air Force Institute of Technology, an MS in Reliability Engineering from the University of Arizona, an MS in Engineering Management from the University of Dayton, and a BS in Electrical Engineering from Boston University. Ed is the coeditor of the Journal of Engineering Management on the Editorial Board of the IEEE Transactions on Technology and Engineering Management Society. Ed is a Fellow of IISE, a Fellow of the Society of Reliability Engineers, a Fellow of the American Society of Engineering Management, a Senior Member of IEEE and ASQ, and a member of INCOSE, INFORMS, ASEE, MORS, and AHRMM.

Matt Phillips is a Graduate Research Assistant at the University of Arkansas. He received a Bachelor of Science in Industrial Engineering with minors in Mathematics and Spanish. He will receive an MS in Operations Management in spring of 2021. During his senior year, he researched scheduling patterns in children's hospitals to reduce wait time. The focus of his interests are statistical analysis and management.

Pattern-Based Methods and MBSE

6

William D. Schindel

Contents

Introduction	152
MBSE Pattern Concept	152
Expanded Perspective and Organization of Chapter	153
State-of-the-Art	154
The Most Important Pattern: What Is the Smallest Model of a System?	154
Introduction to the S*Metamodel	156
S*Models and S*Patterns	162
Distillation and Representation of Learning; Accessibility and Impact of Learning	165
Tooling and Language Issues for MBSE Patterns	167
Best Practice Approach	173
INCOSE Innovation Ecosystem Reference Pattern	173
Model Characterization Pattern: Universal Model Metadata Reference Pattern	177
Illustrative Examples	182
Chapter Summary	183
Impact on Practice, Education, and Research	183
Impact on the Theoretical Foundations of Systems Engineering	187
Cross-References	191
References	191

Abstract

Patterns are recurring regularities, having fixed and variable parts, across engineered systems, systems of engineering, production, distribution, and sustainment, as well as the natural world. Ranging from concrete patterns of engineered product lines to abstract patterns behind architectural frameworks, reference models, ontologies, and general or domain-specific languages, patterns are implicitly involved in all MBSE practice. Methods reported in this chapter exploit the power of explicit MBSE patterns, using the leverage of acquired knowledge to speed processes, reduce rediscovery and error, and lower risk.

W. D. Schindel (✉)
ICTT System Sciences, Terre Haute, IN, USA
e-mail: schindel@ictt.com

Although MBSE has begun exploiting them, model-based patterns have a much longer history in the foundations of the physical sciences and their support for engineering disciplines. Advanced primarily over the last 300 years, all the laws of the physical sciences are represented by model-based patterns describing and organizing observable phenomena. Indeed, that advance was largely centered on the rise of more effective representational models and modeling frameworks, mathematical and otherwise.

In the same fashion, the theoretical foundations of systems engineering are being strengthened by similar observation of system phenomena, their representation as configurable model-based patterns, and validation of credibility of such models for prediction and understanding. Properly harnessed, this creates more effective learning by groups (e.g., teams, enterprises, supply chains, industry groups), in which reusable, configurable model-based patterns are the informational proxy for learning; the techniques of pattern recognition and model verification and validation provide recognized support for formalized learning and model credibility management; and pattern-based processes are automated to enhance generation or checking of models supported by patterns.

Keywords

MBSE patterns · Pattern-based engineering · Pattern-based models · System patterns · S*Metamodel · S*Patterns · SE foundations · Ecosystem pattern

Introduction

MBSE Pattern Concept

This chapter reports on the concepts, practices, and impacts of pattern-based methods on Model-Based Systems Engineering (MBSE). In the narrowest perspective on pattern-based methods, one may view an MBSE model as a reusable, configurable data structure, and configure it for multiple specific applications (as in setting values of parameters or populated/depopulating model elements), thereby taking advantage of past efforts (perhaps by others) to assemble and validate the general model-based pattern. Pattern governance efforts are also included to curate the pattern over time, reflecting ongoing learning. A single MBSE Pattern thus supports multiple SE projects, each with its own configured model, including learning derived to improve the pattern. Figure 1 illustrates this basic perspective.

A general vehicle MBSE Pattern, such as referenced in Fig. 1, might be expected to contain a variety of configurable MBSE elements, including but not limited to Stakeholder Features, System Requirements, Functional Interactions, Interfaces, Modes and States, Functional Roles, Design Components, Parametric Couplings, Failure Modes and Effects, relationships between then, and other information. This is further discussed in the "Illustrative Examples" section [58].

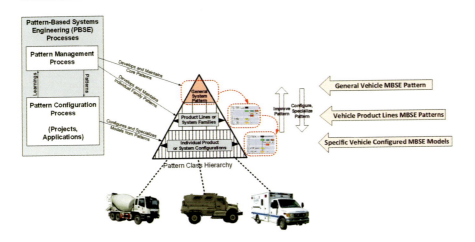

Fig. 1 Elementary perspective on MBSE patterns: A product lines example. (From Ref. [58], permission granted to reproduce with attribution)

Expanded Perspective and Organization of Chapter

The practice and full impact of pattern-based MBSE is best understood not just through the above narrow perspective – performing MBSE using model-based patterns as configurable data structures found in architectural frameworks, ontologies, and product line engineering (PLE) structures (ISO 26580 2021). Instead, the larger perspective encouraged here harnesses the unifying role of patterns across engineering, science, and business, embedding pattern-based MBSE methods in a larger framework. This wider perspective allows the consumer, practitioner, and leader of MBSE methods and the enterprise to more fully exploit the potential impacts that pattern-based methods provide to MBSE, the enterprise, supply chain, and industry groups. This is also discussed further in the "Best Practice Approach" section later below.

Descriptions of systems engineering (Walden et al. 2015) [27, 60] historically emphasize processes and the information they consume and produce (whether model-based or otherwise). One might view these process descriptions as rather thorough outlines of all the things that should be done in order to assemble and coordinate the information necessary to describe a system over the course of its life cycle management – needs, requirements, designs, analyses of failure modes and risks, processes of production, logistics, support, and otherwise. These detailed process descriptions generally spend relatively little of their content on distinguishing between *new information* (not yet known, to be discovered, analyzed, validated) versus *what is already known*. Traditional SE process descriptions were simply not intended to emphasize the difference, leaving it to the project team.

Pattern-based methods of MBSE are concerned with not just the traditional systems engineering needs as addressed by MBSE models, but additionally with optimizing the integration and balance of previously known higher-confidence information with newly acquired information, to be further validated. Those with a background in statistical methods may recognize in this description something that sounds Bayesian [4] in its nature: how do we optimally combine a priori information with new information, and what is our uncertainty about the combination? Indeed, the ecosystem-level description of the "Best Practice Approach" section below resembles a giant adaptive Kalman filter, optimally combining new information with known patterns, themselves improving, all viewed through a lens of uncertainty management.

The use of engineering patterns for leverage predates their use in MBSE models. The formalization of design patterns in civil architecture and software development [1, 17] encouraged a large following and extensive literature in parts of the engineering community, based on use of pre-MBSE prose templates that were configured as forms. These should be appreciated as part of the background of current MBSE Patterns methods [6]. However, the more recent combination of explicit technical models with recurrent patterns has had the same kind of profound impact that it had in the still earlier histories of mechanics, electrical science, and other engineering disciplines based on explicit models of observable phenomena.

Early efforts to apply MBSE models in the above context revealed demands on the capabilities of MBSE modeling languages, frameworks, and tooling that were found to be satisfiable but not always familiar to the MBSE community, since they were closer to mathematical physics [15, 43]. These needed model characteristics were formally defined in such a way that they could be addressed using the broad range of existing standards-based or other third-party commercial off-the-shelf (COTS) modeling tools and languages, and are discussed in the "State-of-the-Art" section below.

The resulting practices in pattern-based MBSE led to a range of domain patterns and applications reported in the "Illustrative Examples" section below. Those advances included formation of the INCOSE MBSE Patterns Working Group (2016) [20], which has over the years collaborated with other technical societies and working groups on additional pattern-based MBSE topics in the References. From these collaborations emerged the INCOSE Ecosystem Reference Pattern, describing the larger descriptive (not prescriptive) framework in which pattern-based MBSE occurs in any ecosystem. The Model Characterization Pattern of universal model metadata also resulted. These reference patterns are discussed in the "Best Practice Approach" section below and references cited there.

State-of-the-Art

The Most Important Pattern: What Is the Smallest Model of a System?

As suggested by Fig. 1, model-based patterns exist at different levels of abstraction; some patterns are specializations (or generalizations) of other patterns, resulting in taxonomies of patterns such as Fig. 2. At the lower levels of Fig. 2, one sees more

6 Pattern-Based Methods and MBSE

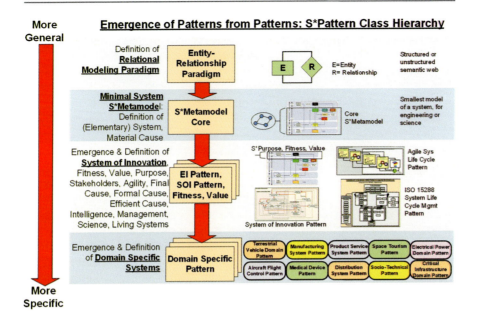

Fig. 2 Example taxonomy of MBSE patterns. (From [44], permission granted to reproduce with attribution)

specific domain patterns that effectively establish domain ontologies and domain-specific languages. Each more specific model level below may be understood as expressing configurations of models in the language introduced by the model at the level above it.

Understanding that the upper patterns each establish the conceptual modeling frameworks for the patterns below them, we can see that a very important pattern near the top of the hierarchy of Fig. 2: the S*Metamodel (an abbreviation for Systematica Metamodel). It establishes the framework/ontology/language in which other models will be expressed. The question it addresses is "What is the smallest model of any system, for purposes of engineering and science, across the life cycle of systems?" [54]. Since engineers and scientists have long been creating models for various engineering and scientific purposes over life cycles, we can expect that this language should be the familiar language structure of engineering and science, and indeed that is the case. In more recent years, this has also been the focus of attention of modeling language and modeling tool standards development, and we want to be able to take advantage of the wealth of third-party COTS tools and languages in our MBSE patterns work. The extra burden of representing patterns over the life cycle of systems introduced a few formalities that can be addressed using available third-party COTS offerings, but the overall MBSE environment has been in enough state of change over the last two decades that it has been useful to specify a neutral reference framework of the minimal capabilities that any of those COTS frameworks must include for purposes of MBSE Patterns work. That framework is the S*Metamodel, described further in Refs. [22, 52, 54, 58] (Schindel 2011, 2019b).

Introduction to the S*Metamodel

Engineering disciplines such as ME, EE, ChE, CE are based upon underlying models of phenomena (mechanical, electrical, chemical, etc.) that are the fruits of physical sciences, mathematics, and philosophy. Newton's laws of motion, Maxwell's equations, and other underlying models describe aspects of the nature of subject systems, not engineering procedures for those systems, while opening up many procedural avenues that operate within the constraints of those underlying models of Nature. In a similar fashion, the S*Metamodel describes the underlying "systemic" aspects of systems of interest, based upon the fruits of science and mathematics. In the tradition of those same physical sciences [19], these underlying models (whether specific to one technical discipline or systems in general) seek the "smallest model" capable of (verifiably) describing, predicting, or explaining the phenomena of interest (Schindel 2011).

The rise of a number of MBSE methodologies and system representation standards [25] has provided many of the needed elements of that underlying "smallest model" framework, and the S*Metamodel builds on those, adding some important missing and compressing other redundant aspects. Throughout, this is in the spirit of seeking out the smallest (simplest) verifiable model necessary to describe systems for purposes of engineering and science over life cycles.

Figure 3 is a simplified summary of some of the key portions of the S*Metamodel. This diagram is not the sort that is produced in a related engineering project (illustrated in Fig. 4), but instead is an entity-relationship representation of the underlying classes and relationships upon which those project-specific models are based. Those project-specific models may be in any modeling language (including but not limited to SysML, IDEF, or otherwise) and supported by any engineering tool or information system, as in Fig. 4.

Whereas the formal S*Metamodel is represented in OMG UML, the conceptual awareness extract of Fig. 3 is a simplified representation of selected S*Metaclasses and key relationships that connect them. It summarizes some of the most important S*Metamodel concepts (discussed further below) leading to some of the above benefits. Additional references are listed later below.

In reading Fig. 3:

The color coding provides an informal reminder of stereotypes mapping in formal modeling languages (e.g., SysML), and related views, especially for nontechnical views and viewers.

- A *Stakeholder* is an entity having a value stake in the behavior or performance of a system.
- A *Feature* is an aspect of the behavior or performance of a system that has stakeholder value, described in the concepts and terminology of that stakeholder,

6 Pattern-Based Methods and MBSE

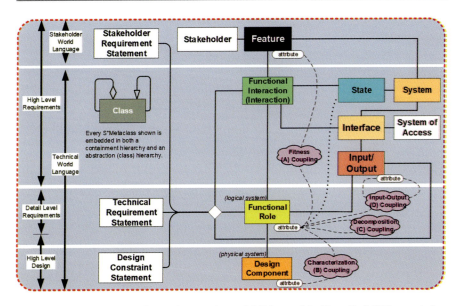

Fig. 3 Informal summary of some key portions of S*Metamodel. (From Ref. [22], permission granted to reproduce with attribution)

and serving as the basis of selection of systems or system capabilities by or on behalf of the stakeholder. Features are parameterized by *Feature Attributes*, which have subjective stakeholder valuations.

- (*Functional*) *Interaction* means the exchange of energy, force, mass, or information between system components, each of which plays a *(Functional) Role* in that Interaction.
- Functional Roles are described solely by their behavior, and parameterized by *Role Attributes* which have objective technical valuations.
- The *State* of a system determines what behavior it will exhibit in future Interactions. The State of a system may be changed by those Interactions.
- *Input-Outputs* are exchanged energy, force, mass, or information between system components, during an Interaction.
- An *Interface* is an association of a System (which has the Interface) with one or more Input-Outputs (which flow through the Interface), Interactions (which describe behavior at the Interface), and *Systems of Access* (which provide the medium of transport between Interfaces).
- A *Design Component* is described solely by its identity or existence, not its behavior; that behavior is described by the functional role(s) allocated to that Design Component. Design Components are parameterized by *Design Component Attributes* that have only identity or existential valuations, but no behavioral aspects.
- *Requirements Statements* are prose or other descriptions of the behavior of Functional Roles during Interactions. They are the prose descriptive equivalent

158 W. D. Schindel

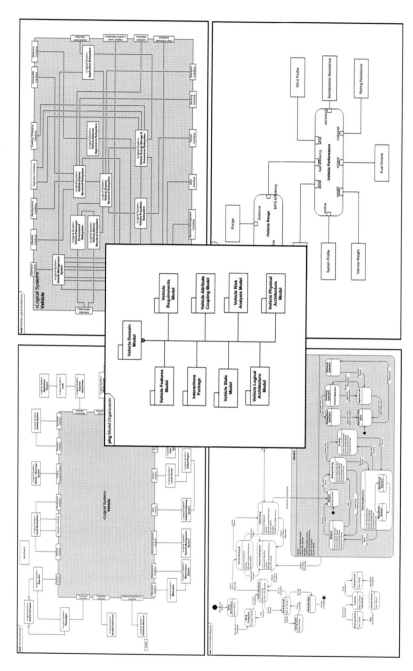

Fig. 4 Example application of S*Model extracts (details in section "Illustrative Examples"). (From Ref. [22], permission granted to reproduce with attribution)

to the Roles they describe, and are parameterized by *Requirements Attributes* which are identical to the related Role Attributes [39].

- *Fitness (A) Couplings* describe the quantitative value dependencies (parametric couplings) between Stakeholder Feature Attributes and Functional Role Attributes, quantifying fitness space or trade space in relation to technical performance space.
- *Characterization (B) Couplings* describe the quantitative value dependencies (parametric couplings) between Functional Role Attributes (technical performance space) and Physical Design Component Attributes (design component identity space). These characterize technical behaviors as a function of identities of design components.
- *Decomposition (C) Couplings* describe the quantitative value dependencies (parametric couplings) between Attributes of decomposed (smaller) Functional Roles and emergent technical behavior Attributes of larger parent Functional Roles.
- *Input-Output Transfer (D) Couplings* describe the quantitative value dependencies (parametric couplings) between values of dynamical Inputs and Outputs, representing the quantitative aspect of input-output transformations performed by functional roles during Interactions.

It is important to remember that the above are underlying *metamodel* concepts. Practical S*Models are represented using whatever modeling languages and tools may be of value. Figure 4 illustrates selected SysML views of a vehicle S*Model.

Interactions, Requirements, and States

Three examples of the type of strengthening discussed above are (1) the S*Metamodel use of Interactions, (2) the related way that Requirements formally enter into the model, and (3) the States of the system. The following conceptual framework, illustrated in Fig. 5, leads to the insight that all System Technical Requirements are manifest as behavior occurring during physical interaction of a subject system with its external environment:

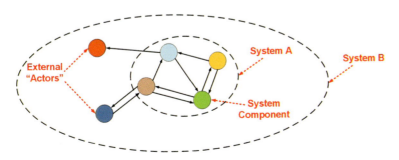

Fig. 5 The system perspective. (From Ref. [22], permission granted to reproduce with attribution)

- A *system* is a collection of interacting components. (A component can itself be a system.)
- By *interact*, we mean that one component exchanges energy, forces, mass flow, or information with another component, resulting in component changes of state.
- By *state* of a component, we mean the condition of the component that determines its input-output behavior.
- The behavior of an interacting component during an interaction, visible only externally to the other component(s) with which it is interacting, is referred to as the *functional role* of the component in the (functional) interaction. This is the case whether the behaviors of that component and the other components are known or not.
- The only behavior that a functional role can exhibit is its *input-output behavior*, exhibited during interactions that occur in different modes or states.
- For linear systems, external behavior can be entirely characterized by mathematical transfer functions, relating inputs to outputs in a particular mathematical form. Systems in general are not linear, so that mathematical form is in general not available. However, for all systems we can still retain the idea that requirements can only describe *relationships between inputs and outputs* (quantitative, temporal, probabilistic, or other behavior, and often expressed as prose *Requirements Statements,* which can now be recognized as describing nothing more or less than that input-output relationship) [39].
- The totality of the externally visible behavior of a system, added up over all its interactions with its environment, is the set of Requirements that describe the input-output relationships of that system during interactions with its external environment.
- The problem of finding all the requirements for a system can thus be shifted to finding all the interactions of the system (Schindel 2013b). (It is understood that "all the requirements" is subject to pragmatic limitations on how much fidelity is needed for engineering purposes. However, business stories continue to arise of negative impact of important missed requirements in time-constrained projects. The approach describes here provides a demonstrated means of more quickly arriving at "enough" of the interactions and requirements to significantly improve on earlier real-world experiences, in both completeness and speed.)

Interactions provide a powerful way to analyze systems using MBSE models [41]. This includes both "hard" manufacturing production process, equipment, and material transformations [49], and "soft" psychological and emotion-laden human interactions [47] important in human-system interaction analysis and integration.

Selectable System Features and Stakeholder Value

The S*Feature model subset of the overall S*Metamodel illustrates the integrative and compressive impacts of the S*Metamodel, through the different parts played by Features in these models, as follows.

For human-engineered systems, and for systems in nature for which selection processes occur, selectable system features are described by the S*Metamodel.

6 Pattern-Based Methods and MBSE

These describe the value landscape of stakeholders for the subject system. In the case of natural systems, not in the setting of human-engineered and used systems, it is frequently found that there are nevertheless selection processes at work in Nature, so that the framework is still applied based on the selectable Features that are "valued" by such selection processes [48, 53].

S*Models are intended to identify all the classes of Stakeholder for systems of interest, not just direct users or customers, and to establish modeled Feature sets for all those Stakeholders. This (Features) portion of an S*Pattern is then used to configure the pattern for individual applications, product configurations, or other instances. It turns out that the variation of configuration across a product line is always for reasons of one stakeholder value or another, so Feature selection becomes a proxy for configuring the rest of an S*Pattern into a specifically configured instance model.

Because S*Features and their Feature Attributes (parameters) characterize the value space of system stakeholders, the resulting S*Feature Configuration Space becomes the formal expression of the trade space for the system. It is therefore used as the basis of analysis and defense of all decision-making, including optimizations and trade-offs. The S*Feature Space also becomes the basis of top-level dashboard model views that can be used to track the technical status of a project or product. All "gaps" and "overshoots" in detailed technical requirements or technologies are projected into the S*Feature Space to understand their relative impact.

Because the S*Stakeholder and Feature model subset is intentionally comprehensive across stakeholder issues, Features play a direct role in modeling failure mode Effects, as discussed in the next section.

Failure Modes and Effects

Models based on the S*Metamodel leverage Failure Impacts (negating aspects of Features), Counter Requirements (negating aspects of Requirements behavior), Failure Modes (off-nominal behavior states), and modeled relationships between them to generate high-quality FMECA table drafts or other risk management views with reduced effort and increased coverage. This approach deeply integrates the information and processes of other parts of the engineering cycle with the risk analysis process, as illustrated in Ref. [46].

Attributes and Attribute Couplings

Several classes of the S*Metamodel include modeled attributes, which are variables (parameters, characteristics) that further parameterize S*Models – these may be numerically valued, or discrete valued, or enumerated list valued, or of other type. In principle any S*Metaclass can have attributes, but three are particularly emphasized: attributes of Features, Roles, and Physical Components.

Feature Attributes parameterize the value space of stakeholders, in the language and conceptual framework of those stakeholders; as such, they often describe subjective stakeholder variables, such as Comfort, Risk, or Responsiveness. They include all stakeholder Measures of Effectiveness (MOEs). *Role Attributes* parameterize the space of technical behavior specification, and exactly these same attributes are associated with the Requirements Statements. They parameterize objective, testable technical descriptions of behavior, such as Thermal Loss, Reliability, or Maximum Speed, and include all technical measures of performance. *Physical Component Attributes* describe nothing about behavior (which is focused on the two previous attribute types), but instead describe identity and existence, such as Product Model, Part Number, Serial Number, Material of Composition, Department, or Employee ID.

Attribute Couplings are part of S*Models, describing how the values of these different attribute types vary with respect to each other. For example, Fig. 3 shows that A-Couplings describe how the stakeholder values of Feature Attributes vary with respect to change in the values of technical Role Attributes. B-Couplings describe how technical behavior-parameterizing Role Attributes values vary as the values of Physical Component Attributes vary. In S*Models and S*Patterns, these modeled Attribute Couplings distill and integrate into the model what has been learned quantitatively by physical sciences, experiment, stakeholder observation, experience, and first principles. They include formulae, graphical curves, data tables, or other representations of parametric interdependencies. They also provide a means of transforming degrees of freedom, as in Dimensional Analysis, to seek out fundamental representations [4] (e.g., Reynolds number in fluid flow, etc.).

S*Models and S*Patterns

*S*Models* are MBSE models conforming to the S*Metamodel (Fig. 3) (that is, they contain Features, Interactions, Roles, States, Design Components, Interfaces, Requirements, Attributes thereof, couplings between them, etc.). *S*Patterns* are S*Models (with all their parts) that have been constructed to cover a system configuration space bigger than single system instances, and are sufficiently parameterized and abstracted to be configurable to more specific S*Models, and thereby reusable, as in Fig. 6 [3, 43, 56, 58].

Like S*Models, S*Patterns may be expressed in any system modeling language (e.g., SysML, IDEF, etc.) and managed in any COTS system modeling tool or repository, by an S*Metamodel mapping or profile.

Patterns compress information to discover essence as well as parametric variation; pattern-based methods accordingly evolve greatest lower bound representations of phenomena – right out of the history of physical science. The most important pattern that emerges is the smallest model of a system necessary for the historical purposes of engineering and science. This turns out to be both smaller and larger than conventional wisdom, but in any case readily mapped into the currently available industry tools and languages.

Architectural Frameworks, Ontologies, Reference Models, Platforms, Families, Product Lines

S*Patterns are concerned with "whole systems," as described above. As such, they are fundamental to supporting the life cycles of Platform Products, Product Families,

6 Pattern-Based Methods and MBSE

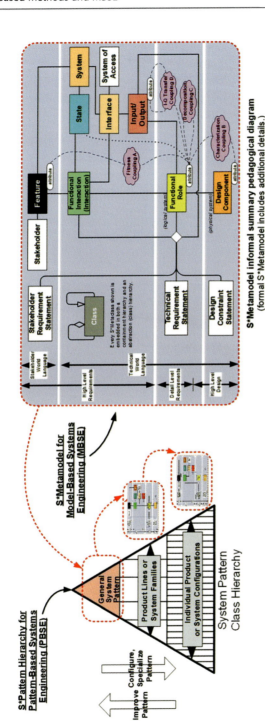

Fig. 6 S*Patterns are S*Models of system families, configurable/reusable as models of individual system types. (From Ref. [22], permission granted to reproduce with attribution)

and Product Lines. They may also be compared to whole system-level descriptions provided by Architectural Frameworks, Ontologies, and Reference Models, as follows.

A *Platform* is a system family abstraction that can be configured to serve the needs of different applications, market segments, customers, regulations, or other specialized requirements that apply in some cases but not others [33]. Platforms leverage the economic value of systems. S*Patterns specify, at both high and detail levels, the requirements, designs, failure modes and risks, verifications, applicabilities, configuration rules, and other aspects of platforms. So, S*Patterns can be used to implement Platform Life Cycle Management [42]. Because they are S*Metamodel-compliant, they include the minimum set of model elements necessary for product life cycle management.

Product Lines and *Families* are terms variably used to describe either the different system configurations of the above Platform families, or else the component subsystems that are variously configured and combined to make up those larger systems. All of these may be described by S*Patterns.

Product Line Engineering (PLE) refers to the engineering processes and support approach used to engineer product line families and the component systems from which they are formed. Emerging approaches such as Product Line Engineering [28] (ISO 26580 2021) describe approaches to certain aspects of PLE. S*Patterns support the implementation of PLE approaches and practices.

Architectural Frameworks are model-based descriptions of repeat-use information frameworks for descriptions of certain aspects of systems, for a given enterprise, domain, or other setting [29]. As such, an Architectural Framework may target less than all the classes of information necessary to describe a system over its life cycle, but could (and in some cases may be intended to) cover all those information classes. S*Patterns cover all the classes for the whole life cycle. Therefore, one could say that an S*Pattern is an Architectural Framework that has been built out sufficiently to cover the S*Metamodel scope, and for which no additional (redundant) information is included.

An *Ontology*, in information science, is a formal model naming and defining the types, properties, and interrelationships of the entities that are fundamental for a particular domain [18]. How are Ontologies related to S*Patterns? A specific ontology could in principle include all the classes and relationships of a specific S*Pattern (e.g., a Vehicle Manufacturing System Pattern), but most ontologies do not try to cover that much detail. Such an Ontology might roughly be said to describe the name space of an S*Pattern, but the *relationships* that practitioners typically include in Ontologies are less parsimonious than those of the S*Metamodel, so that Ontologies can become relationally more complex-looking, even if not as informative in detail. The S*Metamodel itself is certainly an Ontology, for S*Models. One approach to improving the utility of Ontologies for systems can be to start with an S*Pattern and identify certain subset views of it as Ontological Views. In that

6 Pattern-Based Methods and MBSE

approach, the Ontology is a by-product of the S*Pattern, automatically synchronized with it because it is a viewable subset of the S*Pattern.

Patterns, Configurations, Compression, Specialization

As illustrated by the "down stroke" in Fig. 6, a generic S*Pattern of a family of systems is specialized or "configured" to produce an S*Model of a more specific system, or at least a narrower family of systems. Since the S*Pattern is itself already built out of S*Metamodel components, for a mature pattern the process of producing a "configured model" is limited to two transformation operations:

- *Populate*: Individual classes, relationships, and attributes found in the S*Pattern are populated (instantiated) in the configured S*Model. This can include instances of Features, Interactions, Requirements, Design Components, or any other elements of the S*Pattern. These elements are selectively populated, as not all necessarily apply. In many cases, more than one instance of a given element may be populated (e.g., four different seats in a vehicle, five different types of safety hazard, etc.). Population of the S*Model is driven by what is found in the S*Pattern, and what Features are selected from the S*Pattern, based on Stakeholder needs and configuration rules of the pattern, built into that pattern.
- *Adjust Values of Attributes*: The values of populated Attributes of Features, Functional Roles/Technical Requirements, and Physical Components are established or adjusted.

This brings into sharp focus what are the fixed and variable aspects of S*Patterns (sometimes also referred to as "hard points and soft points" of platforms). The variable data is called "configuration data." It is typically small in comparison to the fixed S*Pattern data. Since users of a given S*Pattern become more familiar over time with its fixed ("hard points") content (e.g., definitions, prose requirements, etc.), this larger part is typically consulted less and less by veterans, who tend to do most of their work in the configuration data (soft points). That data is usually dominated by tables of attribute values, containing the key variables of a configuration. Since this is smaller than the fixed part of the pattern, in effect the users of the pattern experience a "data compression" benefit that can be very significant, allowing them to concentrate on what is or may be changing (Schindel 2011).

Distillation and Representation of Learning; Accessibility and Impact of Learning

As also illustrated by the "up stroke" in Fig. 6, discoveries are encountered during projects involving configured S*Models, and some of these cause improvements to be fed back to the S*Pattern, which thereby becomes a point of accumulation of all learning about what is known about the family of systems that pattern represents.

This reduces the amount of "searching" required of future project users to take advantage of what is already known, and in particular reduces the likelihood of re-learning the same lessons by mistake and rework. Notice that this "distillation and abstraction" process is quite different than simply accumulating a lot of separate "lessons learned" in a large searchable space – it is instead translating them into their foundational implications at the pattern level, for future users of the pattern, as a single point of learning well-known and accessible to distributed users. It is the S*Pattern equivalent of representation of scientific learning, and is also a model-based analogy to governance of prose engineering standards.

Patterns provide compression. For MBSE Patterns, this generally means that the variable (parameterized) aspects of a configured pattern, combined with the fixed part of the pattern, together constitute a whole configured MBSE model, but practically only the variable configuration needs to be reproduced, because we "already have" the fixed part. The variable configuration information is usually small compared to the fixed part, since it only describes (binary) presence or absence of elements, and values of configurable variables (attributes, parameters). This "smallness" can be valuable in the sense of the storage or transmission cost of the information, or the mental burden relief for subject matter experts, who typically know the fixed pattern and are interested in a specific configuration's variable aspects.

Patterns also reduce the time, effort, and risk of initial generation of configured model information from validated pattern content, based on input (to the pattern configuration process) of situation-specific stakeholder needs. For example, consider configuration of draft requirements (see Fig. 9). Although this >90% reduction of time and effort is not necessarily the most profound impact of pattern-based MBSE methods, it is often one of the more readily measured aspects, because of past experience with requirements processes, and may in itself significantly justify the use of MBSE Patterns methods.

The Pattern-to-Configured Model transformation (discussed further in later sections and illustrated in Fig. 9) results in (selective and frequently multi-instance) population of a collection of Configured Model class objects from the MBSE Pattern, along with population of relationships between them, and setting of values of attributes (parameters). A misunderstanding to avoid is the notion that this is effectively populating a "smaller" model than the pattern, since not every pattern class is likely to be populated in the model. By far the more typical system situation is that the configured model "expands" from the pattern. S*Patterns are "folded up" in the sense that various model components often appear only once in the S*Pattern but multiple times in the configured S*Model. For example, consider an electrical power distribution circuit for a manufacturing plant facility. The modeled aspects of power transmission interactions, power loads and sources, interfaces, systems of access, and other aspects can be modeled once in the S*Pattern for manufacturing plants, but instantiated tens or hundreds of times in a configured S*Model for a specific site. (The notion of "deletion only" configuration processes may work for a limited class of systems, but is not a general approach usable across systems of higher scope and complexity.)

Tooling and Language Issues for MBSE Patterns

As described by other chapters of this handbook, MBSE practice is closely associated with MBSE modeling languages and automated tools. Different modeling languages arise in response to a number of forces, some of them associated with MBSE-based patterns. Automated tools are used for model authoring, model artifact publishing and distribution, model life cycle management, and more specialized functions, some of them associated with MBSE-based patterns. Although modeling languages and automated tools are often associated with each other, they should be understood as separate, coupled subjects. This chapter's discussion of both is limited to issues associated with MBSE-based patterns. Refer to the other chapters of this handbook for more general MBSE tooling and language issues.

A primary MBSE Pattern emphasis for tooling and languages is the formalized mapping of the S*Metamodel (and as a result, S*Patterns and S*Models) into widely used standards-based modeling languages and third-party commercial off-the-shelf (COTS) automated tooling.

Modeling Languages and MBSE Patterns

The most direct connection between modeling languages and patterns is that a modeling language is itself an expression of a pattern or patterns. Recall this chapter's introduction of patterns as recurrences having fixed and variable parts. A modeling language provides a fixed overall grammar (limited, even if relatively flexible) that facilitates the expression of varying instances of statements (including assertions and queries) about some subject, which in MBSE is most often a modeled system of interest. The "aboutness" part of this capability is supplied by an ontology, which describes what the model's statements may be about; this might be a very specific general subject, such as models of engineered or natural systems in general, or a more specific subject, such as automotive domain systems.

Referring Fig. 2, it can be seen that different languages can be associated with patterns at different levels of specificity. For languages of a more general nature, the underlying ontology is associated with the language's metamodel. For languages of a more domain-specific nature, the "ontology" term is more common than "metamodel." Combinations of these are of particular interest: A general modeling language, supported by its metamodel, may be used to describe models which are further limited by a domain-specific ontology; this case is of particular interest to improve semantic interoperability across a set of people, teams, organizations, models, or automated tools. Modeling languages may also differ from each other for reasons other than a domain-specific ontological constraint, when the abstract metamodel constructs of the language differ.

By "semantic interoperability" of the resulting models, we mean that different people, organizations, or automated tools will interpret the meaning of an explicit model in the same way (refer to Fig. 7). Note that there are forms of interoperability that fall short of semantic interoperability. For example, two model authors may generate their own models that accurately describe overlapping aspects of a single real-world engineered system, using the same generalized modeling language, and

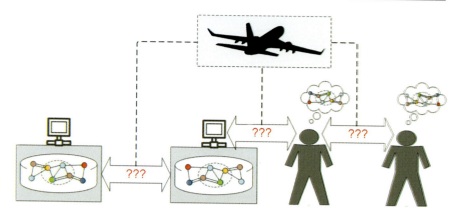

Fig. 7 The challenge of semantic interoperability. (From Schindel (2020), permission granted to reproduce with attribution)

yet produce models that are not semantically interoperable. Indeed, this is common, and is not merely an information technology problem – it can mean the people involved are unable to talk to each other without misunderstanding. A common case of this in systems engineering is the modeling of two subsystems which interact with other, so that they share interactions across shared interfaces. Another common case is found in operational and maintenance models of a single system.

Problems of the above sort are reduced by using shared ontologies or architectural frameworks. The usual interpretation of such ontologies and frameworks is that they frame the general but not detailed framework for the multiple models to be created. Because of this intended lack of detail, it is still possible for two models to conform to the framework and yet be semantically non-interoperable. A prominent example of this is for models of two interacting subsystems to describe differently the set of interactions between them (as to name, scope, partitioning, etc.). By contrast, an S*Pattern, described in earlier sections of this chapter, "fills in" the additional detail of the "smallest possible model" necessary to specify those details for the purposes of engineering and science. This also illustrates why most S*Patterns are complete enough to be applied by configuration (population of existing model components and valuation of existing attributes or parameters) versus specialization (addition of new specialized classes and attributes or properties).

For reasons noted above, as well as other economies, MBSE Patterns methods make use of formal mappings into widely used standards-based modeling languages [26, 35]. Accordingly, the S*Metamodel has been formally mapped into popular standards-based and some other modeling languages the data schema of engineering and life cycle management tools. The resulting mapped profiles permit MBSE Patterns methodology in those standard languages, as in Ref. [23]. This in turn bears on ability to enhance the semantic interoperability of models in those languages [57].

Another area of rapid evolution in languages associated with patterns is associated with semantic technologies (including web-based standard languages) for use in automation of machine reasoning about models [34, 66].

Accordingly, MBSE modeling languages should be understood as still evolving very significantly – a sign of progress but also continuing change to follow, including complete language metamodel replacement now underway [36]. The combination of expanding diversity in domain-specific patterns and languages, along with continuing major change in general modeling languages, illustrates the importance of formal mappings between different languages or between different models, patterns, and ontologies. Gaining competence in use of such mappings is important for protecting investment in model-based Intellectual Property (IP); it is not enough to rely upon the wish for a single stable language at any time in the near future, and perhaps ever (refer to Fig. 8).

The growing diversity of modeling languages and domains inherent to the overall ecosystem is additionally mitigated by the use of configured Model Characterization Pattern (MCP) model "wrappers." This is discussed in section "Model Characterization Pattern: Universal Model Metadata Reference Pattern."

Automated Tooling and MBSE Patterns

If MBSE is understood to be inherently associated with automated tools, then pattern-based MBSE is even more linked to automation. The reason for this

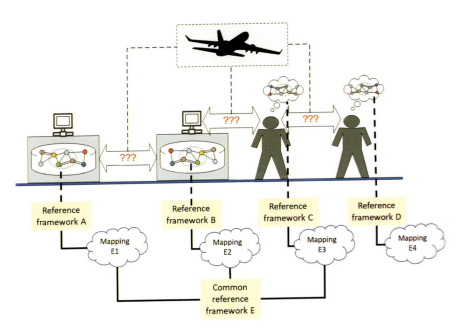

Fig. 8 Mappings between languages, patterns, enhanced by a common reference. (From Schindel (2020), permission granted to reproduce with attribution) [40]

intensified connection to automation is that pattern-based MBSE adds an additional critical automation capability – after an MBSE Pattern has been constructed, with its built-in configuration rules, the stage has been set for subsequent automated use of that pattern in auto-configuration generation of models, auto-checking of models and empirical data against the pattern, and other automated exploitation of the MBSE pattern. Tooling becomes more valuable with MBSE Patterns. For example, Fig. 9 illustrates aspects of automated generation of a configured MBSE model from an MBSE Pattern, based on stakeholder needs expressed through configured Features.

Existing MBSE COTS modeling tools typically permit the addition of automated algorithms such as the pattern configuration algorithm illustrated by Fig. 9 [58]. The information model or schema of MBSE COTS modeling tools permits sufficient flexibility to enable the mapping of the related S*Metamodel (see section "Modeling Languages and MBSE Patterns") into the tool schema, as described in the previous section and in Ref. [23].

Model authoring tools are only part of the ecosystem automated tool chain inhabited by MBSE Patterns. A semantic parallel to the category of numerical simulation tools is the category of semantic reasoning tools. These are particularly associated with pattern-based methods because the reasoning performed by such a tool about a given MBSE model is based on reference to more general rules that the model should satisfy – which is exactly what an MBSE Pattern is. Semantic tools are in turn associated with related languages, discussed in the previous section; such tools include Pellet [61] and others. These lead to the fact that many contemporary semantic reasoning applications are based on the standards-based information technology of the Semantic Web [67]. Other specialized tooling associated with exploitation of MBSE patterns include Product Line Engineering (PLE) tooling and related standards [28] (ISO 26580 2021).

Additional connections of MBSE Patterns to automated tooling at three levels of detail are as follows.

- *Innovation Ecosystem Pattern (ASELCM):* This MBSE Pattern can be configured for a given project, enterprise, or supply chain ecosystem to represent the federation of multiple automated tools, repositories, platforms, and other information technologies supporting the pattern-based ecosystem (refer to section "INCOSE Innovation Ecosystem Reference Pattern") [55].
- *Trusted Model Repository Pattern (TMR):* This MBSE Pattern can be configured to represent the stakeholder feature set(s) of single or multiple federated automated tools (refer to Fig. 10) [64].
- *Model Characterization Pattern (MCP):* This MBSE Pattern can be configured once for each virtual model of any type to represent, access, and manage the diverse capabilities and nature of virtual models of any type, including but not limited to MBSE models (refer to section "Model Characterization Pattern: Universal Model Metadata Reference Pattern") [24].

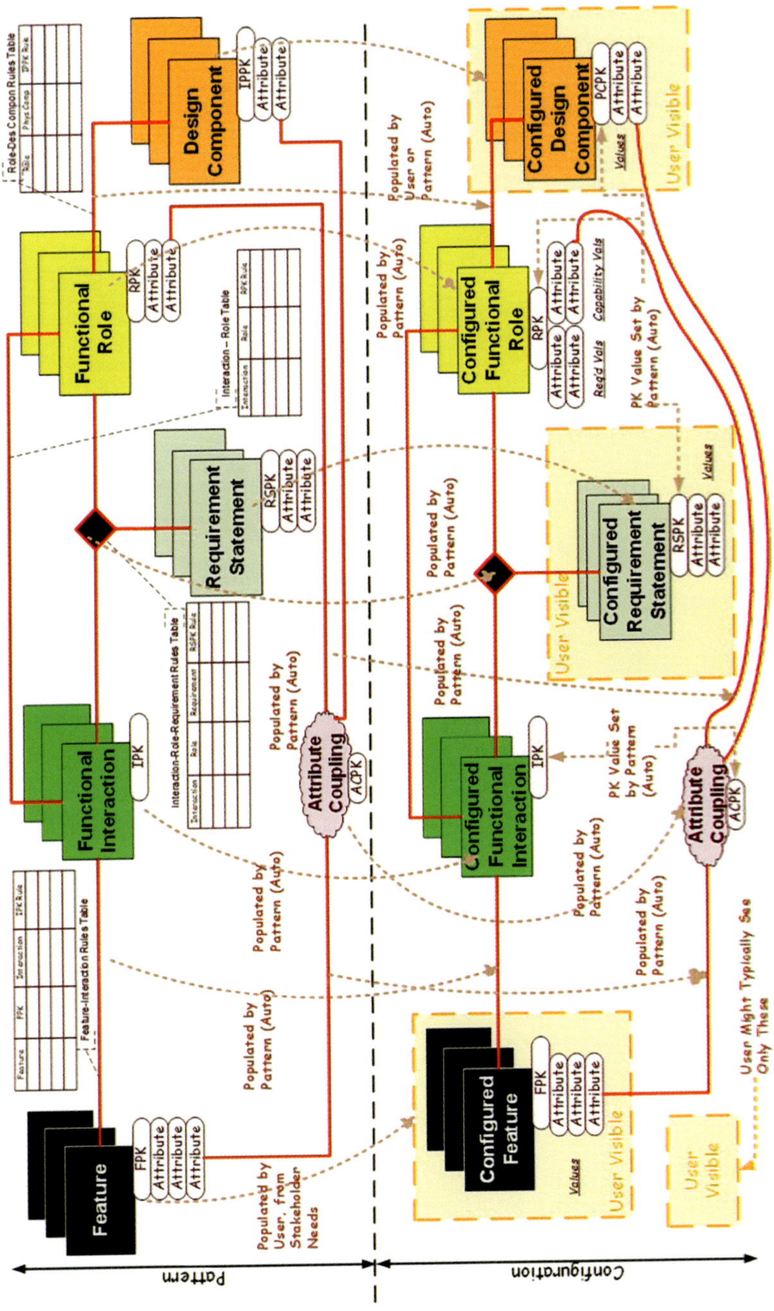

Fig. 9 Illustration of selected aspects of semiautomatic pattern configuration. (From Ref. [22], permission granted to reproduce with attribution)

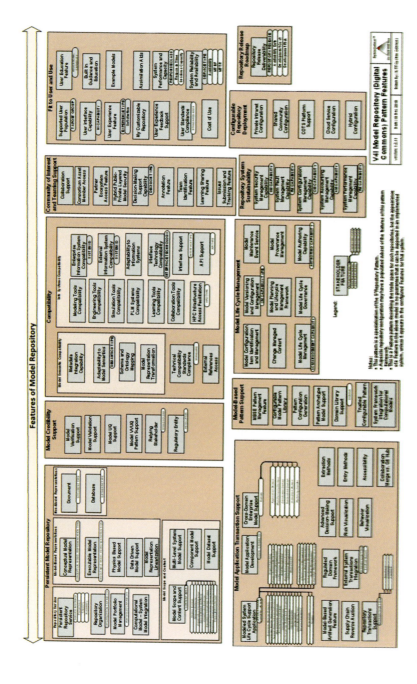

Fig. 10 Trusted Model Repository (TMR) pattern, configurable stakeholder features. (From Ref. [64], permission granted to reproduce with attribution)

Best Practice Approach

Section "State-of-the-Art" introduced "the most important" MBSE foundational pattern, upon which the others are based – the S*Metamodel. Section "Best Practice Approach" moves to two practice-based patterns that can be applied to plan, practice, and advance basic MBSE Pattern practices at any stage of improvement.

INCOSE Innovation Ecosystem Reference Pattern

Digital Engineering in general, MBSE more specifically, and pattern-based MBSE in particular, all offer to improve how we perform systems engineering and the life cycle of innovation. (Likewise, agile systems engineering and other enhancements beyond the scope of this handbook additionally contribute to more effective outcomes.) In planning such improvements, engineers often emphasize the description around models of engineered systems ("System 1") – but it is the system of engineering itself that is being changed, and all the more so with pattern-based methods. Accordingly, we have found that the representation of that "System 2" (the system of engineering) as a first-class modeled system in its own right is essential to convey understanding and represent current and planned methods and capabilities. The INCOSE Agile SE Life Cycle Management (ASELCM) Pattern, summarized in Fig. 11, emerged in our related INCOSE work as a key reference pattern for describing any and all past, current, and future means of engineering and life cycle management, whether agile or not, model-based or not, digital or not. It has been applied extensively in private industry; it was applied by a joint working group team in the public 2-year INCOSE project that created agile systems engineering case studies in four major enterprises [9, 11–13]; it is more recently in use in studies by other technical societies of digital thread and digital twin methods and implementations.

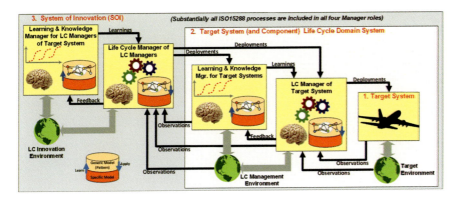

Fig. 11 The ASELCM pattern: Level 1 logical system reference boundaries. (From Schindel and Dove (2016), permission granted to reproduce with attribution)

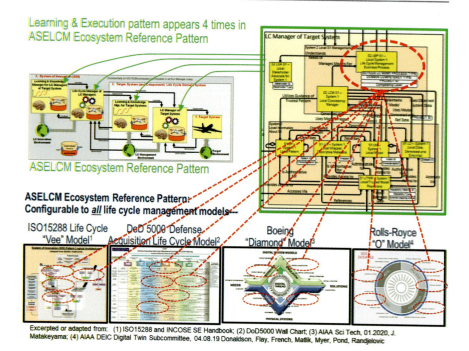

Fig. 12 The systems engineering "Vee" appears four times in the ASELCM pattern. (From [44], permission granted to reproduce with attribution)

The traditional "Vee diagram" view of the ISO15288 model (Fig. 12) focuses on key interdependencies of the life cycle management processes, arising from the nature of developed systems. What we will see below emphasizes certain aspects of the same processes – the *discovery, learning,* and *use of learning* aspects, and how they relate to the very same ISO15288 processes. It is a different *emphasis* on the traditional processes – not a replacement or abandonment of them.

The reference model shown in Fig. 11, the Agile Systems Engineering Life Cycle Management (ASELCM) Pattern, was used by the INCOSE ASELCM Discovery Project (Schindel and Dove 2016). It includes three major subsystems.

System 1: Target system of interest, to be engineered or improved.

System 2: The environment of (interacting with) System 1, including not only its operational environment, but also all the engineering, production, support, and life cycle management systems of System 1, *including learning about and representing System 1 and its environment.*

System 3: The life cycle management systems for System 2, including learning about and representing Systems 2 and its environment.

Note that System 2 is further divided into the following:

Learning and Knowledge Manager for Target System: Discovers, validates, and accumulates new and existing knowledge about System 1 and its operating environment

Life Cycle Manager for Target System: Uses what has already been learned (in A above) about System 1, performing all the necessary life cycle management processes

The same sort of subdivision occurs for System 3, but concerned with discovery and learning about System 2 and its environment, and managing its life cycle. So, System 3 includes all process improvement for System 2.

The ASELCM Pattern of Fig. 11 shows observation and feedback loops. This pattern models Innovation itself, not just the innovated thing – and is nonlinear, iterated, and exploratory as to configuration space. It is a complex adaptive system reference model for system innovation, adaptation, operation/use/metabolism, sustainment, and retirement or replacement. It applies to 100% human-performed or automation-aided innovation, or hybrids thereof, whether performed with agility or not, ISO15288 oriented or informal, and whether performed well or poorly. It includes representation of proactive, anticipatory systems. The rise of a number of newer innovation methods and emphases, in business and technical systems, supports the need for such a combined or integrated reference model:

- Agile engineering of systems and software [8, 38]
- Product Line Engineering of composable, configurable systems (ISO 26580 2021) [28]
- Experiment-based innovation (Schrage 2014) [2, 32, 59]
- Fail fast and recover early [10, 31, 45]
- Lean business start-up, the minimum viable product, and pivoting [37]

ASELCM is neutral descriptive, not prescriptive, reference, configurable to describe less effective as well as more effective innovation processes and information as a complex adaptive system of systems.

Effective Ecosystem-Level Learning: More than "Lessons Learned" Reports

The emerging innovation methods cited above particularly emphasize *learning*, whether it is discoveries about stakeholders and their value space, the evolving environment, competitive alternatives, system concept of operations and technical requirements, designs and technological characterization, or failure modes and design limits. As methodologies couched in agility, experiment, pivot, or fail fast and recover early, the hallmark of these methods is admission that a changing or uncertain world creates risks and opportunities in the form of incomplete knowledge. Of course, this has always been true, at least to some degree, in the world of innovation, traditional or otherwise, but the newer methods particularly emphasize means of accelerating the related discovery and learning process, managing-related risks.

Accordingly, strategies for learning are of particular importance [14, 30]. This learning amounts to filling in more knowledge in the models of the configuration spaces described above. Because these spaces are usually very large, with many degrees of freedom and parametric ranges, and because exploration, experimentation, and learning require expending time and other resources, the *strategy* for picking what to learn about, what to invest experiment and learning resources in, becomes important. The concept of configuration space and trajectories through it can help us see this exploration as "flying through" the space in designated "search patterns." Interest in optimal strategies (i.e., trajectories, routes) for exploration of this space becomes a natural extension of the theory of Design of Experiments [16], and has become the subject of a significant literature on experiment, in its own right (Schrage 2014; Kohavi et al 2009) [2, 5, 32, 62, 63].

For the systems engineering process, there are a number of learning-related implications:

- *How is Continuous, Incremental Learning Represented?* In the approach described above, what is already known about System 1 is represented by the smallest model sufficient for purposes of engineering or science. It follows that what is learned in the future about System 1 would be represented as (incremental) changes to that model.
- *Learning Must be Compressed and Placed "In the Way" of Future Performance:* For learning to be effective, it must impact future behavior. Just "storing" what is learned is not the objective, which is improved future performance about what was learned. So, what was learned must be effectively incorporated in future performance. While the internal means of this are somewhat masked by biology for individual humans, when it comes to teams and enterprises, we must ask how learning is to improve future performance of the group. We suggest that it is not effective to accumulate ever-growing sets of "lessons learned reports," even if searchable as databases. The INCOSE MBSE Patterns Working Group describes S*Patterns as the configurable, reusable models of whole target systems (INCOSE Patterns Working Group 2015). These are subsequently configured as the starting point of future performance, so that whatever has flowed into the Patterns becomes a (configurable, as needed) part of future performance. Think of "muscle memory" in humans.
- *Learning in Each ISO15288 Process:* Figure 12 shows that the ISO15288 life cycle management processes appear twice in System 2 and twice in System 3. Two of those appearances are learning processes – they are the learning aspect of each of the (already defined) ISO15288 processes. They are about learning new things about the subject of those processes – whether they are about stakeholder or technical needs, designs, verifications, or otherwise. Every ISO105288 process potentially has a learning aspect. But each of them also has a "non-learning" execution only aspect, in which what has already been learned is applied. It is not the case that engineering a system requires learning. In the case of Product Line Engineering (PLE) for configurable platforms, there are rapid-execution versions of each of the ISO15288 processes that essentially "configure" what is already

6 Pattern-Based Methods and MBSE

defined in the platform pattern, for a specific case. The platform and its supporting patterns represent what was learned in the past – what we already know.

- *What About What We Already Know?* The traditional description of the systems engineering process actually describes all the things we would do if we knew nothing in advance about a system or its domain. But what about what we already know, which is usually quite a lot? Very little of the traditional life cycle process description addresses that question, nor how it would be blended with new learning processes. So, splitting up processes into the learning-execution pairs of Fig. 12 has the further advantage of explicating this important aspect, essential to agility.
- *Learning About System 2:* These same points, concerning System 2's learning about System 1, will also apply to System 3's learning about System 2.

Model Characterization Pattern: Universal Model Metadata Reference Pattern

Intended Uses and Origins of the Model Characterization Pattern

The purpose of the Model Characterization Pattern is to provide a common, shareable, configurable, universal framework of information to describe a virtual model of any type (e.g., numerical simulations, system models, artificial neural networks, architectural databases, empirical datasets, summarized by Fig. 13).

The MCP effectively "characterizes" any model, and that is why it is called the "Model Characterization Pattern." The MCP is not the virtual model it describes, but is instead the description of that virtual model. Virtual models are very diverse, as to

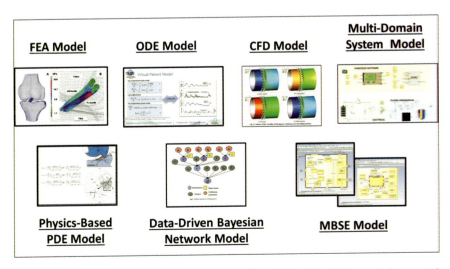

Fig. 13 Diverse virtual model types, technologies, styles, origins. (From [24], permission granted to reproduce with attribution)

the technology of the model, the subject of the model, the purpose of the model, the modeling styles of the model author, the model credibility, and many other aspects. The single MCP data structure is able to describe all these different types and instances of virtual models because the MCP is a configurable S*Pattern data structure (Fig. 14).

When the information characterizing a specific virtual model is inserted into the generic MCP's data structure, the result is referred to as a "configured MCP," because it now describes a specific virtual model. The configured MCP is sometimes also referred to as a "Model Wrapper," because it is in many ways analogous to the printed labels and bar codes that are found on packaged food products on a grocery shelf.

In the language of the systems community, the MCP is a "configurable system pattern" that is in fact a model of all virtual models, ready to be configured to describe specific model instances. In the language of the modeling community, the MCP is a "metadata pattern," which simply means that it is information that describes a virtual model and other related subjects about that virtual model.

The MCP has arisen over several years of collaborative activities by the INCOSE MBSE Patterns Working Group, the ASME VV50 Model Life Cycle Working Group, and the V4 Institute. Configured MCPs (model wrappers) may be used for many different purposes, across the life cycle of the virtual models they describe. Among these MCP applications are the following:

- More effectively plan new or improved virtual models, and to know whether you need them, versus making use of existing model assets you can more readily discover by their "wrappers"

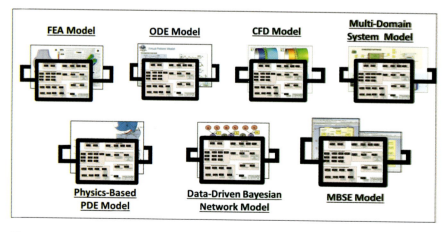

Fig. 14 Configurations of universal Model Characterization Pattern (MCP) metadata. (From [24], permission granted to reproduce with attribution)

6 Pattern-Based Methods and MBSE

- Improve access to collections of models by exposing their characteristics to potential users more effectively, whether individually or through managed directories of characterization data
- Rapidly generate very systematic model requirements for new or existing models, for use in model development, verification, validation, and life cycle management
- Lower the experience threshold needed to plan and manage virtual models, including model VVUQ as well as broader credibility assessment of models
- More effectively manage large collections of diverse virtual models and related information
- More effectively share models across supply chains and regulatory or industry domains
- Lower cost and time necessary to obtain trusted/credible models in regulated or other domains
- Use or manage models that were generated by others; increase the range of others who can effectively use models that you generate; reduce the likelihood of model misuse
- Improve the accumulation and effective use of model-based enterprise knowledge
- Improve the integration of model-related work across specific engineering disciplines and overall systems engineering
- Increase ability to manage the integration of multiple computational models (e.g., using FMI), including their integrated VVUQ

Because of the wide range of possible MCP applications indicated above, it is not necessary, or even desirable, to try to accomplish all of these at one time – a single application from the above list could be so valuable to justify use of the MCP for only that purpose, saving the other applications for possible future consideration. Note that the individual stakeholders likely to find these applications of value will also vary from one application to another. Usually no single individual will be a "customer for" or an expert on all the different applications listed above.

Summary of the MCP Model Stakeholder Feature Groups
The top-level interface to the MCP is the Stakeholder Feature Pattern portion of the MCP. Its purpose is to describe, at a summary level associated with stakeholders of a virtual model of interest, the "degrees of freedom," "parameters," "characteristics," or "stakeholder requirements" for a given virtual model being characterized by the MCP. Six different Feature Groups organize a collection of different Stakeholder Features addressing a wide variety of model stakeholder issues, summarized by Fig. 15.

The generic MCP is configured to describe a given virtual model, whether it is being planned or already exists, by filling in the "answers" to the questions implied by the various parameters of the Feature Groups. What is the scope and content of the model? What is the intended use of the model? How trusted is its credibility for

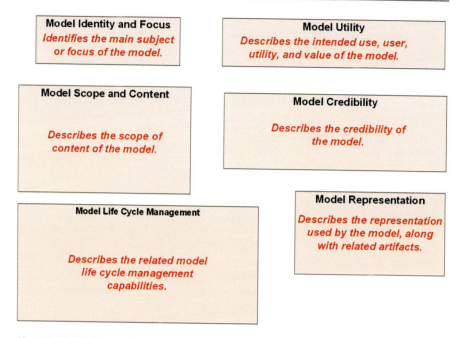

Fig. 15 Model characterization pattern feature groups. (From [24], permission granted to reproduce with attribution)

that use? What is the technical scope of the model? How is it represented? Is the model a long-term enterprise asset or a one-time quick model unlikely to be used again?

Detailed Reference on the MCP Model Stakeholder Features

Individual Stakeholder Features are populated or depopulated based on their relevance for a given model of interest. The MCP collection of approximately 35 MCP Feature types is shown in Fig. 16. Some Features may be depopulated when they are not applicable for a given model of interest (e.g., the Failure Modes and Effects Feature). Some Features are nearly always of interest (e.g., the Modeled System of Interest Feature). Some Features may be populated multiple times (e.g., the Model Intended Use feature may be multiply populated to represent different uses). Many Features also have Feature Attributes, which are additional parameters whose value may be set to represent different configurations.

A detailed description of each of the MCP Features and their Feature Attributes is beyond the scope of this introduction, but may be found in INCOSE MBSE Patterns Working Group (2019e). That reference also provides additional background on configurable patterns in general, as well as an example configuration of MCP.

6 Pattern-Based Methods and MBSE

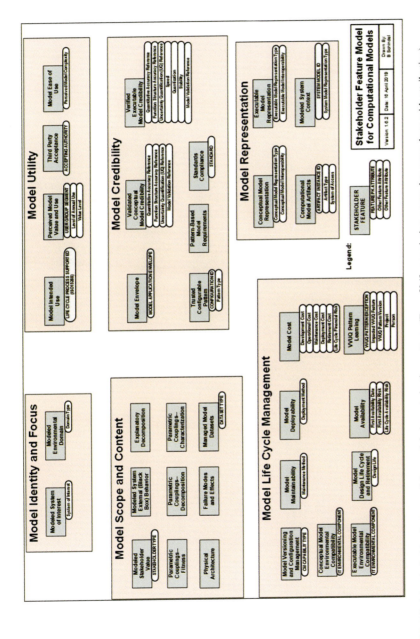

Fig. 16 Model stakeholder features of the model characterization pattern. (From [24], permission granted to reproduce with attribution)

Illustrative Examples

MBSE Pattern methods have been applied since around the year 2000, across a variety of domains in commercial, defense, and institutional environments. Table 1 lists some of these, the following section summarizes example aspects, with the References providing further detail.

When these applications are classified by the System 1–2–3 boundaries of the Innovation Ecosystem Pattern (Fig. 11), it can be seen that most application patterns fall into the following groups:

(1) *Products and Services (System 1):* The most frequently assumed uses of patterns are for product line engineering applications.
(2) *Systems of Product Research, Engineering, Production, Marketing and Distribution, Service and Sustainment, Operations (System 2):* These patterns are additionally vital for enabling the framework in which the patterns of (1) above are applied.
(3) *Systems of Improving System 2 (System 3):* Engineering Education, Engineering Methods Research and Development, Automated Engineering Tooling Development, Engineering Performance Monitoring, Engineering Methods Deployment, all of which are to observe, analyze, understand, and improve (2).

Table 1 Examples of PBSE applications to date

Medical devices patterns	Construction equipment patterns	Terrestrial surface vehicle patterns	Space tourism pattern
Manufacturing process patterns	Production inspection vision system patterns	Packaging systems patterns	Lawnmower product line pattern
Embedded intelligence (EI) patterns	Innovation ecosystem pattern	Consumer packaged goods patterns	Orbital satellite pattern
Product service system patterns	Product distribution system patterns	Plant operations and maintenance system patterns	Oil filter system family pattern
Life cycle management system patterns	Production material handling patterns	Engine controls patterns	Military radio systems pattern
Model characterization metadata pattern	Transmission systems pattern	Precision parts production, sales, and engineering pattern	Higher education experiential pattern
Flight primary control actuator pattern	Automated verification system pattern	Market penetration performance pattern	Clinical diagnostics pattern

6 Pattern-Based Methods and MBSE

The scope of these patterns depends upon their intended uses, across narrow or large parts of the system life cycle. Depending upon that purpose, key sections of application patterns may include:

- Stakeholder Feature model, describing the stakeholder value selection trade space, product line segmentation and configuration, and effects of fault-induced failure modes
- External (black box) environmental interactions with domain actors, external interfaces, domain architectural relationships or input-outputs, system black box interface requirements
- System life cycle and operational states and modes
- System logical and design architecture, decompositions, and design allocations
- Parametric couplings representing quantitative relationships of vertical and horizontal nature, often subject of virtual simulations
- Failure modes and their connections to failure effects, risk analyses
- Pattern configuration rules for the above segments and their relationships, driven by selection patterns within the Stakeholder Features segment listed first above

Some of the above are illustrated by the sample extracts of Figs. 17, 18, 19, 20, 21, 22, and 23, and by additional detail of Ref. [58].

For examples of other MBSE Patterns, refer to:

- Oil Filter Family Product Line Pattern (INCOSE MBSE Patterns Working Group 2018)
- Global Pharmaceutical Packaging Family Pattern [3]
- Construction Vehicles Control Systems Family Product Line Pattern [56]
- Automated Product Test Systems Pattern [7]
- Manufacturing Systems Pattern [49]
- Innovation Ecosystem (ASELCM) Pattern [55]
- Higher Education Experiential Pattern [30]

Chapter Summary

Impact on Practice, Education, and Research

Pattern methods for MBSE, along with key foundational patterns, tooling, and instructional support, have advanced during two decades as illustrated by this chapter.

For individuals, teams, and enterprises that have harnessed MBSE Patterns, dramatic impacts are visible. Some of the least profound impacts are nevertheless easiest to measure, as in the case of system requirements cycles, reduced by a factor of ten to first draft time, quality, and completeness. More profound but also harder to measure are impact through avoided relearning of the same error correction cycles, time to market, and quality. A significant literature has accumulated across a variety

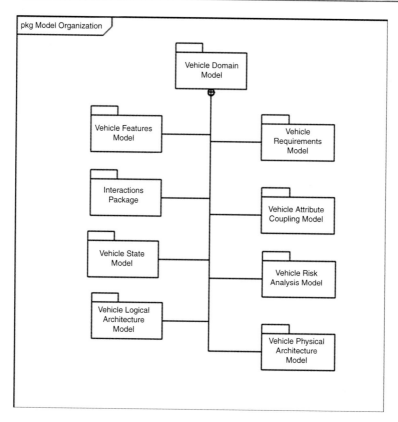

Fig. 17 Pattern model organization into packages. (From Ref. [58], permission granted to reproduce with attribution)

of practices and domains. Nevertheless, current expansion in instruction and use of model-based methods often continues to emphasize "from-scratch" modeling of products in something of a free-for-all set of individual styles and idioms. This means that large competitive improvements are still available.

Model-based system patterns offer a special leverage in engineering education. The majority of engineering discipline courses have at their root implicit, if not yet explicit, patterns of technologies, designs, methods, materials, and processes. The domain or abstract patterns of steam plants, building structures, vehicles, production facilities, consumer products, logistics networks, information technologies, energy systems, and others all exemplify course areas in which more explicit patterns can sharpen the effectiveness of educational investments by students and practitioners. When these are made more explicit in the educational process, the student gains.

The impact of patterns on research runs deep. Schindel (2019) argues an increased emphasis on the domain-specific side, versus generalized systems engineering, is needed to fully exploit this potential.

6 Pattern-Based Methods and MBSE

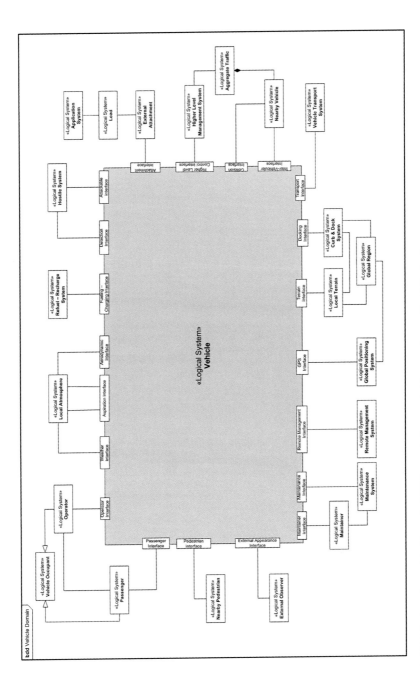

Fig. 18 Vehicle domain model. (From Ref. [58], permission granted to reproduce with attribution)

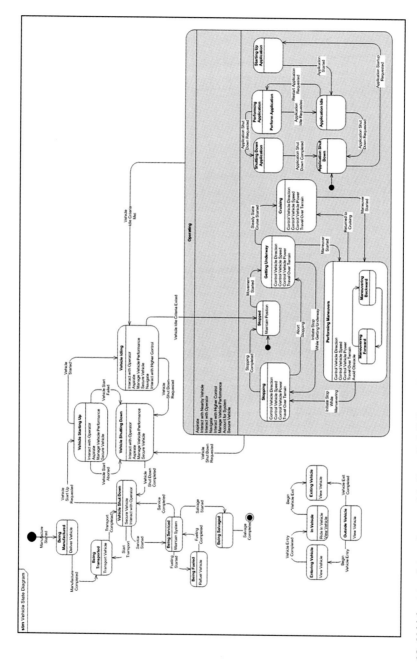

Fig. 19 Vehicle states: Interactions model. (From Ref. [58], permission granted to reproduce with attribution)

6 Pattern-Based Methods and MBSE

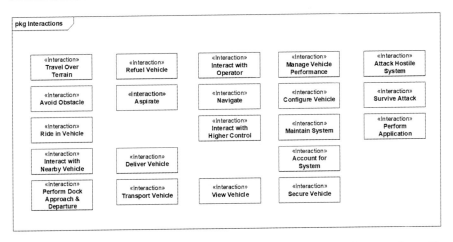

Fig. 20 Vehicle interactions package. (From Ref. [58], permission granted to reproduce with attribution)

Impact on the Theoretical Foundations of Systems Engineering

The theoretical foundations of systems engineering have a different kind of history than the foundational histories of the other engineering disciplines. Electrical, Mechanical, Civil, and Chemical Engineering are supported by physical sciences based on observable phenomena and the theoretical foundations that have grown up with them. Because of its differently motivated emergence in the middle of the twentieth century, systems engineering for its first 50 years particularly emphasized unifying processes and procedures, integration across the other disciplines, connection to stakeholder interests, management of risks, and other important issues that are nevertheless not of the same nature as the observable phenomena and related theoretical edifices created of the last three centuries with the rise of physical sciences and engineering disciplines. Various efforts to establish more theoretically based foundations for systems engineering have been evident, but not with the same effect as the dramatic impacts of the other phenomena-based foundations of EE, ME, CE, or ChE [51].

In the interest of strengthening the theoretical foundations of systems engineering (not in place of the above, but in addition), the International Council on Systems Engineering has undertaken an effort to rally increased focus on observable phenomena-based theoretical foundations for systems engineering. Like the earlier cases of the other engineering disciplines and their related physical science disciplines, this foundation is primarily based on patterns of observable phenomena, placing patterns squarely at the center of the future strength of the theoretical foundations of systems engineering – the same place the discipline-specific phenomena patterns occupy in the older engineering disciplines.

1			Actors																					
2	Interaction Name	Interaction Definition	Vehicle	Operator	Passenger	Vehicle Occupant	Nearby Pedestrian	External Observer	Maintainer	Maintenance System	Local Atmosphere	Refuel System	Hostile System	External Attachment	Load	Application System	Higher Level Management	Nearby Vehicle	Vehicle Transport	Curb & Dock System	Local Terrain	Global Region	Remote Management System	Global Positioning System
3	Account for System	The interaction of the vehicle with its external managers, in which it accounts for vehicle utilization.	x	x					x	x							x						x	
4	Aspirate	The interaction of the vehicle with the Local Atmosphere, through which air is taken into the vehicle for operational purposes, and gaseous emissions are expelled into the atmosphere.	x								x													
5	Attack Hostile System	The interaction of the vehicle with an external hostile system, during which the vehicle projects an attack onto the hostile system's condition.	x										x											
6	Avoid Obstacle	The interaction of the vehicle with an external object, during which the vehicle minimizes contact with or proximity to the object.	x				x																	
7	Configure	The interaction of the vehicle with people or systems that manage its arrangement or configuration for intended use.	x						x	x														
8	Deliver Vehicle	The interaction of the vehicle with the process of its delivery, including manufacture, distribution, and development. This includes delivery of each configured version and update of the vehicle product line or family.																						
9	Interact with Higher Control	The interaction of the vehicle with an external higher level management system, along with the vehicle operator, through which the vehicle is fit into larger objectives.	x														x							
10	Interact with Nearby Vehicle	The intearction of the vehicle with another vehicle, in which information is exchanged to identify one vehicle to another.																						
11	Interact with Operator	The interaction of the vehicle with its operator.																						
12	Maintain System	The interaction of the vehicle with a maintainer and/or maintenance system, through which faults in the vehicle are prevented or corrected, so that the intended qualified operating state of the vehicle is maintained.	x						x	x														
13	Manage Vehicle Performance	The interaction of the vehicle with its operator and/or external management system, through which the performance of the vehicle is managed to achieve its operational purpose and objectives.	x	x																				
14	Navigate	The interaction of the vehicle with the Global Positioning System, by which the Vehicle tracks is position on the Earth.	x																					x
15	Perform Application	The interaction of the vehicle with an external Application System, through which the vehicle performs a specialized application.	x													x								
16	Perform Dock Approach & Departure	The interaction of the vehicle with an external docking system, through which the vehicle arrives at, aligns with, or departs from a loading / unloading dock.	x																	x				
17	Refuel Vehicle	The interaction of the vehicle with a fueling system and its operator, through which fuel is added to the vehicle.	x									x												
18	Ride In Vehicle	The interaction of the vehicle with its occupant(s) during, before, or after travel by the vehicle.	x	x	x	x																		
19	Secure Vehicle	The interaction of the vehicle with external actors that may or may not have privileges to access or make use of the resources of the vehicle, or with actors managing that vehicle security.	x	x																				
20	Survive Attack	The interaction of the vehicle with an external hostile system, during which the vehicle protects its occupants and minimizes damage to itself.	x										x											
21	Transport	The interaction of the vehicle with a Vehicle Transport System, through which the Vehicle is transported to an intended destination.	x																x					
22	Travel Over Terrain	The interaction of the vehicle with the terrain over which it travels, by means of which the vehicle moves over the terrain.	x																		x			
23	View Vehicle	The interaction of the vehicle with an external viewer, during which the viewer observes the vehicle.	x					x																

Fig. 21 Vehicle interactions-actors matrix. (From Ref. [58], permission granted to reproduce with attribution)

6 Pattern-Based Methods and MBSE

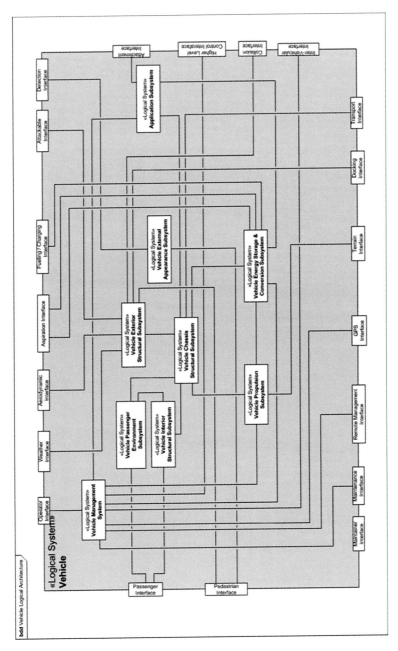

Fig. 22 Vehicle logical architecture. (From Ref. [58], permission granted to reproduce with attribution)

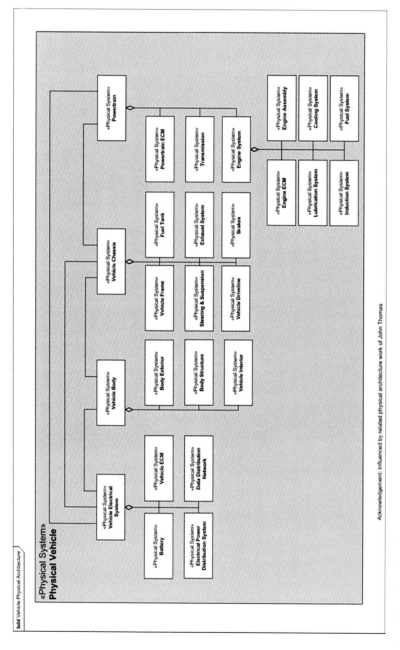

Fig. 23 Vehicle physical architecture. (From Ref. [58], permission granted to reproduce with attribution)

Cross-References

▸ Adoption of MBSE in an Organization
▸ Managing Model-Based Systems Engineering Efforts
▸ MBSE for Acquisition
▸ MBSE Methods for Inheritance and Design Reuse
▸ Semantics, Metamodels, and Ontologies
▸ Transitioning from Observation to Patterns: A Real-World Example

References

1. Alexander, C., Ishikawa, S., Silverstein, M., Jacobson, M., Fiksdahl-King, I., and Angel, S. A Pattern Language. Oxford University Press, New York, 1977.
2. Anderson, Eric T. and Simester, Duncan, 2011, "A Step-by-Step Guide to Smart Business Experiments", Harvard Business Review, March 2011, Retrieve from https://hbr.org/2011/03/a-step-by-step-guide-to-smart-business-experiments
3. Bradley, J, Hughes, M., and Schindel, W., "Optimizing Delivery of Global Pharmaceutical Packaging Solutions, Using Systems Engineering Patterns" Proceedings of the INCOSE 2010 International Symposium, 2010. Download from – https://www.omgwiki.org/MBSE/lib/exe/fetch.php?media=mbse:patterns:bradley_hughes_and_schindel%2D%2Dse_in_packaging_v1.3.1.pdf
4. Buckingham, E., On Physically Similar Systems: Illustrations of the Use of Dimensional Analysis", Phys. Rev. 4(4): 345, Bibcode: 1914.
5. Clarke, B., "Why These Tech Companies Keep Running Thousands of Failed Experiments", Fast Company, Sep 26, 2016.
6. Cloutier, R., Applicability of Patterns to Architecting Complex Systems: Making Implicit Knowledge Explicit. VDM Verlag Dr. Müller. 2008.
7. Cook, D., and Schindel, W., "Utilizing MBSE Patterns to Accelerate System Verification", in Proc. of the INCOSE 2015 International Symposium, Seattle, WA, July, 2015. Download from – https://www.omgwiki.org/MBSE/lib/exe/fetch.php?media=mbse:patterns:insight_v20-1_0317_moog_aircraft.pdf
8. Dove, R. and LaBarge, R., "Fundamentals of Agile Systems Engineering – Part 1, Part 2", in Proc. of INCOSE International Symposium., 2014.
9. Dove, R., and Schindel, W., "Case Study: Agile SE Process for Centralized SoS Sustainment at Northrop Grumman", in Proc. of INCOSE International Symposium, Adelaide, Australia, July, 2017. Download from – https://www.omgwiki.org/MBSE/lib/exe/fetch.php?media=mbse:patterns:is2017%2D%2Dnorthrup_grumman_case_study_dove_and_schindel_bp.pdf
10. Dove, R., et al, 2016, in Proceedings of the INCOSE 2016 Socorro Systems Summit, from: http://www.incose.org/ChaptersGroups/Chapters/ChapterSites/enchantment/library-and-resources/socorro-systems-summit%2D%2D-2016-oct-28-29/proceedings
11. Dove, R., Schindel, W., and Garlington, K., "Case Study: Agile Systems Engineering at Lockheed Martin Aeronautics Integrated Fighter Group", in Proc. of INCOSE International Symposium, Washington, DC., July, 2018. Download from –http://www.omgwiki.org/MBSE/lib/exe/fetch.php?media=mbse:patterns:is2018_-_aselcm_lmc_case_study.pdf
12. Dove, R., Schindel, W., and Scrapper, C., "Agile Systems Engineering Process Features Collective Culture, Consciousness, and Conscience at SSC Pacific Unmanned Systems Group", in Proc. of INCOSE International Symposium, Edinburgh, Scotland, UK, July, 2016. Download from – https://www.omgwiki.org/MBSE/lib/exe/fetch.php?media=mbse:patterns:is2016_%2D%2D_autonomous_vehicle_development_navy_spawar.pdf

13. Dove, R., Schindel, W., Hartney, R., "Case Study: Agile Hardware/Firmware/Software Product Line Engineering at Rockwell Collins", in Proc. of IEEE International Systems Conference, Montreal, Quebec, Canada, April, 2017. Download from – https://www.omgwiki.org/MBSE/lib/exe/fetch.php?media=mbse:patterns:pap170424syscon-casestudyrc.pdf
14. Dyer, J., Christensen, C., Gregersen, H. The Innovators DNA: Mastering the Five Skills of Disruptive Innovators, HBR Press, 2011.
15. Estefan, J. 2008. "Survey of Candidate Model-Based Systems Engineering (MBSE) Methodologies", rev. B. Seattle, WA, USA: International Council on Systems Engineering (INCOSE). INCOSE-TD-2007-003-02. Accessed April 13, 2015 at http://www.omgsysml.org/MBSE_Methodology_Survey_RevB.pdf
16. Fisher, Ronald, The Design of Experiments, Ninth Edition, Macmillan, New York (US), 1971.
17. Gamma, E., Helm, R., Johnson, R., Vlissides, J. Design Patterns: Elements of Reusable Object-Oriented Software. Addison-Wesley Publishing Company, Reading, MA, 1995.
18. Genesereth, M. R., & Nilsson, N., Logical Foundations of Artificial Intelligence, San Mateo, CA: Morgan Kaufmann Publishers, 1987.
19. Gingerich, O., The Book Nobody Read: Chasing the Revolutions of Nicolaus Copernicus, New York: Walker & Co. Publishing, 2004.
20. INCOSE MBSE Patterns 2016. Working Group web site: http://www.omgwiki.org/MBSE/doku.php?id=mbse:patterns:patterns
21. INCOSE MBSE Patterns Working Group, "S*MBSE Patterns: A Small Scale Example", INCOSE Patterns Working Group, 2018. Download from: https://www.omgwiki.org/MBSE/lib/exe/fetch.php?media=mbse:patterns:oil_filter_example_v1.6.2.pdf
22. INCOSE MBSE Patterns Working Group, "MBSE Methodology Summary: Pattern-Based Systems Engineering (PBSE), Based On S*MBSE Models", V1.6.1, March, 2019. Download from –https://www.omgwiki.org/MBSE/lib/exe/fetch.php?media=mbse:patterns:pbse_extension_of_mbse%2D%2Dmethodology_summary_v1.6.1.pdf
23. INCOSE MBSE Patterns Working Group, "S*Metamodel Mapping for MagicDraw/Cameo Systems Modeler Version 19", 2019b. Download from —http://www.omgwiki.org/MBSE/lib/exe/fetch.php?media=mbse:patterns:systematica_mapping_for_magicdraw_csm_v1.9.1a.pdf
24. INCOSE MBSE Patterns Working Group, "The Model Characterization Pattern (MCP): A Universal Characterization & Labeling S*Pattern for All Computational Models", V1.9.3. INCOSE MBSE Patterns Working Group, June, 2020. Download from – https://www.omgwiki.org/MBSE/lib/exe/fetch.php?media=mbse:patterns:model_characterization_pattern_mcp_v1.9.3.pdf
25. ISO 10303-233:2012 "Industrial automation systems and integration – Product data representation and exchange – Part 233: Application protocol: Systems engineering", 2012.
26. ISO, "ISO/PAS 19450:2015, Automation systems and integration — Object-Process Methodology", International Standards Organization, 2015. Download from –https://www.iso.org/standard/62274.html
27. ISO/IEC 15288: "Systems Engineering—System Life Cycle Processes". International Standards Organization (2015).
28. ISO/IEC 26580:2021, "Software and systems engineering – Methods and tools for the feature-based approach to softwared and systems product line engineering", 2021.
29. ISO/IEC/IEEE 42010:2011, "Systems and software engineering - Architecture description" 2011.
30. Kline, W., and Schindel, W., "The Innovation Competencies - Implications for Educating the Engineer of the Future", in Proc. of American Society of Engineering Education 2014 Conference, Indianapolis, Indiana, US, 2014. Download from – https://www.omgwiki.org/MBSE/lib/exe/fetch.php?media=mbse:patterns:asee_2014_paper_on_education_models.pdf
31. Kohavi, R., Crook, T., Longbotham, R., Frasca, B., Henne, R., Ferres, R., Melamed, T., "Online Experimentation at Microsoft", retrieved from: http://www.exp-platform.com/documents/expthinkweek2009public.pdf
32. Manzi, J. Uncontrolled: The Surprising Payoff of Trial-and-Error for Business, Politics, and Society, Basic Books, NY, 2012.

33. Meyer, M., Lehnerd, A., The Power of Product Platforms, Free Press, 2011.
34. NASA JPL, "JPL Ontological Modeling Language Specification Document (OML)", NASA, 2015. Download from – https://github.com/JPL-IMCE/gov.nasa.jpl.imce.oml.doc
35. OMG, "OMG System Modeling Language", Version 1.6, Object Management Group, 2016. Download from – https://www.omg.org/spec/SysML/
36. OMG, "SysML V2.0 Specification Documents", Object Management Group, 2021. Download from – https://github.com/Systems-Modeling/SysML-v2-Release
37. Ries, E. The Lean Startup: How Today's Entrepreneurs Use Continuous Innovation to Create Radically Successful Businesses, Crown Business, New York, 2011.
38. Rigby, D., Sutherland, J., and Takeuchi, H., 2016, 'The Secret History of Agile Innovation', Harvard Business Review, April 20, 2016.
39. Schindel, W. "Requirements statements are transfer functions: An insight from model-based systems engineering", Proceedings of INCOSE 2005 International Symposium, 2005.
40. Schindel, W. "Interoperable Model Semantics: An issue that won't go away". INCOSE Patterns Working Group, April, 2020. Download from – https://www.omgwiki.org/MBSE/lib/exe/fetch.php?media=mbse:patterns:interoperable_semantics_v1.3.1.pdf
41. Schindel, W. "System Interactions: Making the Heart of Systems More Visible", Proc. of INCOSE Great Lakes Regional Conference, 2013.
42. Schindel, W. "The Difference Between Whole-System Patterns and Component Patterns: Managing Platforms and Domain Systems Using PBSE", INCOSE Great Lakes Regional Conference on Systems Engineering, Schaumburg, IL, October, 2014.
43. Schindel, W., "Pattern-Based Systems Engineering: An Extension of Model-Based SE", INCOSE IS2005 Tutorial TIES 4, 2005.
44. Schindel, W., "Consistency Management as an Integrating Paradigm for Digital Life Cycle Management with Learning", INCOSE MBSE Patterns Working Group, 2020. Download from – https://www.omgwiki.org/MBSE/lib/exe/fetch.php?media=mbse:patterns:aselcm_pattern_%2D%2D_consistency_management_as_a_digital_life_cycle_management_paradigm_v1.2.2.pdf
45. Schindel, W., "Fail-Fast Rapid Innovation Concepts", INCOSE Enchantment Chapter Socorro Systems Summit–Collaborative Knowledge Exchange. Socorro, NM, 2017.
46. Schindel, W., "Failure Analysis: Insights from Model-Based Systems Engineering", Proc. of INCOSE International Symposium, 2010.
47. Schindel, W., "Feelings and Physics: Emotional, Psychological, and Other Soft Human Requirements, by Model-Based Systems Engineering", Proc. of INCOSE International Symposium, 2006.
48. Schindel, W., "Innovation, Risk, Agility, and Learning, Viewed as Optimal Control & Estimation", in Proc of INCOSE 2017 International Symposium, Adelaide, Australia, July, 2017.
49. Schindel, W., "Integrating Materials, Process & Product Portfolios: Lessons from Pattern-Based Systems Engineering", Proc. of Society for the Advancement of Material and Process Engineering, 2012. Download from – https://www.omgwiki.org/MBSE/lib/exe/fetch.php?media=mbse:patterns:sampe_baltimore_2012_v1.3.6_.pdf
50. Schindel, W., "Interactions: Making the Heart of Systems Visible", in Proc. of the INCOSE Great Lakes 2013 Regional Conference", 2013. Download from – https://www.omgwiki.org/MBSE/lib/exe/fetch.php?media=mbse:patterns:system_interactions%2D%2Dmaking_the_heart_of_systems_more_visible_v1.2.2.pdf
51. Schindel, W., "Implications for Future SE Practice, Education, Research: SE Foundation Elements, Discussion Inputs to INCOSE Vision 2035 Theoretical Foundations Section", V2.3.2, 2020. Download from – https://www.omgwiki.org/MBSE/lib/exe/fetch.php?media=mbse:patterns:science_math_foundations_for_systems_and_systems_engineering%2D%2D1_hr_awareness_v2.3.2a.pdf
52. Schindel, W., "S*Metamodel", Metamodel Version 1.7, July, 2019. Download from – http://www.omgwiki.org/MBSE/lib/exe/fetch.php?media=mbse:patterns:systematica_5_metamodel_v7.1.6a.pdf
53. Schindel, W., "Systems of Innovation II: The Emergence of Purpose", Proceedings of INCOSE 2013 International Symposium, 2013.

54. Schindel, W., "What Is the Smallest Model of a System?", in Proc. of INCOSE 2011 International Symposium, 2011. Download from – https://www.omgwiki.org/MBSE/lib/exe/fetch.php?media=mbse:patterns:what_is_the_smallest_model_of_a_system_v1.4.4.pdf
55. Schindel, W., and Dove, R., "Introduction to the ASELCM Pattern", Proc. of INCOSE IS2016, Edinburg, UK, 2016. Download from – https://www.omgwiki.org/MBSE/lib/exe/fetch.php?media=mbse:patterns:is2016_intro_to_the_aselcm_pattern_v1.4.8.pdf
56. Schindel, W., and Smith, V., "Results of applying a families-of-systems approach to systems engineering of product line families", SAE International, Technical Report 2002-01-3086, 2002. Download from – https://www.omgwiki.org/MBSE/lib/exe/fetch.php?media=mbse:patterns:sae_tr_2002-01-3086_v1.2.25.pdf
57. Schindel, W., Lewis, S., Sherey, J., Sanyal, S., "Accelerating MBSE Impacts Across the Enterprise: Model-Based S*Patterns", to appear in Proc. of INCOSE 2015 International Symposium, July, 2015.
58. Schindel, W., Peterson, T., "Introduction to Pattern-Based Systems Engineering (PBSE): Leveraging MBSE Techniques", in Proc. of INCOSE 2013 International Symposium, Tutorial, June, 2013.
59. Schrage, M., The Innovator's Hypothesis: How Cheap Experiments Are Worth More Than Good Ideas, MIT Press, Cambridge, MA (US), 2014.
60. SEBoK Editorial Board. 2020. The Guide to the Systems Engineering Body of Knowledge (SEBoK), v. 2.3, R.J. Cloutier (Editor in Chief). Hoboken, NJ: The Trustees of the Stevens Institute of Technology. Accessed [DATE]. www.sebokwiki.org.
61. Sirin, E., et al. "Pellet: A Practical OWL-DL Reasoner", in J. of Web Semantics, June 2007, 5(2):51–53.
62. Teller, A., Ted Talk by Astro Teller, Apr 14, 2016, Alphabet X, retrieved from: www.ted.com/talks/astro_teller_the_unexpected_benefit_of_celebrating_failure
63. Thomke, S., Experimentation Matters: Unlocking the Potential for New Technologies for Innovation, Harvard Business Review Press, Boston, MA (US), 2003.
64. V4 Institute, "V4 Institute Framework, Assets, Support", Conference Poster, in INCOSE 2018 Great Lakes Regional Conference, Indianapolis, IN, USA, September, 2018.
65. Walden, D. et al, INCOSE Systems Engineering Handbook, Fifth Edition, International Council on Systems Engineering, San Diego, CA, 2015.
66. W3C, "OWL 2 Web Ontology Language, Document Overview (Second Edition), W3C Recommendation 11 December 2012. Download from – https://www.w3.org/TR/2012/REC-owl2-overview-20121211/
67. W3C, "OWL Reasoners", download 03.01.2021 from – https://www.w3.org/2001/sw/wiki/OWL/Implementations

William D. (Bill) Schindel is president of ICTT System Sciences. His engineering career began in mil/aero systems with IBM Federal Systems, included faculty service at Rose-Hulman Institute of Technology, and founding leadership of three systems enterprises. An INCOSE Fellow, Schindel serves as chair and co-founder of the INCOSE MBSE Patterns Working Group. He co-led a project on Systems of Innovation in the INCOSE System Science Working Group, was a member of the lead team of the INCOSE Agile Systems Engineering Life Cycle Discovery project, and is a contributor to the INCOSE Vision 2035 and INCOSE Systems Engineering Fifth Edition Handbook.

Overarching Process for Systems Engineering and Design

A. Terry Bahill and Azad M. Madni

Contents

Introduction to Modeling	197
Uncertainty Is Ubiquitous	197
Model-Based System Engineering	198
Purpose of Models	199
Kinds of Models	199
Types of Models	199
Tasks in the Modeling Process	200
Model for a Baseball-Bat Collision	200
Checklist for Tasks Necessary in a Modeling Project	202
Requirements Discovery Process	208
Where Do Requirements Come From?	209
A Use Case Template	209
A Use Case Example from a Chocolate Chip Cookie-Making System	211
A Test Plan for This System	213
Tradeoff Study Process	214
Tradeoff Study Example from a Chocolate Chip Cookie Acquisition System	216
Risk Analysis Process	220
Risk Analysis Example from a Chocolate Chip Cookie Making System	220
Comparing the Requirements, Tradeoff, and Risk Processes	222
Comparing the *Activities* of the Requirements, Tradeoff, and Risk Processes	222
Comparing the *Products* of the Requirements, Tradeoff, and Risk Processes	223

A. T. Bahill (✉)
Systems and Industrial Engineering, University of Arizona, Tucson, AZ, USA
e-mail: terry@sie.arizona.edu

A. M. Madni
Systems Architecting and Engineering, Astronautical Engineering Department, University of Southern California, Los Angeles, CA, USA
e-mail: azad.madni@usc.edu

© Springer Nature Switzerland AG 2023
A. M. Madni et al. (eds.), *Handbook of Model-Based Systems Engineering*,
https://doi.org/10.1007/978-3-030-93582-5_16

A Requirement from the Chocolate Chip Cookie Acquisition System	224
An Evaluation Criterion from the Chocolate Chip Cookie Acquisition System	226
A Risk from the Chocolate Chip Cookie Acquisition System	227
Comparing a Requirement, an Evaluation Criterion, and a Risk	228
The SIMILAR Process	229
State the Problem	230
Investigate Alternatives	232
Model the System	232
Integrate	233
Launch the System	233
Assess Performance	234
Reevaluate	234
The Overarching Process	235
Effects of Human Decision-Making on the Overarching Process	237
Confirmation Bias	237
Severity Amplifiers	239
Framing	240
The Overarching Process	242
Uncertainty in Stating the Problem for the Overarching Process	244
A System for Handling Uncertainty in Models and Documentation	244
Chapter Summary	249
Cross-References	253
References	253

Abstract

This chapter presents three key processes central to systems engineering: requirements discovery, tradeoff studies, and risk analysis. It compares and contrasts these three processes and then combines them into a single *Overarching Process*. The three original processes can then be viewed as specific tailorings of the Overarching (superset) Process. Similarly, the Overarching Process can be viewed as a top-level process (a superset) for model-based system engineering (MBSE) implementations. The Overarching Process itself is not an example of model-based systems engineering, except at a high level. This chapter also identifies the activities in the Overarching Process that contribute to uncertainty. All of these activities involve human decision-making. Therefore, most mistakes caused by uncertainty are found in the system models and documentation. These mistakes often arise from confirmation bias, severity amplifiers, and framing. The two key examples used in this chapter are the Cookie Acquisition System and the BaConLaws model for baseball-bat collisions.

Keywords

Uncertainty · Requirements · Tradeoff study · Trade study · Risk analysis · Baseball · Decision analysis and resolution · Sensitivity analyses · Multi-objective decision-making

Introduction to Modeling

A model is a simplified representation of some aspect of a real system. Models are ephemeral: They are created, they explain a phenomenon, they stimulate discussion, they foment alternatives, and then they are replaced by newer models. Engineers know how to construct a model, but quite frequently, they miss a few steps. This recognition provides the impetus for this chapter that presents a succinct description of the modeling process.

Requirements discovery, tradeoff studies, and risk analyses are three distinct systems engineering activities. Even though they have the same underlying process structure, they appear different because they employ different vocabularies, inputs, and outputs. To convey the similarity of these processes to the reader, we abstracted and grouped the common activities in these three processes. The incipient development of this approach was presented in *Tradeoff Decisions in System Design* [8]. In the interest of clarity, the processes, shown in our figures, suppress the explicit representation of temporal sequences. Also, in the interest of clarity, we suppress the multitude of feedback loops that arise when several of these activities are performed in parallel.

In this chapter, we discuss these three processes, along with their sources of uncertainty, and present existing ad hoc methods and mechanisms for *identifying* uncertainties. We dealt with *handling* uncertainties in Madni and Bahill [38]. Next, we present the Overarching Process as a superset of these three processes. Finally, we present a system for ameliorating uncertainty in the Overarching Process. In the approach presented, uncertainty is consistently and uniformly addressed using the Overarching Process as a reference model.

Uncertainty Is Ubiquitous

Uncertainty is ubiquitous in our environment, and occasionally, people deliberately create uncertainty, for example, in games of chance such as playing card games. In card games, one or more decks of cards are shuffled to create a random ordering of cards. Thus, when a card facing down is about to be turned over, neither its suit nor its rank is known for sure (ideally).

In the field of metrology, measuring uncertainty is a core concept that quantifies the measurement error that one should reasonably expect. Uncertainty is involved in every measurement and is represented as significant figures using the number system. Numbers are restricted to only the physically meaningful digits. This quantification of uncertainty is then propagated throughout the calculations so that the uncertainty in the calculated values depends on the uncertainties associated with the measured values and the calculation algorithm.

As important, our understanding of nature is incomplete. Therefore, our models of natural phenomena have uncertainty. The Heisenberg uncertainty principle states that we cannot simultaneously measure both the position and velocity of an electron (or any number of other subatomic particles). As we measure the position ever more accurately, the estimate of the velocity becomes more inexact. Astrophysicists have proposed dark matter and dark energy to patch the holes in our laws of physics that are used to model nature. Even so, our models remain incomplete. Therefore, we still cannot predict the future. ("It's tough to make predictions, especially about the future." Impishly attributed to Yogi Berra.)

In optimization models, uncertainty is used to describe situations where the user does not have full control. Likewise, it is common to include estimates of uncertainty in economic and weather forecasts. Similarly, pollsters employ uncertainty in their models for polling and predicting political elections. In decision science, we employ probabilistic models for human decision-making under uncertainty.

In a systems engineering process, there cannot be a block that says "manage uncertainty" because uncertainty is ubiquitous: It is everywhere. Uncertainty must be managed where and when it occurs. Handling uncertainty is like making a peanut butter and jelly sandwich – just as the peanut butter must be spread over the surface of a slice of bread, so also uncertainty needs to be spread across the whole systems engineering process.

Model-Based System Engineering

System design can be component-based (e.g., WWII battleships), function-based (e.g., 1970s, MIL-STD 499A), requirements-based (e.g., 1990s, MIL-STD 499B), or model-based (e.g., 2000s, OMG, and Estefan [23]). Model-based system engineering and design has the advantage of executable models that improve efficiency and rigor. It also provides a common terminology (ontology), explicit representations, and a central source of truth from which views can be extracted for the needs of particular stakeholders [40]. The earliest development of this technique was in Wayne Wymore's [54] book entitled *Model-Based System Engineering*, although the phrase *Model-Based System Design* was in the title and topics of Jerzy Rozenblit's [50] PhD dissertation. One of the first model-based systems engineering process models was that of Bahill and Gissing [7]. A good summary of the Wymorian process is given in Estefan [23]. Model-based systems engineering depends on having and using well-structured models that are appropriate for the given problem domain [9, 40]. An ancient Chinese proverb that was invented by a New York City journalist a century ago says, "A picture is worth a thousand words." In engineering design, this phrase has morphed into "a model is worth a thousand pictures." This means that models greatly reduce the complexity of a system description. This is akin to design elegance. The complexity is there, but the ability to create views for specific viewpoints enables focusing on issues relevant to different stakeholders. Model elements that are not important to a particular stakeholder can be abstracted or elided.

In this chapter, we derive the Overarching Process that can be used as an enveloping process on top of a model-based system engineering (MBSE) methodology. Other chapters in this handbook (e.g., [23]) show various implementations of

MBSE processes. But this chapter is at a higher level of abstraction. It shows an Overarching Process for MBSE.

The Overarching Process is a top-level, front-end process. Rather than sitting down with the customer on the first day of the project and filling out SysML diagrams, the customer would be better served by starting with the Overarching Process.

"All the really important mistakes are made the first day," Eb Rechtin, *The Art of System Architecting*, p. 28, [49].

Purpose of Models

Models can be used for many reasons, such as guiding decisions, understanding an existing system, improving a system, creating a new design or system, controlling a system, improving operator performance, suggesting new experiments, guiding data collection activities, allocating resources, identifying cost drivers, increasing return on investment, helping to sell the product, and reducing risk [34]. Running business process models clarifies requirements, reveals bottlenecks, reduces cost, identifies fragmented activities, and exposes duplication of efforts [35].

Kinds of Models

There are different kinds of models in systems engineering. These models address different system perspectives: behavioral, structural, performance, and analysis. *Behavioral models* describe how the system responds to external excitation, that is, how the system functions transform the inputs into outputs. The BaConLaws model [3] is a model of behavior. It describes the linear and angular velocity of baseballs and softballs and baseball and softball bats after the collision in terms of these same parameters before the collision. *Structural models* describe the components and their interactions. Three-dimensional CAD/CAM images check the buildability of structures. *Performance models* describe units, values, and tolerances for properties such as weight, speed of response, available power, etc. These might be captured in requirements. Typical baseball performance measures include batting average, slugging average, and On-base Plus Slugging (OPS). *Analysis models* are used to calculate the properties of the whole system from the properties of its parts. For example, the time for a car to accelerate from 0 to 60 mph can be calculated from the mass of the car, the torque transmitted through the drive train, the aerodynamic drag coefficients, and the friction between the tires and the pavement.

Types of Models

There are many types of models [36]. People generally use only a few and erroneously believe them to be the totality of models because they tend to think of models from their narrow perspectives. Examples of most commonly used types of models include models based on physiological and physical laws and principles, differential equations, difference equations, algebraic equations, geometric representations of physical structure, computer simulations and animations, Laplace transforms, transfer functions, linear systems theory, state-space models (e.g., $\dot{x} = \mathbf{A}x + \mathbf{B}u$), state machine diagrams, charts, graphs, drawings, pictures, functional flow block diagrams, object-oriented models, UML and SysML diagrams, Markov processes, time-series models, physical analogs, Monte Carlo simulations, optimization algorithms, statistical distributions, mathematical programming, financial models, PERT charts, Gantt charts, risk analyses models, tradeoff analyses models, mental models, computer-based story representations, scenario models, and use case models. The appropriate type of model depends on the particular system being studied, the question being asked of the model, the operational context, and the modelers' background.

For example, to understand how people make decisions, at least three phenomena should be accounted for: confirmation bias, attribute substitution, and representativeness. For the biological domain, we must first choose the subject, that is, a virus, a bacterium, a plant, or an animal. Once we have chosen our subject, we can then derive its genome. To model something in the social domain, we might use a novel, an encyclical, a song, a poem, or possibly even a joke.

Most models of real-world phenomena require a combination of these types. For example, Bahill [3] uses Newton's principles, the conservation laws of physics, algebraic equations, spreadsheets, figures, tables, simulations, an optimization package, design of experiments, and statistics. Hence, the BaConLaws model comprises many different types of models.

Tasks in the Modeling Process

In this section, we provide a checklist of the principal tasks or steps that should be performed in a modeling study [2]. Modelers should look at each item on the list to determine if they have done that task. If not, then they should explain why they did not do it. But before explaining our checklist, we must present an example that can be used to illustrate the items on the checklist.

Model for a Baseball-Bat Collision

An effective way to understand these tasks in the modeling process is through an example. A suitable example is one in which the details are quantitative, publicly available, and readily accessible and whose principles are commonly understood by engineers. We decided to use the BaConLaws model for baseball from Bahill [3], Chap. 4. A *very* brief synopsis of this model is presented below. The reader may skip the equations without loss of continuity.

7 Overarching Process for Systems Engineering and Design

The following equations comprise the BaConLaws model for bat-ball collisions. First, the kinetic energy lost (transformed into heat) during the collision is

$$KE_{lost} = \frac{1}{2} \frac{m_{ball} m_{bat} I_{bat} \left(v_{ball\text{-}before} - v_{bat\text{-}cm\text{-}before} - \omega_{bat\text{-}before} d_{cm\text{-}ip}\right)^2 \left(1 - CoR^2\right)}{m_{ball} I_{bat} + m_{bat} I_{bat} + m_{ball} m_{bat} d_{cm\text{-}ip}^2}, \quad (1)$$

where $d_{cm\text{-}ip}$ is the distance between the bat's center of mass and the impact point, and CoR is the coefficient of restitution, which also models the energy lost.

The linear velocity of the ball after the collision is

$$v_{ball\text{-}after} = v_{ball\text{-}before} - \frac{\left(v_{ball\text{-}before} - v_{bat\text{-}cm\text{-}before} - \omega_{bat\text{-}before} d_{cm\text{-}ip}\right)(1 + CoR) m_{bat} I_{bat}}{m_{ball} I_{bat} + m_{bat} I_{bat} + m_{ball} m_{bat} d_{cm\text{-}ip}^2}$$

where $v_{ball\text{-}before} < 0$.

$$(2)$$

The linear velocity of the bat after the collision is

$$v_{bat\text{-}cm\text{-}after} = v_{bat\text{-}cm\text{-}before}$$
$$+ \frac{\left(v_{ball\text{-}before} - v_{bat\text{-}cm\text{-}before} - \omega_{bat\text{-}before} d_{cm\text{-}ip}\right)(1 + CoR) m_{ball} I_{bat}}{m_{ball} I_{bat} + m_{bat} I_{bat} + m_{ball} m_{bat} d_{cm\text{-}ip}^2} \quad (3)$$

The angular velocity of the bat after the collision is

$$\omega_{bat\text{-}after} = \omega_{bat\text{-}before}$$
$$+ \frac{\left(v_{ball\text{-}before} - v_{bat\text{-}cm\text{-}before} - d_{cm\text{-}ip} \omega_{bat\text{-}before}\right)(1 + CoR) m_{ball} m_{bat} d_{cm\text{-}ip}}{m_{ball} I_{bat} + m_{bat} I_{bat} + m_{ball} m_{bat} d_{cm\text{-}ip}^2} \quad (4)$$

Our most succinct presentation of this BaConLaws model is

$$\boxed{\begin{aligned}
CoR &= -\frac{v_{ball\text{-}after} - v_{bat\text{-}cm\text{-}after} - d_{cm\text{-}ip} \omega_{bat\text{-}after}}{v_{ball\text{-}before} - v_{bat\text{-}cm\text{-}before} - d_{cm\text{-}ip} \omega_{bat\text{-}before}} \\
&\text{where } 0 < CoR < 1 \\
A &= \frac{\left(v_{ball\text{-}before} - v_{bat\text{-}cm\text{-}before} - d_{cm\text{-}ip} \omega_{bat\text{-}before}\right)(1 + CoR)}{m_{ball} I_{bat} + m_{bat} I_{bat} + m_{ball} m_{bat} d_{cm\text{-}ip}^2} \\
&\text{and } A < 0 \\
v_{ball\text{-}after} &= v_{ball\text{-}before} - A m_{bat} I_{bat} \\
v_{bat\text{-}after} &= v_{bat\text{-}before} + A m_{ball} I_{bat} \\
\omega_{bat\text{-}after} &= \omega_{bat\text{-}before} + A m_{ball} m_{bat} d_{cm\text{-}ip} \\
\omega_{ball\text{-}after} &= \omega_{ball\text{-}before}
\end{aligned}} \quad (5)$$

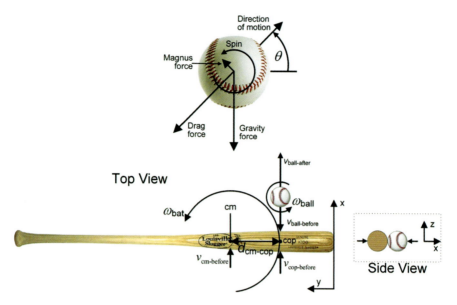

Fig. 1 Forces on a ball-in-flight (top) and a schematic of a head-on bat-ball collision (bottom). The center of mass is cm, and the center of percussion is cop. (© 2019, Bahill. Used with permission)

The numerical value for A is unique for each bat-ball collision. Of course, more complicated models exist, for example, those not described by Fig. 1 and therefore where $\omega_{\text{ball-after}} \neq \omega_{\text{ball-before}}$.

If you do not want to follow the equations of this model, then just imagine watching a baseball game. Your thoughts should fill the squiggly braces in the following checklist. We describe {in squiggly braces} the parts of the BaConLaws model that implement the individual tasks.

Checklist for Tasks Necessary in a Modeling Project

- Describe the system to be modeled. {The BaConLaws model describes head-on collisions between bats and balls, that is, when the bat is going upward at about 10° and the ball is coming downward at about 10° and there is no offset between the bat displacement vector and the ball displacement vector. It gives the velocity and spin of the bat and ball before and after collisions. It does not describe the dynamics *during* the collision nor the swing of the bat.}
- State the purpose of the model. {The purpose of the BaConLaws model was to explain bat-ball collisions with precise, correct equations, without jargon. This included defining the performance criterion function. If the model were being

asked a different question, say about a player's salary contract, an entirely different type of model would have been used, for example, a financial investment model [18, 19].} Here are some baseball models created by physic professors that are not equation-based:

Al Nathan
https://www.scientificamerican.com/article/baseball-physics-opening-day/
David Kagan
https://physics.csuchico.edu/baseball/talks/AAPT(Nov-2012)/slides.pdf
Rod Cross
http://www.physics.usyd.edu.au/~cross/baseball.html
Bob Adair
https://www.amazon.com/Physics-Baseball-3rd-Robert-Adair/dp/0060084367

- Determine the level of the model [11]. {The level for the BaConLaws model encompasses the ball velocity, the bat velocity, and the bat angular velocity after the collision in terms of those same parameters before the collision. The timescale is in milliseconds.}
- State the assumptions and, at every review, reassess the assumptions. {Our assumptions were stated (on pages 24 and 36 of Bahill [3]), and they were reviewed repeatedly.}
- Investigate alternative models. {Many bat swing models were presented in Chap. 1 [3]. Alternative collision configurations were explained in Chaps. 2 and 3. Chapter 3 also presented nine alternative definitions for the sweet spot of the bat. The BaConLaws model was given in Chap. 4, and alternative models were given in Chaps. 5 and 9. Having alternative models helps ensure that you understand the physical system. No model is more correct than another. Alternative models just emphasize different views of the physical system. They are not competing models; they are synergetic.}
- Select tools for the model and simulation. {We used the What'sBest! optimizer, the Pascal language, the Excel spreadsheets, the Math Type equation editor, and MS Word.} This should not be a casual decision. One should not merely accept the default. Tradeoff studies should be used to help select the best tools.
- Make the model. {The BaConLaws model is shown in Eqs. (1) to (5).}
- Integrate with models for other systems. {The outputs of the BaConLaws model became inputs to the ball-in-flight model of Chap. 7 and the Probability of Success model in Chap. 9 of Bahill [3].}
- Gather data describing system behavior. {We used data from our internal databases, peer-reviewed journal papers, and the following online databases:
http://mlb.com/statcast/
https://baseballsavant.mlb.com/statcast_search
https://www.baseball-reference.com/.}
- Show that the model behaves like the real system. {The outputs of the simulations were compared to the data listed in the above paragraph.}

- Verify and validate the model. {*Verification* means, Did you build the system right? For the BaConLaws model, the outputs of the simulations agree with the data listed in the above paragraph. The double checks in the simulation ensured the correctness of the spreadsheets. For example, the kinetic energy lost was computed with equations and also by summing individual kinetic energy components. The conservation laws were used in the derivations, and the final outputs of the simulation were inserted into the conservation law equations to ensure consistency of the spreadsheet. The main output of the BaConLaws model was compared to the output of the Effective Mass model in Chap. 5 of Bahill [3]. The physics was peer-reviewed by two anonymous physics professors. Each of the main BaConLaws equations was derived using at least two techniques. Finally, the equations were checked by an independent mathematician. *Validation* means, Did you build the right system? Our customer wanted a system that described head-on collisions between bats and balls. They wanted a system that would give bat and ball velocity and the bat angular velocity after the collision in terms of those same parameters before the collision. This is what our system does: See Eqs. (2) to (5) above. Finally, we performed a sensitivity analysis, which is a powerful validation tool [30, 53]. It warns if something is wrong with the model. It might also define the boundary conditions for parameters, discover potential brittleness, impact recommended operating procedures, find quirks in how the system must be used, etc.} Enough details should be given to allow other users to replicate your results. If other people cannot replicate your experiments and analysis, then your model fails validation.
- Perform a sensitivity analysis of the model as follows.

Sensitivity Analysis of a Bat-Ball Collision Model

The batter in a game of baseball or softball would like to obtain the maximum batted ball velocity. The larger the batted ball velocity, the more likely the batter will get on base safely. Therefore, we made the batted ball velocity our performance criterion. (Our equations are vector equations. In our analysis, we represented both the magnitudes and directions of the vectors. However, in the book and in this chapter, we only present the magnitudes.)

The linear velocity of the ball after the collision is

$$v_{\text{ball-after}} = v_{\text{ball-before}} - \frac{(v_{\text{ball-before}} - v_{\text{bat-cm-before}} - \omega_{\text{bat-before}} d_{\text{cm-ip}})(1 + CoR) m_{\text{bat}} I_{\text{bat}}}{m_{\text{ball}} I_{\text{bat}} + m_{\text{bat}} I_{\text{bat}} + m_{\text{ball}} m_{\text{bat}} d_{\text{cm-ip}}^2} \tag{6}$$

In a simple sensitivity analysis, an input is changed by a small amount, and the resulting change in the output is recorded. For example, in Table 1, when $v_{\text{bat-cm-before}}$ (the velocity of the center of mass of the bat before the collision) was increased by 1%, $v_{\text{ball-after}}$ increased by 0.62%. This was the most sensitive input listed in Table 1.

Table 1 Relative sensitivity analysis

Inputs	Nominal values, SI units	Nominal values, baseball units	$v_{ball-after}$ when the input was increased by 1%. The nominal value was 91.894 mph	Percent change in $v_{ball-after}$
$v_{ball-before}$	−37 m/s	−83 mph	92.066	0.19
$\omega_{ball-before}$	209 rad/s	2000 rpm	91.894	0
$v_{bat-cm-before}$	23 m/s	52 mph	92.463	0.62
$\omega_{bat-before}$	32 rad/s	309 rpm	92.000	0.12
CoR	0.465	0.465	92.450	0.60

Next, the coefficient of restitution, CoR, was set as a constant, and we computed the partial derivatives of the batted ball velocity, $v_{ball-after}$, with respect to the eight model inputs and parameters. Finally, we used partial derivatives and computed the semirelative sensitivity functions [53].

Table 2 gives the nominal values, along with the range of physically realistic values for collegiate and professional baseball batters, and the semirelative sensitivity values computed analytically. The bigger the sensitivity value, the more important the variable or parameter is for maximizing batted ball velocity.

The equations of the BaConLaws model have the variable d_{cm-ip} for the distance between the bat center of mass and the impact point. To get numerical values for Table 2, we needed a particular impact point. For this, we used the center of percussion, hence d_{cm-cop}. The variable $vt_{bat-cop-before}$ is the total velocity, meaning the sum of the linear and angular velocity, of the center of percussion of the bat before the collision.

The right column of Table 2 shows that the most important property (the largest absolute value), in terms of maximizing batted ball velocity, is the linear velocity of the center of mass of the bat before the collision, $v_{bat-cm-before}$. This is certainly no surprise. The second most important property is the coefficient of restitution, CoR. The least important properties are the angular velocity of the ball, $\omega_{ball-before}$; the distance between the center of mass and the center of percussion, d_{cm-cop}; and the moment of inertia of the bat, I_{bat}. For our analysis, the sensitivities to the distance between the center of mass and the center of percussion, d_{cm-cop}, and the mass of the ball, m_{ball}, are negative, which merely means that as they increase, the batted ball velocity decreases. The second-order interaction terms, which are not shown, are small, which is good. The results shown in Tables 1 and 2, for two different sensitivity analysis techniques, agree.

{The most important parameters, in terms of maximizing batted ball speed, are the velocity of the center of mass of the bat before the collision and the coefficient of restitution, CoR. The least important parameter is the angular velocity of the pitched ball.}

- Explain a discovery that was not planned in the model's design. {(1) We were surprised when the equation for the kinetic energy lost in the collision fell right

Table 2 Typical values and first-order analytic semirelative sensitivities with respect to the batted ball velocity for the BaConLaws model

Inputs and parameters	Nominal values SI units	Baseball units	Range of realistic values SI units	Baseball units	$\bar{S}_a^F = \frac{\partial F}{\partial a}\big|_{NOP} a_0$ semirelative sensitivity values
Inputs					
$v_{\text{ball-before}}$	−37 m/s	−83 mph	−27 to −40 m/s	−60 to −100 mph	8
$\omega_{\text{ball-before}}$	209 rad/s	2000 rpm	209 ± 21 rad/s	2000 ± 200 rpm	0
$v_{\text{bat-cm-before}}$	23 m/s	52 mph	23 ± 5 m/s	52 ± 10 mph	28
$\omega_{\text{bat-before}}$	32 rad/s	309 rpm	32 ± 11 rad/s	300 ± 100 rpm	5
$vf_{\text{bat-cop-before}}$	28 m/s	62 mph			
Parameters					
CoR	0.465	0.465	0.465 ± 0.05	0.465 ± 0.05	25
$d_{\text{cm-cop}}$	0.134 m	5.3 in	0.134 ± 0.05 m	5.3 ± 2 in	−2
m_{ball}	0.145 kg	5.125 oz	0.145 ± 0.004 kg	5.125 ± 0.125 oz	−14
m_{bat}	0.905 kg	32 oz	0.709 to 0.964 kg	25 to 34 oz	10
$I_{\text{bat-cm}}$	0.048 kg-m²	2624 oz-in²	0.036 to 0.06 kg-m²	1968 to 3280 oz-in²	3

out of the BaConLaws set of equations. (2). Before writing that book, we did not expect to prove that cupping the barrel end of the bat does little good. (3) Although it seems intuitive, we were surprised when the mathematics showed that a baseball could be thrown farther than a tennis ball.}

- Perform a risk analysis. {*Risk to our publisher.* The biggest risk is that people might be reluctant to buy a book with equations in it. Also, Springer would be disappointed if sales were low. Therefore, by writing with the reader in mind, we tried to ensure that sales would not be below expectations. We expect no copyright problems because most of the material is original, and we have permissions for the two figures that are not. *Risk to our reader.* Someone could modify their bat and hurt him or herself by working with tools, or they could be thrown out of a game for using an altered bat. *Risk to the authors.* If our equations were wrong, or if important assumptions were omitted, then we would confuse our readers and tarnish our reputations. *Risk to quality.* The book is produced in India. Typographical and editing mistakes that occur are hard to correct because of poor communication channels. *Risk to baseball managers, general managers, and umpires.* It will put a burden on these people to understand the results of mathematical modeling. *Risk to Major League Baseball (MLB).* It could embarrass MLB into disclosing their algorithms. Some of these risks may seem unlikely. However, one of the most important parts of a risk analysis is exploring unlikely risks.}
- Analyze the performance of the model. {This was described above in the verification paragraph.}
- Reevaluate and improve the model. {In the future, we will explain why the curveball curves. We will also investigate the cognitive processing and decision-making of the batter [4, 5, 8, 10, 42]. We will describe the thrust and parry of the pitcher and the batter.}
- Suggest new experiments and measurements for the real system that might challenge existing models. MLB is providing copious amounts of new data. Next, scientists need MLB's actual algorithms and measurements for the spin on the batted ball, particularly for the home run trajectories that are so popular on television. Another proposed area of measurement and display involves the erratic meandering of fielders trying to catch pop-ups. This behavior and the paper by McBeath et al. [42] show that the ball's trajectory often has bizarre loops and cusps. MLB should show these trajectories on the television screen to help laypeople understand the fielders' wanderings. In the third edition of this book, once we build a gold standard input data set for swings of the bat, we will directly compare the BaConLaws model and other bat-ball collision models.}

In this section, we presented a checklist that should be used to ensure that the most important modeling tasks have been performed. The checklist was exemplified with a model for baseball-bat collisions.

Requirements Discovery Process

We must avoid hearing the common customer complaint, "You gave me what I asked for, not what I wanted!" (Norm Augustine, personal communication)

Figure 2 presents the requirements discovery process. Uncertainty exists in several aspects of this process. The activities that have the most uncertainty are marked with a ▼. To begin with, not all stakeholders can be identified with certainty at the start of the project [33]. The identification of customer needs is a step that

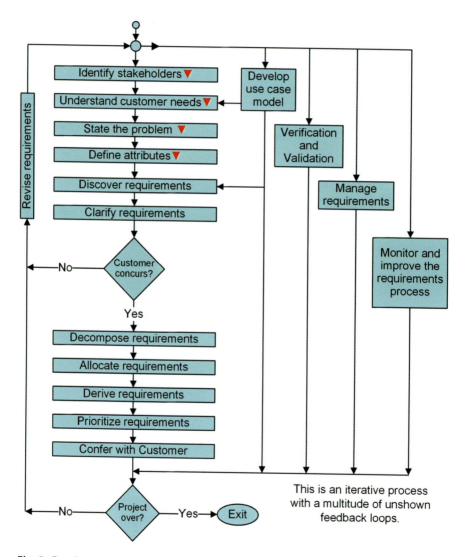

Fig. 2 Requirements discovery process. (Based on Bahill and Madni [8])

requires iterations because uncertainty exists in the initial identification of needs. The same goes for the problem statement. The initial statement of the problem can be expected to be reused and refined as customer needs are more clearly articulated. Therefore, requirements discovery is an iterative process in which each iteration reduces uncertainty in the identification of requirements. Upon customer consensus, the requirements are decomposed, allocated, further refined (derived), prioritized, validated, and finalized. Iteration is the primary means of reducing uncertainty in the requirements discovery and derivation task. Query reformulation is the primary means for reducing uncertainty in the problem statement task. Uncertainty in program schedule and technology maturation is addressed through incorporating schedule buffers and safety margins.

The requirements discovery process has a multitude of unshown feedback loops. For example, the Manage Requirements Activity has inputs to all of the mainline activities, such as discover requirements, clarify requirements. decompose requirements, allocate requirements, derive requirements, and prioritize requirements.

Where Do Requirements Come From?

Requirements come from project stakeholders [6]. Stakeholders include, among others, end users, operators, surrogate customers, managers, sponsors, staff members, testers, maintainers, bill payers, regulatory agencies, potential victims, and systems that will interact with your system [17, 27]. Many requirements can be derived from previous systems. And if you are lucky, requirements can come from the use cases, as shown in the following sections.

A Use Case Template

While the use case diagram is simple, the use case package is complex. It is inadequately explained in most books and papers. Therefore, we start our system development with a formal use case template. Less formal descriptions are called stories.

A *use case* is an abstraction of the required functions of a system. A use case usually produces an observable result of value to the user. Each use case describes a sequence of interactions between one or more actors and the system [47]. Our design process is use case-based.

Name: A use case should be named with a verb phrase in the active present tense form. It should not relate to any particular solution.

Iteration: This is configuration management. Sometimes, we just number them.

Derived from: Explain the source for the use case. For example, it might be the mission statement, the concept of operations (ConOps), a business use case, or a customer requirement.

Brief description: Describe the general sequence that produces an observable result of value to the user.

Level: The amount of detail required in the use case. Do not mix classes of different levels in the same use case.

Priority: The importance of this use case relative to other use cases

Scope: This defines the boundary of what the use case applies to.

Added value: Describe the benefit (usually) for the primary actor. This is an important slot.

Goal: The goal is the behavior that the primary actor expects to get from the use case. You should have a goal or an added value, but probably not both.

Primary actor: Actors are named with nouns or noun phrases. Actors reflect the roles of things outside the system that interact with the system. Primary actors initiate the functions described by use cases.

Supporting actor: Supporting (or secondary) actors are used by the system. They are not a part of the system and thus cannot be designed or altered. They often represent external systems or commercial-off-the-shelf (COTS) components. If your system changes them, then those effects are unintended consequences.

Frequency: How often is the use case likely to be used? When this slot is helpful, it is *very* helpful. When it is not, do not use it.

Precondition: The precondition should contain, among other things, the state of the system and values for pertinent attributes before the main success scenario starts.

Trigger: The trigger should contain the event that causes a transition from the preconditioned state to the first step in the main success scenario.

Main success scenario:

1. This numbered set of steps illustrates the usual, successful interactions of actors with the system. Usually, the first step states the action of the primary actor that starts the use case.

2a. The last step tells you where to go next (e.g., exit use case).

Alternate flows:

2b. Alternate flows describe failure conditions and unsuccessful interactions (exit use case).

The main success scenario and the alternate flows can contain diagrams, such as sequence and activity diagrams.

Postcondition: Describes the state of the system after exit from the use case no matter which flows were executed. This is hard to write.

Specific Requirements

The steps in the main success scenario should suggest the functions that the system is supposed to perform. From these, we should be able to write system requirements.

Functional requirements: Describe the functional requirements with shall statements.

Nonfunctional requirements: Describe the nonfunctional (often performance) requirements with shall statements.

Requirements are sometimes quantified with scoring functions ([54] pp. 385–397; [8], [21], pp. 246–258).

Author/owner: This is an important field. It tells you whom to talk to if you want to change the use case.

Last changed: Use some form of configuration management such as the date of the last change or the revision history.

No standard specifies which slots should be in a use case description. Your minimal set should be based on your company requirements template. The number of slots and the detail in each slot increase as the design progresses from the requirements model to the analysis model to the design model to the implementation model. Other useful slots contain rules, assumptions, and extension points. Do not use slots that do not help you. If you find that the trigger, precondition, or postcondition does not help you to create state machine diagrams, then do not use them. A use case description is also called a use case report, a use case narrative, and stories. A use case package contains a use case description, sequence diagrams, activity diagrams, supplementary requirements, and other UML stuff [47]. Other chapters in this handbook use other templates.

A Use Case Example from a Chocolate Chip Cookie-Making System

Imagine that while reading this book, you experience an irresistible urge for chocolate chip cookies. Frantically, you rummage your kitchen. Lo and behold, you find a tube of chocolate chip cookie dough in your refrigerator! Assume that you have a typical kitchen – a stove, an oven, a timer, pots and pans, utensils, and, of course, cookie sheets. And you have the all-important tube of Pillsbury's Chocolate Chip Cookie dough, with these instructions printed on the label:

- Preheat oven to 350 °F.
- Spoon heaping teaspoons of well-chilled dough about 2 in. apart onto a cool ungreased cookie sheet.
- Bake at 350 °F for 10 min.

You are in business!
Write a use case that will describe how your system should work.
Name: Bake my Cookies
Note: For clarity, we set use case names in a different font, probably Verdana.
Iteration: 3.1
Derived from: Problem statement
Brief description: The cookie-making system bakes cookies to perfection. It is named Cookie.
Level: High
Priority: This use case is of the highest priority.
Scope: A typical home kitchen with pots, pans, utensils, etc.
Added value: Students' brains always work better with a tummy full of cookies.

Goal: To produce stupendous freshly baked cookies
Primary actor: Student
Supporting actors: A tube of Pillsbury's Chocolate Chip Cookie dough
Frequency: Once a month
Precondition: All ingredients and cookware are available.
Trigger: Student gets "hungry" for cookies.
Main success scenario:

1. Student decides to bake cookies.
2. Student turns on the oven and sets the desired oven temperature to 350 °F.
3. Cookie increases the temperature in the oven.
4. Student gets a tube of Pillsbury's Chocolate Chip Cookie dough out of the refrigerator and spoons heaping teaspoons of well-chilled dough about 2 in. apart onto an ungreased cookie sheet. Your Mom would probably worry about the expiration date of the product, which is a possible risk.
5. Cookie signals that the oven has preheated to 350 °F.
6. Student puts the cookie sheet full of cookies into the oven and sets the timer for 10 min.
7. Cookie signals that the baking time is over. If this signal is erroneously too late, the cookies could burn.
8. Student takes the cookies out of the oven. The cookie sheet will be hot. Student must wear an oven mitt. Student puts the cookies on a cooling rack and turns the oven off. Failing to turn the oven off creates a risk.
9. Student eats the cookies and notes their quality (exit use case).

    Unanchored alternate flow: At any time, Student can abort the process and turn off the oven (exit use case).
    Postcondition: The kitchen is a mess, but the oven is off.
    The steps in the main success scenario suggest the functions that the system is supposed to perform. From these steps, we can write the following system requirements. This process was developed by Daniels and Bahill [20].

Specific Requirements
Functional requirements:

ReqF1: Cookie shall provide a mechanism for Student to enter the desired baking time. The abbreviation ReqF* means a functional requirement.
ReqF2: Cookie shall display the desired baking time entered by Student.
ReqF3: Cookie shall heat the oven from room temperature to 350 °F in less than 5 min.
ReqF4: Cookie shall calculate and display the remaining baking time.
ReqF5: Cookie shall emit an audible signal when the oven is preheated and when the baking time is over.
ReqF6: Cookie shall visually indicate when the oven is preheated and when the baking time is over.

ReqF7: Cookie shall execute Built-in Self-Tests (BiST) (derived from company policy).

ReqF8: Cookie shall have a hard upper limit on oven temperature at 550°, even during self-cleaning (derived from the risk analysis). Rationale: This will help prevent fire.

ReqF9: Cookie shall turn off the oven when it is no longer being used (derived from the risk analysis). Implementation could be (1) turning the oven off 20 min after the end of the timer interval, (2) turning on an alarm 20 min after the end of the timer interval, and (3) turning the oven off 20 min after the end of the timer interval if there is no food inside of the oven.

Nonfunctional requirements:

ReqNF1: The remaining baking time displayed by Cookie shall be visible to a Student with 20/20 vision standing 5 ft from the oven in a room with an illuminance level between 0 and 1000 lux. The abbreviation ReqNF* means a nonfunctional (usually performance) requirement.

ReqNF2: Cookie shall raise the temperature of food in the oven so that temperatures at two distinct locations in the food differ by less than 10%.

ReqNF3: Cookie shall update the remaining baking time display every minute.

ReqNF4: The audible signal emitted by Cookie shall have an intensity level of 80 ± 2 decibels (dB) at a distance of 30 cm and a frequency of 440 Hz. Note: "Goalposts" like this, where all values inside the limits are accepted and all values outside the limits are rejected, are no longer fashionable for requirements since Taguchi (see [48]) scoring functions like that shown in Fig. 6 are preferable ([8], pp. 386–389).

ReqNF5: Cookie shall comply with section 1030 of Title 21, Food and Drugs, Chapter I – Food and Drug Administration, Department of Health and Human Services, Subchapter J: Radiological Health.

ReqNF6: The desired baking time shall be adjustable between 1 min and 10 h.

    Author/owner: Hungry Student
    Last changed: January 5, 2021

Requirements must be necessary, verifiable, unambiguous, etc. Bahill and Madni ([8], pp. 379–386) list, with explanations, 28 such characteristics of good requirements.

A Test Plan for This System

An important part of the incipient system design is describing how the system will be tested. Fortunately, with use cases, this is a simple task. To test means to apply inputs, measure and record outputs, compare outputs to requirements, and finally indicate passing status.

This test plan is based on the main success scenario of the Bake My Cookies use case.

1. Tester turns on the oven and sets the desired temperature to 350 °F.
2. Tester waits until Cookie signals that the oven has preheated to 350 °F.
3. Tester stands 5 ft from the oven and observes the visual display. He measures the sound intensity and the frequency of the auditory signal from a distance of 30 cm. He measures the actual temperature inside the oven. He records the results.
4. Tester sets the timer for 10 min.
5. Tester waits until Cookie signals that 10 min is over.
6. Tester stands 5 ft from the oven and observes the visual display. He or she measures the sound intensity and the frequency of the auditory signal from a distance of 30 cm. He or she measures the actual temperature inside the oven. He or she notes the desired and actual elapsed time (10 min) and records all of the results.
7. Tester turns the oven off.
8. Tester notes that the oven temperature is decreasing (end of test).

This series of steps can easily be converted into an activity diagram or a sequence diagram. Some chapters in this handbook may skip the text and go directly to an activity diagram or a sequence diagram.

This section has presented a high-level or front-end process for discovering system requirements. It has not shown how to document requirements at a low level: For this, relational databases or SysML diagrams can be used. Friedenthal et al. [25] and Madni and Sievers [41] show examples of using Use Case Diagrams (uc), Requirements Diagrams (req), Sequence Diagrams (sd), Activity Diagrams (act), Block Definition Diagrams (bdd), Package Diagrams (pkg), and requirements tables and matrices to show the implementation of requirements documentation at a low level.

Tradeoff Study Process

When a decision is important, a formal tradeoff study may be in order [22, 31]. Decisions that may require formal tradeoff studies include bid/no-bid, make-reuse-buy, formal inspection versus checklist inspection, tool and vendor selection, incipient architectural design, hiring and promotions, and helping your customer to choose a solution from among various alternatives.

A tradeoff study is not something that is done once at the beginning of a project. Throughout a project, you are continually making tradeoffs such as creating team communication methods, selecting components, choosing implementation techniques, designing test plans, and maintaining the schedule. Many of these tradeoff decisions should be formally documented.

Companies should have criteria for when to do formal decision analysis, such as:

- When the decision is related to a moderate- or high-risk issue
- When the decision affects work products under configuration management
- When the result of the decision could cause significant schedule delays or cost overruns
- On material procurement of the 20% of the parts that constitute 80% of the total material costs
- When the decision is selecting one or a few alternatives from a list
- When a decision is related to major changes in work products that have been baselined
- When a decision affects the ability to achieve project objectives
- When the cost of the formal evaluation is reasonable when compared to the decision's impact
- On design-implementation decisions when technical performance failure may cause a catastrophic failure
- On decisions with the potential to significantly reduce design risk, engineering changes, cycle time, or production costs

Killer trades are used to eliminate a large number of possible alternatives in one fell swoop. When evaluating alternatives is expensive, then early in the tradeoff study, you should identify important requirements that can eliminate many alternatives. If these requirements are performance related, then they are called key performance parameters (KPP). These requirements produce killer criteria. Subsequent killer trades can often eliminate, if you are lucky, maybe 90% of the possible alternatives.

In the Cookie Acquisition System, a killer criterion is that the cookies must be chocolate chip. Gingerbread, oatmeal, bonbons, rum balls, animal crackers, biscotti, ladyfingers, macarons, etc. will not do. This eliminates maybe 99% of possible cookies.

Alternative solutions should be suggested by stories, use cases, and the concept of operations (ConOps). Everybody should suggest alternatives and criteria during brainstorming sessions and in private contemplation. It is important to get many alternative solutions and criteria and then eliminate most. Bizarre alternatives should suggest new requirements.

Figure 3 presents the tradeoff study process. The activities that have the most uncertainty are marked with a ▼. The first step has several sources of uncertainty. It is inevitably the case that the initial problem statement is imprecise and the tradeoff space (alternative solutions) initially defined is incomplete. These sources of uncertainty need to be addressed before proceeding with tradeoff studies. Probing the statement of the problem, reformulating queries, and identifying new variables that need to be included in the tradeoff space are the means for reducing uncertainty. Thereafter, the steps are relatively straightforward.

In this chapter, we have assumed that the reader is familiar with the requirements discovery process, the tradeoff study process, and the risk analysis process. However, the tradeoff study process has a few subtleties that some readers may not be cognizant of. Other chapters in this handbook on MBSE do not recognize these

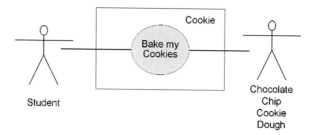

Fig. 3 A use case diagram for the Cookie Acquisition System. The chocolate chip cookie dough is shown as a secondary actor. More formally, the refrigerator might be the secondary actor, and it stores the chocolate chip cookie dough until it is requested

subtleties in doing tradeoff studies. So we offer a simplified example of a tradeoff study in this next section.

Tradeoff Study Example from a Chocolate Chip Cookie Acquisition System

Imagine that while reading this chapter, you experience an irresistible urge for chocolate chip cookies and a glass of milk. You can simply state this as, "I want chocolate chip cookies." You begin to explore how to get hold of chocolate chip cookies to satisfy this urge. You quickly discover that there are no chocolate chip cookies in your home. You do have yogurt, but that does not help. You begin to explore your options. You could head over to a bakery and buy chocolate chip cookies. But wait! That would cut into valuable study time. You simply cannot afford to do that! How about having a pizza delivered instead? No! You want *chocolate chip cookies*. Frantically, you start rummaging the kitchen. Lo and behold, you find a tube of chocolate chip cookie dough in your refrigerator! You are going to *make* chocolate chip cookies! However, is that the best alternative? Perhaps you need to do a tradeoff study.

The following is a tiny excerpt of a tradeoff study for the Chocolate Chip Cookie Acquisition System. The complete example is given in Bahill and Madni [8], pp. 15–30, 469–474, and 616–618). The question that this tradeoff study will answer is, "What is the best way for our student to get chocolate chip cookies?" First, is formal evaluation necessary? Our customer, the student, says that this is important. Therefore, we *will* do a tradeoff study. We will use the following three evaluation criteria which are derived from use case descriptions. Evaluation criteria are often called measures of effectiveness.

Name of criterion: Audible signal indicating cookies are ready
Description: An audible signal shall indicate when the cookies are ready. This signal should have a nominal intensity level of 80 ± 2 decibels.
Weight of importance: 9
Basic measure: Intensity level of an audible signal

Measurement method: During design and construction, the proposed device will be mounted on a test bench and will be activated following test instructions. The sound intensity level in decibels (dB) will be measured at a distance of 30 cm. At the final test and during operation, an actual oven will activate the audible signal, and the sound intensity level in decibels will be measured at a distance of 30 cm.

Units: Decibels (dB)

Trace to functional requirement, ReqF5, in the Bake My Cookies use case ([8], pp. 19–21)

Owner: Engineering

Date of last change: 12/25/2020

Name of criterion: Lost study time

Description: While the Student is making cookies, driving to the bakery to buy cookies, or bargaining with his mother to get her to make cookies for him, he will not be studying. This lost study time is the criterion. **Comment:** For optimal learning, students do need breaks.

Weight of importance: 7

Basic measure: Amount of study time lost in getting the cookies

Measurement method: During design, this lost study time will be calculated by analysis. At the final test and during operation, this lost time will be measured.

Units: Minutes

Scoring function: This criterion requires a scoring function that changes "lost study time" into a "more is better" situation.

Trace to the concept of operations

Owner: Student

Date of last change: 7/8/2020

Name of criterion: Nutrition

Description: Four cookies (2 oz) should contain less than 520 calories, 24 g of fat, and 72 g of carbohydrates.

Weight of importance: 5

Basic measure: Calories, grams of fat, and grams of carbohydrates

Measurement method: Use data from the Internet, for example, http://www.pillsbury.com/products/cookies/refrigerated-cookies/chocolate-chip.

Units: Calories, grams of fat, and grams of carbohydrates

Trace to the concept of operations

Owner: Student

Date of last change: 7/8/2020

Next, we consider these three alternatives:

- Ask your mother to bake cookies for you.
- Use a tube of Pillsbury refrigerated chocolate chip cookie dough.
- Go to the bakery to buy chocolate chip cookies.

Evaluation data (weights and scores) come from expert opinion and measurements during trips to a grocery store and a bakery.

The method for combining the data will be the sum of weighted scores combining function or simply the sum combining function. It is the simplest method for combining data.

The *sum combining function* is

$$f = \sum_{i=1}^{n} w_i \times x_i. \qquad (7)$$

In this equation, n is the number of evaluation criteria to be combined, x_i is the output of a scoring function (with values from 0 to 1) for the ith evaluation criterion, and w_i is the normalized weight of importance for the ith evaluation criterion. Weights of importance are expected to be normalized and vary from zero to one. The resulting scores are multiplied by their corresponding weights. The output of the function is the sum of the weight-times-score for each evaluation criterion. This sum combining function is commonly used, for example, when computing the grade point average of a student.

In a simple specific case, where there are only two evaluation criteria,

$$\text{named } y \text{ and } z \text{ with } n = 2 \text{ then}$$
$$f = w_y y + w_z z. \qquad (8)$$

This function is used when the evaluation criteria show perfect compensation, that is, when both criteria contribute to the result and when more of y and less of z is just as good as less of y and more of z. Stated formally, the sum combining function is appropriate when the decision-makers' preferences satisfy additive independence, which is the case for most industry examples that we have seen.

The question to be answered by the tradeoff study is, "What is the best way for our student to get chocolate chip cookies?"

These evaluation criteria and alternatives were put into a spreadsheet as shown in Table 3. We think the organization is clear enough that we do not have to explain each cell. The exceptions are rolling up the subcriteria into the criteria. So we will explain one of those.

The following numbers are in the blue-shaded cells of Table 3. Calories for Mom's cookies were given a relatively high score of 0.7 (more is better) because Mom won't make unhealthy cookies. The subcriteria calories was given a weight of 9, which became a normalized weight of 0.47. This weight was then multiplied by the score of 0.7 to produce a product of 0.33. Fat and carbohydrates were treated similarly to give numbers of 0.11 and 0.19. Those three numbers were added together to give the criteria nutrition a score of 0.63. This score was multiplied by its normalized weight of 0.24 to give a final score of 0.15.

This tradeoff study shows that Mom's cookies are the preferred alternative. The do-nothing alternative is ranked high, which is worrisome. This probably happened because we do not have any performance criteria, such as Anticipated Tastiness.

7 Overarching Process for Systems Engineering and Design

Table 3 Tradeoff study matrix for the Chocolate Chip Cookie Acquisition System

Hierarchal tradeoff study matrix for the Cookie Acquisition System

Evaluation criteria	Weight of importance	Normalized evaluation criteria weights	Subcriteria weight of importance	Normalized subcriteria weights	Alt 1 Do nothing sc	Alt 1 w × sc	Alt 1 w × sc	Alt 2 Ask Mom sc	Alt 2 w × sc	Alt 2 w × sc	Alt 3 Pillsbury's Chocolate Chip Cookie Dough sc	Alt 3 w × sc	Alt 3 w × sc	Alt 4 Buy cookies at bakery sc	Alt 4 w × sc	Alt 4 w × sc
Audible signal for cookies are ready	9	0.43			0.00		0.00	0.65		0.28	0.50		0.21	0.40		0.17
Lost study time	7	0.33			1.00		0.33	0.80		0.27	0.50		0.17	0.30		0.10
Nutrition	5	0.24			1.00		0.24	0.63		0.15	0.48		0.11	0.35		0.08
Calories			9	0.47	1.00	0.47		0.70	0.33		0.50	0.24		0.40	0.19	
Fat, grams			4	0.21	1.00	0.21		0.50	0.11		0.40	0.08		0.30	0.06	
Carbohydrates, grams			6	0.32	1.00	0.32		0.60	0.19		0.50	0.16		0.30	0.09	
Alternative ratings							0.57			0.70			0.50			0.35
Column sum		1.00		1.00												

These are the abbreviations used in this table: sc stands for the score, wt stands for weight, × is the multiplication sign, norm means normalized, and Alt means alternative

This study was criticized because (1) of its lack of performance and cost evaluation criteria, (2) the customer's desire for milk was ignored, (3) nothing was stated about being healthy! That would have posed an interesting contradiction in the analysis!

This tradeoff study was reviewed by the authors of this chapter and was stored in the process assets library (PAL). Some chapters in this handbook call this the model repository.

Risk Analysis Process

Figure 5 presents the risk analysis process. Activities with the highest uncertainty are marked with a ▼. Once again, preparing for risk requires the identification of risk events. Risks are explored by the risk analyst working with domain experts using storytelling, use cases, mental exploration, and envisioning methods to identify risk events. (This, of course, is not a complete list of the types of analyses used in risk assessment. There are many whole books filled with risk assessment techniques and disaster analyses. Most of the MBSE papers in this handbook devote pages to such analyses. An interesting problem with typical statistical analyses used in risk analyses is that the results may point to incorrect correlations that throw off risk determination. One use for a good model is that it could be useful in poking at causality which makes risk predictions more reliable. Furthermore, likelihood is almost always wrong – it may be high or low but generally wrong (Mike Sievers, personal communication).) Once a comprehensive set of risk events is identified, the risk analysis process proceeds following the steps laid out in Fig. 5.

Risk Analysis Example from a Chocolate Chip Cookie Making System

Risk management starts with the incipient system design. In this section, we (1) construct a risk register identifying and evaluating the major risks, (2) identify early tests or other measures that could be used to mitigate at least some of the major risks, (3) describe a few potential unintended consequences of this system, and (4) describe a few BiST for this system.

Table 4 shows some of the risks to the system and its primary actor. Likelihood expresses our feelings about how likely this failure event is, on a scale of 0 to 1. Severity expresses our feelings about how severe this failure event is, on a scale of 0 to 1. Risk is the product of likelihood and severity.

The biggest risk is incorrect ingredient quantities, which is why people prefer to use a tube of Pillsbury's Chocolate Chip Cookie dough instead of starting from scratch. Next, we identify the most likely outcome and the most severe outcome. Gaining weight is the most likely event that needs to be mitigated, if convenient. A fire in the oven is the most severe event; therefore, we need to keep an eye on the oven.

7 Overarching Process for Systems Engineering and Design

Table 4 A risk register

Risk no	Potential failure event	Consequences	Likelihood per system usage	Severity	Risk
1	Incorrect ingredient quantities when baking from scratch	Off-taste; bad consistency	0.09	0.5	**0.045**
2	Product is out of date	Possible illness	0.002	0.9	0.018
3	Oven temperature too high	Burned cookies	0.009	0.09	0.0008
4	Baking time too long	Burned cookies	0.009	0.09	0.0008
5	Student could gain weight	Student would be unhappy	0.1	0.1	0.01
6	Failure to turn oven off	The oven could start a fire	0.01	0.95	0.0095
7	Student does not wear an oven mitt and burns his or her hand	Student is unhappy	0.04	0.8	0.032

Risk is defined as likelihood times severity

Now, we should identify early tests or other measures that could be used to mitigate the major risks.

Possible early mitigation measures include:

1. Check "use by" dates on ingredients.
2. Use an independent (not a part of the oven) thermometer.
3. Use an independent (not a part of the oven) timer.
4. Cutting back on the quantity or quality of the butter or sugar is not a reasonable mitigation strategy.
5. Build in a governor that will restrict the maximum oven temperature to 550 °F.

It is important to also identify potential unintended consequences of the system. Two possible negative unintended consequences are (1) the aroma could attract undesirable neighbors to crash into your home and (2) heat generated by the oven could heat the house (particularly undesirable in the summer). This could ultimately contribute to global warming!

All systems should have Built-in Self-Test (BiST). Amazingly, a simple oven system has at least five BiSTs: (1) The clock indicates that the oven is connected to electricity, (2) the oven-on light indicates that the function select knob is operable, (3) the preheated indicator shows that the oven heats up, (4) an internal temperature probe indicates that the oven is functioning, and (5) the Student being vigilant by looking for smoke and smelling for cookies burning protects against catastrophic failures.

Figure 5 presented the risk analysis process. Risk events are identified by the risk analyst working with domain experts using storytelling and use cases. Once a comprehensive set of risk events is identified, they are prioritized and stored in the risk register.

Comparing the Requirements, Tradeoff, and Risk Processes

As noted earlier, requirements discovery, tradeoff studies, and risk analyses employ the same underlying structure in their processes. They appear different because they employ different vocabularies, inputs, and outputs. In this section, we explore the commonalities among these three processes. These are the actual phrases that were used in Figs. 2, 4, and 5. The processes that are shown in these figures were developed years before the Overarching Process was conceived.

Comparing the *Activities* of the Requirements, Tradeoff, and Risk Processes

Table 5 compares the requirements discovery, tradeoff studies, and risk analysis processes in terms of the specific vocabularies used in each activity. It is important to note that we did not expect a one-to-one mapping between requirements discovery, tradeoff studies, and risk analysis because they are distinctly different processes employed in engineering design and systems engineering. However, it is striking how closely they correspond to each other from the perspective of process structure. The comparison of these three well-established processes has an added benefit — it facilitates process improvement. For example, comparing the rows in Table 5 might suggest that the requirements process should include an activity for expert review and that all processes should start with an understanding of customer's needs. In other words, activities in each row can be evaluated in terms of their relevance (or not) for the other processes. This activity is intended to introduce consistency and uniformity into the three processes, making it easier to address uncertainty sources and uncertainty handling mechanisms in a consistent, uniform way.

Many chapters in this handbook do not use the verify relationship. In MBSE, verification methods "satisfy" a requirement or possibly "trace" to a requirement but importantly are maintained separately. Consider that we modify a verification method in a real project. Doing that will mean a number of steps including stakeholder reviews and actions by a configuration control board (CCB). Since verification methods typically come later in the development life cycle and not all at once, we are constantly control boarding the requirements document. Rather, we want the verification methods to be kept somewhere else and managed separately so that we can make changes as needed without changing requirements. Another point is that the verification method does not add value for designers, especially since there are usually separate teams dealing with requirements and verification and validation (personal communication Mike Sievers, 2021).

Table 6 shows the arguably most important row of Table 5.

7 Overarching Process for Systems Engineering and Design

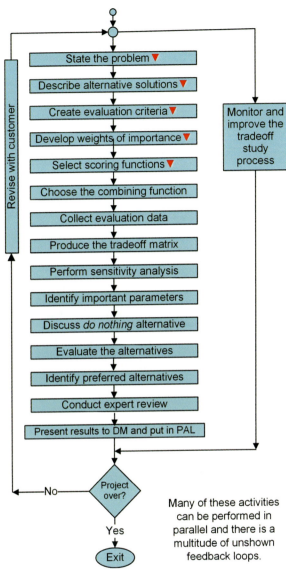

Fig. 4 Tradeoff study process. (© 2017, Bahill and Madni. Used with permission)

Comparing the *Products* of the Requirements, Tradeoff, and Risk Processes

Next, we compare the key products of the requirements discovery, tradeoff studies, and risk analysis processes. These *products* appear to be quite different because they employ different vocabularies and structures.

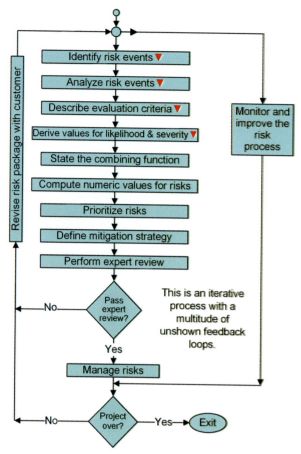

Fig. 5 Risk analysis process. (© 2017, Bahill and Madni. Used with permission)

In the following paragraphs, we discuss the foregoing within the context of a Chocolate Chip Cookie Acquisition System [8]. The Chocolate Chip Cookie Acquisition System is a system that allows a student on an afternoon study break to acquire chocolate chip cookies using one of several approaches. This illustrative example is used to convey the key terms associated with our three processes.

A Requirement from the Chocolate Chip Cookie Acquisition System

Table 7 presents a requirement from the Chocolate Chip Cookie Acquisition System.

7 Overarching Process for Systems Engineering and Design

Table 5 Comparison of the vocabularies used in the activities of these three processes

Terms used in the requirements discovery process, Fig. 2	Terms used in the tradeoff studies process, Fig. 4	Terms used in the risk analysis process, Fig. 5
Identify stakeholders, understand customer needs, state the problem, and develop use case models	State the problem	Identify risk events
Discover requirements	Describe alternative solutions	Analyze risk events
Define attributes	Create evaluation criteria	Derive values for likelihood and severity
Prioritize requirements	Develop weights of importance	Adjust the range of the criteria
Choose utility functions	Select scoring functions	Choose utility functions
Imply the Boolean AND function	Choose the combining function, usually the weighted sum	State the combining function, usually the product
	Collect evaluation data	
	Produce tradeoff matrix	
Identify cost drivers	Perform sensitivity analysis and identify important parameters	Perform sensitivity analysis
	Discuss the *do-nothing* alternative	Identify acceptable risks
	Evaluate the alternatives	Compute numerical values for risks
Clarify, decompose, allocate, and derive requirements		
Revise requirements with the customer	Revise with customer	Revise risk package with customer
		Manage risks
Prioritize the requirements set (find the most important requirements)	Identify preferred alternatives	Prioritize risks (find the greatest risks)
Test, verify, and validate		
Review with the customer and perform formal inspections	Conduct expert review	Perform expert review
	Track the marketplace for new alternatives	Manage risks
Manage requirements (put results in a requirements database)	Present results to decision-maker (DM) and put in PAL	Put results in the risk register
	Choose the combining function	Track outliers (both high frequency but low severity and low frequency but high severity)
Monitor and improve the requirements process	Monitor and improve the tradeoff study process	Monitor and improve the risk process

(continued)

Table 5 (continued)

Terms used in the requirements discovery process, Fig. 2	Terms used in the tradeoff studies process, Fig. 4	Terms used in the risk analysis process, Fig. 5
Inputs		
Requirements come from use cases and stakeholders	Alternatives, evaluation criteria, weights, and scores that come from use cases and the ConOps	Risks come from use cases and are identified by the risk analyst
Outputs		
Requirements specification	Preferred alternatives	Risk register

Table 6 This row from Table 5 might be the most important row

The term used in requirements discovery	The term used in tradeoff studies	The term used in risk analysis
Prioritize the requirements set (find the most important requirements)	Identify preferred alternatives	Prioritize risks (find the greatest risks)

An Evaluation Criterion from the Chocolate Chip Cookie Acquisition System

Name of criterion: Audible signal indicating cookies are ready

Description: An audible signal shall indicate when the cookies are ready. This signal should have a nominal intensity level of 80 ± 2 decibels.

Weight of importance 9

Basic measure: Intensity level of an audible signal

Measurement method: During design and construction, the proposed device is mounted on a test bench and is activated following test instructions. The sound intensity level in decibels (dB) is measured at a distance of 30 cm. At the final test and during operation, an actual oven activates the audible signal, and the sound intensity level in decibels is measured at a distance of 30 cm.

Units: Decibels (dB)

Scoring function input: The measured sound intensity will probably lie between 70 and 90 dB.

Scoring function: SSF5 (76, 78, 80, 82, 84, 0.5, −2, RLS (70–90)) ([54], pp. 385–397; [8], p. 470). Here, we used the mandatory requirement thresholds of 78 and 82 dB as the baseline values because we expect the values to improve through the design process.

Scoring function output: 0 to 1 (Fig. 6)

Trace to functional requirement, ReqF5, in the Bake My Cookies use case ([8], pp. 19–21).

Owner: Engineering

Date of last change: 12/25/2020

7 Overarching Process for Systems Engineering and Design

Table 7 A requirement from the Chocolate Chip Cookie Acquisition System

Attribute	Explanation
Identification tag (ID)	ReqNF4
Name	Audible signal for cookies are ready
Text	An audible signal shall indicate when the cookies are ready. This signal shall have an intensity level of 80 ± 2 decibels at a distance of 30 cm and a frequency of 440 Hz
Priority	9
Verification method	During design and construction, this requirement will be verified by test. The proposed device will be mounted on a test bench and will be activated per test instructions. The sound power level in decibels (dB) will be measured at a distance of 30 cm. At the final test and during operation, this requirement will be verified by demonstration. An actual oven will activate the audible signal, and the sound power level in decibels (dB) will be measured at a distance of 30 cm
Verification difficulty	It will be easy to verify this requirement
Refined by technical performance measure (TPM) [46]?	No
DeriveReqt:	This requirement refines ReqF5: Cookie shall emit an audible signal when the timer has elapsed
Owner	Pat the engineer
Date of last change	January 26, 2021

A Risk from the Chocolate Chip Cookie Acquisition System

Failure event: Audible signal for "cookies are ready" is too loud.
   Potential effects: Someone's hearing could be damaged.
   Relative likelihood: Noise-induced hearing loss affects 10% of Americans. So we assess this likelihood at 0.1.
   Severity of consequences: 1.0
   Estimated risk: 0.1
   Priority: This should be described with a risk management chart.
   Mitigation method: During design, the proposed device is mounted on a test bench and activated following test instructions. The sound intensity level in decibels (dB) is measured at a distance of 30 cm. The device is put under configuration control to ensure that it is not replaced or altered. This design is conservative in that the device should produce only 80 dB. Exposure to 120 dB or less for only a few seconds is unlikely to cause permanent hearing loss.
   Status: Active
   Trace to: ConOps
   Assigned to: Pat the engineer
   Date: Tracking started on April 1, 2020.

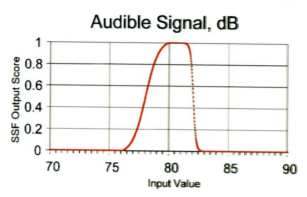

Fig. 6 The audible signal scoring function of Chocolate Chip Cookie Acquisition System

Comparing a Requirement, an Evaluation Criterion, and a Risk

Table 8 compares the three descriptions of the auditory output given above. Once again, it is not surprising that these products are similar because they come from similar processes. However, we might compare these three products and determine whether changes to the templates for each of them are warranted.

For example, the tradeoff study process requires scoring functions ([54], pp. 385–397; [8], pp. 246–258). Should the requirements process and the risk process also *require* scoring functions? Some requirements have scoring functions, but we do not want to require scoring functions for *all* requirements. We have not found risk analyses that used scoring functions.

The risk analysis process has an attribute named *status*. This attribute should change frequently as the design develops. On the other hand, requirements are not likely to come and go like risks. But it would be easy to add a column in the requirements database to cover this possibility. In a tradeoff study, it is not likely that evaluation criteria would come and go. But the whole tradeoff study could have status, and possible values would be under construction, gathering data, alternatives being evaluated, and the decision has been made.

Requirements have an attribute listing the difficulty of satisfying and verifying the requirement. Should the tradeoff study process and the risk process also have such an attribute? First, we do not think that the final tradeoff study should. However, it might be useful to state confidence in the results. This difficulty typically stems from uncertainty in the measurements and evaluation criteria and qualitatively from how different the options are. For example, is therapy better than drugs to treat depression? These are very different alternatives, and thus, it is hard to do a fair comparison. Your confidence in the result should be low. Second, one school of risk analysis has three columns in their risk tables: relative likelihood, severity of consequences, and *difficulty of detection*. We did not use difficulty of detecting the failure event because we found that it added complexity without comparable added value.

Both the requirements process and the tradeoff study process have an attribute to trace where the item came from, for example, from a particular use case or review. It

7 Overarching Process for Systems Engineering and Design

Table 8 Comparison of a requirement, an evaluation criterion, and a risk for the Chocolate Chip Cookie Acquisition System

Requirement	Evaluation criterion for a tradeoff study	Risk
Name: Audible signal indicating cookies are ready	Name of criterion: Audible signal indicating cookies are ready	Failure event: Audible signal indicating "cookies are ready" is too loud
Text: This audible signal shall have an intensity level of 80 ± 2 dB	Description: This audible signal should have an intensity level of 80 ± 2 dB	Potential effects: Someone's hearing could be damaged
Priority: 9	Weight of importance: 9	Relative likelihood: 0.1 Severity of consequences: 1.0 Estimated risk: 0.1 Priority: High
Verification method: The sound power level in decibels (dB) will be measured at a distance of 30 cm	Measurement method: Sound intensity level in decibels (dB) is measured at a distance of 30 cm Units: dB	Mitigation method: Sound intensity level in decibels (dB) is measured at a distance of 30 cm
Difficulty: Easy to satisfy and verify		
	Scoring function SSF 5 (76, 78, 80, 82, 84, 0.5, −2, RLS (70–90)	
		Status: Active
Refined by TPM? No DeriveReqt: This requirement refines Functional requirement ReqF5 in the Bake My Cookies use case	Trace to functional requirement ReqF5 in the Bake My Cookies use case ([8], pp. 19–21)	Trace to ConOps
Owner: Engineer	Owner: Pat the engineer	Assigned to: Pat the engineer
Date of last change: 4/1/20	Date of last change: 12/25/20	Date of last change: 4/1/20

might be hard to implement, but this would also be nice for the risk process. For example, if the risk factor was derived from an FMEA or industrial experience, it could be marked as such.

This section compared the requirements discovery, tradeoff studies, and risk analyses processes. It did this by comparing the activities and products of these processes and by comparing example requirements, evaluation criteria, and risks.

The SIMILAR Process

The SIMILAR process of Fig. 7 is based on Bahill and Gissing [7]. It comprises seven key activities: **S**tate the problem, **I**nvestigate alternatives, **M**odel the system, **I**ntegrate, **L**aunch the system, **A**ssess performance, and **R**eevaluate. These seven activities are conveniently summarized using the acronym SIMILAR. We use this process to

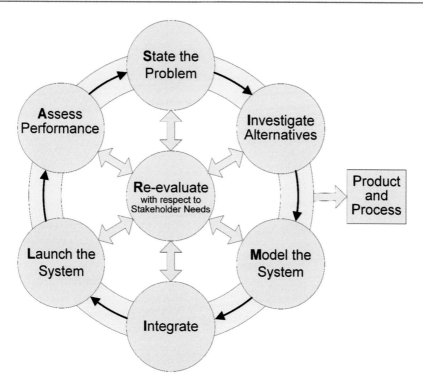

Fig. 7 The SIMILAR process. (Based on Bahill and Gissing [7])

provide the overall context for problem-solving during system design and every other human activity. At the outset, we want to clarify that the activities in the SIMILAR process are performed iteratively and in parallel with many unshown feedback loops. Each activity in the SIMILAR process is described next.

State the Problem

> "The beginning is the most important part of the work" Plato, *The Republic*, 4th century BC.
> "Begin at the beginning," the King said gravely, "and go on 'til you come to the end; then stop." From Lewis Carroll, *Alice's Adventures in Wonderland*.

The problem statement contains many tasks that are performed iteratively, many of which can be performed in parallel. We examined Figs. 2, 4, and 5 and studied Tables 5 and 8 and found the following listed tasks that fit into the problem statement activity.

7 Overarching Process for Systems Engineering and Design

- Understanding customer needs is the first and foremost task.
- Identify stakeholders such as end users, operators, maintainers, suppliers, acquirers, owners, customers, bill payers, regulatory agencies, affected individuals or organizations, victims, sponsors, manufacturers, etc. Our stories that explain how the system will work are a source for identifying stakeholders. Stories have villains. Some villains that the system designer might encounter are competitors, the IRS, the EPA, Mexican drug cartels, and their associated gangs like MS-13.
- Where do the inputs come from? Requirements come mainly from the use cases and the stakeholders. They are presented by the customer and the systems engineer. Evaluation criteria and proposed alternatives for tradeoff studies come from use cases, meetings, and reviews and are presented by the design engineer. Risk events are identified in use cases, brainstorming, meetings, and reviews and are described by the risk analyst (Fig. 8).

- Describe how the system works using stories and use case models. The use case models provide requirements and test cases.
- State the problem in terms of *what* needs to be done, not *how* it must be done. The problem statement may be in prose form or the form of a model.
- Develop the incipient architecture.
- Define the scope of the project. This shows the boundary between what is inside the system and the external world.
- Initiate risk analysis. Yes, the risk analysis of the system should begin at the same time as the requirements discovery and tradeoff study processes.

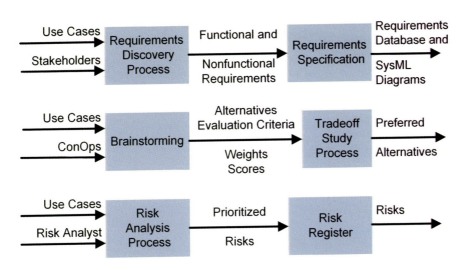

Fig. 8 Inputs and outputs

An interesting aside: By law, with minor room for excursions, the role of a US CEO has historically been defined as maximizing long-term (sic!) shareholder return. Recently, there has been a movement to redefine this requirement as "benefitting owners, employees, customers, communities, society, etc." This is going to pose a huge tradeoff problem (albeit perhaps an appropriate one) for CEOs and boards! It will at least assure lifetime employment for lawyers (Norm Augustine, personal communication).

Investigate Alternatives

We examined Figs. 2, 4, and 5 and studied Tables 5 and 8 and found the following listed tasks that fit into the Investigate Alternatives activity.

- One should investigate alternative requirements, designs, and risk events using evaluation criteria such as performance, cost, schedule, and risk.
- For quantitative analyses, identify attributes of requirements, evaluation criteria for tradeoff studies, and the likelihood of occurrence and severity of consequences for risk events. Assign them weights of importance to show priorities.
- Scoring (utility) functions are mandatory for tradeoff studies but are optional for requirements and risks.
- Select methods for combining the data. State the combining function that will be used. Usually, this will be the Boolean AND function for requirements, the sum of weighted products for tradeoff studies, and a chart or a matrix for risks.
- Finally, one must collect evaluation data and use it to assign values to attributes for requirements, weights and scores for tradeoff studies, and likelihoods and severities for risk analyses.

Model the System

We examined Figs. 2, 4, and 5 and studied Tables 5 and 8 and found the following listed tasks that fit into the Model the System activity.

- Models are typically created for most requirements, alternative designs, and risk events. These models are *consistently elaborated* ([54], pp. 178–180) (that is, expanded) throughout the system life cycle. A variety of models can be used.
- Requirements can be modeled with use case models, textual shall statements, tables, spreadsheets, and specialized databases. Friedenthal et al. [25] and Madni and Sievers [41] model requirements with Use Case Diagrams (uc), Requirements Diagrams (req), Sequence Diagrams (sd), Activity Diagrams (act), Block Definition Diagrams (bdd), and Package Diagrams (pkg). Subsequently, the requirements must be clarified, decomposed, allocated, and derived.
- Tradeoff studies are usually modeled with tradeoff matrices implemented with spreadsheets. The alternative designs within them are modeled with UML

diagrams, SysML diagrams, analytic equations, computer simulations, and mental models.
- Risks are modeled with tables containing values for the likelihood of occurrence and severity of consequences and figures displaying these data.
- Everything must be prioritized. The requirements set should be prioritized to find the most important requirements. For tradeoff studies, the preferred alternatives are identified with a tradeoff matrix. The ranges for likelihood and severity are adjusted for risk events to find the greatest risks.
- The results of a sensitivity analysis can be used to validate a model, flag unrealistic model behavior, point out important assumptions, help formulate model structure, simplify a model, suggest new experiments, guide future data collection efforts, suggest accuracy for calculating parameters, adjust numerical values of parameters, choose an operating point, allocate resources, detect critical evaluation criteria, suggest tolerance for manufacturing parts, and most importantly identify cost drivers.

Integrate

We examined Figs. 2, 4, and 5 and studied Tables 5 and 8 and found the following list of tasks that fit into the Integrate activity.

- Integration means bringing elements together so that they work as a whole to accomplish their intended purpose and deliver value. (A new systems engineering buzzword is emergent behavior. It suggests that, in terms of behavior, the result might be greater than the sum of its parts.) Specifically, systems, enterprises, and people need to be integrated to achieve desired outcomes. To this end, interfaces need to be designed between subsystems. Subsystems are typically defined along natural boundaries in a manner that minimizes the amount of information exchanged between the subsystems. Feedback loops between individual subsystems are easier to manage than feedback loops involving densely interconnected subsystems.
- Evaluation criteria should trace to requirements. Risks should trace to requirements or particular meetings or reviews. Requirements should refine higher-level requirements and should link to risks. Requirements and risks might be refined by technical performance measures (TPMs) [46]. TPMs are evaluated continually during the design process as a way of detecting and mitigating risk.

Launch the System

We examined Figs. 2, 4, and 5 and studied Tables 5 and 8 and found the following listed tasks that fit into the Launch the System activity.

- Launching the System means either deploying and running the actual system in the operational environment or exercising the model in a simulated environment to produce necessary outputs for evaluation. In a manufacturing environment, this might mean buying commercial off-the-shelf hardware and software, writing code, and/or bending metal. The purpose of system launch is to provide an environment that allows the system or its model to do what it is being designed to do.
- The outputs of these processes are a requirements specification, preferred alternatives, and the risk register. One should continually monitor the requirements (in the requirements database), alternative designs (in the process assets library, PAL), and risks (in the risk register) looking for possible changes and bring these to the attention of the decision-makers. One should continually monitor the marketplace looking for new requirements, products, designs, and risks and bring these to the attention of the decision-makers.

Assess Performance

We examined Figs. 2, 4, and 5 and studied Tables 5 and 8 and found the following listed tasks that fit into the Assess performance activity.

- Test, validation, and verification are important tasks for all processes.
- There should be regularly scheduled and performance-initiated expert reviews. The results of these reviews are presented to the decision-maker (DM) and are put in the process assets library (PAL).
- Evaluation criteria, measures, metrics, and TPMs are all used to quantify system performance. Evaluation criteria are used in requirements discovery, tradeoff studies, and risk analyses. Measures and metrics are used to help manage a company's processes. TPMs are used to mitigate risk during design and manufacturing.

Reevaluate

The distinction between an engineer and a mathematician is arguably the use of feedback in design. For two and a half centuries, engineers have used feedback to control systems and improve performance. It is one of the most fundamental engineering concepts. Reevaluation is a continual feedback process with multiple parallel loops. Reevaluation means observing outputs and using this information to modify the inputs, the system, the product, and/or the process.

The SIMILAR process (Fig. 7) shows the distributed nature of the reevaluate function in the feedback loops. However, it is important to realize that not all loops will always come into play all of the time. The loops that are used depend on the problem to be solved and the problem context. Reevaluation includes formal inspections, expert reviews, and reviews with the customer.

A very important and often neglected task in any process is monitoring and improving the process itself. This self-improvement process is shown explicitly in Figs. 2, 4, and 5 with the following tasks: Monitor and improve the requirements process, monitor and improve the tradeoff study process, and monitor and improve the risk process. These processes use a different timescale than the mainline processes. For example, the monitor and improve the X process tasks run with timescales of months to years, whereas the mainline processes, like revise with the customer and prioritize X, run with timescales of days to weeks.

This section presented the SIMILAR process that was developed by Bahill and Gissing [7]. It has served as a general model for doing everything.

The Overarching Process

We used Figs. 2, 4, 5, and 7 and Tables 5 and 8 and created the Overarching Process of Fig. 9 that can be used for system design, requirements discovery, decision analysis and resolution, tradeoff studies, and risk analysis. The processes in these figures were created many years before the Overarching Process was conceived. This Overarching Process is a superset of the three processes: requirements, tradeoffs, and risks. Viewed another way, each of the three processes is a tailoring of the Overarching Process [32]. A new feature in the Overarching Process is identifying and handling uncertainty. In Fig. 9, we have marked with a ▼ those activities that have the most uncertainty.

It is important to note that this is not a waterfall process. One activity does not have to wait for the previous activity to end before it can start. Also, it is an iterative process with a multitude of unshown feedback loops.

In the Investigate Alternatives block, it looks like discover requirements and describe alternative solutions are performed in parallel. Well, some of them are. However, this whole figure is very iterative. The systems engineer gets a few requirements and then gets a few alternatives and criteria. Then he gets a few more requirements and a few more alternatives and criteria, etc.

Some tasks listed in Figs. 2, 3, and 4 might seem to be missing in the Overarching Process of Fig. 9. However, these tasks have been subsumed in the activities shown. For example, the task of "examining the shape of the data" is included in the "choose combining function activity." The task of studying the do-nothing alternative is in the Create Tradeoff Matrix and Evaluate Alternatives activity. The identify important parameters task is in the perform a sensitivity analysis activity. The Develop Incipient Architecture task is included in the Describe Alternative Activities activity. Finally, the task to track outliers (both high frequency but low severity and low frequency but high severity) is in the Monitor and Manage risks activity.

Sometimes key stakeholders impose system-level constraints, requirements, goals, etc. The Overarching Process can include collecting and evaluating these before getting started with stating the problem. That is, we can have work before

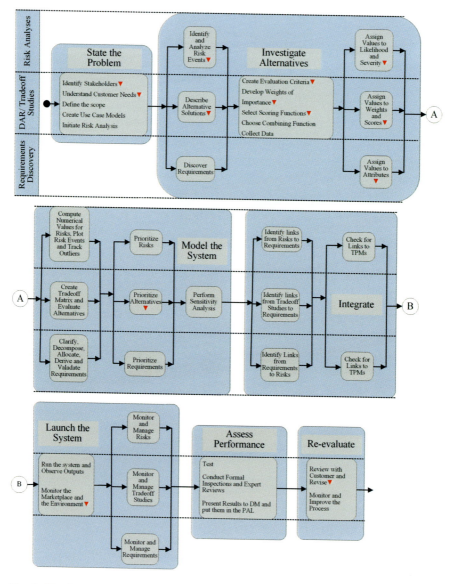

Fig. 9 The Overarching Process. DAR is decision analysis and resolution. DM is the decision-maker, PAL is the process assets library, and TPM is a technical performance measure

Fig. 9 the goal of which is to establish feasibility. When no solution is evident, then there must be iterations that add or eliminate the top-level desires so that the option space is increased. When there is no set of options agreeable to the primary stakeholders, then we do not continue to the next phase (personal communication Mike Sievers, 2021).

7 Overarching Process for Systems Engineering and Design

Effects of Human Decision-Making on the Overarching Process

The study of human decision-making reveals that the presence of cognitive biases can never be ruled out [43, 52]. This is also the contention of the economic school of heuristics and biases, which produced Prospect Theory [29], a theory that describes how people respond to choices under risk and uncertainty. Innate human biases, and external circumstances, such as the framing or the context of a question, can compromise decisions. It is important to note that subjects maintain a strong sense that they are acting rationally even when they are exhibiting these biases [28].

Other chapters in this handbook do not show MBSE processes, diagrams, or viewpoints that help ameliorate human biases in decision-making. So this section is unique in this handbook.

In Fig. 9, we have marked with a ▼ the actions that are the biggest contributors to uncertainty. They all deal with human decision-making rather than uncertainty in the weather, climate, solar variability, geology, political actions, or experimental data. In the upcoming paragraphs, we will explain how the activities of Fig. 9 are affected by uncertainty. Most reasons involve confirmation bias, severity amplifiers, and framing. Therefore, we will first discuss these three decision modifiers.

Confirmation Bias

Arguably, the most important cause of fallibility in human decision-making is confirmation bias. Humans hear what they want to hear and reject what they do not want to hear. Humans filter out information that contradicts their preconceived notions and remember things that reinforce their beliefs. Confirmation bias causes decision-makers to actively seek out and assign more weight to evidence that confirms their hypotheses and ignore or under weigh the evidence that could disconfirm their hypotheses. For example, mothers emphasize the good deeds of their children and de-emphasize their bad deeds. This is why we often hear the mother of a terrorist crying out, "My boy is innocent. He could never have killed all those people." People who think that they have perfect memory and perfect recall tend to ignore instances when they forgot something and tend to secure in long-term memory instances when they correctly recalled events and facts. Senior citizens often believe that they are good drivers despite tests that show that they have poor vision, fading cognitive processes, and slow reflexes. Thirty years ago, most cigarette smokers were in denial about the hazards of smoking. Some people say, "There must be a storm coming because my arthritic joints are hurting."

Social media is making this worse. Not only do you filter what you see and hear, but also Facebook filters what you are exposed to. They present to you things from the friends you care about. These friends are probably ideologically like you, which accentuates the filtering process.

Nickerson [44] reported many common instances of confirmation bias. In one, the subjects were given a triplet such as (2, 4, 6) and were asked to guess the rule that was used to generate the triplet and then try to prove or disprove that rule by giving examples. After each guess, they were told if they were right or wrong. For example, if the subject's mental model for the rule was "successive even numbers," they might guess (10, 12, 14) or (20, 22, 24), triplets that would confirm their mental model, but they would seldom guess (1, 3, 5) or (2, 4, 8), triplets that might disprove their mental model. He also presented another example of confirmation bias – witches.

The execution of 40,000 suspected witches in seventeenth century England is a particularly horrific case of confirmation bias functioning in an extreme way at the societal level. From the perspective of the people of the time, belief in witchcraft was perfectly natural, and sorcery was widely viewed as the reason for all ills and troubles that could not otherwise be explained. In one test of a woman being a witch, the mob tied the suspect to a chair and threw her into a river. If she floated, it was proof that she was a witch, and she was executed. If she sank, well, too bad.

Until the nineteenth century, physicians often did more harm than good because of confirmation bias. Virtually anything that could be dreamed up for the treatment of disease was tried and, once tried, lasted decades or even centuries before being given up. It was, in retrospect, the most frivolous and irresponsible kind of human experimentation. They used bloodletting, purging, infusions of plant extracts and solutions of metals, and every conceivable diet including total fasting. Most of these were based on no scientific evidence. How could such ineffective measures continue for decades or centuries without their ineffectiveness being discovered? Probably, because sometimes patients got better when they were treated, sometimes they did not, and sometimes they got better when they were not treated at all. Peoples' beliefs about the efficacy of specific treatments seem to have been influenced more strongly by those instances in which treatment was followed by recovery than by those instances in which there was no recovery. A tendency to focus on positive cases could explain why the discovery that diseases have a natural history and people often recover from them with or without treatment was not made until much later.

Most people react to news articles with confirmation bias. If a left-wing liberal reads a news story about a scientific study that showed how effective it was to give money to poor people, he might think, "That's an insightful article. I'll remember it." However, if one of those same people reads about a new study showing that giving people money when they are unemployed just makes their lives worse, then he might start looking for flaws in the study. If a person has a long-felt belief that the income gap between the rich and the poor in America is too large and is growing too fast, then a new study that challenges this belief might be met with hostility and resistance. However, if that person readily accepts a study that confirms his belief, then that is confirmation bias.

Before a person participates in an activity that involves evaluating requirements, alternatives, evaluation criteria, weights, scores, or risks, they should be reminded about confirmation bias. During the evaluation process, people should be on the lookout for instances of confirmation bias exhibited by other people and politely suggest that it might be influencing their evaluations.

Most people do not think like scientists: They think like lawyers. They form an opinion and then emphasize only evidence that backs up that opinion.

In filling out the tradeoff study matrix for the Cookie Acquisition System of Table 3, a physical fitness pundit might want to add more (possibly dependent) subcriteria such as protein, fiber, antioxidants, and unsaturated fats to the nutrition evaluation criterion. The system engineer must explain to the team why some people might want more subcriteria and the effects of adding dependent subcriteria.

Severity Amplifiers

Interpersonal variability in evaluating the seriousness of a situation depends on the circumstances surrounding the event. An evaluation may depend on factors such as how the criterion affects that person, whether that person voluntarily exposed himself to the risk, how well that person understands the alternative technologies, and the severity of the results. The following are severity amplifiers: lack of control, lack of choice, lack of trust, lack of warning, lack of understanding, being man-made, newness, dreadfulness, fear, personalization, ego, recallability, availability, representativeness, vividness, uncertainty, and immediacy.

The following paragraphs explain some severity amplifiers. *Lack of control*: A man may be less afraid of driving his car up a steep mountain road at 55 mph than having an autonomous vehicle drive him to school at 35 mph. *Lack of choice*: We are more afraid of risks that are imposed on us than those we take by choice. *Lack of trust*: We are less afraid while listening to the head of the Centers for Disease Control explaining anthrax than while listening to a politician explain it. *Lack of warning*: People dread earthquakes more than hurricanes because hurricanes give days of warning. People in California follow strict earthquake regulations in new construction. People in New Orleans seem to ignore the possibility of hurricanes. *Lack of understanding*: We are more afraid of ionizing radiation from a nuclear reactor than of infrared radiation from the sun. In the 1980s, engineers invented nuclear magnetic resonance imaging (NMRI). When the medical community adopted it, they renamed it magnetic resonance imaging (MRI). They dropped the adjective *nuclear* to make it sound friendlier. *Man-made*: We are more afraid of nuclear power accidents than solar radiation.

Newness: We are more afraid when a new disease (e.g., swine flu, SARS, MERS, Ebola, Zika, and COVID-19) first shows up in our area than after it has been around a few years. *Dreadfulness*: We are more afraid of dying in an airplane crash than of dying from heart disease. *Fear*: If a friend tells you that a six-foot rattlesnake struck at him, how long do you think the snake was? We suspect 3 ft. But of course, the length of the snake is irrelevant to the harm it could cause. It is only related to the fear it might induce. *Personalization*: A risk threatening us is worse than that same risk threatening you. *Ego:* A risk threatening our reputations is more serious than one threatening the environment.

Recallability: If something can be readily *recalled*, it must be more important than alternatives that are not as readily recalled. We are more afraid of cancer if a friend has recently died of cancer. We are more afraid of traffic accidents if we have just observed one. Recallability is often called *availability*. Something readily *available* to the mind must be more important than alternatives that are not as readily available. *Representativeness:* The degree to which an event is similar in essential characteristics to its parent population increases its importance. In the dice game of craps, rolling a seven would be typical of random rolls, and therefore, it would be representative of the parent population and would therefore be important. *Vividness of description*: An Edgar Allen Poe story read by Vincent Price will be scarier, than one that either of us reads to you. *Ambiguity or uncertainty:* Most people would rather hear their ophthalmologist say, "You have a detached retina. We will operate tonight" than "You might have a detaching vitreous, or it could be a detaching retina, or maybe its cancer. We will do some tests and let you know the results in a week." *Immediacy*: A famous astrophysicist was explaining a model for the life cycle of the universe. He said, "In a billion years, our sun will run out of fuel, and the earth will become a frozen rock." A man who was slightly dozing awoke suddenly, jumped up, and excitedly exclaimed, "What did you just say?" The astrophysicist repeated, "In a billion years, our sun will run out of fuel, and the earth will become a frozen rock." With a sigh of relief, the disturbed man said, "Oh, thank God. I thought you said in a *million* years."

In filling out the tradeoff study matrix for the Cookie Acquisition System of Table 3, *personalization* was an important severity amplifier. It created variability in the responses of the team making the evaluations. For example, a diabetic or someone on a strict diet will certainly want to give a much higher weight to the nutrition evaluation criterion and will give different weights to the nutrition subcriteria. Similarly, a dietitian or a nutritionist will also give different weights to the nutrition subcriteria. When these large variabilities occur, the systems engineer should explain possible causes to the team.

Framing

In the human decision-making community, *utility* is a subjective measure of happiness, satisfaction, or reward a person gains (or loses) from receiving a good or service. Utility is considered not in an absolute sense (from zero), but subjectively from a reference point, established by the decision-maker's (DM) perspective and wealth before the decision, which is his frame of reference [28]. (Kahneman and Tversky's [29] utility functions show a human's subjective utility as a function of its objective value ([8], pp. 167–176). Economists use utility functions to show consumer preference of one product over another and assign a numerical value to that preference. Systems engineers use utility functions in tradeoff studies to relate different evaluation criteria that use different units of measure. Despite the incompatibility of the measures, these diverse evaluation criteria (measures of effectiveness) may nevertheless contribute to the same overall goal. Utility functions therefore convert the different input evaluation criteria (physical characteristics) into output quantities (called utility values) which are mutually and completely compatible. Wymore's scoring functions ([8], pp. 246–257) are eloquent mathematical elaborations of such utility functions.) Framing (the context of a question) could affect his decision. The section on severity amplifiers stated that interpersonal variability in evaluating the seriousness of a situation depends on framing. That is, the circumstances surrounding the event will affect how a DM responds to it. An evaluation may depend on factors such as how the criterion affects that DM, whether that DM voluntarily exposed himself to the risk, how well that DM understands the alternative technologies, and the severity of the results. In the previous section, we gave over a dozen severity amplifiers that would affect the framing of a problem.

In contrast to defining framing in passing, as we have done so far, we will now explain *framing* directly, based on Beach and Connolly [12]. The DM has a vision, a mission, values, morals, ethics, beliefs, evaluation criteria, and standards for how things should be and how people ought to behave. Collectively, these are called *principles*. They are what the DM, the group, or the organization stands for. They limit the goals that are worthy of pursuing and acceptable ways of pursuing these goals. These principles are difficult to articulate, but they powerfully influence the DM's behavior. They are the foundation of the DM's decisions and goals; actions that contradict them will be unacceptable. The utility of the outcomes of decisions derives from the degree to which these decisions conform to and enhance the DM's preconceived principles.

Goals are what the DM wants to accomplish. The goals are dictated by the principles, the problem, the problem statement, opportunities, desires, competitive issues, or gaps encountered in the environment. Goals might seed more principles. Goals should be SMART: specific, measurable, achievable, realistic, and time-bound.

The DM has *plans* for implementing the goals. Each goal has an accompanying plan. Each plan has two aspects: (1) Tactics are the concrete behavioral aspects that deal with local environmental conditions, and (2) forecasts are the anticipation of the future that provides a scenario for forecasting what might result if the tactics are successful. The plans for the various goals must be coordinated so that they do not interfere with each other and so that the DM can maintain an orderly pursuit of the goals. The plans are also fed back to the principles; therefore, they might foment more principles.

Framing means embedding observed events into a context that gives them meaning. Events do not occur in isolation; the DM usually has an idea about what led up to them. This knowledge supplies the context, the ongoing story that gives coherence to experiences, without which things would appear random and unrelated. A frame consists of the principles, goals, and plans that are deemed relevant to the decision at hand and that fixes the set of principles that influence that decision.

The DM uses contextual information to probe his or her memory. If the probe locates a contextual memory that has similar features to the current context, then the current context is said to be recognized. *Recognition* defines which principles, goals, and plans are relevant to the current context and provides information about the goals and plans that were previously pursued in this context. If a similar goal is being pursued this time, then the plan that was used before may be used again.

In summary, framing means describing all aspects of the problem, the problem statement, and the DM's mind that will affect decisions.

In filling out the tradeoff study matrix for the Cookie Acquisition System of Table 3, *framing* was important. The student's frame of mind includes his or her present grade in the class and the scheduled occurrence or not of an exam the next day. These will affect his or her weights and scores for the lost study time evaluation criterion. The tradeoff study team must be aware of their teammates' frames of mind.

Because of these human mental mistakes, and many more [14, 52], weights and scores in tradeoff studies and other values that depend on human judgments are subjective and have large variations. This is the reason for performing sensitivity analyses: to identify simple judgments that have a large effect on the outcome.

The Overarching Process

Figure 9 shows a diagram for the Overarching Process. We have marked with a ▼ those activities that are the biggest contributors to uncertainty. We will now examine these activities. Specifically, we identify human psychological factors that can adversely influence human decision-making when dealing with uncertainty. Here, we only give short phrases listing these factors. Three are described in detail above. The others are explained in detail in Smith et al. [52] and Bohlman and Bahill [14].

- State the problem. This activity tends to be affected by severity amplifiers and framing. Additionally, it is affected by incorrect phrasing, attribute substitution, political correctness, and feeling invincible.
- Identify stakeholders. This activity is affected by framing.
- Understand customer needs. This activity is affected by confirmation bias, severity amplifiers, and framing.
- Define the scope, which is given in a high-level use case.
- Create use case models.
- Initiate risk analysis.
- Investigate alternative solutions. This activity is affected by confirmation bias, severity amplifiers, and framing.
- Identify and analyze risk events. This activity is affected by confirmation bias and severity amplifiers.
- Create evaluation criteria. This activity is affected by severity amplifiers. Additionally, it is affected by dependent evaluation criteria, relying on personal experience, the Forer Effect, and attribute substitution.
- Develop weights of importance. This activity tends to be affected by severity amplifiers. Additionally, it can be affected by whether the weights are the result of choice or calculation.
- Select scoring functions. Mistakes here include mixing gains and losses, not using scoring functions, and anchoring. The biggest mistake is stating output scores with false precision.
- Choose combining functions. Lack of knowledge is the key problem in this activity. There are several appropriate combining functions. One of the oldest and most studied means for combining data under uncertainty is the certainty

factor calculus employed by the Mycin expert system at Stanford University in the 1980s [16]. It is now called the sum combining function.
- Assign values to (1) attributes, (2) weights and scores, and (3) likelihood and severity. All three of these activities can be adversely affected by confirmation bias, severity amplifiers, relying on personal experience, magnitude and reliability, and judging probabilities poorly. In addition to these human decision-making errors, we also have metrology errors. No measurement is exact: You can always do better (that is, until we get to subatomic particles). Therefore, you decide how much uncertainty you will allow in your measurements and budget appropriately.
  - Our understanding of the laws of physics is not accurate. This precludes precise models for nature. Likewise, our models for dark energy and dark matter are inaccurate. But these physics problems do not manifest in most of our design problems.
  - If models are being used to compute values for risks, evaluation criteria, etc., then statistical measures like mean, standard deviation, and correlation coefficients can be used to identify the range of expected values. For some problems, for example, calculating the time for an asteroid to impact the earth, the best we can do is give uncertainty ranges.
- Prioritize alternatives. This activity is affected by confirmation bias, severity amplifiers, and framing. This activity can be degraded by serial consideration of alternatives, isolated or juxtaposed alternatives, conflicting evaluation criteria, adding alternatives, maintaining the status quo, and uneven level of detail. The order in which the alternatives are listed has a big effect on the values that humans give for the evaluation data. Therefore, a tradeoff study matrix should be filled out row by row with the status quo being the alternative in the first column. This makes the evaluation data for the status quo the anchors needed for estimating the evaluation data for the other alternatives. This is a good choice because the anchoring alternative is known and is consistent, and you have control over it. Prioritization also depends on the algorithm being used to combine the data.
- Perform sensitivity analyses. Done right, there should be no problems. Otherwise, lack of training and the Hawthorne effect can potentially confound the study.
- Monitor the marketplace and the environment. This activity is typically affected by severity amplifiers. Additionally, tunnel vision can throw off the analysis. Therefore, to avoid tunnel vision, the environment must be a part of the framing.
- Conduct formal inspections [24] and expert reviews. These inspections and reviews are done entirely by humans. Therefore, every human limitation such as cognitive biases, misconceptions, and preconceptions must be addressed.
- Review with stakeholders and revise. The most common mistake in design projects is failing to engage stakeholders and consult with experts in universities and local industries [14]. It is imperative to engage all stakeholders, especially in upfront engineering, to avoid the likelihood of extraneous design iterations and rework.
- The out arrow at the lower right feeds back to all of the boxes in Fig. 9.

Most of these areas of uncertainty involved human decision-making, and our models for this are imprecise. Overall, the biggest cause of uncertainty is simply that we cannot predict the future. A total reversal of the Earth's magnetic field is imminent, but we cannot predict when it will occur.

Uncertainty in Stating the Problem for the Overarching Process

Now that we have *identified* sources of uncertainty in the Overarching Process, we will present examples of some techniques for *handling* uncertainty in the tradeoff study process. The first and most important step in performing a tradeoff study is stating the problem. Uncertainty can cause mistakes in the problem statement. This section is based on Diogenes [1]. These are some of the tasks that were described in the state the problem paragraph of the SIMILAR process section of this chapter:

- Using stories and use cases, explain what the system is supposed to do.
- Understand customer needs.
- Identify stakeholders.
- Discover the inputs and their sources.

At the beginning of any system design, we do not know exactly what the finished product will do. Functions, requirements, and desirements may have been stated, but incomplete understanding, mistakes, unknown technology, and improvement opportunities usually change the preconceived functioning of any system. To understand and explain what the system is supposed to do and how it works, we use a multitude of stories and use case models. An example of a use case model for handling uncertainty is coming up shortly.

It turns out that all of these activities involve human decision-making. Therefore, most of the mistakes caused by uncertainty will be found in the system models and documentation.

Understanding customer needs, identifying stakeholders, and discovering the system inputs are all affected by uncertainty, confirmation bias, severity amplifiers, framing, and many mental mistakes [52].

The primary reason that these mental mistakes are so important is that people do not realize that they exist. And the people that know of their existence believe that these mistakes do not affect *their* decision-making. However, when the results of these mistakes are pointed out, most people are willing to rewrite to eliminate their undesirable effects. So the best way to get rid of such mistakes is to bring them out in the open.

A System for Handling Uncertainty in Models and Documentation

We will now present our process for ameliorating such mental mistakes. To handle uncertainty, all of the work products must be available for public view, must be subjected to formal reviews, and must be approved in expert reviews, and all of these activities must be in a feedback control loop with frequent small iterations.

The first step in our process is to prepare a document that explains confirmation bias, severity amplifiers, and framing, as well as the mental mistakes of poor problem stating, incorrect phrasing, attribute substitution, political correctness, and feeling invincible [52]. Portions of this chapter could serve this purpose. All people involved in the process must read this document in advance. We have been using this approach for over a decade [1], but it may seem new to the systems engineering community. The following use case is our process for identifying and ameliorating mistakes in a system design.

The system described by this use case can be used to *handle* uncertainty in the models and documentation of the Overarching Process. It is an abstract that included use case.

Name: Perform Formal Inspection to Find Mistakes Caused by Uncertainty

Iteration: 4.6

Derived from the concept exploration document, Diogenes [1]

Brief description: A formal inspection is a structured group review process used to find defects and mistakes in requirements, programming code, test plans, models, and designs [24]. The flow of this use case is called from the handling uncertainty use case, which is not described in this chapter. When this sub-flow ends, the use case instance continues from where this included use case was called.

Level: Medium

Priority: Medium

Scope: The Inspection Team, the work products to be inspected, and the process assets library (PAL).

Added value: The company will be able to look for unresolved uncertainties, mental mistakes, risks, opportunities for Built-in Self-Test (BiST), and unintended consequences of the system being designed all at the same time. This should increase efficiency. Furthermore, discovering *positive* unintended consequences could provide additional revenue.

Goal: Find defects caused by uncertainty and mental mistakes. Find unidentified risks, opportunities for BiST, and unintended consequences of the system being designed.

Primary actors: The Inspection Team is comprised of the moderator, systems engineer, author/designer, reader, recorder, and additional inspectors

The **moderator** leads the inspection, schedules meetings, distributes inspection materials, controls the meetings, reports inspection results, and follows up on rework issues. Moderators should be trained in how to conduct inspections. Risk or quality assurance managers often serve in this role.

The **systems engineer** coordinates the inspection with the overall design process. The systems engineer delivers the lists of unresolved uncertainties, mistakes, risks, opportunities for BiST, and unintended consequences to risk management, test engineering, marketing, management, and legal. He or she also puts these lists in the project PAL.

The **author/designer** creates and/or maintains the work products being inspected. The author/designer answers questions asked about the work products during the inspection, looks for defects, and fixes defects. The author/designer, or other members of the design team, cannot serve as moderator, reader, or recorder.

During the meeting, the **reader** leads the Inspection Team through the work products being inspected, interprets sections of the artifact by paraphrase, and highlights important parts.

The **recorder** classifies and records unresolved uncertainties, mental mistakes, risks, opportunities for BiST, unintended consequences of the system being designed, and issues raised during the inspection. The moderator might perform this role in a small Inspection Team.

The **inspector** attempts to find errors in the work products. This role is filled by several people. All participants act as inspectors, in addition to any other responsibilities. The following may make good inspectors: the person who wrote the specification for the work products being inspected; the people responsible for implementing, testing, or maintaining the work product; a quality assurance representative; a representative of the user community; and someone who is not involved in the project but has infinite experience and impeccable wisdom.

Secondary actors: The process assets library (PAL)

Frequency: Once a month or before specified reviews

Precondition: An author/designer has requested an inspection of his work product.

Trigger: This use case will be included from the handling uncertainty use case.

Main Success Scenario:

1. **Planning activity:** The moderator selects the Inspection Team, obtains the problem statement and the work products to be inspected from the author/designer, and distributes them along with other relevant documents to the Inspection Team. As a rule of thumb, the work products to be inspected typically comprises 200 lines of code or 2000 lines of text.
2. **Overview meeting:** The moderator explains the inspection process to the Inspection Team. This will take from 10 min to 3 h depending on the backgrounds of the team members. The author/designer may describe the key features of the work products.
3. **Preparation:** Each member of the Inspection Team examines the work products before the actual inspection meeting. Each member should be looking for unresolved uncertainties, mental mistakes, risks, opportunities for BiST, and unintended consequences of the system being designed. (Often, risks are handled by a separate department isolated from design. But there is no reason why it has to be this way. And it may be more efficient to include it with these other activities.) Typically, this will take 2 h for each member. The amount of time each person spends will be recorded. This time would be substantially increased for an inspector running models and simulations to verify the system.
4. **Inspection meeting:** The moderator and reader lead the team through the work products. The issues are brought up one by one, and each one is discussed in a

round-robin fashion where each member comments on each issue. (Although time constraints may try to prevent this, the moderator should ensure that each participant says something. This promotes the sense of inclusion and most importantly is the best mechanism for discovering unknown unknowns.) During the discussion, all inspectors can report unresolved uncertainties, mental mistakes, risks, opportunities for BiST, and unintended consequences of the system being designed, all of which are documented by the recorder. The meeting should last no more than 2 h.

How can we prime our inspectors to look for unresolved uncertainties and unintended consequences?

If they are looking at an activity, action, process, procedure, or another verb phrase (an active verb followed by a measurable noun), then tell them to ask, "What problems could this activity create for other systems?" "How could doing this activity hurt other systems?" "If this activity failed, how could that hurt other systems?"

If they are looking at an object, component, model, or another noun phrase, then tell them to ask, "How could this object hurt other systems?" "How could this object fail?" For each failure event, ask, "How could this failure event hurt other systems?"

If they are looking at a risk, then tell them to ask, "How could this failure event hurt other systems?"

If they are looking at a use case scenario or other sequences of events, then tell them to ask, "What-if?" For example, when the document states, "The user does this and the system does that." Ask, "What if it doesn't?"

Inspectors should look for common mental mistakes that people make [52], particularly for attribute substitution, which is the most common mental mistake [51].

Inspectors should look to see if the designers used fundamental principles of good design [5], including design for resiliency [39, 45].

But we really want the mindset of looking for unresolved uncertainties and unintended consequences to become a part of company culture.

5. **Databases:** The team creates and maintains five databases that contain newly resolved and unresolved uncertainties, mistakes, risks, opportunities for BiST, and unintended consequences of the system being designed.

6. **Prioritized. lists:** The moderator and the systems engineer consolidate and edit the five databases to create five prioritized [15] lists.

The list of newly resolved and unresolved uncertainties is given to the systems engineer.

The prioritized list of mistakes is given to the author/designer for rework and resolution.

The prioritized list of risks that could adversely affect the system being designed is given to risk management.

The prioritized list of opportunities for Built-in Self-Test (BiST) is given to test engineering.

The prioritized list of positive unintended consequences that could beneficially affect other systems is given to marketing.

The prioritized list of negative unintended consequences that could adversely affect other systems is given to management and the legal department.

7. **PAL:** The systems engineer puts these prioritized lists in the project PAL.
8. **Rework:** The author/designer fixes the mistakes. Each of the other owners will know what to do with his or her list.
9. **Follow-up:** The moderator must verify that all fixes are effective and that no additional mistakes have been created. The moderator checks the exit criteria for completing an inspection.
10. **Update PAL:** The team updates the project PAL (exit use case).

Postcondition: The project PAL has been updated, and the systems engineer is ready to schedule a new inspection.

Specific requirements can be derived from this use case [20] as follows. These requirements came directly from the main success scenario.

Functional Requirements:

FR3-1 The moderator shall form the Inspection Team.

FR3-2 The moderator shall collect the inspection work products and other relevant materials and distribute them to the Inspection Team To Be Determined (TBD) days before the inspection.

FR3-3 The moderator shall chair the overview meeting.

FR3-4 Each member of the Inspection Team shall examine the work products before the actual inspection meeting looking for unresolved uncertainties, mental mistakes, risks, opportunities for BiST, and unintended consequences of the system being designed.

FR3-5 Each member of the Inspection Team shall record and report the number of hours he or she spent inspecting the materials. Typically, this will be 2 h.

FR3-6 The moderator shall chair the inspection meeting.

FR3-7 The recorder shall create and maintain the five databases that contain newly resolved and unresolved uncertainties, mistakes, risks, opportunities for BiST, and unintended consequences of the system being designed.

FR3-8 The moderator and the systems engineer shall consolidate and edit the databases to create prioritized lists.

FR3-9 The systems engineer shall deliver the lists to their respective owners.

Stipulation: Each owner will know what to do with his or her list.

FR3-10 The systems engineer shall put these prioritized lists in the project PAL.

FR3-11 The moderator shall verify that all fixes are effective and that no additional defects have been created. The moderator shall check the exit criteria for completing an inspection.

It is often said that we can impose requirements on our system, but we cannot impose requirements on operators, pilots, and other secondary actors. This is still true. However, here, we are imposing requirements on members of the Inspection Team. That is all right because they are a part of our system.

7 Overarching Process for Systems Engineering and Design   249

Nonfunctional Requirements:

NFR3-1 The moderator shall schedule the inspection meeting for 2 h. The moderator shall prepare about two-dozen pages of models and documentation for each inspection.

Author/owner: Terry Bahill
Last changed: January 29, 2021

This is the end of the section describing our process for ameliorating uncertainty in the models and documentation caused by human mental mistakes.

Chapter Summary

In the year 2020, the COVID-19 pandemic devastated the world. Scientists continually produced new scientific results and new models using uncertain knowledge of the virus and the pandemic. Inevitably, predictions made with these models were uncertain. Politicians, supposedly using these scientific "facts" and uncertain models, frequently strutted out new policies – policies that often contradicted previous policies. Because of this flip-flopping, people lost faith in their governments' abilities to understand and interpret scientific results from models.

In the next few pages, we will show how governments could have used the principles in this chapter to help manage this pandemic. The third activity in the Overarching Process is to model the system. As shown in this chapter, making models is affected by confirmation bias, severity amplifiers, and framing. These phenomena affected the interpretation of scientific results and the assumptions made in models of this pandemic.

The early COVID pandemic models needed improvements. These early models were created and used by scientists, the media, government bureaucrats, and politicians. We will refer to this group collectively as *they*. Their models needed additional knowledge about the COVID virus, such as transmission by asymptomatic carriers, the impact of super-spreaders, a diminished role for children, the role of airborne aerosol-mediated transmission, and the usefulness of wearing masks. And most importantly, they should have predicted human behavioral responses to government-imposed policy interventions and the mental health problems created by social isolation and the closing of schools.

They form a pyramid. Around 10^6 to 10^7 peer-reviewed journal papers are published each year. Of course, not all of them are about COVID or even about science. Only 10% are scientific. These scientific papers are not read by the public. Then there are on the order of 10^5 science writers who convert these papers into layman articles. These articles are given to maybe 10^4 people in the media, who transform them into summaries and snippets fed to the public. The media also report the number of COVID cases and deaths both geographically and temporally. But these numbers are not science. They are just a collection of numbers. They were not derived using the scientific method. They are not even displayed scientifically (with mean, variance, and sources). To further explain this point, consider the Watson-Crick DNA double helix model published in 1953 that won the Nobel Prize in 1962. Now *that* was science. Compare that to the dozens of DNA ancestry tests on the market today, which do their analyses based on cheek swabs and spit. These companies collect those data to build their databases that include millions of people. They have lots of numbers. But we see little scientific usefulness in those numbers. Just collecting lots of numbers does not make it science. Now, back to the pyramid. Around 10^3 bureaucrats (government bureau chiefs and department heads) filter this information to around 100 politicians making policies for the COVID pandemic. Finally, there is one US president who makes the final decisions. All of these people make up what, in this chapter, we call *they*.

As the year went on, they gathered some of the missing knowledge mentioned above and improved their models. However, they never stated that their old models were based on incorrect assumptions and that these old models were being replaced by new models based on new data and new assumptions. At the beginning of this chapter, we wrote,

> A model is a simplified representation of some aspect of a real system. Models are ephemeral: They are created, they explain a phenomenon, they stimulate discussion, they foment alternatives, and then they are replaced by newer models.

Therefore, old models are *supposed to be* replaced by new models: It is the nature of modeling. It is not something to be ashamed of, something to be covered up. It is *good* to state that because of new knowledge and insights, an old model is being replaced by a new model. It is *very bad* to cover up incorrect knowledge and bad assumptions of an old model.

People infected with the coronavirus (COVID) were generally contagious for around 4 days before they showed symptoms, and some infected people never showed symptoms at all. Because people with influenza (the flu) are most contagious for only 1 day before symptoms appear, this contradicted the decision-makers' preconceived notions about how the virus spread. It took a long time for them to overcome this cognitive bias (tunnel vision).

Their risk analyses were puzzling. Elderly people were at the highest risk; however, elderly people did not get the highest priority for prevention, treatment, or vaccinations. Indeed, New York sent elderly infected COVID patients back into nursing homes! In covering up this fiasco, they never explained why, in their risk analyses, the estimated likelihoods and severities gave such a bizarre recommendation.

Initially, there were frantic efforts to accelerate the manufacturing of ventilators. After they were manufactured and delivered, many were not used. It turned out that high-flow oxygen was a better treatment alternative than ventilators. A tradeoff study might have revealed this mistake earlier.

Their tradeoff studies were faulty or nonexistent. In trying to determine how the virus was transmitted, instead of doing a tradeoff study, they jumped to a single-point solution. They assumed that the virus, like the flu, was spread by touching people and contaminated surfaces. They recommended that people disinfect highly touched surfaces often and wash their hands frequently. Initially, they only looked at evidence that confirmed their bias for this mode of transmission. They ignored the alternative that the virus was spread by aerosol-mediated transmission through the air. As a consequence, in March, they told the public that masks were not useful and people should *not* wear them. (Some have suggested that their secret ulterior motive was to save the good masks for medical personnel.) Several months later, they realized that the coronavirus was transmitted through the air. Then they mandated the wearing of masks for everyone, everywhere, all the time.

To prioritize their alternatives, throughout the year, bureaucrats and politicians made decisions without the benefits of documented tradeoff studies. Thus, they missed golden opportunities to gain public trust. One of the prime benefits of a tradeoff study is that making the evaluation criteria and the scores public produces transparency and greatly increases public trust in the decision-makers.

In prioritizing vaccine distribution, they failed to explain and justify their evaluation criteria. Sometimes, this was a deliberate attempt to disguise their true intent, such as with the contrived evaluation criterion "equity." As a result, their decisions lacked transparency, and people did not trust them. Notably, some vaccination prioritization schemes might have put politicians first (because they were *essential* workers), but they never told that to the public. In fact, the evidence of this that the public saw was media photographs of high-ranking politicians getting the first vaccinations.

They created a divisive fiction that Americans were either science believers (those that agreed with scientific results filtered by bureaucrats and summarized by the media) or science deniers (those who were not convinced by those media summaries). The science deniers were shamed as bad people. Actually, neither of these groups existed [26]. However, Americans were deeply divided. There seemed to be a group that fearfully followed orders and did whatever the government told them to do and another group that did not question scientific results but rather questioned government policies that could hardly have been based on these scientific results, policies such as wearing masks; obeying lockdowns of restaurants, bars, and churches; accepting job losses; closing schools; and letting loved ones die alone. The people labeled science deniers were mostly people who accepted scientific results but disagreed with government policies to ameliorate the problem. The most important activity of the Overarching Process is stating the problem. The invention of science believers and science deniers caused political delight and media churning. But what problem was it supposed to solve?

The biggest problem remaining in 2021 may have been side effects of vaccinations. The scientists probably did risk analyses of the two-dozen admitted vaccine side effects, but the media did not publish their estimated likelihoods and severities. So the public did not understand the risks of the vaccines' side effects.

The most outstanding success during the COVID pandemic was Investigating Alternatives. The search for a vaccine started with dozens of alternatives. Scientific researchers and companies kept a half-dozen alternatives alive throughout the entire year. Developing a half-dozen vaccines and inoculating millions of people in less than a year was the fastest vaccine development and deployment in history.

Note added in proof: This summary was written in the Spring of 2021. This note is being inserted in the Summer of 2022. The predictions of this section have held up well over the last year and no changes were made to this section.

But that is enough about the COVID pandemic. Let us now summarize uncertainty in systems engineering and design in general.

Uncertainty arises from factors that are both external and internal to the system. Examples of factors that contribute to external uncertainty are changes in market conditions, the operational environment, new competitors or threats, emerging requirements, partial observability, changes in priorities, and delays in maturation times of promising new technologies. By far, the greatest uncertainty is coping with unknown futures. This problem requires designing for alternate futures – the hallmark of resilient design [39, 45]. Internal uncertainties stem from human behavior and unanticipated challenges that surface during program/project execution, system design and implementation, and creating performance requirements.

There are several traditional approaches to dealing with uncertainty depending on the context. Uncertainty may stem from incomplete/fuzzy requirements or technology maturation rate. If requirements and technologies are both stable in a project, then it is relatively straightforward to plan ahead and execute the plan because there is very little uncertainty to further ameliorate. On the other hand, when a project intends to capitalize on new or emerging technologies, uncertainty is best handled by placing "smart bits" or incorporating real options in both system architecture/design and program schedule [37]. Finally, when the technology aspect is relatively stable, but the requirements continue to evolve, then an incremental commitment approach needs to be pursued [13].

This chapter presented a requirements discovery process, a tradeoff study process, and a risk analysis process. It compared and contrasted these three processes and then combined them into one *Overarching Process*. The three original processes could then be viewed as tailorings of the resulting Overarching (superset) Process. This Overarching Process can also be a top-level or a precursor process for MBSE implementations. The Overarching Process itself is not an MBSE implementation. For example, it does not even use SysML diagrams. This Overarching Process is not a subset of MBSE: It is a superset.

In conclusion, the requirements discovery, tradeoff studies, and risk processes shown in the figures of this chapter were developed many years before the Overarching Process was conceived.

This chapter identified activities in the Overarching Process that were responsible for creating uncertainty. Then it addressed uncertainty handling for the three core processes (requirements discovery, tradeoff studies, and risk analysis) and the general Overarching Process. The approach presented reduces overall complexity in uncertainty management by first creating an enveloping Overarching Process, then systematizing the process of uncertainty handling for this Overarching Process, and then instantiating the approach for the three core processes. This approach should appeal to both engineering and management professionals engaged in modeling, analysis, and design of complex systems in the presence of uncertainty.

Cross-References

▶ MBSE Methodologies

References

1. Bahill AT (2012) Diogenes, a process for finding unintended consequences, *Systems Engineering*, 15(3):287–306. https://doi.org/10.1002/sys.20208
2. Bahill AT (2016) Model for absorption of perfluoropropane intraocular gas after retinal surgeries, International Journal of Medical and Health Sciences Research, PAK Publishing Group, 3(5):50–76, 2016. https://doi.org/10.18488/journal.9/2016.3.5/9.5.50.76. http://www.pakinsight.com/journal/9/abstract/4546
3. Bahill AT (2019) *The Science of Baseball: Batting, Bats, Bat-Ball Collisions and the Flight of the Ball*, Springer Nature, NY, NY, second edition, 2019, ISBN 978-3-030-03032-2
4. Bahill AT, Baldwin DG (2004) The rising fastball and other perceptual illusions of batters. In: Hung G, Pallis J (eds) Biomedical engineering principles in sports. Kluwer Academic, pp 257–287
5. Bahill AT, Baldwin DG (2008) Mechanics of baseball pitching and batting, Chapter 16. In: Ghista D (ed) Applied biomedical engineering mechanics. CRC Press and Taylor & Francis Asia Pacific, pp 445–488
6. Bahill AT, Dean F (2009) Discovering system requirements, Chapter 4 In Handbook of Systems Engineering and Management, second edition, (eds.) AP Sage, WB Rouse, John Wiley & Sons, pp. 205–266
7. Bahill AT, Gissing B (1998) Re-evaluating systems engineering concepts using systems thinking. IEEE Trans Syst, Man, Cybern C. 28(4):516–27
8. Bahill AT, Madni AM (2017) Tradeoff Decisions in System Design, Springer International Publishing, Switzerland
9. Bahill AT, Szidarovszky F (2009) Comparison of dynamic system modeling methods, Systems Engineering, **12**(3):183-200
10. Bahill AT, Baldwin DG, Venkateswaran J (2005) Predicting a baseball's path. Am Sci 93(3):218–225
11. Bahill AT, Szidarovszky F, Botta R, Smith ED (2008) Valid models require defined levels, International Journal of General Systems, 37(5):533–571
12. Beach LR, Connolly T (2005) The Psychology of Decision Making: People in Organizations. second edition. Thousand Oaks, CA: Sage Publications
13. Boehm B, Lane J, Koolmanojwong S, Turner, R (2014) The Incremental Commitment Spiral Model, Addison-Wesley, Pearson Education, Inc.
14. Bohlman J, Bahill AT (2014) Examples of Mental Mistakes Made by Systems Engineers While Creating Tradeoff Studies, Studies in Engineering and Technology, 1(1): 22–43, https://doi.org/10.11114/set.v1i1.239
15. Botta R, Bahill AT (2007) A Prioritization Process. Engineering Management Journal, 9(4): 20–7.
16. Buchanan BG, Shortliffe EH (1984) Rule-based Expert Systems. Reading, MA: Addison-Wesley
17. Buede DM (2009) The engineering design of systems: Models and methods. Second edition. New York: John Wiley and Sons, Inc.
18. Cohen RB (2003) Teaching Note on A-Rod: Signing the best player in baseball, *Harvard Business School Case Study* 5-203-091, April 16, 2003.
19. Cohen RB, Wallace J (2003) A-Rod: Signing the best player in baseball, *Harvard Business School Case Study* 9-203-047, January 27, 2003.
20. Daniels J, Bahill AT (2004) The hybrid process that combines traditional requirements and use cases, *Systems Engineering* 7(4) pp. 303-319.

21. Daniels J, Werner PW, Bahill AT (2001) Quantitative methods for tradeoff analyses, *Systems Engineering*, **4**(3):190-212, 2001, with a correction published in **8**(1):93, 2005
22. Edwards W (1977) How to Use Multiattribute Utility Measurement for Social Decision making. IEEE Trans Syst, Man, Cybern. 7(5):326-40
23. Estefan JA (2021) Survey of Model-Based Systems Engineering (MBSE) Methodologies, to be published in the *Handbook of Model-Based Systems Engineering*, edited by Azad M. Madni and Norman Augustine
24. Fagan M (2020) Improved Fagan Inspections and Continuous Process Improvement http://www.mfagan.com/index.html
25. Friedenthal S, Moore A, Steiner R (2012) A practical guide to SysML: the systems modeling language, Elsevier, Morgan Kaufman, OMG
26. Hilgartner S, Hurlbut JB, Jasonoff S (2021) Was 'science' on the ballot? *Science*, 371, 6532, 893–894.
27. Hooks IF, Farry KA (2001) Customer-centered products: Creating successful products through smart requirements management, AMACOM, New York
28. Kahneman D (2011) Thinking, fast and slow. Farrar, Straus and Giroux, New York. ISBN: 978-0-374-27563-1
29. Kahneman D, Tversky A (1979) Prospect theory: an analysis of decision under risk. Econometrica, 46(2):171–85
30. Karnavas WJ, Sanchez PJ, Bahill AT (1993) Sensitivity analyses of continuous and discrete systems in the time and frequency domains. IEEE Trans Syst Man Cybern 23(2):488–501
31. Keeney RL, Raiffa H (1976) Decisions with multiple objectives: Preferences and value tradeoffs. New York: Wiley
32. Madni AM (2000) Thriving on Change through Process Support: The Evolution of the Process Edge Enterprise Suite and Team Edge. Information Knowledge Systems Management 2.1 7–32
33. Madni AM (2014) Generating novel options during systems architecting: psychological principles, systems thinking, and computer-based aiding. Syst Eng 17(1):1–9
34. Madni AM (2020) Models in Systems Engineering: From Engineering Artifacts to Source of Competitive Advantage, Conference on Systems Engineering Research (CSER 2020), Redondo Beach, October 8–10
35. Madni AM (2020) Minimum Viable Model to Demonstrate the Value Proposition of Ontologies for Model Based Systems Engineering, 2020 Conference on Systems Engineering Research, Redondo Beach, CA, October 8–10
36. Madni AM (2020) Exploiting augmented intelligence in systems engineering and engineered systems. INSIGHT Special Issue, Systems Engineering and AI, March 2020
37. Madni AM, Allen K (2011) Systems thinking-enabled real options reasoning for complex socio-technical systems programs. In: Conference on systems engineering research
38. Madni AM, Bahill AT (2021) Handling Uncertainty in Engineered Systems, Handbook of Systems Sciences, Gary S. Metcalf et al. (eds), (Chapter 22-1), Springer
39. Madni AM, Jackson S (2009) Towards a conceptual framework for resilience engineering. Syst J IEEE 3.2:181–191
40. Madni A, Sievers M (2014) Systems integration: key perspectives, experiences, and challenges. Syst Eng 17(1):37–51
41. Madni AM, Sievers M (2018) Model-based systems engineering: motivation, current status, and research opportunities. *Systems Engineering*, 21: 172–190
42. McBeath MK, Nathan AM, Bahill AT, Baldwin DG (2008) Paradoxical pop-ups: why are they difficult to catch? Am J Phys 76(8):723–729
43. Mohanani R, Salman I, Turhan B, Rodriguez P, Ralph P (2020) Cognitive biases in software engineering: a systematic mapping study. IEEE Trans Softw Eng 46(2):1318–1339. https://doi.org/10.1109/TSE.2018.2877759
44. Nickerson RS (1998) Confirmation Bias: A Ubiquitous Phenomenon in Many Guises. Review of General Psychology. Vol. 2, No. 2, 175-220

45. Neches R, Madni AM (2013) Towards Affordably Adaptable and Effective Systems, Systems Engineering, Vol. 16, No. 2, pp. 224–234
46. Oakes J, Botta R, Bahill AT (2006) Technical Performance Measures. This paper was presented at the National Defense Industrial Association, *NDIA Systems Engineering Conference,* October 2005. It was also presented at the INCOSE Symposium and was published in the *Proceedings of the INCOSE Symposium,* July 2006
47. Overgaard G, Palmkvist K (2005) *Use Cases: Patterns and Blueprints,* Addison-Wesley, Indianapolis, IN.
48. Pignatiello JJ, Ramberg JS (1985) Off-line quality-control, parameter design, and the Taguchi method-discussion. *J Qual Technol* 17(4):198–206
49. Rechtin E, Maier MW (1997) *The Art of System Architecting,* CRC Press, Boca Raton
50. Rosenblit JW (1985) A conceptual basis for model-based system design, Ph.D. dissertation in Computer Science at Wayne State University (published by University Microfilms International, Ann Arbor)
51. Smith E, Bahill AT (2010) Attribute substitution in systems engineering. Syst Eng 13(2):130–148
52. Smith ED, Son YJ, Piattelli-Palmarini M, Bahill AT (2007) Ameliorating mental mistakes in tradeoff studies. Syst Eng 10(3):222–240
53. Smith ED, Szidarovszky F, Karnavas WJ, Bahill AT (2008) Sensitivity analysis, a powerful system validation technique. Open Cybernetics & Systemics Journal, 2:39–56
54. Wymore AW (1993) Model-based systems engineering. CRC Press, Boca Raton

Terry Bahill is Professor Emeritus of Systems and Industrial Engineering at the University of Arizona in Tucson. He served 10 years in the United States Navy leaving as a lieutenant. He received his PhD in electrical engineering and computer science from the University of California, Berkeley, in 1975. He is the author of six engineering books and 250 papers; over 100 of these are in peer-reviewed scientific journals. Bahill has worked with dozens of technology companies presenting seminars on systems engineering, working on system development teams, and helping them to describe their systems engineering process. He holds a US patent for the Bat Chooser™, a system that computes the Ideal Bat Weight™ for individual baseball and softball batters. He was elected to the Omega Alpha Association, the systems engineering honor society. He received the Sandia National Laboratories Gold President's Quality Award. He is an elected Fellow of the Institute of Electrical and Electronics Engineers (IEEE), of Raytheon Missile Systems, of the International Council on Systems Engineering (INCOSE), and the American Association for the Advancement of Science (AAAS). He is the Founding Chair Emeritus of the INCOSE Fellows Committee. His picture is in the Baseball Hall of Fame's exhibition "Baseball as America." You can view this picture at http://www.sie.arizona.edu/sysengr/. His research interests are in the fields of system design, modeling physiological systems, eye-hand-head coordination, human decision-making, and systems engineering application and theory. He has tried to make the public appreciate engineering research by applying his scientific findings to the sport of baseball.

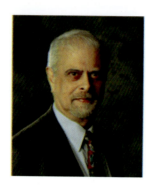

Azad M. Madni is a Professor of Astronautical Engineering and the Technical Director of the Systems Architecting and Engineering Program in the University of Southern California's Viterbi School of Engineering. He is also a Professor (by courtesy) in USC's Schools of Medicine and Education. He is the founder and Chairman of Intelligent Systems Technology, Inc., a high-tech R&D company specializing in game-based educational simulations and methods, processes, and tools for complex systems engineering. He received his BS, MS, and PhD degrees from the University of California, Los Angeles. His research has been sponsored by both government research organizations such as DARPA, OSD, ARL, RDECOM, ONR, AFOSR, DHS S&T, DTRA, NIST, DOE, and NASA and aerospace and automotive companies such as Boeing, Northrop Grumman, Raytheon, and General Motors. He is an elected Fellow of the American Association for Advancement of Science (AAAS), the American Institute for Aeronautics and Astronautics (AIAA), the Institute for Electrical and Electronics Engineers (IEEE), the Institution for Electronics and Telecommunications Engineers (IETE), the International Council on Systems Engineering (INCOSE), and the Society for Design and Process Science (SDPS). His recent awards include the *2016 Boeing Lifetime Accomplishment Award* and *2016 Boeing Visionary Systems Engineering Leadership Award* for contributions to industry and academia (awards received in Boeing's 100th anniversary), the 2016 INCOSE RMC Special Award for pioneering industry-relevant contributions to Transdisciplinary Systems Engineering, the *2016 Distinguished Engineering Educator Award* from the Engineers' Council, the *2016 Outstanding Educator Award* from the Orange County Engineering Council, the 2014 *Lifetime Achievement Award* from the International Council on Systems Engineering, the 2013 *Innovation in Curriculum Award* from the Institute of Industrial Engineers, the 2012 *Exceptional Achievement Award* from INCOSE, and the 2011 *Pioneer Award* from the International Council on Systems Engineering. He serves on the USC's Council of the Center of Cyber-Physical Systems and the Internet of Things (CCI) and the Steering Committee of USC Provost's STEM Consortium. His research interests include formal and probabilistic methods in systems engineering, model-based architecting and engineering, engineered resilient systems, cyber-physical systems, and exploiting disciplinary and technology convergence in systems engineering. He is listed in the Who's Who in Science and Engineering, Who's Who in Industry and Finance, and Who's Who in America.

Problem Framing: Identifying the Right Models for the Job

James N. Martin

Contents

Introduction	258
Background	258
Why Use a Model-Based Approach?	258
Architecture Frameworks	259
Modeling Methodology	262
Model Development	264
Six-Step Method for Model Development	265
Example Using the Method	265
Agile Development of Models Using Sprints	273
Evolution of the Model Development Method	274
Problem Framing	275
Overview of the Problem Framing Approach	275
Problem Framing Steps	277
Problem Framing Execution	280
Summary	284
References	284

Abstract

In the world of modeling, we are now confronted with a new problem. The richness of the tools and methods available to us now have led to an "agony of abundance" when it comes to available options for modeling a system. The Unified Architecture Framework, for example, has specified 82 possible architecture views that can be created. How can systems engineering cope with all the choices? How do you effectively choose the right models? How do you determine the appropriate contents of the models you do choose? This chapter presents a Problem Framing approach that facilitates selection of the right models for the job, the ones that can add the greatest value and have the greatest utility for the

---

J. N. Martin (✉)
The Aerospace Corporation, Chantilly, VA, USA
e-mail: James.N.Martin@aero.org

various potential users of these models, those that can provide the greatest insight into system trade-offs and successful outcomes, and the views that can best inform enterprise and program decision makers.

> **Keywords**
>
> Architecture · Architecture frameworks · Problem Framing · Metamodel · Methodology · Model development

Introduction

Modeling is becoming more and more important in the course of doing systems engineering. Modeling technologies are becoming more sophisticated, and architecture frameworks are becoming more expansive and mature. This leads to the difficulty of knowing what to model for a highly complex system or a large-scale enterprise. The recently released Unified Architecture Framework has over 82 view specifications to choose from. How are we to choose which views are relevant for our situation? Which models do we need to build to generate these various views? How do we populate the models to provide the best insights into the architecture?

This chapter examines the issues with our current situation and presents a Problem Framing approach to help the modeling team determine the most useful and cost-effective models to build to get the greatest return on in-vestment from the modeling efforts. This approach can be a key enabler for successful transition to a fully model-based systems engineering (MBSE) way of doing engineering on our projects.

Background

Why Use a Model-Based Approach?

Good models can help focus the engineering effort for a system of interest by doing the following:

- Inform senior leaders in the enterprise and program managers about the best way to achieve organizational objectives and mission imperatives
- Assess impacts across the enterprise when things are about to change
- Identify alternative Courses of Action (COA) along with pros and cons of each COA
- Inform trade-offs between alternative COAs and selection of the best COA
- Communicate intended way forward (and changes to the plan)

Models can help identify dependencies throughout the enterprise and highlight impacts to team members, mission partners, and other enterprises, as well as to

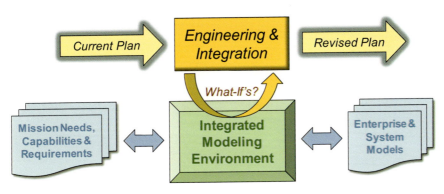

Fig. 1 An Integrated Modeling Environment is a key enabler for creating models and supporting a unified modeling methodology

enabling support organizations. The MBSE approach can help build a truly integrated modeling environment that serves as a single "source of truth" for the program by tying together the mission needs, capabilities and requirements with models of the relevant systems and enterprises, and all their various elements and connections, and connect all of this with associated program elements. This "integrated modeling environment" illustrated in Fig. 1 ties together the disparate bits of information about all the various entities that are associated with the architecture.

Architecture Frameworks

The Nature of Frameworks

Architecture frameworks are designed to help you organize models and identify which views are needed to sufficiently characterize the architecture of a system [1]. It is common for a framework to organize the models and views by use of a "grid," often with the columns representing "aspects" (which are the various ways you can characterize something) and rows representing "perspectives" (which are ways to understand how aspects relate to each other and to the whole).

Aspects are the different ways in which something can be viewed by the mind with respect to the architectural solution's trade space. Perspectives provide insights across the spectrum of the problem space and help reveal the diversity of concerns of affected stakeholders. The Unified Architecture Method illustrated in Fig. 2 uses this approach in organizing the various models to be developed [10].

NATO Architecture Framework (NAF)

The NATO architecture framework (NAF) was developed [8] using a similar grid approach with its rows being about "subjects of concerns" while the columns are representing "aspects of concerns." The question arises as to which aspects and subjects (i.e., perspectives) from this framework are relevant for a particular architecture development effort. Of these 45 views in NAF (Fig. 3), which of them should

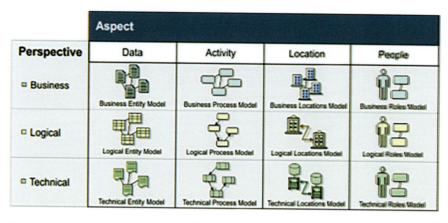

Fig. 2 The Unified Architecture Method (UAM) represents a typical way of organizing models and views of an architecture using aspects and perspectives

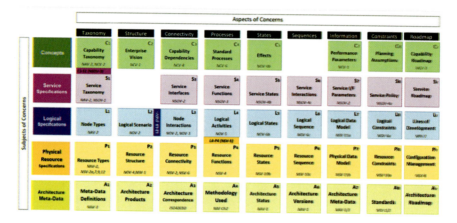

Fig. 3 The NATO Architecture Framework (NAF) organizes its views by aspects of concern and by subjects of concerns

be modeled? The Problem Framing approach is designed to help answer such questions.

DOD Architecture Framework (DODAF)

The architecture framework developed by the US Department of Defense [2] uses a similar grid (Fig. 4) although it is not organized by so-called aspects and perspectives, unless you consider the capability-to-systems stack to be the four main "perspectives" of interest. These perspectives are called "viewpoints" in this framework, resulting in 52 models to choose from thereby making it difficult to know which ones are most relevant and useful. The Problem Framing approach discussed

8 Problem Framing: Identifying the Right Models for the Job 261

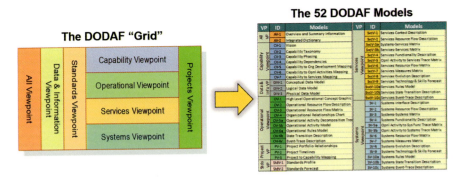

Fig. 4 The DOD Architecture Framework (DODAF) organizes its views by viewpoint that correspond to stakeholder perspectives

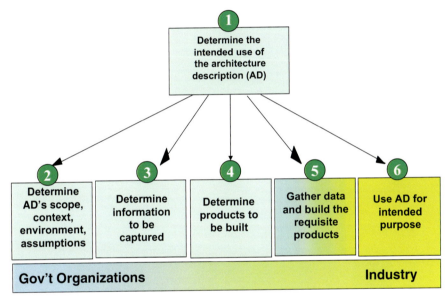

Fig. 5 The 6-step process provided by DODAF conducts a "problem framing" approach in the first four steps prior to building the products (models and views) in step 5

herein can help identify the most relevant and useful models to develop for a given situation.

DODAF did provide a process to facilitate the determination of "products" to be built which happens in step 4 of the process (Fig. 5). In practice, many projects will dictate which products to build regardless of the "intended uses" to be identified in step 1. This was a problem which led to many modeling efforts becoming a "check the box" exercise rather than an effort to make the most useful and insight-provoking models to address the challenges in engineering complex systems.

In practice, many people start with step 5 by immediately building products, either the ones mandated for the project or the ones that have commonly been used in the past. As a result, the modeling products often do not get used as much as expected and some are never even finished due to the lack of acceptance by the intended users. The Problem Framing approach was adapted from this DODAF process while adding refinements over the years as the approach was used on various programs.

OMG Unified Architecture Framework (UAF)

The latest, most mature, framework is the one developed by the Object Management Group (OMG) called the Unified Architecture Framework [9]. Unfortunately (or fortunately, depending on your perspective), it provides 82(!) view specifications to choose from, making the problem of proper selection of views (and related models) even more challenging (Figs. 6 and 7).

Modeling Methodology

In the early days, most people tried to go from the architecture framework directly into modeling without much thought given to how the models would be built to maximize their utility and effectiveness, to enhance model development and reuse, and to minimize difficulties in comprehension and change management. UML as a language was becoming more complex and difficult to use. And then SysML came along with even more ways to model a system. A modeling methodology is needed to help organize the modeling efforts and to standardize the approach for the various modelers involved (Fig. 8).

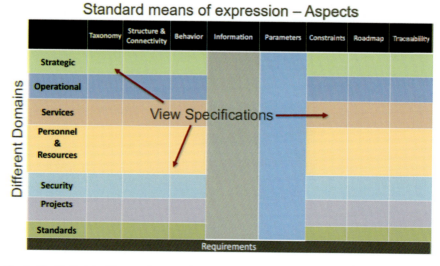

Fig. 6 The Unified Architecture Framework (UAF) provides a two-dimensional grid using aspects and viewpoints for its coordinates for organizing view specifications

8 Problem Framing: Identifying the Right Models for the Job

Fig. 7 UAF specifies 82 different standardized views that can be used in creating a comprehensive architecture description with a variety of architecture views

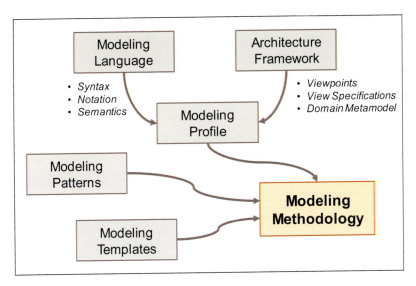

Fig. 8 A modeling methodology incorporates various modeling tools and techniques

This gave rise to the need for modeling profiles (such as the Unified Profile for DODAF and MODAF (UPDM)) that specified how a particular modeling language should be used in a consistent and coherent manner in creating the various views specified by the framework. The modeling profile could provide more guidance on how the models should be built for that particular domain since UML, SysML, and other modeling languages are quite general in nature.

Domains and programs started creating their own particular modeling patterns and templates to ease the burden when creating models, and this also had the benefit of encouraging greater model reuse and enabling more rapid and agile model development practices. These modeling patterns are similar to those already in use by the specialty disciplines such as software development, electronics and mechanical design, reliability and maintainability, security and safety, etc.

The architecture frameworks over time became more robust with additional views added to expand the aspects of a system and perspectives of the problem domain that could be modeled to better understand the system trade space and evaluate the architecture alternatives. Figure 7 illustrates the variety of views that are standardized in the Unified Architecture Framework (UAF).

Model Development

Many projects build particular models because the "process says so." Or the company handbook dictates certain system diagrams and other depictions. Or the customer wants to receive a specific set of deliverables to satisfy their mandates. However, it has been found that more useful models can be produced if they are based on important questions that are be asked about the system by its various stakeholders. Examples of such questions are shown in (Fig. 9). The more often the same questions keep getting asked, the more likely these are good candidates for building a model to answer them. Hence, this is the reason that the Problem Framing

❑ **Architecture**
- Is the architecture aligned with strategic goals & objectives and with leadership priorities?
- Is the architecture aligned with community architectures and visions?
- Are system architectures aligned with enterprise architecture and requirements?
- Are program plans aligned with architectures?

❑ **Requirements & Test**
- How do enterprise capability requirements relate to capabilities and gaps/shortfalls?
- How do system level requirements relate to capabilities and gaps/shortfalls?
- Where and Where to add/modify requirements to address gaps/shortfalls?
- How to show traceability from requirements to architecture elements and to mission needs?
- How and When to verify requirements?

❑ **Enterprise Integration**
- What are capabilities needed to satisfy the mission needs and expectations?
- Where are the key capability gaps and shortfalls?
- How should capabilities be sequenced to best fill the gaps and address the shortfalls?
- How do capabilities relate to community capability documents and other commitments?
- How are systems and services affected when filling these gaps and addressing shortfalls?
- How are programs and projects affected when filling these gaps and addressing shortfalls?

❑ **Mission Resilience**
- What capabilities are needed for identification and mitigation of threats?
- How should the enterprise handle threats?
- How to assess protection strategies?

Fig. 9 Models and views should be based on the important questions about the architecture that address various stakeholder concerns and other key considerations

8 Problem Framing: Identifying the Right Models for the Job

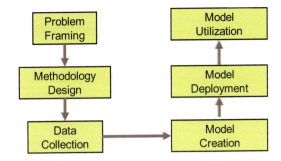

Fig. 10 Six-step method for model development provides a structured way of building the right models and getting maximum utility from them

approach is heavily focused on the questions that can readily be answered using the models.

Six-Step Method for Model Development

Illustrated in Fig 10 is a commonly followed pattern for developing models. This provides a structured way of building the right models and getting maximum utility out of them. However, more often than not, people tend to start with step 4 (Model Creation) and immediately start building models without a good understanding of why particular models are needed, or what kinds of data need to be put into those models.

It has been found that many modeling projects end up floundering and not producing the models that were deemed to be worthy of the effort expended. Some of these projects get canceled before they even finish. The purpose of Problem Framing (step 1 in this method) is to increase the likelihood of having useful models and focusing the efforts on the most important questions that need to get answered (and those that can possibly *only* be answered using a good set of models!).

This method for model development shown above was used as the basis for the Architecture Elaboration process in the ISO 42020 international standard on architecture processes [5]. The purpose of the Architecture Elaboration process is to "describe or document an architecture in a sufficiently complete and correct manner for the intended uses of the architecture" (ibid.). The Problem Framing approach described in this chapter was incorporated into the early steps of the Architecture Elaboration process in the ISO standard (Fig. 11).

Example Using the Method

To illustrate what happens during these steps in the method, an example is provided from the project that developed the NOAA Observing System Architecture (NOSA). The NOSA model [6] was built originally to support the annual budget planning process at the National Oceanic and Atmospheric Administration (NOAA). The

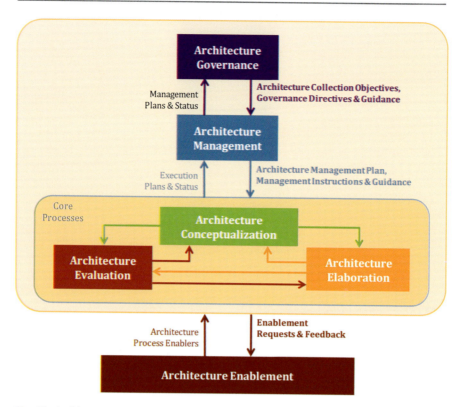

Fig. 11 Architecture processes in the ISO/IEC/IEEE 42020 international standard

NOAA Administrator needed a better way to justify the budget submissions being provided to Congress. The NOSA model served this purpose well and ended up being used for other things as well, such as requirements development, capability and capacity planning, technology investment decisions, program assessment and evaluation, etc.

The NOSA model (or at least a variant of it) is still in use today at NOAA for doing program assessment and evaluation that informs the annual budget planning process [7]. It is being adapted now to support enterprise level portfolio management and enterprise integration activities.

The NOSA reference model shown in (Fig. 12) was instrumental in being able to properly model (and understand) the great diversity of observing systems that support the wide variety of environmental observation missions for NOAA and its mission partners (illustrated in Fig. 13).

Step 1: Problem Framing

Problem Framing starts primarily with the "business questions" that are often asked by managers and leaders in the organization: program managers, chief engineers, middle managers, senior executives, and others (Fig. 14).

8 Problem Framing: Identifying the Right Models for the Job

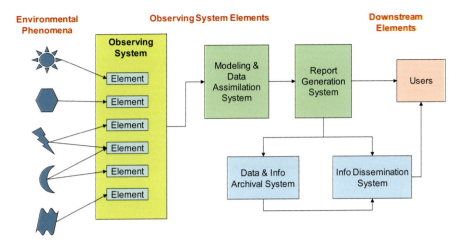

Fig. 12 Generic reference model of an observing system architecture

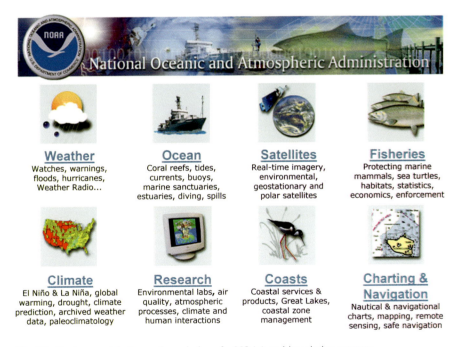

Fig. 13 Environmental observation missions for NOAA and its mission partners

A few of the thirty business questions for NOSA are illustrated in (Fig. 15). Notice that these are usually at a higher level than those questions asked inside a program or project, although many of these enterprise-level questions are still of interest to program and project managers (and the systems engineers who work

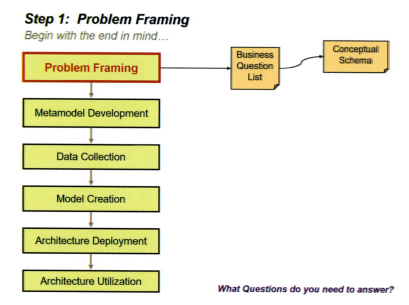

Fig. 14 Problem Framing starts with identification of the key business questions and results in a conceptual schema of the problem space

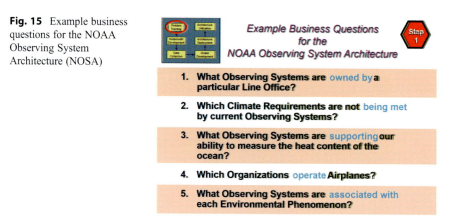

Fig. 15 Example business questions for the NOAA Observing System Architecture (NOSA)

therein). Calling these "business" questions emphasizes that fact that we are mainly interested in transformation of the enterprise as a whole and not as much about changing individual systems under the purview of that enterprise.

The NOSA model spanned the entire NOAA organization, hence it was an enterprise architecture rather than an architecture for a single system. This is because the purpose of the architecture was to inform seniors on how their programs were

8 Problem Framing: Identifying the Right Models for the Job

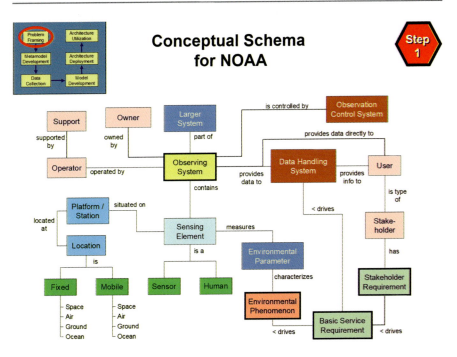

Fig. 16 Conceptual schema for the NOAA Observing System Architecture (NOSA)

doing in helping to achieve strategic objectives, and hence drive the organizational budget (and program plans) to better align with corporate imperatives.

The primary output from Step 1 is a conceptual schema (Fig. 16) that specifies the basic kinds of entities and relationships in the enterprise. This was developed by examining the business questions to identify each entity (i.e., the nouns) and each relationship (i.e., the verbs). This conceptual schema will be used as the basis for developing the metamodel in Step 2.

Step 2: Metamodel Development

A metamodel is the underlying set of concepts that are used to generate consistent and complete models (Fig. 17). It defines the semantics and syntax of the modeling notation to be used. Even though a modeling language will necessarily come with its own semantics and syntax, the metamodel is adapting these basic language properties to a particular domain. So, the "domain" metamodel is intended to capture the key concepts for the domain which can serve to enforce some standardized approach for that domain and to enable better integration of models across that domain.

The metamodel and architecture framework for the NOSA model are illustrated in (Figs. 18 and 19). Existing architecture frameworks were examined, but none were

Step 2: Metamodel Development

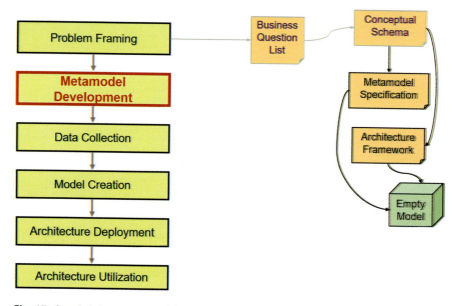

Fig. 17 Step 2 defines a metamodel specification and the selection or development of an architecture framework

found suitable, so a custom framework was created. Likewise, existing metamodels in commonly used frameworks were examined and were found lacking in the right concepts and model element types that would be relevant to the NOAA missions and their associated concepts of operations.

Notice that Step 2 in the early days was called Metamodel Development since that was what was needed back then. But as time went on, we came to realize that we also needed to be concerned about designing the overall modeling methodology rather than just creating or modifying the metamodel (to be used as the basis for the architecture models to be built).

The modeling methodology (see section "Modeling Methodology") included not only the metamodel, but also the modeling language, modeling profile, modeling patterns, modeling templates, and other job aids that could help in the modeling effort [7]. Therefore, we eventually changed the name of Step 2 to Methodology Design.

Step 3: Data Collection

The metamodel and framework, along with the business questions, were used to develop a set of detailed questions to be used as the basis for collecting data to populate the NOSA model (Fig. 20). From the original 30 business questions identified in Step 1, we derived 150 detailed questions that were then used as the

8 Problem Framing: Identifying the Right Models for the Job

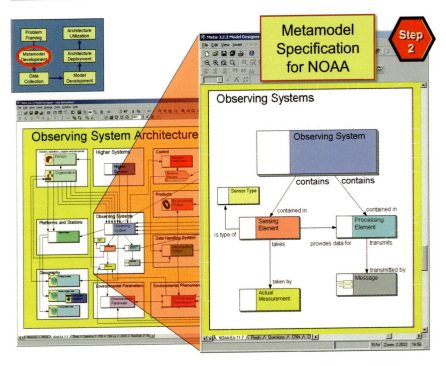

Fig. 18 Metamodel specification for the NOAA Observing System Architecture (NOSA)

basis for several survey forms posted on the NOAA network to help collect data from subject matter experts around the world (Fig 21).

Step 4: Model Creation
The survey forms were built on top of a database which facilitated the collection of data. The data fields were mapped to the model structures in the metamodel (defined in Step 2) to enable the models to be auto-generated from the data (Fig. 22). Data was maintained by the subject matter experts by changing their answers to the survey questions, and then the modified data was ingested into the model on a periodic basis. The architecture framework provided a useful way to organize the model information for easy viewing and exploration. Below you can see a partially populated model with some of the connections highlighted (Fig 23).

Step 5: Architecture Deployment
The "raw" model view shown above was suitable for architects and modelers, but it was not very useful nor intuitive for the intended users of the architecture. Therefore, special user-friendly views were developed that provided only the relevant information particular to a question of interest. In many cases, these views were constructed like "dashboards" that readily highlighted key information of interest for the

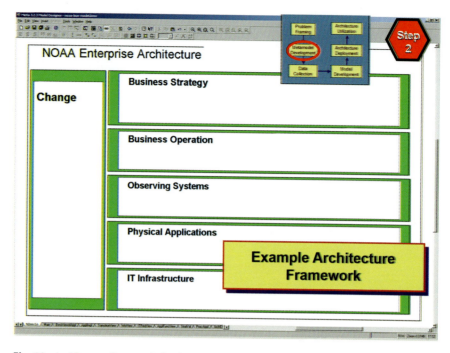

Fig. 19 Architecture framework for the NOAA Observing System Architecture (NOSA)

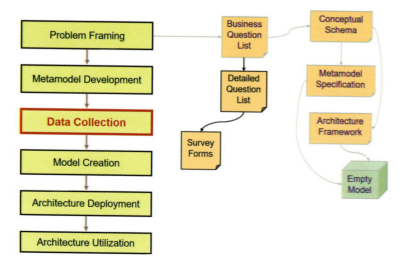

Fig. 20 Step 3 collects the data needed to populate the architecture models

8 Problem Framing: Identifying the Right Models for the Job

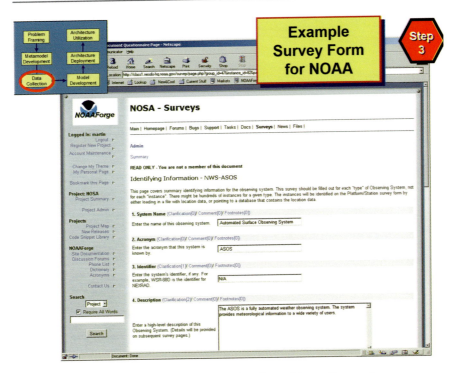

Fig. 21 Example survey form used to collect the data from subject matter experts

intended uses that had been identified in Step 1 (Problem Framing). Example of a such a view is shown in (Fig 24).

These "user friendly" views were carefully created to use the language of the stakeholder to avoid confusing them with modeling jargon and extraneous metadata that was only of interest to the modelers and systems engineers. This is where many of the modeling tools fall short in that they are often only able to provide the "modeldy-gook" that engineers know and love. Often these user views needed to be created in some other tool than the modeling tool itself (Fig. 24).

Step 6: Architecture Utilization
The final step puts the architecture models out on the network in a way that was accessible to everyone at NOAA around the world. We made this as web-like as we could to make it more intuitive and easier to use, maximizing use of "point and click" features and simplified views (Figs. 25 and 26).

Agile Development of Models Using Sprints

The six steps for model development are not usually used in a one-pass operation. It is better to use an incremental approach, similar to the way agile development is

Step 4: Model Creation

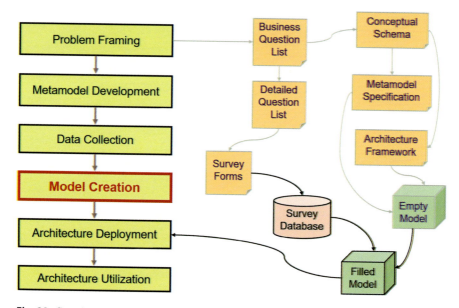

Fig. 22 Step 4 creates the model from collected data to be used as basis for the views

done with multiple "sprints" to quickly get results and learn from each cycle. As the model gets used, more potential uses will be found, and Problem Framing should be repeated to prepare for the next cycle. At each cycle, the models become more complete and more useful to various stakeholders. Lessons learned can be collected and folded into the models and views to be built in the next sprint. Additional use cases and questions that the model can address will be identified and dealt with in subsequent increments (Fig. 27).

Evolution of the Model Development Method

This time-tested approach has been used for over twenty years in a couple of dozen customer engagements. It has evolved as we have learned what works and what does not. Also, the modeling tools and techniques are now more mature, and we have adapted the approach to take account of SysML, UPDM modeling profile, more modern tools and methods, digital engineering and product line management techniques, and so on. The more evolved approach is illustrated in (Fig. 28).

Originally, we were focused on metamodel development in step 2 but later expanded this to design of the modeling methodology (where the metamodel is just one element in the methodology). We also changed the last three steps to focus on models in general rather than on architecture since MBSE was coming into its

8 Problem Framing: Identifying the Right Models for the Job

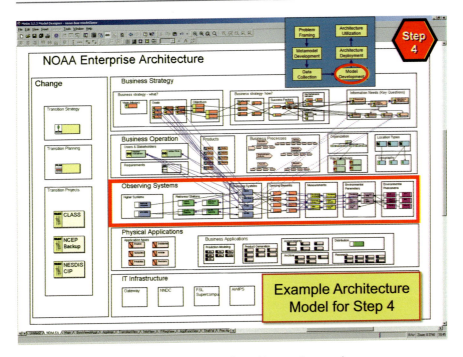

Fig. 23 Model elements populate the views in the architecture framework

own and taking a broader view of models just those used for architecture development and evaluation. The concept of modeling profiles was further advanced with the advent of SysML and the UAF Modeling Language (UAFML), and this idea was also incorporated into the six-step method outlined in this chapter.

Problem Framing

> Begin with the end in mind... (Plato, ~400 BC)

As we gained more experience with this method, it became apparent that we needed to focus on doing a more thorough job on the Problem Framing aspects. We found that the first four steps of the DODAF architecture description process were essentially all about problem framing, so we adapted these steps into our own Problem Framing approach (Fig. 29).

Overview of the Problem Framing Approach

It is important to properly "set the stage" when modeling the enterprise to ensure maximum usefulness and impact of the resulting models that are developed to

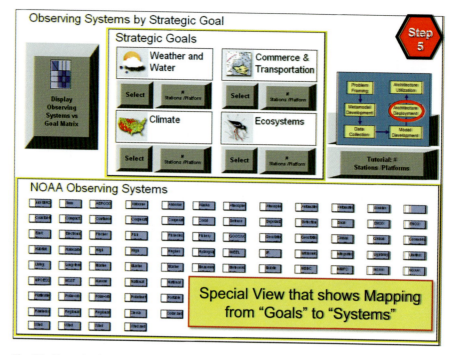

Fig. 24 Example view that answers one of the key business questions from Step 1

support decisions for the program or project. The Problem Framing approach (outlined herein) has proven to be a very good approach for laying the groundwork for models to be used in systems engineering, system integration, and program management. Problem Framing focuses on:

1. Identifying intended uses and users of the models
2. Focusing effort on the questions that can be answered with the models
3. Specification of the views and products that will answer these questions in support of the intended uses and users

This approach is based on established industry best practices in MBSE and in enterprise architecture development and management. The Problem Framing steps illustrated in Fig. 30 highlight the fact that preparing to do the modeling and collecting relevant data should only be done after you have a clear understanding of the problem to be addressed.

It is not uncommon for many modeling efforts to start with step 5 (building the models!) without first stepping back and understanding why you are building the models in the first place.

Step 6: Architecture Utilization

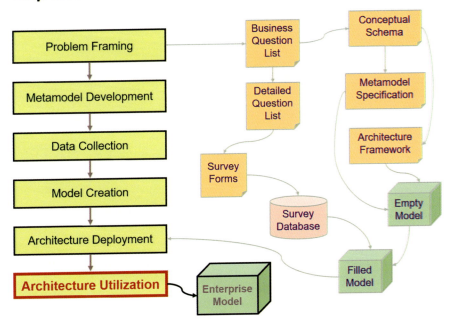

Fig. 25 Step 6 is where the architecture views and models are used in the ways identified in Step 1 of problem framing

Problem Framing Steps

Details of each step in the Problem Framing approach are outlined below. These of course must be tailored for the situation based on the goals and objectives of the modeling effort.

Step 1.1: Intended Users and Uses of the Models

This step is focused on intended users and uses of the models (Fig. 31). Decision makers often have need of the information that can be obtained from examination or analysis of system and enterprise models. The issues to be explored and the questions to be answered with the models are identified in this step. Analytical models often rely on information from descriptive models of the system. The interests and perspectives of intended audience for the models must be understood.

The system life cycle processes in the ISO 15288 standard [3] can be examined for potential uses of the models. Models, to be most effective, should not be restricted merely to technical processes such as requirements, architecture, integration, and verification. They could also be of potential use in the technical management processes (e.g., project planning, decision management, and risk management)

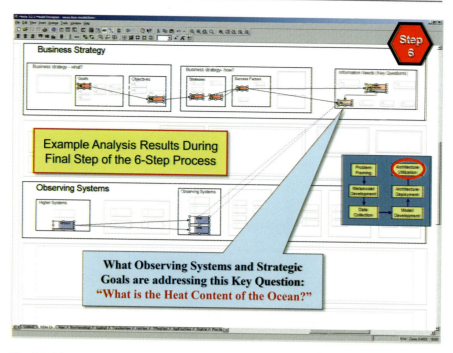

Fig. 26 Views are created (as reflected from models) primarily to answer key questions from important stakeholders

and in the organizational project-enabling processes (e.g., life cycle model management, portfolio management, and knowledge management) that are specified in the ISO 15288 standard. Likewise, models can be of great use in the agreement processes to help manage flow of information between contractors and acquisition agencies.

Step 1.2: Scope and Context of the Models

It is not uncommon to experience "scope creep" when modeling a system. People come along and ask if you could generate yet another view, could you enhance the model with this new data, and could you incorporate these additional scenarios in the architecture models? Without a clear idea of which models and views are of the greatest value for the program, it is easy to succumb to the wishes of those who ask for "just a little bit more." And modelers tend to be an optimistic lot and are often eager to please, thinking to themselves that it will just take a few minutes.

The scope and context need to be carefully laid out during this step (Fig. 32), while considering the environmental situation and the expected operational scenarios. There may be particular constraints to consider such as mandated modeling environments and engineering practices, contractual deliverables, investments in tools and methods, and less than ideal knowledge and experience of modelers and engineers, as well as limited time and budget for such efforts.

8 Problem Framing: Identifying the Right Models for the Job

Repeating the Process (ie, Incremental Development)
Creating useful results early and often…

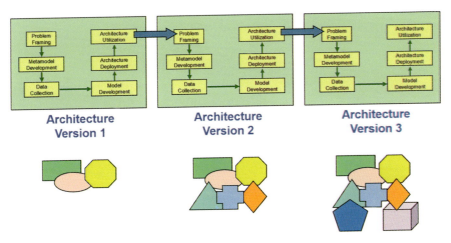

Fig. 27 Incremental development of the model set is a common best practice

Step 1.3: Information and Data Needs
Given the intended uses and users, and keeping in mind the scope and context, the kinds of information and data that are needed can now be defined (Fig. 33). Sources of these items should be identified and validated. There could be mandated or expected ways to present the information coming from the views, or particular formats for the data to be collected and ingested. You need to identify and track down related architectures, designs, and other system technical data.

Step 1.4: Model Views and Products
After the information and data needs are understood, then you can identify the views and products to be generated (Fig. 34). Even if there is already an architecture framework you are using, the best views might not be specified by that framework, so you should consider either extending that framework or choosing particular views from other frameworks. The various engineering disciplines also may have particular views they need for their work and in some cases already have a specific modeling template or format that must be adhered to. Map these views and products back to the intended uses and users, the questions to be answered, issues to be explored, and the analyses to be performed, as well as the decisions to be supported. This mapping will help prioritize the sequencing of model development and the urgency of getting access to specific datasets. The questions, issues, analyses, and decisions should have been prioritized during step 1 to ensure that models of the greatest value will be developed based on the greatest need.

Problem Framing: First step in the journey...

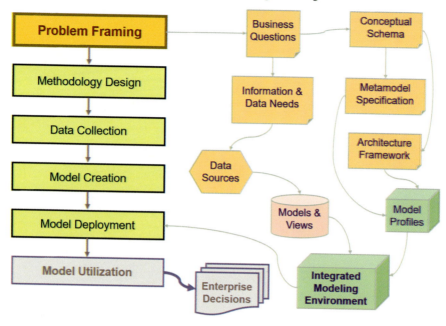

Fig. 28 Methodology design (in addition to Problem Framing) is key to having an efficient and effective modeling approach

Problem Framing Execution

Problem Framing is often done best in a workshop setting with key participants in attendance who have a stake in the outcome of the modeling effort. The workshop leader should have good facilitation and team leading skills and should have significant experience in modeling systems. A workshop can be fast and easy, typically held over a one to two-day period. It can be an excellent opportunity for team building, bringing together various stakeholders on a project that do not normally have much time to work with each other.

It can be done in less than half a day, but experience has shown that this is often not enough time to get the team warmed up and really working well as a group. However, if schedule does not permit full-day sessions, this can be done in half-day sessions spread out over a few days or weeks. However, it is important to not put too much time between sessions to avoid losing momentum.

Advance Planning. It has been found helpful to have a planning meeting with the sponsor ahead of time to help set expectations and to understand their goals and objectives. Think of this planning meeting as a chance to do some good, old-fashioned "purpose analysis." Often the customer will say "we want to implement MBSE" without stating the reasons why or what they are trying to accomplish through such an initiative. Have this planning session about one or two weeks before

8 Problem Framing: Identifying the Right Models for the Job

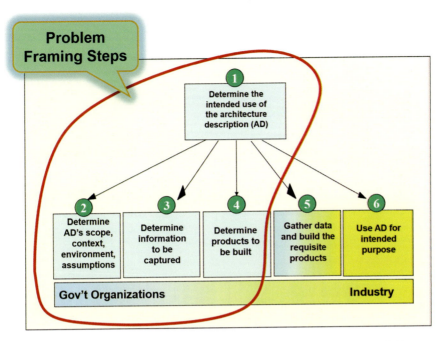

Fig. 29 First four steps of the DODAF process comprise the key Problem Framing activities

the workshop event. This allows enough time to gather relevant source material, identify the right people to attend the workshop, and assign some "homework" to the participants so they can come in with a "strawman" set of ideas with regards to what they think the problem is they are trying to solve.

Homework Assignment. This homework can be stated in terms of having the workshop participants do Step 1 of Problem Framing on their own ahead of time and bring these results with them to the workshop. Ideally, this should be shared with all the other participants before they arrive so they can start pondering other people's conception of the problem areas and the expected outcomes of the modeling effort. The homework should at a minimum ask them to identify the decisions that models can support, the users and uses of these models, the questions that can be answered with such models, and the analyses that will rely on information from the models.

Sponsor Participation. It is good to start the workshop with the sponsor giving the team their expectations for the workshop and for the entire initiative. Often the sponsor cannot stay for the entire duration of the workshop, but it is best if the sponsor can at least come back at the end so the team can give the sponsor an out-briefing on what happened at the workshop, discuss key findings and expected challenges, and establish plans and commitments for next steps.

Storyboarding. During the workshop, it is a good idea to create "storyboards" of the key ideas, along with what the models and views will look like, and what kinds of information they will contain. Put these sketches on large sheets of chapter and hang

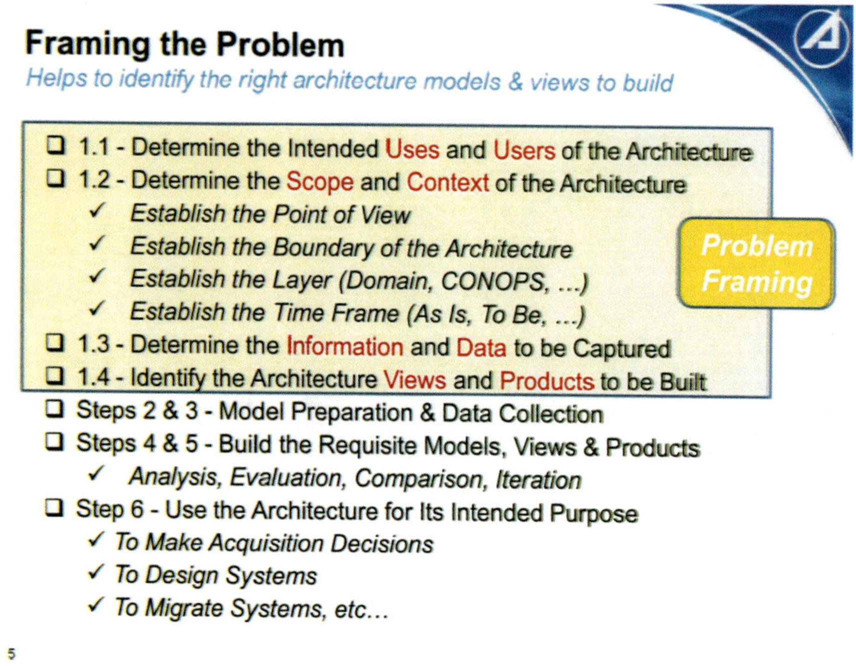

Framing the Problem
Helps to identify the right architecture models & views to build

- 1.1 - Determine the Intended Uses and Users of the Architecture
- 1.2 - Determine the Scope and Context of the Architecture
  - ✓ Establish the Point of View
  - ✓ Establish the Boundary of the Architecture
  - ✓ Establish the Layer (Domain, CONOPS, ...)
  - ✓ Establish the Time Frame (As Is, To Be, ...)
- 1.3 - Determine the Information and Data to be Captured
- 1.4 - Identify the Architecture Views and Products to be Built
- Steps 2 & 3 - Model Preparation & Data Collection
- Steps 4 & 5 - Build the Requisite Models, Views & Products
  - ✓ Analysis, Evaluation, Comparison, Iteration
- Step 6 - Use the Architecture for Its Intended Purpose
  - ✓ To Make Acquisition Decisions
  - ✓ To Design Systems
  - ✓ To Migrate Systems, etc...

Problem Framing

Fig. 30 Summary of the Problem Framing steps prior to building the models

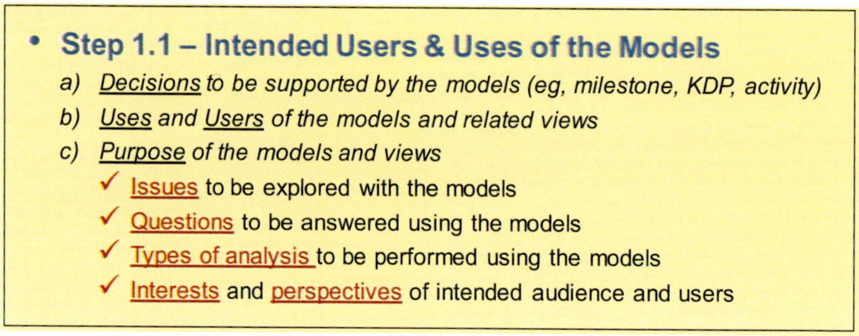

- **Step 1.1 – Intended Users & Uses of the Models**
  - a) <u>Decisions</u> to be supported by the models (eg, milestone, KDP, activity)
  - b) <u>Uses</u> and <u>Users</u> of the models and related views
  - c) <u>Purpose</u> of the models and views
    - ✓ <u>Issues</u> to be explored with the models
    - ✓ <u>Questions</u> to be answered using the models
    - ✓ <u>Types of analysis</u> to be performed using the models
    - ✓ <u>Interests</u> and <u>perspectives</u> of intended audience and users

Fig. 31 Identification of intended users and uses of the architecture models

them on the wall. Having these visible to everyone will serve to stimulate further ideas and to give the team a sense of accomplishment as the walls get plastered. Some of these ideas will end up on the cutting room floor, but that is okay. The important thing is to allow a free flow of ideas, and to quickly capture these fleeting thoughts before they are forgotten. At the end of each day, take pictures of these charts and paste them into a PowerPoint or Word file with notes added to augment these pictures. This will become a nice record of the workshop proceedings without a lot of effort.

8 Problem Framing: Identifying the Right Models for the Job   283

- **Step 1.2 – Scope & Context of the Models**
    a) Scope (ie, Activities, functions, organizations, timeframes, boundaries, layers, etc)
    b) Context (ie, What is the bigger picture? Who are the mission partners?)
    c) Points of view (eg, EA, SE, PM, program office, end user, operator, maintainer, etc)
    d) Environment (eg, Technology, budget, programmatic)
    e) Operational scenarios, situation(s), geographical areas
    f) Major constraints (eg, mandated products/formats, frameworks/tools)
    g) Other key assumptions

Fig. 32 Determination of relevant scope and context of the models

- **Step 1.3 – Information & Data Needs**
    a) Information to be collected for use in generating the products
    b) Precision and granularity of needed information
    c) Expected presentation form or method
    d) Previous or related architectures that can be "mined" for information or data
    e) Potential sources of this information or related data

Fig. 33 Determination of information and data needs

- **Step 1.4 – Model Views & Products**
    a) Types of views and products needed (that serve the intended uses from Step 1)
    b) Identify specific views and products that address the needs
        - Existing views and products (if applicable)
        - New views and products
    c) Contents, structure and form of each item
        - Questions addressed by each
        - Activities support by each
    d) Tools, models, data and other resources needed to develop these views and models
    e) Frameworks and modeling approaches to be used

Captured in a Conceptual Schema and Storyboards

Fig. 34 Identification of required model views and products

Afterwards. Following the workshop, it is important to build rapid prototypes of the models to be developed. Spend only one or two weeks to produce the first (partial) model of the system, enterprise, situation, or whatever is the focus of the modeling effort. Use "fictional" data if necessary to get quick results. The purpose of these prototypes is twofold – first, to find out if the models and views are going to answer the questions identified in Step 1, and second to determine if they will be of value for the intended users and provide the right information for the planned analyses.

Prototyping. Spend some time doing a walkthrough of these prototypes with key stakeholders to ensure they understand what is going to be produced and get their feedback on how well they will serve their needs. Another benefit is that you can start to get buy-in from these stakeholders and gain their confidence that this modeling effort will be a worthy investment and something they can look forward to. You will need to rely on many of the stakeholders to provide you with data to populate your models. To enlist their support, they need to see that there is something in it for them.

Incremental Development. Continue to develop additional prototypes to incorporate ideas from the stakeholders and intended users. Replace the temporary fictional data with real data as it becomes available. You will be surprised how often you can gain really significant insights into the system you are going to develop even when the model is only partially built (and even when it has some fake data in it!). You may find that you need to revisit the Problem Framing, perhaps even having a mini-workshop to reexamine the intended users and uses, and possibly defining new questions to be answered.

Reframing the Problem. You may need to revise the conceptual schema to accommodate things you have learned about the problem domain. Keep doing multiple spirals until you are all finished... Of course, modeling is never really done, but at least if you have done a good job in getting the right questions to be answered, you are more likely to have models that are deemed worth the investment in time and effort.

Summary

Choosing the right models will increase the positive effect that systems engineering will have on project success. The Problem Framing approach presented here has proven to be a good way to get the modeling effort started in the right direction, and to keep it focused on the models that will have the greatest value to the various potential users. If you did not start out this way, it never hurts to stop for a bit and go back to do some problem framing!

References

1. Bernus, P., N. Laszlo, and G. Schmidt, eds. 2003. *Handbook on enterprise architecture*, eds. L. Nemes, G. Schmidt. Heidelberg, Germany: Springer Berlin Heidelberg.
2. DoD. 2009. *DoD architecture framework (DODAF)*, version 2.0. Washington, DC: U.S. Department of Defense (DoD).
3. ISO/IEC/IEEE 15288-2015. *Systems Engineering — System Life Cycle Processes*. International Standards Organization (ISO).
4. ISO/IEC/IEEE 42010-2022. *Enterprise, Systems and Software — Architecture Description*. International Standards Organization (ISO).
5. ISO/IEC/IEEE 42020-2019. *Enterprise, Systems and Software— Architecture Processes*. International Standards Organization (ISO).

6. Martin, JN, 2003a, "On the Use of Knowledge Modeling Tools and Techniques to Characterize the NOAA Observing System Architecture." *INCOSE Proceedings*, INCOSE Symposium.
7. Martin, JN, 2003b, "An Integrated Tool Suite for the NOAA Observing System Architecture." *INCOSE Proceedings*, INCOSE Symposium.
8. NATO 2018, *NATO architecture framework (NAF)*, version 4. Brussels, Belgium: North Atlantic Treaty Organization.
9. OMG 2022, *Unified Architecture Framework*, Object Management Group.
10. UAM 2018, *Unified Architecture Method (UAM)*, http://www.unified-am.com, Accessed 10/19/18.
11. Zachman, JA 1992, "Extending and formalizing the framework for information systems architecture." *IBM Systems Journal* 31 (3): 590-616.
12. Zachman, JA 1987, "A framework for information systems architectures." *IBM Systems Journal* 26 (3): 276-92.

James N. Martin is an Enterprise Architect and a Distinguished Systems Engineer affiliated with The Aerospace Corporation developing solutions for information systems and space systems. Dr. Martin is developer of architecture and modeling methodologies for various clients in the Defense and Aerospace domains. He was a key author on the BKCASE project in development of Enterprise Systems Engineering articles for the SE Body of Knowledge (SEBOK). Dr. Martin led the working group responsible for developing ANSI/EIA 632, a US national standard that defines the processes for engineering a system. He previously worked for Raytheon Systems Company and AT&T Bell Labs on airborne and underwater systems and on communication systems. His book, Systems Engineering Guidebook, was published by CRC Press in 1996. Dr. Martin is an INCOSE Fellow and was leader of the Standards Technical Committee. He was founder and leader of the Systems Science Working Group. He received from INCOSE the Founders Award for his long and distinguished achievements in the field.

Part III

Technical and Management Aspects of MBSE

Model-Based System Architecting and Decision-Making

9

Yaroslav Menshenin, Yaniv Mordecai, Edward F. Crawley, and Bruce G. Cameron

Contents

Introduction	290
Model-Based System Architecting: Crossing a Mental Grand Canyon	292
A Tango of Conceptualizations and Decisions	293
Model-Based Concept Representation	296
System Architecture Framework	296
The Stakeholder Domain (D1)	297
The Solution-Neutral Environment (D2)	298
The Solution-Specific Environment (D3)	299
The Integrated Concept (D4)	301
The Concept of Operations (D5)	303
The Scope of an MBSA Application	304
MBSA and Architectural Decision-Making	305
What is a Decision and Which Decisions are Architectural?	306
Concept Attributes, Metrics, and Decision-Supporting Criteria	309
Capturing Stakeholder Needs	310
Capturing and Discovering Possible Architectures	315
Capturing the Architectural Decision-Making Process Alongside the Resulting Architecture	318
Solution-Specific Architecture Decisions	321
Conclusion	325
References	327

"The straight line, a respectable optical illusion which ruins many a man." – Victor Hugo, Les Misérables

---

Y. Menshenin (✉)
Skolkovo Institute of Science and Technology, Moscow, Russia
e-mail: y.menshenin@skoltech.ru

Y. Mordecai · E. F. Crawley · B. G. Cameron
Massachusetts Institute of Technology, Cambridge, MA, USA
e-mail: yanivm@mit.edu; crawley@mit.edu; bcameron@mit.edu

© Springer Nature Switzerland AG 2023
A. M. Madni et al. (eds.), *Handbook of Model-Based Systems Engineering*,
https://doi.org/10.1007/978-3-030-93582-5_17

Abstract

We explore the application of MBSE for conceptual system architecting. Choosing an architecture is a fundamental activity. Our Model-Based System Architecting (MBSA) framework facilitates the specification of an architecture as a reasoning process – a series of conceptualization and decision-making activities, backed-up by an MBSE environment. Our framework captures both the ontology of a stakeholder-driven and solution-oriented system architecture, and the process of growing the architecture as a series of conceptualization steps through five ontological domains: the stakeholder domain, the solution-neutral environment, the solution-specific environment, the integrated concept, and the concept of operations. Our MBSA approach shifts the modeling focus from recording to conceptualizing, exploring, decision-making, and innovating. In comparison to an "offline" architecting process, our approach may initially require a bigger effort but should enable stronger stakeholder engagement, clearer architectural decision point framing, quicker exploration, better long-term viability, and increased model robustness.

Keywords

Model-Based System Architecting · Model-Based Systems Engineering · Architectural Decision-Making · Object-Process Methodology · Concept Representation

Introduction

Choosing an architecture for a complex system, sometimes called the "fuzzy front end" of design, is a task rife with ambiguity. Traditional approaches have relied on a federated mixture of informal, semiformal, and formal methods. The growing challenge systems face today has made these "offline" approaches largely obsolete. Model-Based Systems Engineering (MBSE) [1] is gradually becoming a mainstream approach for practicing systems engineering. However, while traditional systems engineering works to capture missing connections between subsystems, MBSE today is focused on the descriptive recording of concepts in models [2]. This concept representation is essential for further processing, analysis, and presentation, but it is only one aspect of systems engineering. Current practice and research are overweight with the representational effort in MBSE, and underweight on analysis and decision-making. Similarly, software engineers are expected to deliver operational, functional, secure, and efficient software regardless of the programming languages and software development environment they use; mechanical engineers are expected to deliver valid, verified, buildable, and maintainable part and component designs, regardless of the design technology they design with, etc. Nevertheless, in the current landscape of digital engineering [3], no one imagines that software, hardware, or mechanical engineers will not employ the latest software to manage, design, implement, test, and deploy their deliverables. Systems engineering should be no exception.

We explore the ways in which MBSE can be used to support system architecting, and to ensure that the process remains rigorous and insightful. Reaching a good system architecture must be inherent in any MBSE approach. Accordingly, our model-based system architecting (MBSA) approach uses models and analysis of MBSE to choose an architecture. It is not a detached adaptation or variation of MBSE to system architecting.

Much has been written about the descriptive aspects of MBSE, whether it be in cataloging functional flow or in defining potential system states. However, this documentation does not necessarily support architectural decision-making unless it presents decision points. A decision point could be an opportunity to choose a solution from at least two options. We define what we consider architectural, in order to evaluate how and where MBSE supports decision-making about architecture.

The effort involved in building an MBSE environment and the associated cultural transformation imply that the scope and purpose must be crisply defined, so as to rationalize the investment in MBSE. One of these purposes (but by no means the only one) is to support architectural decision-making. MBSA includes the following cycle of activities in scope:

1. representing potential architectures with models,
2. identifying architectural decisions,
3. conducting analysis in support of emerging architectural decisions,
4. making architectural decisions based on model analysis results, and,
5. capturing decisions in the model for the next architecting iteration.

Model-Based Conceptual Design (MBCD) resembles MBSA. MBCD is the application of MBSE to tradespace exploration during the conceptual stages of systems engineering [4]. The activities performed during the conceptual stages of system engineering are defined as architecting, and their main outcome is an architecture – a holistic view of the entire system. By contrast, activities performed to realize the architecture, particularly planning solutions with engineering and scientific knowhow – are considered as designing – where the main outcome is the design: a blueprint for developers to implement or build the system. A complex component's design may constitute architecting for that component as a bona fide system, e.g., the jet engine in an airplane, or a communication network that connects many sensors and controllers.

MBSA has also been used as an acronym for Model-Based System Architecture [5], in a framework which uses the Systems Modeling Language (SysML) [6]. That approach focused on providing a repository of artifacts, which facilitate communicating with stakeholders, assuring requirements traceability, and specifying systems and subsystems. We employ the MBSE paradigm as a reasoning mechanism, and not only as a documentation approach, because we believe that it generates additional value to stakeholders.

Previously, the Model-Based System Architecting and Software Engineering (MBASE) approach [7] advocated a holistic process for software architectures, software lifecycle guidance. The MBASE approach was in-fact document-centric. The model-based ecosystems were not yet mature enough to accommodate a complete system architecting, design, development, deployment, and operation thread. Therefore, MBASE started with an Operational Concept Description (OCD), but focused on generating documents like the System and Software Requirements Definition (SSRD), System and Software Architecture Description (SSAD), Life Cycle Plan (LCP), and Feasibility Rationale Description (FRD) [8]. It also concerned some critical aspects in software deployment such as iterative development, following the Spiral paradigm [9], transition, and software support.

The discussion on the necessity, relevance, and sufficiency of system architecting, particularly in software-intensive systems, has been ongoing especially with the appearance of short-cycle iterative and continuous development and deployment approaches [10]. The key argument remains, and gets validated in many famous failures [11] that holistic system architecting increases confidence in the ability to meet stakeholder needs, develop a robust architecture that can adapt to changes, and reduce the amount of technical debt as the system evolves [12].

Many of the building blocks described in this chapter originate from previous holistic frameworks for system architecting [13], in which modeling played a key role in concept description, but could not yet be regarded as a fully model-based approach.

Bahill and Madni introduce a model-based approach known as the SIMILAR process, which stands for: (a) Stating the Problem, (b) Investigating Alternatives, (c) Modeling the System, (d) Integrating Components, (e) Launching the System, (f) Assessing Performance, and (g) Re-evaluating the System [14]. The MBSA approach that we proposed focuses and extends on the early stages in the SIMILAR framework and especially on early iterations in which the conceptual architecture is the main artifact, and little or no physical components are available.

Model-Based System Architecting: Crossing a Mental Grand Canyon

MBSA often begins with concept brainstorming in response to some need or set of needs, and ends with a formalized review and sign off of a well-defined and buildable architecture. In between, there is a series of conceptualizations and decisions: The leap from stakeholder needs to a well-defined organization of structural and behavioral elements does not happen overnight. This mental 'Grand Canyon' is simply too wide to jump all at once, and a series of intermediate steps is necessary. The question is: how can we wisely plan these steps that will lead us safely to the other side? This idea is illustrated in Fig. 1.

A system architecture is a description of the structure and behavior of a system that jointly provide one or more functions to serve the needs of system stakeholders. MBSA relies on a formal modeling language to capture, present, and reason about

9 Model-Based System Architecting and Decision-Making

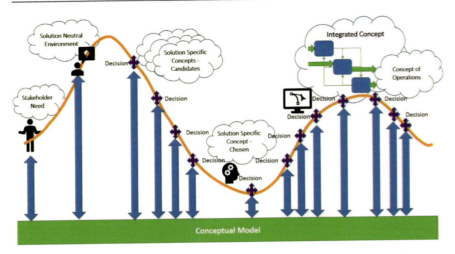

Fig. 1 Crossing the mental Grand Canyon from needs to operationally-viable solutions, through a series of system architecture decisions, using a model as the knowledge base and primary reasoning engine

the system architecture, but the deliverables are essentially the same as those of the traditional (not model-based) process: a specification of the system architecture, which can serve as the basis for further requirement specification, design, development, testing, and operation. This high-level concept of MBSA is illustrated in Fig. 2. We shall be using Object-Process Methodology (OPM) [1] as a model for this chapter, due its relative simplicity (using OPCloud, and its automatically generated text specifications [15]). (The title of Fig. 2 is directly drawn from the text specification that OPCloud generates for this diagram, making it an unambiguous description of the diagram.). The complete reference model for our MBSA framework is included in [16]. In Fig. 2 the objects (such as "Stakeholder" and "System Architect") are denoted by rectangles, whereas the process ("Model-Based System Architecting") is denoted by oval. The filled in black triangle inside a triangle means that the "Need" is the attribute of "Stakeholder." The link with arrow informs about the consumption of the attribute "Need" by the "Model-Based System Architecting" process. The link with filled in circle at the end is the agent link ("System Architect"), whereas the link with the open circle at the end is the instrument link ("Model-Based Systems Engineering Environment"). The full description of the OPM symbols can be found in [1].

A Tango of Conceptualizations and Decisions

A concept is an initial mapping of what we want to accomplish to the form that will be used to accomplish it. For example, "the rocket will land upright using stabilizer fins," "the vehicle will work on both fuel and electrical power," or "all the communications will go through the central message hub." The concept is part of the system

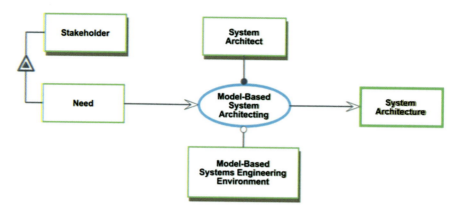

Fig. 2 Model-Based System Architecting: Stakeholder exhibits Need. Model-Based System Architecting consumes Need of Stakeholder. System Architect handles Model-Based System Architecting. Model-Based System Architecting requires Model-based Systems Engineering Environment. Model-Based System Architecting yields System Architecture

architecture and should be specified appropriately within the scope of the MBSA process.

A concept maps function to form [13]. The function of a system is a process (an activity), which typically affects one or more operands (the objects that are changed by the activity). The form is a set of elements that support this function. This is analogous to the three core parts of all languages being the noun (instrument of the action), verb (activity that describes the action), and noun (the object of the action) [17]. Figure 3 illustrates the basic pattern of a concept and the association among the concept, function, form, and architecture.

The highest-level concept of the entire architecture should be a short phrase or sentence. For example, "SelfDriving Car handles Transporting of up to 4 Passengers to a distance of 500km." In this short example, we clearly see (a) the process: Transporting, (b) the form: Self-Driving Car, (c) the operand: Passenger (up to 4). Additionally, this statement includes an optional attribute: Distance (up to 500 km), which may be drawn from some need.

The long journey from needs to solutions passes through a series of steps and is by no means a straight line. Many of these steps go back and forth in what could be imagined as a tango dance. Many mental models have been proposed for this series of steps, most notably the V model and other examples [18], which mostly advocate a procedure of activities. We present a generic classification. We argue that at each point, architecting is either one of two cognitive tasks: conceptualizing or deciding. Conceptualizing is describing or specifying concepts, while deciding is selecting concepts from the available candidate pool. After deciding, the decision becomes part of the solution. Conceptualizing and deciding are collectively referred to as *reasoning*. Each architecting step is a reasoning step, and MBSA is a series of reasoning steps, as shown in Fig. 4. Both types of reasoning – conceptualization and deciding – can benefit from a model-based approach.

9 Model-Based System Architecting and Decision-Making

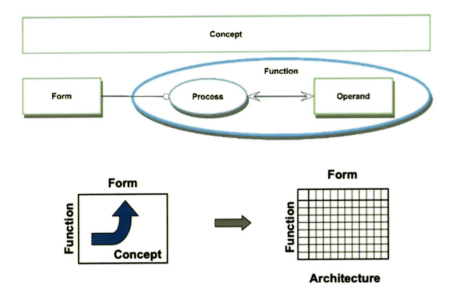

Fig. 3 **Concept**: Form enables Function; Function = Process that affects an Operand. The allocation of Function to Form leads to an Architecture [13]

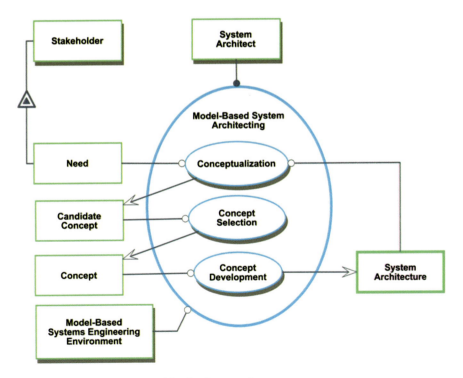

Fig. 4 Model-Based System Architecting in-zoomed

Model-Based Concept Representation

In this section, we discuss a conceptualization process, built around a concept representation framework, supported by OPM modeling language. In subsection "System Architecture Framework," we define an ontology for system architecting with five domains. Subsections "The Stakeholder Domain (D1)," "The Solution-Neutral Environment (D2)," "The Solution-Specific Environment (D3)," "The Integrated Concept (D4)," "The Concept of Operations (D5)" sequentially reveal each one of the ontological domains through a conceptual reference model. The iterative nature of the system design process is demonstrated through the interplay between the domains. The Scope of an MBSA Application is explained in subsection "The Scope of an MBSA Application." We then use this framework to examine how MBSA changes the system architecting process.

System Architecture Framework

An ontology is a formal vocabulary of domain concepts. It is critical to adopt and formalize an ontology for a coherent discussion about concept representation, particularly for system architectures. Our ontology is illustrated in Fig. 5. This ontology underpins a framework that introduces the core entries within the system architecture concept representation, and proposes ways to encode these entries,

	Domain 1 (D1) Stakeholders				
1	Stakeholder				
2	Need				
	Domain 2 (D2) Solution-Neutral Environment		**Domain 3 (D3) Solution-Specific Environment**		**Domain 4 (D4) Integrated Concept**
3	Solution-neutral operand (SNO)	8	Solution-specific operand (SSO)	17	Internal Operands (IO)
4	SNO value attribute	9	SSO value attribute	18	IO value attribute
5	SNO other attribute	10	SSO other attribute	19	IO other attribute
6	Solution-neutral process (SNP)	11	Solution-specific process (SSP)	20	Internal Processes (IP)
7	SNP attribute	12	SSP attribute	21	IP attribute
		13	Generic Form	22	Internal Elements of Form (IEoF)
		14	Generic Form attribute	23	IEoF attribute
		15	Specific Form	24	Structure
		16	Specific Form attribute	25	Interactions
					Domain 5 (D5) Concept of Operations
				26	Concept of Operations
				27	Operator
				28	Context

Fig. 5 A System Architecture Ontology and a Concept Representation Framework, adapted from [19, 20]

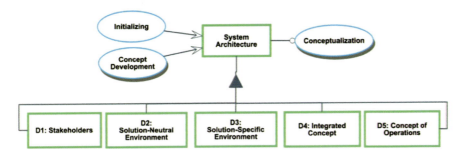

Fig. 6 A System Architecture ontology with five concept domains

preferably in a modeling environment. The ontology consists of five domains: Stakeholders (D1), Solution-Neutral Environment (D2), Solution-Specific Environment (D3), Integrated Concept (D4), and Concept of Operations (D5). A concept domain is a subset of the ontology, which focuses on a specific aspect of the architecture, and has a mapping to other. Domains are distinguished by color. We list 28 entries within these domains, based on a concept representation framework introduced in [19, 20]. We reference these 28 entries using {EXX}, such as {E15} referring to Specific Form.

The first three domains, D1–D3, represent the simplest formulation of a concept. The fourth and fifth domains lie downstream to reflect a latent termination, i.e., extending D1–D3 as long as it is still appropriate to continue detailing the architecture. The exact timing for terminating varies with solution types and contexts [19, 20]. D1–D3 are distinguished from D4–D5 in the abstract vs. specific levels of discussion. The system architect should be comfortable with switching from the abstract discussion (D1, D2, and D3) to a more concrete level of detail (covered by D4 and D5). Moreover, iterative system architecting means that D5 can impact D1, in a cycle of revising, diverging, and converging. Figure 6 shows the system architecture as a composition of the domains.

The Stakeholder Domain (D1)

The Stakeholder domain (D1) captures stakeholders and their needs. A stakeholder {E01} is "any group or individual who can affect or is affected by the achievement of the system's objectives" [21]. In other words, many groups and individuals can be stakeholders in the broadest sense, depending on the context. This emphasized the importance of a broad operational context in which the system is intended to operate. Consider the European "Green Deal" [22]: according to the given definition, all humans on Earth are Green Deal stakeholders.

Stakeholder needs {E02} are defined as answers to the question "What problems are we trying to solve?" [23]. Needs should be problem-oriented and not solution-oriented. Needs are often specified before the system architect gets involved. Needs are often fuzzy, ambiguous statement by stakeholders. This fuzziness challenges

Fig. 7 The stakeholders domain (D1): stakeholder {E01} exhibits need {E02}

system architects to clearly formulate the essence of the need, e.g., what is the expected capability, or expected outcome, or expected change to the current state. The special importance of the stakeholder needs is that they are used to formulate functional requirements in a solution-neutral environment.

The stakeholder needs might come from the variety of the sources. The first of them is the stakeholders themselves: this is the task of the system architect to frame the discussion with stakeholders in such a way that would help formulating those needs. Another potential sources of needs are the use cases, constraints, requirements that might come from extensive literature review.

The system architect's goal is to formulate the functional intent in each stakeholder need. Needs are associated with the problem statement first, expressed in the solution-neutral environment and realized through the process. Figure 7 illustrates the stakeholders domain (D1) in which the stakeholders are denoted by rectangle and need is defined as an attribute (denoted by a black triangle inside a triangle) of stakeholders.

The Solution-Neutral Environment (D2)

The need for a solution-neutral environment is a fundamental design principle [24]. The solution-neutral environment (D2) facilitates the elicitation of functional requirements, which must be free of any bias toward prospective solution approaches, specific technical disciplines, or implementation strategies [25]. Therefore, the system architect specifies the essential information about the solution-neutral process before solution concept development. The functional intent (such as "transporting passengers," "transferring money," or "playing music") should be formulated before the possible alternative solutions are set up.

Figure 8 encodes the solution-neutral environment (SNE) and its entries. The solution-neutral operand (SNO) {E03} is an object of interest that will undergo some transformation by the solution-neutral process (SNP) {E06}. The solution-neutral process manifests the dynamic nature of the function: it reflects the action. The SNP and SNO should be abstract so that a variety of alternatives will emerge and an

9 Model-Based System Architecting and Decision-Making

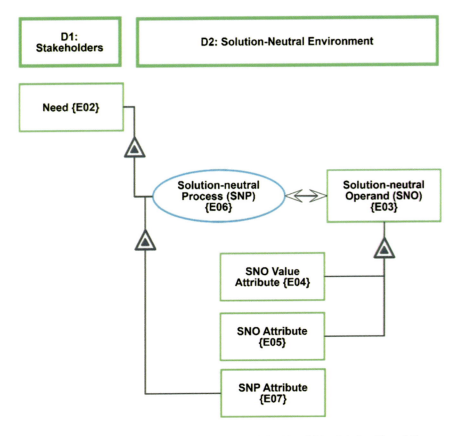

Fig. 8 The Solution-Neutral Environment (D2): Need {E02} exhibits Solution-Neutral Process (SNP) {E06}. Solution-Neutral Process (SNP) {E06} affects Solution-Neutral Operand (SNO) {E03}. Solution-Neutral Operand (SNO) {E03} exhibits SNO Value Attribute {E04} and SNO Attribute {E05}. Solution-Neutral Process (SNP) {E06} exhibits SNP Attribute {E07}

informed decision-making process will take place. The SNE entries may have attributes, which are appropriate to start elaborating at this stage.

SNP {E06} maps to the need {E02}, which is specified in D1, as shown in Fig. 8. Need is realized via the performance of some process – the SNP and the consumption, transformation, or generation of some operand – the SNO. Changes in need are likely to entail changes in SNP.

The Solution-Specific Environment (D3)

The solution-specific environment (D3) encodes alternative architectures. A solution entry defines how the system is going to perform the solution-neutral functions. Solution-specific processes (SSPs) and solution-specific operands (SSOs) are mapped to their solution-neutral counterparts (which were presented in D2), such that it is clear which solution-specific entry attempts to realize each solution-neutral one. For example, "Transporting by Air (Flying)," "Transporting by Land (Rolling)," and "Transporting by Sea (Sailing)" are SSPs refining the SNP "Transporting."

The solution-specific environment is derived from the solution-neutral one via the generalization-specialization relation (drawn as a blank triangle in OPM, as shown in Fig. 9). The SNO "person" generalizes the SSO "passenger" in a transportation context, "patient" in a medical context, and "user" in a technological context. Domain jargon can better describe the artifacts, entities, and human roles (e.g., the SSO "exoplanet" in deep space exploration).

The number of possible solutions is a product of the number of D3 entries per D2 entries, therefore it increases with every additional solution-specific entry. However, the specification of solutions also narrows down the funnel of possible solutions. While the solution-neutral environment leaves room open for as many solutions as

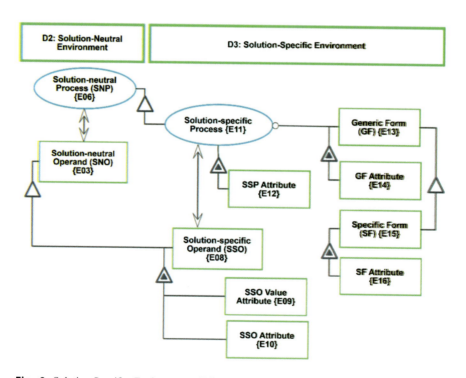

Fig. 9 Solution-Specific Environment (D3): Solution-Specific Process (SSP) {E11} affects Solution-Specific Operand (SSO) {E08}. Generic Form (GF) {E13} enables Solution-Specific Process (SSP) {E11}. Solution-Specific Operand (SSO) {E08} exhibits SSO Value Attribute {E09} and SSO Attribute {E10}. Solution-Specific Process (SSP) {E11} exhibits SSP Attribute {E12}. Generic Form (GF) {E13} exhibits GF Attribute {E14}. Specific Form (SF) {E15} exhibits SF Attribute {E16}

possible, solution-specific entries identify specific ways that realize the solution-neutral intent to choose from, and close the door to other unlisted ideas.

The solution-specific environment may also be associated with the principal solution – the deliverable of conceptual design [26]. A principal solution is a concept, and the early outline of an architecture. The key purpose of the solution-specific environment is to discover the architecture by specifying those forms as principal solutions. This is achieved by specifying Generic Form entities {E13} and associating them with the SSOs they enable or support. Every SSP has several optional Generic Forms that may implement it. This is a fundamental conceptual design principle, which, to some extent, further extends the solution space. For example, the SSP "Flying" can be implemented by several Generic Forms, e.g., Airplane, Helicopter, and Drone.

Each Generic Form can be specialized into Specific Forms (SFs) {E15} within the scope of the Generic Form. For example, "Jet Airplane," "Turbo-Prop Airplane," and "Propeller Airplane" are three SFs of the Generic Form "Airplane." The Vertical Take-off and Landing (VTOL) Aircraft concept, which is featured by Lockheed Martin's V-22 "Osprey," is a form with lineage to two Generic Forms: Airplane and Helicopter. Therefore, a Specific Form which is a combination of several Generic Form can be a valid concept. In fact, converging multiple dimensions of Generic Form into a minimal set of specific forms is desired, if it reduces the tradespace into a smaller set of comprehensive, integrated solutions.

The Integrated Concept (D4)

An integrated concept fuses multiple concepts into a cohesive architecture. Two integrated concepts should be distinguishable from each other at a relatively high-level of abstraction (i.e., following a relatively small number of abstraction steps, such as the listing of internal processes or the breakdown into components). The integrated concept must also reach a sufficient level of granularity that allows for the critical transition from system architecture to subsystem design [27]. The integrated concept actually results from decomposition, rather the composition, i.e., by increasing the granularity of the architecture.

The Integrated Concept Domain (D4) encodes the internal processes, operands, structures, and relations, as illustrated in Fig. 10. Digital thread flows from the specific form {E15} in D3. Each function is compounded from an Internal Process {E20} and an Internal Operand {E17}. Internal Element of Form {E22} enables the functions. The structure (physical interaction of elements of form) and interactions (functional relationship of elements of form) between the system concept's entities are demonstrated at the bottom of Fig. 10.

A system architecture captures vertical and horizontal relations [28]. The vertical relations capture the decomposition or breakdown of systems into subsystems. The horizontal relations capture interactions between elements, such as flows of material, energy, or information. D4 caters to both the vertical relations – encoded in the upper part of Fig. 10, and the horizontal ones, encoded in the lower part of Fig. 10.

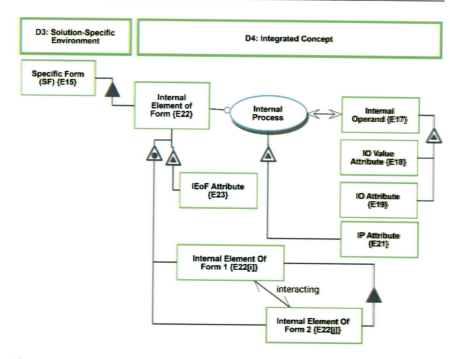

Fig. 10 The Integrated Concept Domain (D4): Internal Process {E20} affects Internal Operand {E17}. Their attributes are specified {E21}, {E18}, {E19}, respectively. The internal elements of form {E22} is used to execute the function. IEoF's attribute is {E23}. The lower part specifies structural and interaction relations {E24} and interactions {E25} among instances of IEoFs {E22}

The multiplicity of potential vertical breakdowns and horizontal interactions gives rise to the concern that the architecture of the integrated concept will become too complicated and messy. We should therefore set bounds on the amount of information to specify at this stage. Miller's Law states that an average human can hold 7 ± 2 objects in short-term memory [29]. It has since become common to assert that 7 ± 2, Miller's Magical Number, is a good limit for complexity, because constructs that include more than 7 ± 2 items are likely to become difficult to grasp.

Completeness and complexity go together in our approach. That is to say: a complete integrated concept at the first level of decomposition (from a specific concept to the set of internal structures), is complete in the sense that it *utilizes its complexity quota*, so to speak: A view that comprises no more than 7 ± 2 elements make a good candidate for completeness of specification. That is not to say that there cannot be more elements. More elements should be clustered with the existing 7 ± 2 elements. Thus, 7 ± 2 is in fact an estimate for sufficiency and a constructive measure of complexity, in the sense that it encourages the architect to converge on this range for complexity management. We can therefore say that a problem that does not converge on a 7 ± 2 element scale at any given level of hierarchy, may not qualify for this approach.

The Concept of Operations (D5)

The Concept of Operations (ConOps) domain (D5) specifies the overall high-level idea of how the system will be used to meet stakeholder expectations [23]. The Department of Defense Architecture Framework (DoDAF) refers to the ConOps as a high-level abstraction graphic that captures how the system will operate, how it will work out together to help the operational stakeholders achieve their goals [30]. We have shown a similar model-based approach for analyzing the DoDAF Operational Viewpoint, which covers the ConOps [31]. The ConOps ties the system concept with the environment, and over time. The ConOps is important as it informs all stakeholders with the context and integrative operation of the system: what processes are to be performed, in which sequence, and how they will be executed by components of the architecture. Eventually, the purpose of the ConOps is to illustrate how the architecture delivers value. ConOps should include both the system of interest, and the accompanying systems that are necessary to consider during the system design process.

D5 focuses on the context represented through the whole product system {E28}, as shown in Fig. 11. It includes the accompanying systems, system enterprise which is responsible for the system of interest {E15}, and operator {E27}. An operator is a

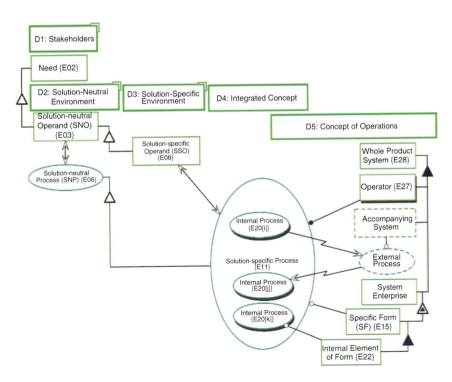

Fig. 11 The Concept of Operations Domain (D5)

person or group of people who operate the system. There is always one higher level in which an architecture resides, unless we aim to architect a universe, which, to the best of our knowledge, is the most inclusive architecture of all.

The context defines how the system interacts with its environment. The same architecture can perform perfectly in one context and poorly in another. For example, a Formula 1 racer will be amazing on the racing tarmac but less adept on the loose surfaces of the Dakar Rally. Even if the architecture remains the same, the context provides the boundaries and constraints in which the solution architecture must operate successfully.

While it might make sense to consider the ConOps earlier in the process, it can also be harmful because setting too many constraints and restrictions limits our ability to come up with good solutions. Consider, for example, that operational stakeholders will impose a ConOps that heavily relies on manual or cognitive actions, while the whole solution can be autonomous or semi-autonomous. We would rather explore multiple options and validate autonomous solutions by finding an appropriate ConOps, rather than try to fit it into a human-intensive environment. Thanks to a paradigm shift in the automotive industry, many more functions are delegated to automation and relieve the driver of the cognitive load rather than intensify the burden.

It is still possible to specify ConOps upfront for an already well-defined operational architecture in which the architect has to integrate a new capability or functionality. The specification of legacy elements, platforms, and reusable assets can be helpful both for solution-specific concept viability and for further-up elicitation of needs. In some cases, a critical analysis and challenging of existing operational concepts may help elicit the true underlying needs of operational stakeholders, which may open up the door to other major architectural enhancements.

ConOps has connections with the other domains, which is illustrated in Fig. 11. The iterative nature of the system design process is embodied in the clear digital thread that starts with stakeholders and their needs and culminates in D5. Figure 11 demonstrates the role of D5 in context representation, as well as inclusion of the system design process in a coherent way in which the domains are interwoven to deliver a value from system operation to meet stakeholders needs.

The Scope of an MBSA Application

The scope of an MBSA project may be a subset of our framework. The system architect may choose to focus only on some domains, depending on how broad or narrow an exploration they desire. Ideally, we would try to model just enough to have a reasonable evaluation of our architectural options, identify evaluation criteria, and move forward with an architecture. MBSA should capture sufficient detail to support the detailed design. Unfortunately, the broader the architectural decisions under consideration, the more general the models must be to account for the breadth of options. The presented framework assumes that the fixed effort available in the architecting phase is a tradeoff between breadth and depth of architectures evaluated.

MBSA is designed to minimize unnecessary effort. If, for instance, a solution-specific environment (D3) is already defined due to various constraints (for instance, implementing some functionality using specific hardware type), we may skip the divergence from solution-neutral environment (D2) and attempt to match the solution-specific environment (D3) with stakeholder needs (D1). Solution-neutral and solution-specific functionalities are defined in a way that clarifies and simplifies the MBSA effort. This approach also helps systems architects focus on those functionalities that are most critical to constitute decision points that would direct the architecture one way or another.

System architecture, like civil architecture, is both science and art [32]. This scope should answer the following questions, including explicitly specifying what lies outside the boundary of MBSA:

1. Which functions must, should, should not, and must not be captured?
2. Could introspecting on the functions of interest yield a re-formulation of the problem? If so, how general must the model be?
3. Which components should and should not be captured?
4. How do we determine if someone is a stakeholder and whether they should be included?
5. How do we evaluate synergies or conflicts in a given architecture?
6. What are the insights derived from the process of architecting beside the outline of a selected architecture, and how do we preserve those insights in order to further inform the design process?
7. At which level of granularity is it sufficient to decompose the system of interest in relation to context and specific needs of stakeholders?

MBSA allows for recording the answers to these questions within the architecture model, and within the context of our concept representation framework – thus extending and empowering the cognitive process done by the system architect in order to consider and answer these questions. Indeed, just like a painting is an artifact of the artistic process, a model can record the emergent propositions that we include in a system architecture, such as elegance, empowerment, holism, and inspiration – all of which are subjective perceptions that we hope stakeholders will experience when presented with the selected system architecture.

MBSA and Architectural Decision-Making

In this section we focus on the value of a model-based process for architectural decision-making. Architectural decisions are those reasoning steps that affect the direction in which an architecture evolves. Decisions are made throughout the process, and some are based on the model. We focus on how model-based system architecting would be different from a traditional "offline" system architecting process. It has been generally asserted that models promote easier design reuse, evaluating more options, and automating design space exploration. We are interested in a deeper question of how we might expect decision-making to change. The mere availability of models has not broadly changed the decision-making process. We ask what is it about MBSE capabilities that would lead us to believe that the process of choosing among potential architectures is different, and better?

R&D stakeholders often make incremental decisions on large programs, before detailed solution comparison tables are available. The front end of design is often ambiguous, and there is a gulf to be crossed between the concept-as-a-napkin-sketch and detailed specifications. We need more rigorous bridges than the traditional ones – slideshow illustrations and overly complex draft box-and-line drawings.

The ability to visualize decision-supporting information as discussed in the previous section had taken a back seat while the MBSE community had focused on the modeling environment and the user experience of the modeler or analyst. Product managers and engineers have been somewhat neglected and have lost their ability to look at a model, recognize a dilemma, understand the options, and make or at least advise a decision. In this section we try to remedy this situation by focusing on decisions rather than on excellence in modeling.

Following a brief discussion of some decision-theoretic concepts (subsections "What is a Decision and Which Decisions are Architectural?" and "Concept Attributes, Metrics, and Decision-Supporting Criteria"), we study several ways in which MBSE enhances architectural decision-making: Capturing stakeholder needs as cost and benefit manifestations of architectural decisions (sub-section "Capturing Stakeholder Needs"); Capturing and discovering the tradespace of possible candidate conceptual architectures, and highlighting decision points and inviting the architect and stakeholder to resolve them; Specifying architectural decisions by detailing the solution-specific architecture in the context of the problem domain; and highlighting the decisions that were made or will have to be made throughout the architecting process, their impact on the evolving architecture, and the trace of justification and rationalization of the emerging architecture (sub-section "Capturing the Architectural Decision-Making Process Alongside the Resulting Architecture").

We consider driver behavior tracking, an issue that vehicle owners are familiar with, as they want to ensure the safe and lawful behavior of those who drive their vehicles. This issue is well known to vehicle fleet operators, rental companies, insurers, and parents of adolescent children. We would like to find a solution for this problem.

What is a Decision and Which Decisions are Architectural?

Decisions are the choices that one makes about something after considering several possibilities [33]. This definition emphasizes that a) each decision emerges from several alternatives, and b) the choice should be made after reasoning and consideration. A decision is the outcome of a decision-making process.

Architectural decisions are those important and critical-to-make decisions that have a significant impact on the *concept* – i.e., a significant transformation of system structure and behavior [13]. Our cognitive and mental abilities and subjective biases may make the most important decisions indiscernible from less important ones. The model-based approach helps place stakeholders on the same page and ensure that priorities, impacts, and implications are clear to all, conventionalized, and objective, as part of the decision-making process.

Decision-making is the process of reaching a decision. It generally consists of three phases: Decision Problem Definition, Deciding, and Decision Execution. A more detailed description of the canonical outline of decision-making is provided by [34].

The system architect's primary role is decision-making, and decision making is the essence of architecting, however, more and more architectural decisions are made in groups, and the architect's role becomes one of facilitating, moderating, informing, and recording architectural decisions [35]. This notion highlights the importance of a suitable platform that would assist the system architect throughout the architectural decision-making process. Decision support capabilities include information management, formulation, recommendation, selection, execution, and learning [36].

The relevance of several alternatives is natural to humans. From the most trivial chore to the most pressing and fundamental issues of our lives, there are always at least two options, and even when there is one visible option, there is also a shadow, or default option of "doing nothing" (DN). When we consider medical treatment, we identify alternative clinics, physicians, medical approaches, available days and hours, healthcare coverage, and the risk of worsening our medical condition. Complex system architecting is no different: When we design a new aircraft, we evaluate the desired capacity, fuel consumption, range, piloting automation capabilities, situational awareness, etc. Alternatives emerge from key attributes, relevant values, and feasible combinations.

Decision analysis is the scientific foundation of decision-making. It is rooted in both the exact and social sciences, giving rise to two DM paradigms: the analytical and the behavioral. Analytical, model-centered approaches emerged from classical probabilistic and utility-theoretic approaches and focused on rational choice [37–39]. Behavioral decision theories view decision-making as a non-normative, human-centered process, with all the issues it raises [40]. According to the behavioral school, a person or group of people make decisions under various constraints, uncertainty, and bias [41]. The behavioral approach deals with heuristics, rationality and rationalization, analysis paralysis, and a host of other aspects and phenomena of human cognition. Managerial aspects like decision tracking and assurance are considered mainly behavioral.

Two primary handbooks on systems engineering (INCOSE's and NASA's [23, 42]) discuss decision-making as an engineering process, concerning both programmatic and architectural aspects. Programmatic decisions are made at decision gates, to simplify project and risk management. Architectural decisions concern aspects like functionality, design, technology, and vendor selection.

Trade Study, or Tradespace Exploration, is the process of analyzing various architectural alternatives, and trading-off figures of merit until a balanced solution is obtained [43, 44]. The primary phases of a trade study are: problem scoping, communicating with stakeholders, defining evaluation criteria and weights, defining and filtering alternatives, evaluating candidate alternatives based on measures of merit, selecting the best alternative, reviewing and re-evaluating the selection, assessing impact, and validating assumptions.

Reasoning about the alternatives is a fundamental part of our actions. There is a need to support the reasoning process through the recording and visualization of alternatives, compositions of alternatives, and comparisons of alternatives. The model-based approach empowers reasoning by reducing all the alternatives into a common formal language and representation. The representations of alternatives are not always straightforward, and the need to formulate them under the syntax and semantics of the modeling language makes the alternative concepts and their attributes emerge through an interactive cognitive-computational process, rather than through a deterministic mechanistic process of converting ideas to models.

Additional concerns related to decision making involve the consideration of uncertainty, risk, and subjective bias. Being able to integrate requirement prioritization, solution architecture selection, design exploration and decision making, and risk analysis is the essence of informed decision-analytic system architecting [14]. We would add here that the ability to do all of this in a model-driven way and in a model-based environment, particularly during early conceptual architecting iterations, is the essence of this chapter.

While the concept is a function-form mapping at the highest level of abstraction, a series of architectural decisions maps a set of functions to their respective forms. Consider an example of the functional intent of Money Moving, as illustrated in Fig. 12. In this case the stakeholder {E01} is defined as a *Person*, who has a need {E02} *Change the location of money from point A to point B*. The solution-neutral process *Moving* {E06} affects the solution-neutral operand *Money* {E03} by changing its state from *Point A* to *Point B*. The solution-neutral process *Moving* {E06} can be interpreted as either *Transporting* {E11.1} (moving money in the form of bills and coins in a physical way, e.g., using a secure courier), or *Transferring* {E11.2} (offsetting sums of money electronically between two accounts, sometimes commonly called *wiring*, after an ancient legacy dating back to the days of the telegraph). Both the physical *Transporting* and the digital *Transferring* are solution-specific processes that lead to completely different alternative architectures for delivering money to or for stakeholders: *Flying Vehicle* {E13.1.1} and *Land Vehicle* {E13.1.2} in case of *Transporting* {E11.1}, and *Wire Transfer* {E13.2.1} and *Check Deposit* {E13.2.2} in case of *Transferring* {E11.2}.

The distinction among SSPs is not always obvious. *Transferring* can also be interpreted in a physical sense, for example, through a *Check Deposit*. Further decomposition of SSPs can quickly reveal the distinctions between alternative concepts for realizing the original solution-neutral functional intent. The Internal Processes of *Transporting* and *Transferring* reveal the differences between these two conceptually distinct solution-specific architectures. The physical steps are distinctly different from the digital steps needed to complete the Money Moving procedure under each architecture.

9 Model-Based System Architecting and Decision-Making

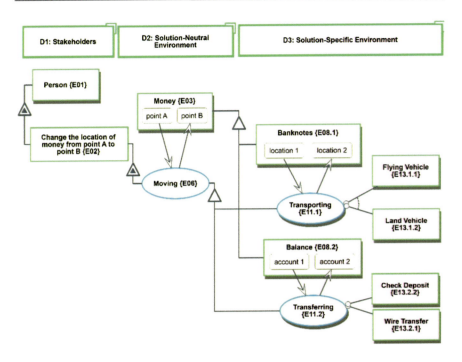

Fig. 12 Architectural choices for the solution-neutral functional intent of Money Moving

The common representation and formulation of the two distinct approaches using the same modeling framework enables reasoning about these approaches. Placing these abstract concepts together and showing how they are both derived from the same solution-neutral functional intent, facilitates the cognitive process of comparing alternatives. The system architect can go back and forth between the solution-neutral and solution-specific environments. The common representation facilitates discussion with stakeholders about the alternatives to validate the functional intent, to discover new alternatives, and to eliminate the irrelevant ones. This iterative process leads to continuous improvement and validation of the emerging architecture.

Concept Attributes, Metrics, and Decision-Supporting Criteria

We now discuss ways to include decision-supporting criteria for concept evaluation as part of the MBSA process. The terms criterion and metric may be used interchangeably, but while a metric is typically perceived as a general-purpose quantitative index (e.g., the Dow-Jones, the outside temperature, or the second moment of inertia), a criterion is defined in the context of a decision problem and is usually weighed against at least one other criterion. Both criteria and metrics can be qualitative or quantitative. Qualitative criteria must be ordinal or ranked, such that it is clear which value is better. That said, two stakeholders could aspire for opposite trends of the same criterion. For example, airports want to maximize the volume of air traffic while nearby residents want to minimize it. A binary criterion is either high or low, met or unmet, true or false, success or failure. A ternary criterion has three levels, e.g., high-medium-low, red-orange-green (also known as a traffic light criterion, etc.), and so on. Any number of ranks can be applied to a criterion. However, higher separation of ranks, require more precision in the induction of the value such that one can be confident about the suitability of the ranking assigned to an alternative in a specific criterion.

Criteria depend on the context and vary from concept to concept, and across stakeholders. Nevertheless, all criteria can be encoded in the model-based framework and support decision-making about the system architecture throughout the process. We can capture any criterion as an attribute of solution-neutral environment (D2), solution-specific environment (D3), and integrated concept (D4). A value-related attribute is the one which is changed by the associated process.

Consider, for instance, a dilemma between three vehicle architectures: front-wheel drive, rear-wheel drive, and all-wheel drive. The question is not merely about mechanical feasibility – all options are feasible, and all have uses and applications, and therefore the evaluation must be based on the needs. Recognizing this situation as a decision problem is the first stage in the decision-making process. The decision problem should account for metrics that are sufficiently detailed on the one hand but sufficiently design-agnostic on the other hand. This is perhaps the essence of the distinction between architecting and designing. The decision may rely on cost, size, weight, weight distribution, volume, power consumption, maintainability, etc. – as long as we can assign values to such criteria through elaboration of candidate solution architectures. The concept representation framework encodes the key performance metrics (which serve as decision-supporting criteria) as the attributes of the solution-neutral domain (D2), solution-specific domain (D3), and integrated concept (D4). Specifying the attributes is therefore critical for alternative evaluation and comparison and not only as additional information. The MBSA framework enables of any system concept analysis that the system architect and design team may come up with.

Capturing Stakeholder Needs

Any spreadsheet would do for listing stakeholders and needs, but a model-based approach enhances this process in several meaningful ways. Model pattern reusability is useful for identifying stakeholders. A pattern of stakeholder types could include broad stakeholder categories and roles (owner, operator, regulator, etc.), as shown in Fig. 13.

A dependable and reusable pattern of stakeholders – a decomposition and specification of subdivisions and attributes of the stakeholder concept – can be very useful in quickly and efficiently choosing stakeholders, rather than rediscovering and recreating them for each model anew. A stakeholder pattern may include both

9 Model-Based System Architecting and Decision-Making

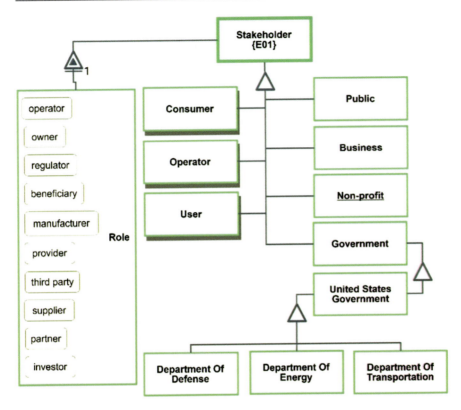

Fig. 13 A reusable pattern of stakeholders

high-level and lower-level roles, units, and profiles, as well as generic or umbrella needs. For instance, a national or regional energy authority obviously wants to provide energy to its residents, while minimizing operating costs and obeying the law. They might also care about clean and sustainable energy generation solutions. Such needs can be encoded in a pattern and utilized for various applications, even such that only consider the energy authorities indirectly, e.g., mass transportation projects.

Stakeholder entries may also be elaborated with domain terminology. For example, a defense stakeholder pattern can draw from the DoD Architecture Framework (DoDAF) [30]. Other aspects could include geographic distribution, available resources, standards, regulations, laws to comply with, strengths, weaknesses, opportunities and threats (SWOT), and so on.

Specifying stakeholders in the model, whether through discovery, documentation, or reuse, informs the system architect about all those parties and people who might need to weigh in on problem and solution domain decisions. Stakeholders who are not the primary customers or beneficiaries of an architecture must also be identified, and their needs must also be well-understood, especially if they are in tension.

A stakeholder set also leads to an emerging stakeholder network. Mapping stakeholder relations can be done in stakeholder value networks [45].

Consider for instance the Israeli missile defense system "Iron Dome," which has the intent of protecting the country against incoming rockets and ballistic missiles. Thanks to the protective umbrella that this system has cast over rural southern Israel, the economy in those areas began to thrive. Industry, commerce, and tourism indirectly became stakeholders. Nationwide franchises have had a chance to lobby on the deployment of Iron Dome batteries, due to the national economic impact these have had on their business. Before the system had been field-tested, it was very difficult to find deployment locations for launchers, sensors, logistic support, command and control outposts, and military encampments to support the massive operation of Iron Dome. Years later, shopping centers with a piece of the system on their outskirts suddenly became attractions. Having all stakeholders and needs in a common model could have led to different dynamics.

Institutional stakeholders often elaborate their needs as so-called stakeholder requirements. We include those as identified stakeholder needs, which simplifies the process, however there is a caveat. Stakeholder requirements are not really requirements. While this assertion may seem rude to some readers who have been stakeholders, this assertion stems from the understanding that both parties – stakeholder and system architect – are interested in getting to the stakeholder's essential needs, prior to establishing any solution that would meet those needs.

A model supports a hierarchy of interrelated needs that serve to justify those bottom-line or most central needs that stakeholders chose to state as their expectations from the system. For example, the Vehicle Owner in our Driver Behavior Tracking example ultimately wishes to maximize vehicle utilization while minimizing the risk to the driver and vehicle. Business owners might also be interested in monitoring schedule compliance by their professional drivers. We can figure out together with the owner or, say, the consumer association as a representative organization, how technology can help, next to other approaches like education or regulation. An in-vehicle technology could focus on collecting, analyzing, reporting, or acting upon drive and driver behavior characteristic data.

The presence of requirements in a model is truly informative when they serve as references for architectural decisions – regardless of how needs and requirements are captured. Systems engineers are responsible for traceability – showing that each requirement (or need) is mapped to pieces of the solution, so that stakeholders (and especially customers) may track and verify the fulfilling of their expectations from the system.

Decision-making is a process of choosing one possible solution out of a population or solution space, according to some criteria by which the solutions are assessed, scored, or ranked. Stakeholder needs constitute such solution assessment criteria because they capture the benefits and costs that stakeholders will reap from any given solution.

In driver behavior tracking, the owner is obviously a stakeholder, but also the driver, the insurer, the vehicle manufacturer, and the regulator. We might say that the public is also a stakeholder, in case the owner is a public entity such as a

9 Model-Based System Architecting and Decision-Making

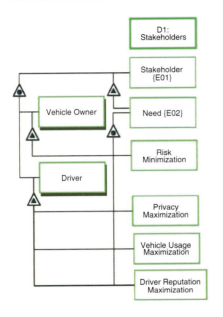

Driver and Vehicle Owner are instances of Stakeholder, E01.

Driver Reputation Maximization, Privacy Maximization, Risk Minimization, and Vehicle Usage Maximization are instances of Need, E02.

Vehicle Owner exhibits Risk Minimization.

Driver exhibits Driver Reputation Maximization, Privacy Maximization, and Vehicle Usage Maximization.

Fig. 14 Stakeholders of the Driver Behavior Tracking System are affected by the main functionality of the system: Driver Behavior Tracking

consumer-serving delivery provider, government agency, law enforcement agency, non-profit organization, etc. For simplicity, we will focus on two stakeholders: Vehicle Owner and Driver. Figure 14 illustrates the instantiation of D1 with our two Stakeholders and their Needs.

The Vehicle Owner needs to minimize the risk to the vehicle, driver, and passengers while the driver wants to maximize her use of the vehicle and maintain a good reputation as a responsible driver. The driver is a central stakeholder in our case. One of those needs, for example, is Privacy. In some settings, drivers' right to privacy exceeds the vehicle owner's right to monitor their vehicle. For instance, car rental companies must not spy on their customers (in most countries), and employers may not be allowed to monitor their employees beyond scheduled work hours. We may not invade the right to privacy, and therefore some practices, e.g., in-vehicle ambient voice recording, may not be permitted. On the other hand, for law enforcement and public safety agencies, operational activity monitoring and debriefing is a critical activity which may significantly benefit from such a capability. It is therefore clear that while such a privacy-violating capability may be useful for driver behavior monitoring, it must account for the circumstances and might not always be applicable. These notions impact our architectural decision-making process by illuminating the tradespace about such aspects, thereby validating the architectural decisions to follow.

We can articulate needs as global metrics with aspired trends, e.g., Risk Minimization or Privacy Maximization. This approach prepares the needs for multi-attribute utility analysis [46] and multi-attribute tradespace exploration (MATE) [47], which aggregates all benefits and costs. This approach also helps in discovering opposite and biased needs and objectives. For example, in the case of privacy, the owner may not directly wish to violate the driver's privacy, but their need to receive as much information as possible about driver behavior may eventually compromise driver privacy. It does not mean that the owner has a need to "Minimize Privacy." If stakeholders had two conflicting needs, i.e., opposite trends on the same metric, we can quickly see how these might give rise to a potential conflict. For example, assuming that driver behavior tracking is not a passive process from the driver's perspective, the driver might want to "Minimize Data Collecting" while the vehicle owner might want to "Maximize Data Collecting" in order to gain as much information as possible. One way of resolving such conflicts is by weighing the alternatives and understand the fundamental needs. In our case the driver's fundamental need is actually to minimize disturbances. Such stakeholder needs revisions may eliminate inherent bias in need specifications, and resolve conflicts.

The specification of solution-neutral functions is mapped to needs and solution-neutral functions (SNPs), as shown in Fig. 15. Once clearly defined, we should have

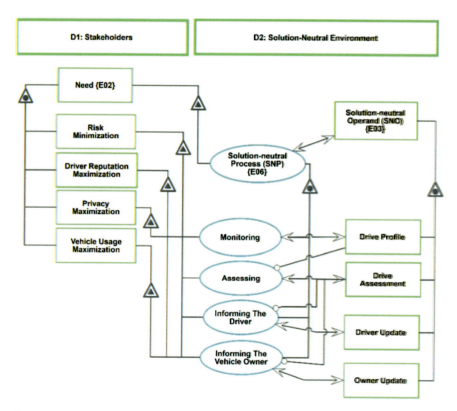

Fig. 15 Mapping stakeholder needs to solution-neutral processes and solution-neutral processes to solution-neutral operands

9 Model-Based System Architecting and Decision-Making

a problem-domain, solution-neutral model, covering the stakeholders, their needs, operational activities, and operands of interest.

Presenting the problem model to stakeholders allows them to validate a formal concept representation and clarify their needs in a common language. This approach is better than any approach in which there is no shared model that ensures consistency, common language, and holism. A stakeholder-driven model of the problem-domain is a necessary but not a sufficient condition for supporting a solution space generation process. In fact, failure to correctly represent the problem domain with a model may result in converging on a limited set of feasible solutions, which raises the risk of an invalid solution. The narrower the problem-domain description is, the likelier the solution architecture to be a bad solution for the wrong problem.

Capturing and Discovering Possible Architectures

The system architecting process can be structured as a series of decisions [13]. Key to this idea is identifying and filtering forward those decisions that have the greatest impact on the system's performance and cost. We would like the model to assist us in confining the scope to the necessary minimum so that we would be able to focus on the critical decisions and the supporting conceptualization to inform them.

The number of potential architectures under consideration is theoretically the product of the numbers of options per decision. For example, if we have two decision variables and each decision variable has two options, then the total number of architectures is 2x2 = 4. If a third decision emerges, with, say 3 options, the number of integrated alternatives becomes 2x2x3 = 12, and so on. Many combinations may be logically infeasible and therefore excluded, but we can still end up with a large number of combinations. Figure 16 illustrates the mapping of architectural options to architecture decisions, with three decision points (A, B, and C), each with two options. Theoretically, we have a total of 2x2x2 = 8 options. We also illustrate three possible system architectures: SA1, SA2, and SA3. We map each architecture to the options that it relies on per each architectural decision.

Table 1 summarizes all possible combinations of decisions to be made. Three out of the eight combinations have been identified as relevant candidates for further exploration. Generating the combination table from the model is an important capability for ensuring consistency between the map of the problem domain and the list of applicable integrated solutions. It is also critical for ensuring that suitable, reasonable, and feasible alternatives are considered and not only the obvious, immediate, or convenient ones. Consider, for instance, that combination 6, that we can encode as a vector [a2 b1 c2] vis a vis the vector of options [A B C], is a truly brilliant combination that has not even been considered. Furthermore, if a new, previously unthought-of, or neglected decision variable is added, it brings the number of combinations to 16. This could be a game changer in many ways.

Maintaining a valid table of options, based on an evolving model, even after an initial conceptual architecture is determined, has tremendous importance in ensuring the validity, sufficiency, and completeness of the tradespace, and consequently the

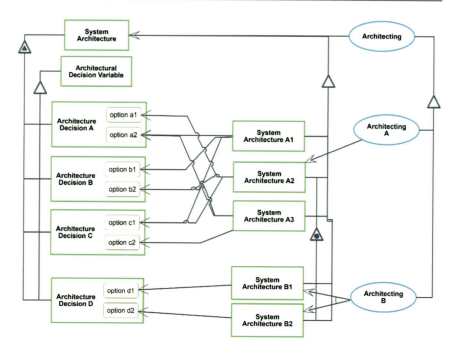

Fig. 16 Combinations of architectural decisions generate as many architectural candidates as the product of the numbers of options per decision variable

Table 1 Three out of eight potential combined architectures based on three decisions with two options each

Combination	Arch. Decision A	Arch. Decision B	Arch. Decision C
1	a1	b1	c1
2	a1	b1	c2
3 - Architecture 2	a1	b2	c1
4	a1	b2	c2
5 - Architecture 1	a2	b1	c1
6	a2	b1	c2
7	a2	b2	c1
8 - Architecture 3	a2	b2	c2

validity of the tradespace exploration effort – both by ensuring a consideration of all feasible options and by continued validation of the tradespace.

A retrospective of the Apollo Program considered nine possible decisions (including Lunar-orbit Rendezvous, fuel type, and other considerations) [13]. If each decision has only two options, then the number of possible combinations is 2^9, which is 512 architectures. Each option must be enumerated and considered, even if only superficially, in order to filter out irrelevant options and converge on a small and manageable set of combined options. We could map five additional

9 Model-Based System Architecting and Decision-Making

candidate architectures to all remaining possible combinations of decision options, but with 512 alternatives, this seems to be impractical.

Enumerating the alternatives is only the first stage. We must determine each candidate solution's feasibility by some study and elaboration of details that will allow us to reach a confident conclusion about each candidate architecture in order to decide if we keep or drop it.

The set of decisions can also be used in generating placeholders for all options and then selecting candidates to populate. We can filter out options which are unlikely, infeasible, or too expensive. We can consider the three architectures illustrated above as the three finalists that survived the filtering process out of the original eight (possibly following thorough analysis using our concept representation framework). We will return to these options later as we elaborate them from the placeholder level to full-blown architecture specifications, relying on structural and functional building blocks.

There are situations in which options are implicit in the choice to regard or disregard an artifact. Consider, for instance, a specific stakeholder's requirement about compliance with some protocol, which is not mission-critical, but potentially a good idea in terms of interoperability, reusability, and risk reduction. We can easily relax the assumption that protocol compliance must be obtained by specifying two optional states for that protocol as either accepted or deferred/rejected. This step will immediately inflate the tradespace by a factor of 2 because each available combination will have to be assessed with and without the protocol. Thus, we can significantly expand the tradespace by referring to any predetermined concept as possibly-redundant. The binary enumeration of option states is therefore critical to comprehensive coverage and enumeration of tradespace options.

We can reduce the tradespace by disqualifying either the accepted or rejected option for binary decisions, or by splitting the decision process into a series of decision points, in which each decision point consists of a subset of the decision variables, resulting in more decision steps but significantly less options to enumerate, assess, and choose from. Listing the options in the model, evaluating a subset of combinations, and selecting those combinations we wish to explore further, makes this process significantly easier, smarter, and more consistent.

The impact of MBSA here is therefore in the ability to enumerate candidate alternatives based on a model of architectural decisions. By specifying options as states of the artifact, we facilitate a process of enumerating the candidate alternatives.

For driver behavior tracking, we might consider, for instance the following aspects as fundamental to determining a conceptual architecture:

1. One size fits all vs. usage-specific variants – this would imply a single product versus a product line – and will significantly affect the architecture of the product platform and the variants.
2. Open vs. closed sensor policy – are we going to allow a variety of sensors to plug into our solution or only one or two specific sensors.
3. Vehicle-integrated vs stand-alone solution – are we going to embed the system in the vehicle and fully integrate it with vehicle systems such as the dashboard

displays and vehicle component bus, or install it separately (possibly with a small connection to the vehicle for basic monitoring or interfacing capability).
4. Driver management vs anonymity – are we going to include driver identification and personalization, such that data and behavior patterns are directly associated with a specific driver, or leave the task of figuring out who drove the vehicle while misbehaving without the assistance of our technology.
5. Driver notification available or not – are we going to alert the driver, or only collect the data and report to the subscribing customer (who could be the owner, the driver, or the insurer but it would not be a real-time indication).

We could continue defining more aspects of an architecture and gradually increase the tradespace. We already have 32 combinations if each decision variable only has two options. We can also refer to these issues as five serial decisions and only consider one variable at a time. This will result in evaluating 10 solutions in total – 2 per step. The risk is in missing potentially preferable solutions hiding among the other 22, by discarding possible combinations by nailing down one variable after another. With 6 variable serial decision-making, the ratio is 64 combinations to 12 inspected solutions and the difference is 52 ignored solution candidates.

For N binary decision variables the ratio between exhaustive search and serial decision-making is $2^N : 2N$. While serial decision-making seems inherently sub-optimal, this is how many of us make architectural decisions - resolving one issue or a couple of issues at a time. A model-based approach allows for both visualizing and analyzing the problem space with clear understanding of the implications of breaking down the problem into a series of smaller problems, as opposed to thoroughly studying the entire state space. Reaching a compromise is often a good idea, but it still requires good understanding of how decision variables can be grouped together into conceivable subspaces of the entire tradespace.

Capturing the Architectural Decision-Making Process Alongside the Resulting Architecture

The analogy between systems architecting and multi-criteria decision-making can be formalized using Category Theory. Category Theory is a branch of mathematics that focuses on the equivalence of representations and transformations of mathematical structures [48]. A category consists of a set of objects, which represent types, and a set of morphisms, which define mappings among types. These mappings can include relations, conversions, mathematical functions, etc. For example, a morphism sign: $R \to S$ converts any real number in R to a value in S, $S = \{-1, 0, 1\}$ according to its sign: a positive number maps to 1, a negative number maps to -1, and zero maps to 0. Morphisms can also act on multiple objects and generate multiple objects. For example, a morphism $R \to S$ converts the sign of a product of two real numbers to a value in S.

Analogous to the mapping of concepts to models [49], architecting is a mapping from the Problem Domain to an Architecture Co-Domain. This mapping should correspond to the notion that Deciding is a mapping from Problem to Decision. Architecting and deciding can be viewed as categorically equivalent if there exists a complete mapping of the decision domain to the architecture domain. In Category Theoretic terms, a mapping between categories is a functor. We would like to show that there exists a functor ADF: A → D such that for every object and morphism in A there exists a mapping to objects and morphisms in D. Similarly, we would like to show that there exists a functor DAF: D → A, such that for every object and morphism in D there exists a mapping to objects and morphisms in A.

In a category of system architectures, the objects are Architectures A, and morphisms are mappings of one architecture to another architecture, i.e., *architecting* steps. We shall initially argue simply that *architecting*: A → A. The morphism *architecting* includes a set of elaborations that change the architecture to address the problems that the previous architecture presents. This mapping also considers options to change the current architecture. Therefore, it would be more correct to state that architecting {A,P,O} → A, where P is a problem object, and O is a candidate operation object.

Each architecture A(n) presents a set of problems P, that has to be solved by the architecture A(n + 1). The sequence of solving the problem may transit through a set of architectural alternatives A(n + 1,1), A(n + 1,2), A(n + 1,3), etc. Therefore, architecting is also a polymorphism on A(n) due to its ability to create multiple sequel architectures. Alternative architectures are generated according to candidate operations on A(n). For example: add/modify/remove a block (structural element), add/modify/remove a function, add/modify/remove operand, add/modify/remove assignment of function to form, add/modify/remove output relation from function to operand, and so on. We can also merge or split items, e.g., break down a function into two or more smaller functions, merge several outputs into one big output entity, etc.

Some revisions of a given architecture model are not recommended, even if they are syntactically valid using a given modeling language. We should follow the careful transition through our concept representation framework, in order to maximize stakeholder value and solution-neutral problem definition, to ensure solution tradespace exploration, appropriately follow architecting guidelines, and minimize solution discrepancy.

While every architecting step changes the architecture, not every operation constitutes a decision problem. A decision point emerges when multiple options are possible. Although any inclusion or exclusion of an item in the architecture could pose a dilemma, constitute a decision point, or incur a discussion, we will usually conclude that it is preferable and worthwhile to include rather than exclude any aspect that enriches the concept and context of the architecture. For instance, we should not refrain from including any stakeholder or stakeholder needs even if they seem far-fetched or infeasible at a specific point in time. In case we wish to consider alternatives with and without a specific feature, capability, or aspect, we should define its state set as a binary existent/non-existent such that it will be taken into account in the definition of the tradespace. We therefore define an architecture decision point for A(n) as a situation in which all the following conditions hold:

1. At least two options are possible regarding a specific aspect of A: $A(n)A^1(n+1)$; $A(n)A^2(n+1)$; ....
2. Choosing one option over another may result in an architecture change, in a different solution, or in a different cost-benefit balance with respect to original stakeholder needs: $A^i(n+1) \neq A^j(n+1)$, for any i, j.
3. The decision cannot be made at a later point in time, i.e., the next iteration must be different from the current: $A^i(n+1) \neq A(n)$ for any i.

The above three conditions allow very specific issues to become decision points, and filter out trivial architecture modifications. However, it is now critical to identify these decision points and separate them from the rest of the architecture modification steps. Tagging model elements as decision variables enables this. For example, a sensor could be on or off from the operational perspective, but could be local or remote from the architectural perspective. Sensor activation (on/off) is obviously not an architectural decision, but determining the sensor location (local/remote) is. Therefore, we would characterize the sensor with two attributes: activation state and location. Only the location of the sensor is an architectural decision. We will tag that as a decision for further filtering and analysis.

Decision problems have been traditionally captured using decision trees [13, 50]. One of the major issues with decision trees is their obvious decision-centricity, as opposed to solution-centricity: they emphasize the decision-making process, but it is sometimes difficult to see how decisions represent solutions. Said otherwise, once a system architecture has been created and built, the architect's intent may vanish. Conversely, system architectures show the result of implemented architectural decisions, but not the decision-making process that effected those decisions.

A computational framework for coordinated design of cyber-physical system components under a given architecture (e.g., power, mass, and capacity optimization for a vehicle, communication system, etc.) attempted to address this decision-to-design discrepancy [51]. The problem is framed as the resource intake required to deliver the assigned functionality relative to an existing architecture. We are interested in extending this approach to develop a tradespace of architectures and select preferable architectures. We need to capture both the design and the designing process, in a coordinated and consistent way. This is where MBSE fits in.

Although MBSE has focused on representing system architectures with diagrams, it is possible to harness the power of modeling notations to capture decision processes. We should therefore strive to encode the process of deciding about architectural options in the conceptual system model, as well as its outcomes (in the form of architecture artifacts), this dual architecture-decision, we may begin to generate a model-based decision-driven system architecture. This assertion has been initially corroborated in [36] where a canonical decision-making process model was introduced.

Modeling the architectural decision-making process also facilitates better planning of the process. By planning ahead multiple architecting iterations, the system architect can clarify the prioritization of needs or the preference of solution technologies along predefined milestones. The model does not have to show only the immediately decision at hand. By deriving architectures from other architectures we will be able to serialize the architectural decision-making process and significantly reduce the computational and cognitive effort that is necessary to reason about a combinatorically exploding tradespace. In the above example, rather than considering 16 candidate combinations of the four binary decision variables, the first iteration considers 8 and the second iteration considers 2, hence the total number of reviewed candidates is 10. While this may result in overlooking 6 candidate architectures, a decision to prioritize decision variables A, B, and C and then decide about D, due to several legitimate considerations, is practically encoded in this model and can even be audited, debriefed, or even revisited - if it will not be too late.

MBSE is an environment that can foster concept discovery. In practice, reaching a viable architecture is a major milestone, short of comparing multiple architectures. MBSE environments could, at minimum, assist the architect in detecting this moment, for example by showing the architect that satisfactory coverage of needs has been achieved.

MBSE should also be able to indicate to the architect that a work-in-process architecture is infeasible or prone to reach a dead end. MBSE environments can fulfill this role if they are able to track the fulfillment of goals, whether those goals are defined within the model or as external evaluations and judgements that the model has to satisfy.

Solution-Specific Architecture Decisions

Architectural decisions constitute a gateway between the problem-oriented decision point formulation of the architecture, and the solution-oriented architecture specification and elaboration. While the decisions are made based on cost and benefit considerations, we recall that the architecting morphism also accounts for the set of available operations on the given architecture model. However, the size of this set of operations is completely arbitrary. Namely, we can carry out any number of operations we feel is sufficient to establish confidence in the architecture's ability to deliver the costs and benefits that we assess for it. In a world of incomplete information, constant change, and limited resources, we must make the call in many cases, and often end up revisiting and revising our former decisions.

MBSE could make this process of elaborating an architecture more structured. For example, we can insist on elaborating any architectural candidate one or two levels down in order to gain confidence in that potential solution. In many cases, these extra levels of detail could help us realize we are about to reach a dead end or come up with a solution that does not make sense.

Following the model-based specification of both the architecture and the architectural decision-making process, and remaining within the same conceptual modeling framework, we can explore candidate architectures by mapping them to what models truly excel in – the architecture's functional and structural specification. MBSE supports the specifying of operational processes and solution-neutral functions to be supported by the architecture, and mapping each operational process to the stakeholders that are involved in it, each operand to a need, and each functionality to the operational process that it contributes to, either by plain membership in the set of activities that compose the process, or through the specific generation of output or outcomes that can be used in the operational process.

A robust MBSE environment provides for both diverging from a problem statement to a space of alternative solutions, as we have seen so far, and for converging on a specific solution or subset of solutions as the architect desires, as we discuss next. The ability to continue using the same environment to grow a conceptual architecture into a comprehensive solution design is a major benefit, and yet, it is not always the case with MBSE environments.

A concept has to include solution-specific operands, processes, form, and the allocation of form to function, to make a good architecture. MBSE can force or advise architecture elaboration in order to capture these aspects. Robust MBSE environments will also make it easy for the architect to compare concepts both within the same model and across multiple models. The former approach helps in seeing the big picture in one place, while the latter helps in letting the solution architect focus on the details of the solution.

For driver behavior tracking, suppose we have chosen an open architecture, we can now incorporate an in-vehicle camera, an environment recording camera, a vehicle-integrated data collection device, or a driver assistance system, to collect information from the vehicle. In fact, any combination of these four solutions could be a candidate solution as well, and while some combinations may potentially provide greater value on some criteria, e.g., those that contribute to fulfilling the operational needs, they may also be too expensive, complicated, big, heavy, etc. All of these factor into the stakeholder needs, but now we can actually start discussing such attributes as mass, volume, power consumption, bandwidth, and performance. This is where architectural modeling both informs the more technical and functional architectural decisions (e.g., where to install a camera, or how to process the data), but, as emphasized, it also helps validate the open sensor architecture that we have chosen in the conceptual architecting phase.

By mapping each solution we propose to a stakeholder need in a model, where everything is traced back to the preference schemes and framed as a decision-driven process, is again a significant game changer. It is always a challenge to come back to a stakeholder with a concept and elicit new requirements and expectations. It is also clear that stakeholders must remain in the loop because the solution directly impacts the concept of operations, as shown in the concept representation framework. Even if the conceptual architecture is converged, this phase opens up a whole new tradespace of options and decision-making remains critical and essential as it previously was in mapping out stakeholders needs to solution concepts.

Transparent, collaborative, model-based stakeholder engagement significantly improves stakeholder need validation by introducing visual, model-driven

9 Model-Based System Architecting and Decision-Making

projections of solution architectures on problem statements. Choosing a vehicle-integrated solution inevitably generates concerns about interference with vehicle control and the potential risk of cyber-security.

The outcome of solution generating vis-à-vis well-defined stakeholder needs is illustrated in Fig. 17. We begin with a functional decomposition of the system's main functionality, while ensuring the nested functionalities are as solution-neutral and as need-oriented as possible. In our case we specify Road Monitoring, Driver Monitoring, and Vehicle Monitoring as SSPs that map to the Monitoring SNP. Real-time Analysis and Off-Line Analysis are two solution approaches for Analysis. Informing the Vehicle Owner and the Driver can be implemented either in Real-Time or after the fact.

Specifying potential forms that may contribute to the execution of each function lays out the tradespace of possible combined solutions. We recall that any

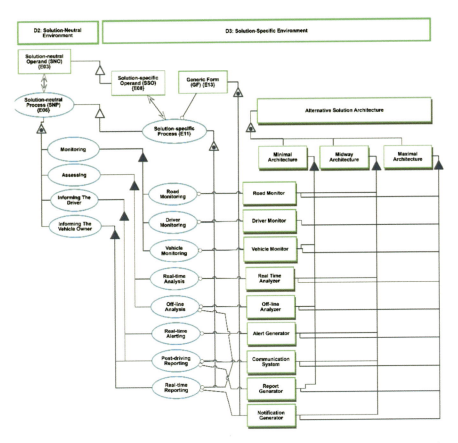

Fig. 17 Deriving solution-specific functions from solution-neutral functions, specifying generic form to support solution-specific functions, and assembling three typical architectures - minimal, midway, and maximal - as combinations of generic form. The selected architecture will be further developed in D4

combination of options constitutes a theoretically possible solution and implies a decision point. We map each SSP to Generic Forms. We then propose three integrated architectures: minimal, maximal, and midway architectures. This is also a way to reduce the problem-space from an exponential combination to a set of functions that must be performed by the system, such that all functions and needs are fulfilled, or at least fulfillable. This representation does not directly address the decisions to be made, but we can decide for each GF artifact whether to implement it or not.

We must provide at least one form to provide each function. In some cases, the same form may perform more than one function, as the technology, or commercial product it relies on allows. In some cases, a solution candidate requires another solution candidate. For instance, in our example, in order to run a software application within the vehicle, we must be able to run it either on a dedicated device or on the vehicle's multimedia system.

In this example we define a Minimal Architecture as the combination of a Vehicle Monitor, Off-Line Analyzer, and Report Generator. We can also refer to this trio as one architectural building-block and define any architecture by adding to this one. By specifying candidate elements of form and function without specifying their states we argue that they can either exist or not exist in the solution. This is another approach to list solution factors that can later co-facilitate combined solution architectures, complementing the approach for listing options as states of the constituent decision units. With three out of nine elements unified, we can now say that the solution space has $2^7 = 128$ options. Any solution which includes more than the Minimal Solution and less than the Maximal Solution (which includes all the eight elements) is a partial solution that we can specify in terms of its constituent elements and evaluate according to our decision criteria. A model is useful here both in visualizing the range of options, and in elaborating any point in this range.

Decision criteria might include overall weight, volume, power consumption, and integration effort. Note that such design parameters may emerge as constraints to a solution rather than pre-conceived stakeholder needs, which is yet another reason for an iterative model-based architectural decision-making approach as we advocate here. We can begin discussing the cost, in financial, energetic, or performance-related terms, after we have secured a solution for stakeholder needs. Another approach would be to combine these benefit and cost factors together and consider them together in the same iteration. Both approaches are possible in an MBSE environment, as the evolvability of the model is an inherent capability of the MBSE process. As explained before, it depends on the decomposability and conceivability of the tradespace and is up to the solution architect to figure out. Physical qualities must be considered together because of the mutual effects (e.g., the combination of mass, power, and capacity). The model can clearly capture bundles of solution-specific attributes as decision variables and facilitate (and in some cases execute) the computation of a Pareto frontier, i.e., a set of combinations that meet all the criteria at the best values. We can similarly analyze software considerations and determine the most appropriate decomposition into digestible and sensible design decisions.

We may argue that the contribution of in-vehicle camera is smaller than that of a road-observing camera, or that the integration with a Driver Assistance System is better for alerting the driver than a multimedia system interface. The purpose here is not necessarily to argue for one approach in favor of another, but to show how a model-based scheme can greatly enhance the visibility of the decision-making process, the understanding of trade-offs and composition of alternatives, and the propagation of value and enablement all the way to stakeholder needs. Communicating such a model to stakeholders is also easier and more intuitive, as it can illustrate the impact on stakeholder needs, which greatly increases clarity and transparency.

Conclusion

We have shown how systems architecting is essentially a reasoning process, which consists of conceptualization and decision-making steps. MBSE facilitates, but also changes the architecting process. It is worth taking a step back to summarize the ways in which MBSA differs and grows from legacy, "off-line" system architecting. The architecting process (whether legacy or model-based) is a source of discovery and insight. It addresses the need to understand, through sufficient specification and completeness of coverage at a given abstraction level of analysis. These drivers enable us to judge when the architectural modeling is concluded.

We have presented a model-based concept representation framework that formalizes the conceptualization process to generate a system architecture. The framework advocates a path to a solution that accounts for all the major concerns in a system architecture. Indeed, this is not a linear process, but an iterative one, rife with revisions, divergence, and convergence. The framework accommodates both ongoing conceptualizations and decision-making when critical decisions must be made.

We conclude with three key MBSA principles:

1. MBSA is an iterative reasoning process, in which the model records and informs the evolving conceptual architecture.
2. MBSA fosters divergence before convergence: allowing for options to emerge from a solution-neutral environment, and converging on a solution after considering multiple solution-specific approaches.
3. MBSA projects can focus on the relevant conceptualizations and not necessarily follow an A-to-Z approach – the amount of modeled information should be just enough to reach a decision.

The "offline" architecting process is based on siloed analysis and discussion, and we find there is substantially more effort in setting up a model to answer architectural questions. Some of this effort is due to the need to translate decision-theoretic concepts into conceptual modeling language. However, once a model is developed and can serve as a baseline, the resulting exploration becomes quicker with each iteration. The model-based approach could provide a return on investment due to the reusability and evolvability of model assets through multiple iterations. We caution, though, that the architecting process should not be conceived as a procedural one: our framework is not just a checklist. Given that no architectural model will be able to capture detailed design information, we find that the process of modeling and sense-making is important to building a shared understanding among the architecture team members. Therefore, the process remains both cognitive and computational, and the two streams augment each other thanks to the shared artifact that the model constitutes, which lends itself both to human reasoning and machine processing.

We summarize the primary benefits of MBSA as follows:

Communication with Stakeholders: Communicating with stakeholders is an essential part of the decision-making process. Engaging stakeholders, getting them involved, committed, and informed, is critical for architecting the solution to their needs. A shared language for discussing concepts, problems, options, and solutions with a reference model is instrumental in facilitating, formalizing, and documenting the discussion.

Knowledge and Architectural Building Block Reusability: A growing concern in enterprises working on evolving systems in an agile world, the reuse of existing knowledge about the domain and design of existing designs, is becoming a critical aspect of MBSE. Reusability of knowledge, architecture, and design artifacts has two major impacts on the decision-making process. Reliance on existing assets reduces uncertainty, ambiguity, and programmatic risk (schedule, budget, quality, etc.). Furthermore, reducing the explorable tradespace by reusing existing component designs is an approach to coping with a combinatorial solution space explosion. System architecture decisions that are based on a formal model that incorporates reusable domain content and building blocks have higher degrees of confidence and may be more likely to survive the emergence of new constraints, issues, or materializations of risk.

Consistent Problem-Space Mapping: MBSA facilitates conceptual mapping of the problem space in a domain-agnostic manner. It is then possible to specify problem-domain needs, key evaluation metrics, and performance indicators, and to layout the architectural solution space that may contain at least one feasible solution to the problem. Evolving problem understanding, some of it in parallel to decision-making, helps reshape and recreate the problem domain. We believe that this consistency in mapping the problem space helps us recognize patterns across designs.

Gradual Transformation of Problem Statements to Architectures: MBSA facilitates a smooth transition from the problem domain to the solution domain by allowing for a traceable mapping of candidate and chosen solutions to problems. While we could describe the problem domain in a variety of less-formal ways, and the solution in a variety of approaches, a model-based approach formalizes the transition, the transformation, and the traceability of solution domain aspects to problem domain ones.

Consistent Comparison and Evolution of Alternative Architectures: Comparing and reasoning about architectures within the modeling frameworks is challenging. MBSA allows this by mapping conceptual architecture decisions to quantifiable metrics of stakeholder utility on the one hand, and to quantifiable metrics of design validity on the other. We can promise a conceptual architecture that helps our stakeholders in many ways, but failing to validate the architecture may be bad for business. MBSA facilitates a smooth transition across modeling, assessment, verification, and validation methods, with a conceptual model as a focal point. The alternative approach rephrases the decision-making problem in mathematical decision-analytic terms that do not rely on the model. There are two risks there: (a) the full architecture scope will not be sufficiently understood when developing a selected solution, and (b) validating the solution architecture by tracing back to the "numbers" will be impossible.

Documenting both the Architecting Process and its Outcome – the Architecture: By documenting our decision-making process in a model, we are creating a mapping of the entire process rather than the outcome. One should not guess why a particular architecture has been selected. We believe MBSE will be much stronger in capturing functional intent. A decision-oriented model can substantiate the solution on the considerations that fed the decision-making process leading to that particular solution. While this is not a common MBSE practice, we assert that it is possible, achievable, and desirable within an MBSE framework to ensure that the process is appropriately documented, and that the traceability of solutions to options to problems to needs to stakeholders becomes clear.

Aspiration to Sufficiency: We have discussed the importance of model sufficiency – knowing that we have modeled enough to make an architectural decision. We reasoned that by applying Miller's Law of $\sim 7 \pm 2$ elements in every set as a threshold for sufficiency. We can measure the compliance of various sets of objects and processes with this criterion and determine our confidence in the model and our willingness to rely on it.

References

1. Dori D (2016) Model-Based Systems Engineering with OPM and SysML. Springer, New York
2. McDermott TA, Hutchinson N, Clifford M, Van Aken E, Slado A, Henderson K (2020) Benchmarking the Benefits and Current Maturity of Model-Based Systems Engineering across the Enterprise. Systems Engineering Research Center (SERC)
3. Hale JP, Zimmerman P, Kukkala G, Guerrero J, Kobryn P, Puchek B, Bisconti M, Baldwin C, Mulpuri M (2017) Digital Model-based Engineering: Expectations, Prerequisites, and Challenges of Infusion. NASA
4. Morris BA, Harvey D, Robinson KP, Cook SC (2016) Issues in Conceptual Design and MBSE Successes: Insights from the Model-Based Conceptual Design Surveys. INCOSE Int Symp 26: 269–282 . https://doi.org/10.1002/j.2334-5837.2016.00159.x
5. Weilkiens T, Lamm JG, Roth S, Walker M (2016) Model-Based System Architecture. In: Model Based System Architecture, First Edi. John Wiley & Sons, Inc, pp 27–33
6. Object Management Group (2019) OMG Systems Modeling Language Version 1.6
7. Klappholz D, Port D (2004) Introduction to MBASE (Model-Based (System) Architecting and Software Engineering). In: Zelkowitz M V. (ed) Advances in Computers. Elsevier, pp 203–248

8. Boehm B, Klappholz D, Colbert E, Puri P, Jain A, Bhuta J, Kitapci H (2004) Guidelines for Model-Based (System) Architecting and Software Engineering (MBASE). 1–159
9. Boehm B (2006) Some future trends and implications for systems and software engineering processes. Syst Eng 9:1–19. https://doi.org/10.1002/sys.20044
10. Boehm B, Oram A, Wilson G (2010) Architecting: How much and when? O'Reilly Media
11. Bahill AT, Henderson SJ (2005) Requirements Development, Verification, and Validation exhibited in famous failures. Syst Eng 8:1–14. https://doi.org/10.1002/sys.20017
12. Lane JA, Koolmanojwong S, Boehm B (2013) Affordable Systems: Balancing the Capability, Schedule, Flexibility, and Technical Debt Tradespace
13. Crawley E, Cameron B, Selva D (2015) Systems Architecture: Strategy and Product Development for Complex Systems. Prentice Hall
14. Bahill AT, Madni AM (2017) Tradeoff Decisions in System Design. Springer International Publishing Switzerland
15. Dori D, Kohen H, Jbara A, Wengrowicz N, Lavi R, Levi-Soskin N, Bernstein K, Shani U (2020) OPCloud: An OPM Integrated Conceptual-Executable Modeling Environment for Industry 4.0. In: Kenett RS, Swarz RS, Zonnenshain A (eds) Systems Engineering in the Fourth Industrial Revolution: Big Data, Novel Technologies, and Modern Systems Engineering. Wiley
16. Menshenin Y, Mordecai Y (2020) Model Based System Architecting Reference Model. V01_20_12
17. Chomsky N (1956) Three models for the description of language. IRE Trans Inf Theory 2: 113–124. https://doi.org/10.1109/TIT.1956.1056813
18. INCOSE (2015) INCOSE Systems Engineering Handbook: A Guide for System Life Cycle Processes and Activities, Fourth Edi. John Wiley & Sons, Inc.
19. Menshenin Y, Crawley E (2020) A system concept representation framework and its testing on patents, urban architectural patterns, and software patterns. Syst Eng 23:492–515. https://doi.org/10.1002/sys.21547
20. Menshenin Y (2020) Model-based framework for system concept - Ph.D. Thesis. Skolkovo Institute of Science and Technology
21. Freeman RE (2001) A Stakeholder Theory of the Modern Corporation. Perspect Bus Ethics 3. https://doi.org/10.3138/9781442673496-009
22. European Commission (2019) The European Green Deal. Brussels
23. NASA (2016) NASA System Engineering Handbook, SP-2016-61. NASA
24. Suh NP (1990) The principles of design. Oxford University Press on Demand
25. Nordlund M, Lee T, Kim S-G (2015) Axiomatic Design: 30 Years After. In: Proceedings of the ASME 2015 International Mechanical Engineering Congress and Exposition IMECE2015. ASME, Houston, Texas
26. Pahl G, Beitz W, Feldhusen J, Grote K-H (2007) Engineering Design A Systematic Approach. Springer-Verlag London
27. Maier JF, Eckert CM, Clarkson PJ (2016) Model granularity and related concepts. In: Proceedings of the DESIGN 2016 14th International Design Conference
28. Eppinger SD, Browning TR (2012) Design Structure Matrix Methods and Applications. Des Struct Matrix Methods Appl. https://doi.org/10.7551/mitpress/8896.001.0001
29. Miller GA (1956) The magical number seven, plus or minus two: Some limits on our capacity for processing information. Psychol Rev 63:81–97. https://doi.org/10.1037/h0043158
30. United States Department of Defense (DoD) (2010) The DoDAF Architecture Framework Version 2.02. https://dodcio.defense.gov/Library/DoD-Architecture-Framework/. Accessed 3 Dec 2020
31. Mordecai Y, James NK, Crawley EF (2020) Object-Process Model-Based Operational Viewpoint Specification for Aerospace Architectures. IEEE Aerosp Conf Proc 1–15. https://doi.org/10.1109/AERO47225.2020.9172685
32. Maier MW, Rechtin E (2000) The Art of Systems Architecting, Second Edi. CRC Press LLC
33. Cambridge Dictionary (2020) Decision. https://dictionary.cambridge.org/us/dictionary/english/decision. Accessed 18 Dec 2020

34. Zeleny M (1982) The Decision Process and Its Stages. In: Zeleny M, Cochrane J (eds) Multiple criteria decision making. McGraw-Hill, Inc., New York, pp. 85–95
35. Weinreich R, Groher I (2016) The Architect's Role in Practice: From Decision Maker to Knowledge Manager? IEEE Softw 33:63–69. https://doi.org/10.1109/MS.2016.143
36. Mordecai Y, Dori D (2014) Conceptual Modeling of System-Based Decision-Making. In: INCOSE Internaional Symposium. INCOSE, Las-Vegas, NV, USA
37. Pratt, Raiffa, Schlaifer (1964) The Foundations of Decision Under Uncertainty. 59:353–375
38. Saaty TL (1990) How to make a decision: The analytic hierarchy process. Eur J Oper Res 48:9–26. https://doi.org/10.1016/0377-2217(90)90057-I
39. Howard R (1968) The Foundations of Decision Analysis. IEEE Trans Syst Sci Cybern 4:211–219. https://doi.org/10.1109/TSSC.1968.300115
40. Kahneman D (2003) A perspective on judgment and choice: mapping bounded rationality. Am Psychol 58:697–720. https://doi.org/10.1037/0003-066X.58.9.697
41. Tversky A, Kahneman D (1974) Judgement under Uncertainty: Heuristics and Biases. Science (80-) 185
42. INCOSE (2015) INCOSE Systems Engineering Handbook: A Guide for System Life Cycle Processes and Activities, Fourth Edi. John Wiley & Sons, Inc., San Diego, CA, USA
43. Haskins C, Forsberg K, Krueger M, Walden D, Hamelin RD (2011) Systems Engineering Handbook, v. 3.2.2. International Council on Systems Engineering
44. Parnell GS, Parnell GS, Madni AM, Bordley RF (2017) Trade-off Analytics: Creating and Exploring the System Tradespace Chapter 2: A Conceptual Framework and Mathematical Foundation for Trade-off Analysis
45. Rebentisch ES, Crawley EF, Loureiro G, Dickmann JQ, Catanzaro SN (2005) Using Stakeholder Value Analysis to Build Exploration Sustainability. Engineering 1–15. https://doi.org/10.2514/6.2005-2553
46. Malak RJ, Aughenbaugh JM, Paredis CJJ (2009) Multi-attribute utility analysis in set-based conceptual design. CAD Comput Aided Des 41:214–227. https://doi.org/10.1016/j.cad.2008.06.004
47. Ross AM, Hastings DE, Warmkessel JM, Diller NP (2004) Multi-Attribute Tradespace Exploration as Front End for Effective Space System Design. J Spacecr Rockets 41:20–28. https://doi.org/10.2514/1.9204
48. Breiner S, Sriram RD, Subrahmanian E (2019) Compositional Models for Complex Systems
49. Mordecai Y, Fairbanks J, Crawley EF (2020) Category-Theoretic Formulation of Model-Based Systems Architecting: The Concept → Model → Graph → View → Concept Transformation Cycle
50. Haimes YY (2009) Multiobjective Decision-Tree Analysis. In: Risk Modeling, Assessment, and Management, Third Edit. John Wiley & Sons, Inc.
51. Censi A (2017) A Class of Co-Design Problems with Cyclic Constraints and Their Solution. IEEE Robot Autom Lett 2:96–103. https://doi.org/10.1109/LRA.2016.2535127

Yaroslav Menshenin, PhD, is a Research Scientist at the Space Center, Skolkovo Institute of Science and Technology – Skoltech (Moscow, Russia). He holds a PhD (2020) from Skoltech and a Specialist Degree (MSc equivalent) (2012) from the National University of Science and Technology "MISIS" (Moscow, Russia). He is also a graduate of the Singularity University located at NASA Ames Research Center (California, USA). Dr. Menshenin is a Member of INCOSE, AIAA, IFIP WG 5.1, and the DESIGN Society. He was also a visiting doctoral candidate at the System Architecture Group, MIT (2016–2017).

Yaniv Mordecai, PhD, is a post-doctoral research fellow at the Engineering Systems Laboratory, Massachusetts Institute of Technology (Cambridge, Massachusetts, USA). He holds a PhD (2016) from Technion – Israel Institute of Technology, (Haifa, Israel); and MSc (2010) and BSc (2002) from Tel-Aviv University (Tel-Aviv, Israel). He is also a senior systems architect with Motorola

Solutions. Dr. Mordecai is a senior member of IEEE and board member of IEEE Israel and of the Israeli Association for Systems Engineering – INCOSE_IL. Yaniv is the recipient of the IEEE Systems, Man, and Cybernetics Society Doctoral Dissertation Award (2017) and the OmegaAlpha Association Exemplary Doctoral Dissertation Award (2017).

Edward F. Crawley, ScD, is the Ford Professor of Engineering and Professor of Aeronautics and Astronautics and Engineering Systems at Massachusetts Institute of Technology (Cambridge, Massachusetts, USA). He holds a Sc.D (1980), M.S. (1978), and B.S. (1976) from MIT. Dr. Crawley was the founding president of Skolkovo Institute of Science and Technology, Moscow (2011–2016), Director of the Bernard M. Gordon – MIT Engineering Leadership Program (2007–2012), Executive Director the Cambridge University-MIT joint venture (2003–2006), and head of the Aeronautics and Astronautics Department at MIT (1996–2003). He is a Fellow of AIAA and RAeS; Member of the National Academy of Engineering (NAE), Royal Academy of Engineering (RAEng UK), Royal Swedish Institute of Engineering Science (IVA) and the Chinese Academy of Engineering (CAE); NASA Astronaut Finalist (1980); and Regional Soaring Champion (1991, 1995, and 2005).

Bruce G. Cameron, PhD, is the Director of the System Architecture Group, Lecturer in System Design and Management, and Faculty Director of the Architecture and Systems Engineering Certificate Program – all in Massachusetts Institute of Technology (Cambridge, Massachusetts, USA) He holds a PhD (2011) and dual MS (2007) from MIT and a B.A.Sc from the University of Toronto (Toronto, Canada). Dr. Cameron is also a co-founder of Technology Strategy Partners.

Adoption of MBSE in an Organization

10

Tim Weilkiens

Contents

Introduction	332
Related Work	333
Change, Resistance, and Bad Practices	335
Aimless Modeling	336
Fast False Start	337
Overmodeling	337
Model Island	337
Ivory Tower	338
Fly Out of the Learning Curve	338
Developing the MBSE Methodology	338
Selection of the Modeling Languages	343
Selection of the Modeling Tools	344
Verification and Validation	345
Running the MBSE Infrastructure	346
Summary	347
Cross-References	347
References	348

Abstract

The adoption of MBSE in an organization is a challenging task. There is no single path to achieve this. It depends on many aspects, a holistic approach, and the people who lead the adoption process. It sounds like the challenge of a typical systems engineering project. In fact, essential steps of systems engineering also fit the implementation of an MBSE process. Here, too, the purpose must be defined, the requirements specified, and the architecture of the MBSE environment developed.

T. Weilkiens (✉)
oose Innovative Informatik eG, Hamburg, Germany
e-mail: tim.weilkiens@oose.de

© Springer Nature Switzerland AG 2023
A. M. Madni et al. (eds.), *Handbook of Model-Based Systems Engineering*,
https://doi.org/10.1007/978-3-030-93582-5_18

This chapter discusses the challenges of introducing MBSE, lists typical pitfalls, and presents an approach to developing the MBSE methodology with MBSE itself.

Keywords

MBSE · Methodology · Adoption

Introduction

The introduction of a new methodology into an organization is always a process of change. Not only the technical issues of the introduction but also the human factors are crucial for success. This chapter sheds marginal light on the human factors and focuses on the technical aspects of the introduction, i.e., the definition of the methodology, the languages, and the tools. The content of this chapter is taken – partly directly – from the book [16].

A model-based approach requires a harmonic triad of methodologies, modeling languages, and modeling tools (Fig. 1).

The three domains must fit together. Compromises must be made in particular areas to achieve this. For example, a given modeling tool that does not fully support the language or methodological steps, conversely, offers proprietary functions that enable methodological steps and lead to a dependency of the methodology on the modeling tool. Often the tool is set, and the languages and methodologies must adapt. This common practice is precisely the opposite of the recommended way of first deriving the methodologies from the purpose, the languages from the methodologies, and, finally, the appropriate tools.

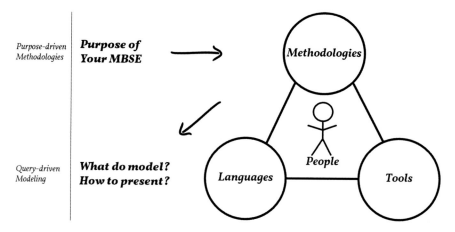

Fig. 1 Triad of modeling

People play a central role. They have to work with the methodologies, languages, and tools. Here, too, the interaction must fit. People must have the ability, and aspects such as culture also play a role.

▶ Chapter 3, "MBSE Methodologies" on MBSE methodologies looks at MBSE from a different perspective and describes the relationships of the elements process, method, tools, environment, technology, and people.

An MBSE methodology should not be used 1:1 out-of-the-box without customization. Every organization has its own specific needs, and none methodology can satisfy them all.

There are several published MBSE methodologies available. They are useful templates and starting points. Nevertheless, it must be analyzed whether all the tasks included are needed: some must be adapted to your needs, some must be executed more intensively and some less intensively, and some necessary tasks are probably not covered by the given methodology.

Before customizing a methodology, the purpose and the requirements of the MBSE adoption must be elaborated. The MBSE methodology is the implementation of these requirements. In Fig. 1, this approach is called purpose-driven methodologies.

Stakeholders of the engineering projects define the desired engineering artifacts and views. The methodology describes how to achieve these outputs. On this basis, you can derive the languages in which the artifacts are expressed and the tools used to create and present them. In Fig. 1, this is called query-driven modeling. The queries to satisfy the stakeholders' requests determine the type of modeling and when the modeling will be completed. When all artifacts are created, the modeling is complete. Each element in the model must serve a purpose.

Related Work

The International Council on Systems Engineering (INCOSE) states in its vision 2025 document that MBSE will become the norm for systems engineering [9]. As systems and the organizations that develop them become more complex, the document-based approach reaches its limits. The information is stored in a distributed manner, is inconsistent, and cannot be presented in different stakeholder-optimized views. Nevertheless, the systems engineering discipline still relies on a document-based approach and is in an early stage of maturity of an MBSE adoption [9].

Mary Bone and Robert Cloutier summarized the results from the OMG SysML Request for Information 2009 survey in a paper for the Conference on Systems Engineering Research [2]. Culture and general resistance to change were identified as the largest inhibitor to the adoption of MBSE. This outcome is not surprising since MBSE is a significant change in the engineering process. The management of change processes is a topic outside of this book's scope but must be considered for an MBSE introduction in an organization.

Another finding of the survey was that MBSE requires a steep learning curve, including SysML, the tool, and the methodology, which fits the triad of modeling presented in the previous section. Taking a learning curve takes time, and you can fly out of the curve if you are too fast, as the anti-pattern "Fly out of the learning curve" in the following section describes.

That the adoption of MBSE is a challenging task is also mentioned by Mohammad Chami and Jean-Michel Bruel in their paper summarizing another MBSE survey [3]. They observed a higher maturity of MBSE over the years. Whereas in the past, the question was "Why modeling?," the question "How modeling?" followed, and now the question "How do I model effectively?" is asked.

Andreas Vogelsang et al. concluded that people have bad experiences and frustration about MBSE adoption due to too high expectations [15]. Therefore, it is crucial to specify the expectations at the beginning clearly and to validate them early in the adoption process to be able to react in time. These steps are part of the approach presented in this chapter.

Vogelsang et al. also highlighted the resistance to change. People do not adopt MBSE because it changes existing processes.

The D3 MBSE Adoption Toolbox presented by Mohammad Chami et al. [4] defines three phases for an MBSE adoption: definition, development, and deployment. They also emphasize the three modeling components, methods, languages, and tools, and add a fourth component personnel.

Jonas Hallqvist and Jonas Larsson describe in their paper how to introduce MBSE into an organization using systems engineering principles based on a real-world MBSE adoption at the aerospace company Saab [7]. They mention that the system engineering principles helped overcome the MBSE introduction problems. This work underpins the approach presented in this chapter to introduce MBSE with MBSE approaches.

The key stakeholder of an MBSE implementation is the management or, more generally, the people responsible for budget and strategic direction. Besides the adoption challenges of change resistance and elaborating on how to introduce MBSE, also the business case for MBSE must be analyzed and presented to the management. For some organizations, MBSE is also the entry point into the systems engineering discipline. We must differentiate the business case for systems engineering and the business case for MBSE.

Joseph P. Elm and Dennis R. Goldenson published a report summarizing the results of a survey quantifying the relationship of systems engineering application and project performance [5]. The result clearly shows that projects that practice systems engineering perform significantly better in terms of time, cost, and requirement satisfaction than projects without systems engineering.

Azad M. Madni and Shatad Purohit analyzed the economic aspects of MBSE [11]. They came to the conclusion that compared to traditional systems engineering, MBSE requires a higher upfront investment but will win in later phases. Other studies report similar findings that MBSE is financially and technically a benefit over traditional or no systems engineering. For example, Joseph Krasner mentioned that the application of MBSE results in a reduction of 55% in total development cost compared to traditional systems engineering [10].

Change, Resistance, and Bad Practices

There are three main interest groups when introducing MBSE into an organization:

1. The initiators and drivers of MBSE
2. The organizational units that should work model-based in the future
3. The organizational units responsible for the time and budget of the engineering projects

All three groups need to pull together to make the introduction a success – another harmonic triad of MBSE adoption. In particular, this means that the people in the organizational unit, who will use MBSE, want and support it. And that the responsible organizational unit is ready to provide the necessary money and time.

Implementing MBSE is a significant investment and typically only unfolds value when you include multiple projects. It is a strategic decision of the organization. Therefore, the management must be convinced and support the MBSE introduction. The previous section listed some related works about the business case of MBSE.

Helpful in convincing the organization, respectively, the management, to implement MBSE, in addition to listing the benefits and the very worthwhile business case with potential cost reductions over 50%, can be the following thought experiment: What would your own market environment look like if a competitor has successfully implemented MBSE, has a consistent model of the product, and, according to the studies, has a high-cost reduction, faster development times, and better coverage of requirements and can use emerging technologies such as artificial intelligence based on this?

This is a realistic scenario. Can you hold your own next to such a competitor? It is rather unlikely that you can find sound arguments against this scenario and justify why an introduction of MBSE is not necessary. At the same time, it should be noted that MBSE is not a silver bullet, and its use is not recommended across the board for any type of project.

Figure 2 shows the percentage distribution of the degree of support among those affected by a change [12]. Often the focus of persuasion is on the small group of opponents because they are usually much louder than the other groups. That ties up valuable resources that should be used to convince the large group of undecided people.

An essential factor for the success of the implementation is also the support of an experienced, external expert in the project to:

- Transfer the new MBSE knowledge quickly and directly into the organization
- Recognize typical pitfalls at an early stage and thus to save time and not to stir up fears of a failure in the organization
- Obtain a neutral external view without operational blindness that means the inability to reflect critically on oneself due to working too routinely

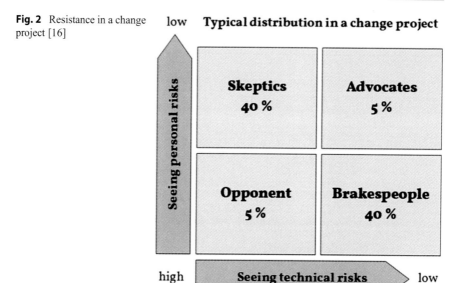

Fig. 2 Resistance in a change project [16]

The expert should only support the introduction but does not take it over. Responsibility and tasks must remain within the organization to anchor the new approach in the long term. After a successful start, the expert should be mostly superfluous and only consulted selectively for specific questions or further developments.

The following are typical pitfalls in the introduction of MBSE. The knowledge of the following list of bad practices can help to avoid errors in the first place.

Aimless Modeling

A project can take a methodology and a SysML tool and start modeling the system. That works apparently very well. After the methodology's guidance, they go from requirements identification to requirements analysis to architecture and so on.

But how detailed should they model the system functions? Don't they anticipate software engineering when it comes to software-related functions? How detailed are the blocks of the architecture? Do the cables, screws, and housing also need to be modeled? And most importantly, what is the value of this modeling to the project?

Modeling is not an end in itself, and only if the purpose is known, the answers to the questions above can be given. If all needs of the stakeholders of the engineering effort are satisfied, the model is complete.

In Fig. 1, this is referred to as query-driven modeling, which is derived from test-driven development (TDD) [1]. TDD is a software approach that, in simple terms, states that test cases are first written and then the software is developed until all test cases are successful. Then one is finished.

The MBSE model is only a source of engineering information that must answer engineering stakeholder questions. If all questions can be answered, the model is complete. These questions are also called competency questions and are used in the ontology domain to set the model's scope based on [6]. It is important to remember that creating the model itself is also a valuable outcome, not just the result model.

Typical engineering stakeholders of an MBSE project are, for example, all kinds of involved engineers, the management, supplier, manufacturing, marketing, quality manager, and the legal department.

Fast False Start

Full-fledged model-based specifications, traceability from requirements through architecture to implementation and back, simulation, document generation, automated analysis and verification, and more, these are all powerful tools in a model-based approach that can make projects successful. But they are also all tools that have to be learned to use correctly. To do this, you have to build up the systems modeling environment and knowledge and, above all, practice a lot and gain experience.

A typical mistake is to introduce too much in one step. The probability of failure is high, and the opinion is quickly anchored in the organization that modeling does not work. Therefore, introduce a model-based methodology in several small steps. If things go well, you can increase your walking speed. If things do not go well, you have a manageable list of issues for improvement. Think big; start small!

Overmodeling

Overmodeling is a common phenomenon in MBSE. With SysML, almost anything can be modeled. The modeling language sets no limits and does not indicate when to stop modeling. Less is more, and just because it works, it should not be modeled.

In addition to creating a model element, there is also a lot of follow-up work for each model element: changes to the definition or representations in diagrams changes through dependencies to other elements, adjustments of simulations, model generators and automated analyses, and so forth. These efforts are usually much higher than the initial effort to create the model element.

An overmodeled model becomes sluggish, and the effort quickly exceeds the benefit.

Model Island

It is a bad practice if the model and associated tasks are on an isolated island and not integrated into the other development tasks.

If the modeling is not integrated, it cannot unfold its full potential and is discarded at the latest when the project comes under pressure, which usually happens at some point.

Modeling must be an integrated and mandatory part of development from the outset, not an optional complement. Optional activities do not withstand project pressure and are soon dropped. In particular, this means that the results of the MBSE activities, which typically existed previously in the engineering process in a document-based format, must be replaced by the new MBSE activities. Specific attention should be paid to this, as resistance from the organization is expected, as shown, for example, in the previous section based on related work [15].

Ivory Tower

The ivory tower is similar to the model island. In the tower, there are a few modelers who create very sophisticated models. The models are very detailed and use unusual and rare elements and rules of the modeling language. Even if the models are correct, they are only understood and used by a few project participants. Accordingly, the models (and also the modelers) are hardly integrated into development processes and have little significance and benefit for the project.

Fly Out of the Learning Curve

To use modeling effectively, the modelers and the organization must also learn and gain experience. That takes time and initially runs contrary to one of the primary goals of modeling, becoming faster and more effective. Introducing MBSE to speed up when the project pressure is already high will not work.

If you do not plan the learning curve's necessary times, it will probably carry you out of the curve at some point, and the introduction of the model-based approach can fail. That, of course, applies to the introduction of any new method and is independent of MBSE.

Developing the MBSE Methodology

The core principles of MBSE methodologies can be used to develop a customized MBSE methodology for a project or organization. In this chapter, the SYSMOD Methodology Adoption Process (SMAP) depicted in Fig. 3 is used. However, it is just an example, and any other MBSE methodology can be used. The logical order of the SMAP steps is depicted by a SysML activity diagram. The process steps are the boxes with rounded corners, and the attached rectangles describe the input and output artifacts of the step. The arrows depict the flow of control and the artifacts.

10 Adoption of MBSE in an Organization

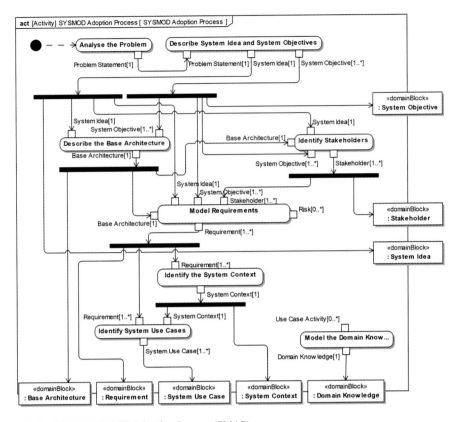

Fig. 3 SYSMOD MBSE Adoption Process (SMAP)

To better illustrate SMAP, this chapter explains how it was applied to a fictional organization that would like to introduce MBSE to become more competitive. Their MBSE introduction is a strategic organization-wide decision and is not limited to a single project.

The step "Analyse the Problem" at the beginning examines whether MBSE really solves the causal problems or just symptoms. The outcome is a problem statement, which then forms the basis for defining the idea and objectives of MBSE adoption in the next step.

Typical results of the problem analysis are too long development times, too high costs, and too low quality. Not to be underestimated is the growing problem in engineering organizations that, of course, also processes, and tools must continue to evolve to continue to create innovative products. We cannot create tomorrow's products with yesterday's tools. To put it bluntly, no one today wants to create software with punch cards or design mechanical components on drawing boards. The introduction of MBSE is just a natural evolution of engineering capabilities.

INCOSE's Systems Engineering Vision 2020 [8] gives a good argument for the introduction of MBSE: "A key driver [for MBSE] will be the continued evolution of

Fig. 4 Objectives of the MBSE adoption

complex, intelligent, global systems that exceed the ability of the humans who design them to comprehend and control all aspects of the systems they are creating."

This argument for MBSE could also be part of a problem statement when viewed from a different perspective. Organizations need to ensure they are sufficiently innovative to avoid being pushed out of the market by competitors.

Since the concept of a "model-based approach" is broad, it is essential to describe clearly what one understands by it, how one wants to introduce it, and what the organization expects from the introduction. These steps are essential to avoid aimless modeling. Four objectives for the approach were derived from the identified problems (Fig. 4).

Figure 4 also lists some of the stakeholders of the MBSE introduction. Of course, there are many more stakeholders that must be identified for requirement elicitation. This work is done in a step after the description of the system idea and objectives.

Typical objectives of an MBSE introduction are cost and time savings. Such objectives need special attention insofar as the introduction of MBSE at the beginning leads to the opposite: it costs time and money. The time component and the introduction effort should be taken into account by the objectives since the initial opposite can easily lead to the conclusion that the MBSE introduction has failed.

It is essential to know all the stakeholders, their concerns, and their needed views on the engineering information. The list of stakeholder views is one input to derive the languages and tools for the MBSE methodologies. Besides modeling tools like SysML tools, an outcome could also be that office document tools are required. They can be used just for viewing modeling information, for example, automatically generated text documents from a SysML model. Still, it is also possible that some information is stored in office documents instead of a model. Although a model should be preferred, an MBSE approach does not have to be 100% model-based.

The black bar in Fig. 3 specifies that the description of the base architecture can be done parallel to the stakeholder identification. The base architecture of an MBSE introduction represents the technical constraint requirements for the MBSE methodology, for example, the systems modeling environment with given concrete tools.

Some typical stakeholder requirements can be seen in Fig. 5.

The MBSE context shows the users of the MBSE and systems to which there are interfaces (Fig. 6). For space reasons, the interface boxes (ports) are not labeled in the diagram. In this example, the MBSE users are typical roles of a systems engineering project. The context also lists the external systems with interfaces to the systems modeling environment, which is part of the MBSE "system" in the

#	Id	Name	Text	Stakeholders
1	MBSE-REQ4	Measurability	KPI's and goals must be defined that objectively represent the success of MBSE.	Head of Engineering
2	MBSE-REQ3	Product Roadmap	The introduction of MBSE must not delay the planned product release roadmap.	Head of Product Management
3	MBSE-REQ1	Training	There must be a training offer for the new methods and tools on the market.	Human Resource
4	MBSE-REQ2	Employee Development	The existing employees must be trained for the new approach.	Human Resource

Fig. 5 Some MBSE stakeholder requirements

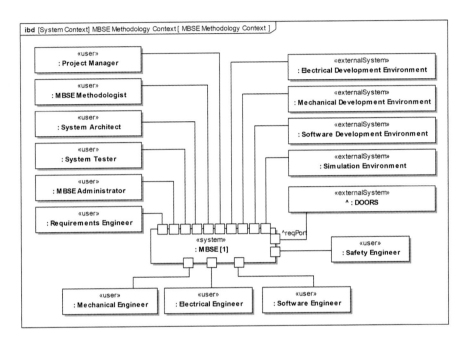

Fig. 6 Example system context for MBSE

middle. The elements with a leading ^ symbol are inherited from the base architecture definition.

To integrate the methodology into the overall engineering landscape, you need to focus on the interfaces, which means the context diagram's ports. The ports to human actors are about views and UI needs. The ports to external systems are about interoperability, i.e., data formats, adapter, and exchange technologies like Open Services for Lifecycle Collaboration (OSLC) [14].

The use case analysis leads to a list of the methodological steps required to achieve requested stakeholder results by the methodology. Figure 7 shows some use cases of the MBSE approach and the actors involved.

In contrast to the typical use case analysis in systems engineering, the use cases of the MBSE methodology are typically not described in detail. Instead, however, the incoming and outgoing objects, a short description of the use case, the actors' interfaces, and the necessary views and other remarkable points are noted. Figure 8 shows an example of the use case "Determine architectural elements for a requirement."

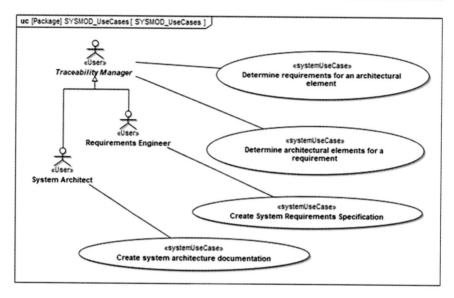

Fig. 7 Methodology use cases

Determine architectural elements for a requirement	System
Description	On the basis of the system model, the list of requirements for an architectural element is determined, which are partially or completely implemented by the architectural element.
Actorsr	Traceability-Manager
Trigger	When changes are made to the architecture, it must be checked whether all requirements have been taken into account.
Result	List of affected architectural elements in Excel format
Precondition	None
Postcondition	None

Fig. 8 Example use case description

The domain knowledge defines the incoming and outgoing objects of the methodological steps. It provides a clean and concise description of the engineering artifacts. The diagram in Fig. 9 shows only the relationships between the objects. There are also properties and a textual description for each domain block defined but not depicted in the diagram. For example, the domain block "Requirement" also defines the list of requirement properties.

The relationship, for example, between the Stakeholder on the left side and the System Objective specifies that every stakeholder has zero to many system objectives and every system objective has exactly one stakeholder.

Having clarity on the language and tool independent aspects of MBSE implementation makes it much easier to make the right decisions on tools and modeling languages. Questions like "Is tool A better than B?" or "Should we rather take the Object Process Methodology (OPM) language instead of SysML" can be answered well on this basis.

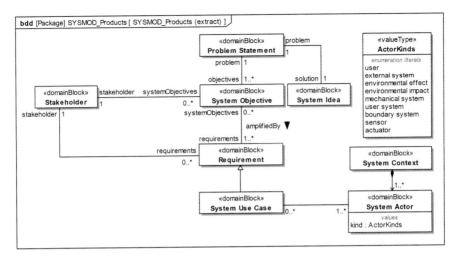

Fig. 9 MBSE methodology domain knowledge

For the implementation of the methodology, among other things, the modeling languages and tools must be selected, which is considered in the following sections.

Selection of the Modeling Languages

The modeling languages must satisfy the needs of the methodology. If the methodology's task is "Create a 3D specification of the mechanical parts," SysML is not the right choice. The system idea of SysML is "SysML is designed to provide simple but powerful constructs for modeling a wide range of systems engineering problems. It is particularly effective in specifying requirements, structure, behavior, allocations, and constraints on system properties to support engineering analysis" [13].

SysML is only one potential candidate for an MBSE modeling language. Typically, it is a good choice to cover the overarching requirements and system architectures. It is a general-purpose language and, therefore, probably not suitable for special tasks. More than one language can be used in an MBSE project. Since they are typically implemented in different tools, these should be designed to be interoperable.

The outcomes of the methodology must be mapped to the modeling languages. The mapping determines which languages are suitable and which subsets of the languages are used. A mapping to SysML is shown below.

For example, the system context is an outcome of the step "Identify the System Context" in the SYSMOD methodology. Figure 10 depicts the system context domain blocks identified during the MBSE methodology analysis described in the previous section.

The system context includes the actors of different kinds; the connections between the system and the actors, including the interfaces and item flows; and the system of interest itself.

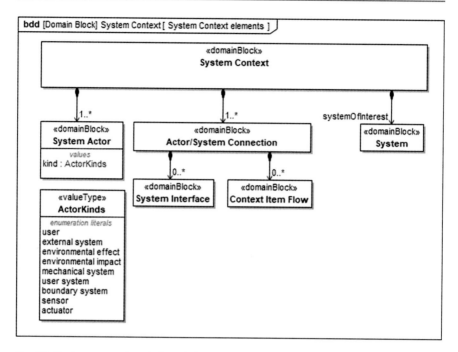

Fig. 10 System context domain knowledge

SysML has appropriate elements to describe these artifacts. Stereotypes can add additional properties and semantics not covered by SysML. In addition to the model elements, you must also define the views requested by the stakeholders.

Figure 11 lists the mapping of the system context concept to SysML.

An outcome of the analysis can also be that a required artifact of the methodology cannot be expressed in the selected modeling language. In that case, another language must be selected for this artifact and probably also another tool, as well as a way how to link this artifact with the other elements to keep the traceability and consistency.

Based on the customized subset of SysML, the modeling tool should be adapted as much as possible to optimize the subset's support, for example, making the model elements easily accessible in the toolbox and menus while removing the unused model elements.

Selection of the Modeling Tools

The selection of modeling tools derives from the selection of the modeling languages. The modeling language elements and views are, so to say, the mandatory functional requirements for the tool. Also, you have a set of mandatory and optional quality requirements. The set of quality requirements is different for each project or organization. Typical requirements are:

Methodology Artifact	SysML Element/View
System Context	Block with stereotype «systemContext»
Actor	Actor or Block with SYSMOD actor stereotypes
Actor/System Connection	Connector
System Interface	Proxy Port, Interface Block
Context Item Flow	Item Flow, Block with stereotype «domainBlock»
View System Context	Internal Block Diagram
Documentation System Context	Automated Document Generation

Fig. 11 Mapping methodology/language

- Interoperability, for example, interfaces to other engineering tools, import/export formats, OSLC support
- Usability
- Support by vendor and community
- Conformance to standards
- License cost
- Extensibility, for example, support of profiles, and adaption of the UI
- Automation of modeling tasks

Ideally, the methodology and the modeling languages are defined before the appropriate tools are selected.

Verification and Validation

As with the development of a product, introducing a new methodology must be verified to ensure that it is implemented correctly and meets or supports the objectives.

For verification, the implementation of the use cases and the requirements are checked, typically, by inspection. For the validation, the methodological approach's objectives must be considered and examined whether they are met or adequately supported.

If not already covered by the objectives, it must also be critically considered whether the MBSE approach really benefits and does not only create additional effort. At least the following questions should be honestly answered:

- Is the benefit higher than the effort minus the initial learning curve?
- Is the methodology lived by the participants or only worked through?
- Did the procedure lead to redundant work because the new ones did not replace existing steps?

From the answers, related analyses, and other verification and validation, it can be deduced whether the methodology is basically successful. The decision can be made whether and how the MBSE approach should be continued. This usually leads to adjustments in methodology. The ongoing projects certainly also regularly generate new requirements.

It is usually challenging to identify objective KPI's that show the success of MBSE. Surveys of project participants are well suited, ideally taking place at regular intervals to uncover changes over time.

Overall, the adaptation and evaluation of the methodology is a continuous improvement process that must be carried out consciously.

Running the MBSE Infrastructure

The previous section presented an approach on how to develop an adapted MBSE methodology, including selecting the languages and tools. It is an iterative process that always needs feedback from ongoing projects and needs to be adapted appropriately.

For MBSE projects to run well, infrastructure is needed in addition to the appropriate methodology. Figure 12 shows the SYSMOD infrastructure process as an example, showing the steps to build the infrastructure for MBSE.

The first step, "Tailor the MBSE Methodology," is covered by the previous section. People must be familiar with the methodology, modeling languages, and modeling tools. You must differentiate between people who must read the models and people who build the models. These are different skills whereby the reading skill is a subset of the building skill. Creating a model requires a much more in-depth knowledge than reading a model or understanding views of the model.

First, you need an overview of the MBSE capabilities of the project stakeholders. Together with the required skills for the MBSE methodology and the roles assigned to people, this results in a list of missing skills. From this, the training needs can then be derived.

When conducting the training, it is important that the three aspects of methodologies, languages, and tools (Fig. 1) are separated from each other so that the people can later orient themselves better in the MBSE world.

In addition to the training, you must provide consulting services to quickly resolve issues, remove hurdles, and get feedback from the projects. In the beginning, this can be done by external experts. In the long run, internal persons should take over continuous consultation. External experts are only used for particular topics, in exceptional cases, and for occasional reviews.

The deployment of the methodology should enable the successful application of MBSE. The methodology should not be introduced in one step, instead, define intermediate goals. Think big and start small.

The systems modeling environment (SME) is the set of tools used for the methodology. It is necessary to enable methodology-specific extensions to use a modeling tool for a tailored methodology. Additionally, you should remove

10 Adoption of MBSE in an Organization

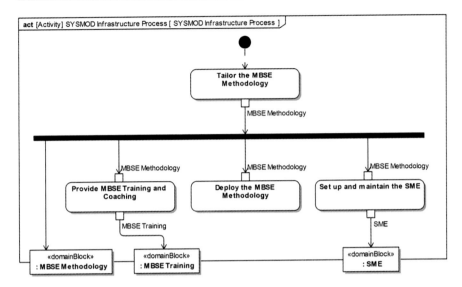

Fig. 12 SYSMOD infrastructure process

superfluous elements from the tools and add convenience functions to provide a more convenient user experience.

If required, model libraries must be made accessible, and toolchains between the tools must be established.

Summary

The introduction of an MBSE methodology is a challenging task. It is much more than just buying a modeling tool and learning the modeling language. It's also not something you can buy from a consulting firm.

The introduction has to be treated like a project with goal definition, requirements, and development of a solution. Many techniques of (model-based) systems engineering can be applied to introduce MBSE.

The human factor in the introduction must not be underestimated and be addressed explicitly. It is a change project, where the human is the biggest challenge.

This chapter has shown, in particular, using SYSMOD as an example, how an MBSE methodology can be used to develop an MBSE methodology. The methodology is then the basis for selecting modeling languages and tools.

Cross-References

▶ MBSE Methodologies

References

1. K. Beck. Test-Driven Development By Example. Addison-Wesley. 2002.
2. M. Bone, R. Cloutier. The current state of model based systems engineering : results from the OMG TM SYSML request for information. Annual Conference on Systems Engineering Research, CSER2010, Hoboken, NJ. 2009.
3. M. Chami, J. Bruel. A Survey on MBSE Adoption Challenges. (2018) In: INCOSE EMEA Sector Systems Engineering Conference (INCOSE EMEASEC 2018), 5 November 2018 - 7 November 2018 (Berlin, Germany).
4. M. Chami, A. Morkevicius, J. Bruel. Towards Solving MBSE Adoption Challenges: The D3 MBSE Adoption Toolbox. INCOSE Internation Symposium 2018. https://doi.org/10.1002/j.2334-5837.2018.00561.x.
5. J. Elm, D. Goldenson. The Business Case for Systems Engineering Study: Results of the Systems Engineering Effectiveness Survey. Carnegie Mellon University. Journal contribution. 2018. https://doi.org/10.1184/R1/6585080.v1.
6. M. Gruninger, M. Fox. Methodology for the Design and Evaluation of Ontologies. In: Proceedings of the Workshop on Basic Ontological Issues in Knowledge Sharing, IJCAI-95, Montreal.
7. J. Hallqvist, J. Larsson. Introducing MBSE by using Systems Engineering Principles. INCOSE Internation Symposium 2016. https://doi.org/10.1002/j.2334-5837.2016.00175.x.
8. Internation Council on Systems Engineering (INCOSE). Systems Engineering Vision 2020. INCOSE-TP-2004-004-02. September, 2007
9. International Council on Systems Engineering (INCOSE). Systems Engineering Vision 2025. 2014.
10. J. Krassner. How Product Development Organizations can Achieve Long Term Cost Savings Using Model-Based Systems Engineering (MBSE). 2015. Ashland, MA: American Technology International, Inc.
11. A. Madni, S. Purohit. Economic Analysis of Model-Based Systems Engineering. Systems. 2019. https://doi.org/10.3390/systems7010012.
12. Niko Mohr, Jens Marcus Woehe. Widerstand erfolgreich managen: Professionelle Kommunikation in Veränderungsprojekten. Campus Verlag. 1998.
13. Object Management Group. OMG Systems Modeling Language (OMG SysML). Version 1.6. formal/19-11-01.
14. Open Services for Lifecycle Collaboration (OSLC). https://open-services.net. accessed October 2020.
15. A. Vogelsang, T. Amorim, F. Pudlitz, P. Gersing, J. Philipps. "Should I stay or should I go? On forces that drive and prevent MBSE adoption in the embedded systems industry". International Conference on Product-Focused Software Process Improvement. Springer, Cham. 2017.
16. T. Weilkiens. SYSMOD – The Systems Modeling Toolbox. 3rd edition. MBSE4U. 2020.

Tim Weilkiens is a member of the executive board of the German consulting company oose, a consultant and trainer, lecturer of master courses, publisher, book author, and active member of the OMG and INCOSE community. Tim has written sections of the initial SysML specification and is still active in the ongoing work on SysML v1 and the next generation SysML v2.

He is involved in many MBSE activities, and you can meet him at several conferences about MBSE and related topics.

As a consultant, he has advised a lot of companies from different domains. The insights into their challenges are sources of his experience that he shares in his books and presentations.

Tim has written many books about modeling, including Systems Engineering with SysML (Morgan Kaufmann 2008) and Model-Based System Architecture (Wiley 2015). He is the editor of the pragmatic and independent MBSE methodology SYSMOD – the Systems Modeling Toolbox.

Model-Based Requirements

11

Alejandro Salado

Contents

Introduction	350
Theoretical Framework	351
Dedicated Classes and Flagged Models	352
Requirements As a Specific Class of Element	352
System Models As Requirements	353
Math-Based Models of Requirements	356
Wymorian Models of Requirements	356
Property-Model Methodology (PMM)	360
Semantic Extensions to Model the Problem Space	364
Elicitation, Derivation, and Trade-Off Analysis	369
Better Problem Formulation Due to Automated Enforcement of Syntactic Rules	369
More Comprehensive Requirements Derivation Due to Increased Semantic Precision	370
Early Verification and Validation	370
Cognitive Assistance to Increase Completeness	372
Chapter Summary	373
Cross-References	374
References	375

Abstract

This chapter presents several approaches to capture requirements using models. The different modeling constructs are presented, followed by examples of how they work when capturing requirements. The approaches are organized in three groups. The first group includes the approaches that are more widely used in practice. They include those that use a dedicated model class to capture requirements and those that flag system models as requirements. The second group includes approaches that leverage mathematical constructs, paying attention to the Wymorian framework and the Property-Model Methodology. The third group

A. Salado (✉)
The University of Arizona, Tucson, AZ, USA
e-mail: alejandrosalado@arizona.edu

addresses those methods that are based on defining dedicated semantics to capture the problem space. The chapter concludes with a discussion of how model-based approaches to capture requirements affect the elicitation, derivation, and trade-off of needs and requirements.

Keywords

Model-based requirements · Requirements engineering

Introduction

Adequately framing and formulating the problem before initiating design and development activities is one of the core principles of systems engineering [1–3]. Solving a wrong problem likely leads to an ineffective solution [4]. Requirements written in natural language (e.g., using shall statements) have traditionally been the most common mechanism to formulate engineering problems [5, 6].

The development of Model-Based Systems Engineering (MBSE) has led to the idea of substituting traditional shall statements by models [7–12]. This transformation is expected to provide several benefits, including enhanced mapping, traceability, and system decomposition [11]. Furthermore, incorporating models of requirements within a central model of the system, instead of leveraging textual statements in natural language, could "facilitate requirement understanding and foster automatic analysis technique" [10].

This chapter presents several approaches to model requirements. These are organized in three groups. The first group includes approaches that are widely employed yet either do not represent real models of requirements or embed poor practices of requirements engineering. In other words, these approaches are claimed to be model-based approaches to capture requirements; however, either they do not do so or do so poorly. The second group includes approaches that are based on mathematical underpinning. These methods are constructed following a rigid mathematical structure. While these may still be descriptive in nature, the rigidity of the structure is used to achieve uniformity and precision. The third group includes approaches that leverage nonmathematical structures to semantically capture problem spaces. All approaches have been selected based on the judgment of the author as to their maturity and/or theoretical and/or practical relevance while collectively covering the major types of approaches described in the literature.

Before the three groups of approaches to model-based requirements are presented, the theoretical framework adopted in this chapter to conceptualize requirements is described. This framework will be used to evaluate the different approaches to model-based requirements and discern which ones yield adequate versus non-adequate models of requirements.

Finally, the chapter concludes with some insights as to how model-based requirements can improve elicitation, derivation, and trade-off analysis during problem definition with respect to using textual requirements.

Theoretical Framework

The theoretical framework presented in [13] is adopted in this chapter. It is assumed that the purpose of requirements is to define, formulate, and/or scope the problem to be solved or the opportunity to be sought [1].

The *problem space* is defined as a representation of a problem. A *problem* is defined as a desired situational change: There is a current situation and a desired future situation that is different from the current one. The notion of *representation* explicitly differentiates the problem space from the problem itself. The *solution space* is a set of elements that solve the problem represented by the problem space. Therefore, a problem space yields a solution space [4]. In systems engineering, the solution space is a set of systems that solve the problem represented by the problem space.

von Bertalanffy's definition of a *system* is adopted: A system is a set of inter-related elements [14], where the type of inter-relations is unrestricted. The notion of system boundary is central to this definition: Some elements form the system and some elements do not.

Two classes of systems are considered: open systems and closed systems. An *open system* is one that exchanges information, matter, or energy with external systems [14]. That is, an open system can be described as an entity that processes inputs into outputs [14]; it executes a function. A *closed system*, to the contrary, is one that does not exchange any information, matter, or energy with any external system [14]. That is, a closed system cannot be described as a process of inputs into outputs. Instead, a closed system may be described by properties of the whole or sequences of *snapshots* of its inside.

Two classes of problem spaces are thus of interest for systems engineering: problem spaces of outcomes and problem spaces of functions.

A *problem space of outcomes* defines the desired state of a universe of discourse. It defines the outcomes desired by the stakeholders. Outcomes are not inputs into a system or outputs provided by a system; an outcome cannot be produced by a system. Rather, outcomes are consequences of interactions (or inter-relations, in von Bertalanffy's terms). A problem space of outcomes necessarily yields a solution space that consists of closed systems. The outcome is, thus, a representation of a condition, state, or property (used indistinctively here) of the whole closed system and, by definition, is not an output of it.

A solution to the problem space of outcomes is a closed system that consists of several open systems in inter-relation. Realizing those (open) systems can be conceived as solving a problem, and hence, this creates the need for a second class of problem spaces in systems engineering: the problem space of functions.

A *problem space of functions* defines a set of desired output trajectories for a given set of input trajectories (that is, a set of desired functions). Since closed systems cannot accept inputs or provide outputs, a problem space of functions necessarily yields a solution space formed by open systems. Therefore, any desired open system that is needed to interact in a certain way within a desired closed system is a response to a problem space of functions.

Solving the problem space of functions requires finding a suitable open system that achieves the desired functions at its boundary. Such an open system will be referred to as a *System of Interest (SOI)*.

Two classes of *requirements* (i.e., mechanisms to formulate the problem space) are sufficient: one class of requirement to formulate each class of problem space. The term *Stakeholder Needs* is used to refer to the mechanisms that formulate the problem space of outcomes, and the term *System Requirements* is used to refer to the mechanisms that formulate the problem space of functions.

As per the definitions of the two classes of problem spaces, *stakeholder needs* capture the desired properties of the interaction or engagement of the SOI with external systems. Therefore, their satisfaction occurs after putting the SOI into its environment so that the engagement of all external systems with it is exercised. *System requirements* capture the desired properties of a system's boundary, which do not depend on the actions of external systems. Therefore, their fulfillment can be demonstrated by assessing the system in isolation. This distinction is essential to assess the adequacy of a given formulation mechanism in the context of the class of problem space that it is intended to formulate.

Dedicated Classes and Flagged Models

This section presents the approaches that are arguably most widespread in practice. The first approach specifies dedicated elements that capture requirements within an MBSE environment. The second approach adds the possibility to flag system models as requirements.

Two aspects are worth noting before presenting the different modeling approaches and constructs. First, the examples used in the chapter are not intended to show a complete set of requirements or a complete use of the modeling constructs. The purpose is to show how different approaches attempt to model requirements. Second, the discussion centers on the modeling of the requirement not on techniques that aid the elicitation, derivation, and/or organization of needs and requirements. For example, Use Case diagrams are very useful in this regard. They capture situations that are relevant to the problem at hand but do not model the problem itself.

Requirements As a Specific Class of Element

Most modeling languages used to support MBSE implementations, such as the Systems Modeling Language (SysML) [15], STRATA® [16], or the Lifecycle Modeling Language (LML) [17], offer specific classes of elements to capture requirements. These elements are defined as text objects with other properties (such as unique identification numbers) that can be linked through various kinds of relationships to other elements in the model. These constructs are used to capture relationships between requirements (for example, those related to derivation and/or

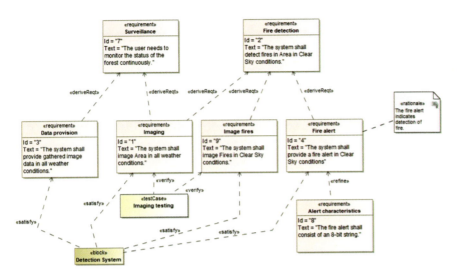

Fig. 1 Example of requirement elements in SysML

refinement), relationships between requirements and verification artifacts (for example, those that identify which verification activity is used to verify which requirement), and relationships between requirements and system components (for example, those that identify which component must satisfy which requirement), among others.

Figure 1 provides an example of this approach to model requirements, using SysML specification. Arguably, this is the most common approach to capture requirements in an MBSE environment [18, 19]. However, while the objectification of the requirement enables effective traceability between different systems engineering artifacts, the requirement remains a text string encapsulated as an object. Therefore, these objects cannot be really considered requirement models different from text-based requirements.

System Models As Requirements

Realizing the modeling limitations of using textual requirements as objects described in the previous section, a different path taken by the MBSE community has focused on flagging parts of a system model as requirements, avoiding the need for *shall* statements in this case [11, 12, 20]. In the case of SysML, for example, an *activity diagram* or a *state machine diagram* would be flagged and considered a requirement. In this case, the system is expected to do what the model represents.

This approach is certainly model based, as it represents a requirement as a model without using textual statements. However, it comes with problems associated with good practices in requirements engineering [21]. Particularly, because a system model is used as a requirement, the model enforces a particular design solution, thus unnecessarily reducing the solution space [5]. Bellagamba [22] articulated this problematic, and Wach and Salado [21] demonstrated it with the following example.

Following the requirements in Fig. 1, consider a need for a system to detect fires in a certain area. Two different operational conditions may be present: *Clear Sky* and *Rain*. Regardless of the operational condition, the system is required to continuously capture images of the surveilled area and send them to a station. In addition, the system is required to observe fires and send alarms to a station, at least when the operational condition is *Clear Sky*. Therefore, the problem space is defined by four conditions:

1. Imaging of surveilled area in all weather conditions
2. Imaging of fires in surveilled area in Clear Sky condition
3. Provision of gathered image data in all weather conditions
4. Provision of a fire alert for detected fires in Clear Sky condition

These four conditions can be modeled as shown in Fig. 2. Certainly, the four conditions are met by the system model. A system developed following this model as the requirement would implement two different states, each of them being active depending on the weather condition. However, the authors show that the system model shown in Fig. 3 also fulfills the four conditions. Yet, such a solution, based on a single state, cannot be derived from Fig. 2 because it is inconsistent with it. A good requirement model would not discard any possible solution to the four conditions previously listed. Therefore, this shows that simply flagging a system model as a requirement can lead to poor requirements engineering practices.

Fig. 2 System model representing the need to detect a fire

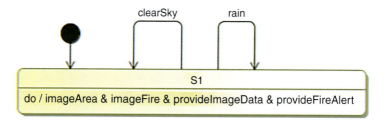

Fig. 3 Alternative system model representing the need to detect a fire

This effect is not limited to using state machines but is inherent to using any kind of system model. A similar example using activity diagrams in SysML is shown in Fig. 4. The system model fulfills the four conditions given above, where the activity *Sense weather* indicates when *Identify fire* must be executed. (Note: assume that the model is complete with all interactions being adequately defined.) The model enforces that any system solution implements those activities and their mutual sequencing. However, the alternative solution shown in Fig. 5 would also be acceptable against the four conditions. Yet, as in the case with state machine diagrams, using any of the two models as the requirement would make it impossible to accept each other as a solution, even if both are acceptable.

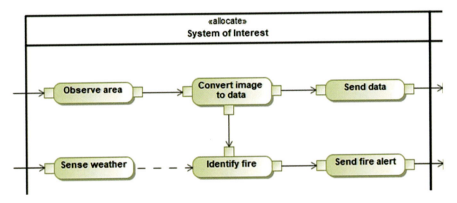

Fig. 4 A system model using an activity diagram representing the need to detect fires

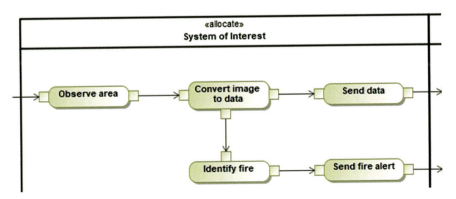

Fig. 5 Alternative system model using an activity diagram representing the need to detect fires

Math-Based Models of Requirements

This section presents two model-based approaches to model requirements that leverage mathematical constructs as the underlying foundation: the Tricotyledon theory of system design (T3DS, which will be referred to in this chapter as the Wymorian models of requirements) [23], and the Property-Model Methodology [24]. These two approaches have been selected for this chapter because their development is solid and they are based on different mathematical frameworks.

The two approaches have been developed by the original authors with varying levels of depth. A full description of the modeling constructs and the capabilities that they enable is outside the scope of this chapter. The readers are encouraged to reach out to the original sources for an in-depth treatment of the material. This chapter provides a summary of the key modeling constructs and shows how they can be applied to model requirements.

Wymorian Models of Requirements

Wymore [23] establishes its modeling constructs for requirements within an overarching mathematical framework for discrete systems. Discrete systems are those defined in temporal scales that are discrete and not continuous. However, it is worth mentioning that he envisioned a path to adapt the framework to continuous systems [23], although never developed it, and other authors have shown this to be feasible [25]. Therefore, while the constructs here are formally restricted to discrete systems, they can be extended to continuous systems.

The basic Wymorian requirements modeling construct is called a system design requirement (or system design problem), is denoted by *SDR*, and is defined as a sextuple of the form:

$$SDR = (IOR, TYR, PR, CR, TR, STR) \qquad (1)$$

Where *IOR* is an input/output requirement, *TYR* is a technology requirement, *PR* is a performance requirement, *CR* is a cost requirement, *TR* is a trade-off requirement, and *STR* is a system test requirement. Note that all these are defined as sets of requirements, not as individual ones. These six types of requirements can be grouped in three categories: requirements that determine which systems fall within the solution space or not (IOR and TYR), requirements that set an order on the solution space (that is, they determine the preferability of one solution over another within the solution space; PR, CR, and TR), and requirements that establish the *observation* conditions under which the other requirements are considered to have been met (STR). Because STR does not influence the solution space but defines procedural conditions of the realization of the system, it will not be covered in this chapter. Each of the other ones will be precisely defined next.

11 Model-Based Requirements

The *IOR* captures the behavior that the SOI must exhibit by specifying:

- The length of the operational life of the SOI
- The set of inputs to be accepted by the SOI
- The set of input trajectories or histories that the SOI might experience
- The set of outputs producible by the SOI
- The set of output trajectories that are possible for the SOI
- The function that matches outputs and inputs or input trajectories and eligible output trajectories

Formally, the modeling construct for the *IOR* is defined as a sextuple denoted by *IOR* = (*OLR*, *IR*, *ITR*, *OR*, *OTR*, *ER*), where:

- $OLR \in \mathbb{N}$.
- *IR* is a nonempty set.
- *ITR* is a subset nonempty of a family of functions $f : TSR \rightarrow IR$, where $TSR = \{n \in \mathbb{N} : n < OLR\}$.
- *OR* is a nonempty set.
- *OTR* is a subset nonempty of a family of functions $g : TSR \rightarrow OR$.
- *ER* is a function defined over the set *ITR* with values in the set of nonempty subsets of *OTR* such that $OTR = \cup \ range(ER)$.

In line with the theoretical framework presented in the previous section, it is important to note that the "IOR is not a system model ..., but it determines a space of system designs/models each of which satisfies the IOR in the sense that each model in the space has the same input and output sets as specified by the IOR and the IO behavior of the model is within the range specified by the IOR" [23]. This is why, in general, *IOR* matches individual input trajectories to subsets of output trajectories, while a system model would only provide one output trajectory for an individual input trajectory.

Consider the previous fire detection problem as an example. The four conditions defining the problem space are listed below again:

1. Imaging of surveilled area in all weather conditions
2. Imaging of fires in surveilled area in Clear Sky condition
3. Provision of gathered image data in all weather conditions
4. Provision of a fire alert for detected fires in Clear Sky condition

OLR would be defined as the required lifetime of the SOI, using a time scale of interest to define the different trajectories. For example, assume a lifetime of 10 years, and a timescale of 1 min, driven by the maximum time that it must take to provide the fire alert from the time the fire is observed. (Note that there are no rules to define the timescale other than enabling defining all necessary requirements. The rule in this example has been chosen for illustrative purposes only.) Then, $OLR = 10 \cdot 365 \cdot 24 \cdot 60 = 5,256,000$.

IR is given by the different inputs that the SOI must be able to accept. For simplicity, *IR* = {Surface, Fire, Sky}, where, for simplicity, each input can take one of two values: Surface = {Forest, No forest}, Fire = {Fire, No fire}, and Sky = {Cloudy, Clear}. Note that these could be further refined with specific properties of fires, clouds, forests, etc.

ITR is given by any possible combination of *IR* for each time point in the lifetime. That is, each element in *ITR* is a vector of length *OLR* where each component takes one value in Surface × Fire × Sky, listed in Table 1, and all elements in *ITR* collectively cover all possible variations.

OR is given by the different output that the SOI must produce. For simplicity, *OR* = {ImageData, FireAlert}, where, for simplicity, each output can take one of two values: ImageData = {Yes, No} and FireAlert = {Yes, No}. Note that these could be further refined with specific properties such as image resolution.

OTR is given by any possible combination of *OR* for each time point in the lifetime. That is, each element in *OTR* is a vector of length *OLR* where each component takes one value in ImageData × FireAlert, listed in Table 2, and all elements in *OTR* collectively cover all possible variations.

ER would be defined by a mapping of each element in *ITR* to a subset of elements in *OTR*. This is difficult, and lengthy, to do because of the size of both sets (defined by *OLR*). A parameterization construct is introduced such that the modeling can be simplified by grouping subsets of functions within *ER* that take the same values based on a subset of parameters in *IOR*. In this case, for example, matching of functions does not depend on history. That is, past states do not play a role in which output trajectories are adequate for an input trajectory, but just the value of the input trajectory does. As a result, it is possible to do a direct mapping between the values in Tables 1 and 2. Specifically:

Table 1 Possible values for each *ITR* element

Value	Surface	Fire	Sky
1	Forest	Fire	Cloudy
2	Forest	Fire	Clear
3	Forest	No fire	Cloudy
4	Forest	No fire	Clear
5	No forest	Fire	Cloudy
6	No forest	Fire	Clear
7	No forest	No fire	Cloudy
8	No forest	No fire	Clear

Table 2 Possible values for each *OTR* element

Value	ImageData	FireAlert
1	Yes	Yes
2	Yes	No
3	No	Yes
4	No	No

11 Model-Based Requirements

- Because the need is to detect fires in forests, Values 5 to 8 in Table 1 must match to Value 4 in Table 2.
- Because fire alerts must only be provided in the case of a fire, but images must be provided always, Values 3 and 4 in Table 1 must match to Value 2 in Table 2.
- Because fire alert must be provided in the case of a fire in clear sky, Value 2 in Table 1 must match to Value 1 in Table 2.
- Finally, because no restriction is defined for the provision of fire alerts when there is a fire in cloudy conditions, Value 1 in Table 1 must match to Values 1 and 2 in Table 2.

It is possible to define an alternative eligibility function ER^* as a subset of matching of values from IR to OR such that $ER^* = \{(1, 1), (1, 2), (2, 1), (3, 2), (4, 2), (5, 4), (6, 4), (7, 4), (8, 4)\}$. Note how, as indicated earlier, this definition of required behaviors does not limit the solution space as the system models in section "System Models As Requirements" did; pairs (1,1) and (1,2) would allow for both solutions in Figs. 2 and 3 to be acceptable.

The TYR captures the constraints imposed into the technological options that are acceptable to design and develop the system. For example, the development of a given system could be restricted from using ITAR-marked components. This requirement is modeled as a nonempty set of systems. Such set would include all technologies and components that are admissible. The set can also be defined as the complement of the set of restricted technologies, which is a more robust definition against technological evolutions [26].

The PR captures the preferability of solutions with respect to IOR and is defined as a partial ordering on the solution space. In essence, the PR defines those conditions that enable comparing "how well" two different system solutions satisfy the IOR. Originally, Wymore defined such requirements as sort of figures of merit that are different from the IOR but applicable to it. In such a construct, the resolution of ImageData, for example, would be part of the PR, whereas the corresponding IOR would be restricted to the dichotomy of providing image data or not. That is, the PR is modeled as a set of variables that are mapped to the IOR and whose values are mapped to a preference ordering.

An alternative interpretation or extension is also possible. Instead of modeling the IOR as a set of dichotomic behaviors, each different value of characteristics of the inputs and outputs is considered different trajectories. The PR is then modeled as a partial ordering on ER^*. Continuing with the previous example, a PR that captures that it is indifferent to detect fires in all weather conditions or just in clear sky would be given by:

$$PR_1 = \{\prec : a \sim b, \forall a, b \in ER^*\} \qquad (2)$$

whereas a PR that captures that detecting fires in cloudy conditions is not as important as the rest of the conditions would be given by.

$$PR_2 = \{\prec : ((1, 1) \prec a) \wedge (a \sim b), \forall a, b \in ER^* \setminus \{(1, 1)\}\} \qquad (3)$$

The *CR* captures the referability of solutions with respect to the cost that it takes to develop and operate the SOI, both financially and nonfinancially. The *CR* establishes, thus, a corresponding partial ordering in the solution space. No restriction is defined on the kind of variables and functions that relate them that may be used. A model of a *CR* may be in this sense as simple as given in Eq. (4) or as complicated as a nonlinear multiattribute utility function [27].

$$CR = \{\prec : a \prec b \leftrightarrow \text{cost}_{total}(a) > \text{cost}_{total}(b)\} \quad (4)$$

The *TR* captures the preferability between the *PR* and the *CR*, establishing an ordering in the solution space with respect to the ordering established by the *PR*, the ordering established by the *CR*, and the trade-off between them. The *TR* is modeled as a scoring function between the two orderings.

Property-Model Methodology (PMM)

The PMM takes a different approach and models any type of requirement using the same structure, based on the semilattice mathematical construct [24]. A basic well-formed requirement on a system Σ is defined as a constraint applied to a property of an object of such a system when a condition occurs (in the form of an event) or is achieved (in the form of a state).

Formally, the requirement is expressed as $\text{Req} : C \Rightarrow val(o.p) = d$, where Req is an identifier that denotes the requirement statement, C is a state condition of Σ, o is an object of Σ, p is a property of o, and d is a domain in which the value of $o.p$ must be located when C occurs or is achieved. All such elements are necessary for a requirement to be well-formed, that is, complete. (Note: The existence of C is formally declared as optional, but its omission can also be interpreted as a *for all conditions*.)

As an example, consider the common requirement to limit the mass of a satellite during launch, and assume the value of 1,000 kg for illustrative purposes. The requirement can be modeled as follows using PMM:

$$\text{Req1} : C_1 \Rightarrow val(o_1.p_1) <= 1{,}000 \text{ kg} \quad (5)$$

Where:

- $C_1 \coloneqq$ During launch
- $o_1 \coloneqq$ Satellite
- $p_1 \coloneqq$ mass

Here, C_1, o_1, and p_1 are not only textual definitions but also objects within the model. By doing so, the semilattice structure of the modeling construct enables operating with the requirements in several ways, including defining composite requirements, automated identification of dependencies between requirements, and hierarchical nesting of requirements. These three operations are described next.

Composite Requirements. To ease requirements management, it is sometimes convenient when formulating requirements to establish an overarching statement that is later refined as a set of more detailed conditions that need to be fulfilled. In this way, one can work with a *general* aspect (and its compliance) of the system to a set of requirements or focus on each one of those individual requirements in the set. PMM uses the *composite* construct to model this relationship between the overarching statement and each requirement it encompasses.

For example, consider a requirement for a satellite to be compliant to a standard, in this case an electromagnetic compatibility one, denoted by EMC_{STD}. The requirement can be modeled as:

$$\text{Re q2} : \text{In orbit} \Rightarrow val(o_1.EMC_{STD}) = \{\text{yes}\} \quad (6)$$

Complying with a complete standard usually requires fulfilling many conditions (that is, requirements). PMM addresses this aspect by enabling defining one requirement as a composition of others. For example, considering the following requirements as capturing aspects of electromagnetic compatibility (simplified for illustrative purposes):

$$\text{Re q3} : \text{In orbit} \Rightarrow val(o_1.Z_{IN}) > 10 \text{ Mohm}$$
$$\text{Re q4} : \text{In orbit} \Rightarrow val(o_1.RE_{0.15-0.50 \text{ MHz}}) < 70 \text{ dB}\mu\text{V} \quad (7)$$
$$\text{Re q5} : \text{In orbit} \Rightarrow val(o_1.RE_{0.50-30 \text{ MHz}}) < 65 \text{ dB}\mu\text{V}$$

Then, the original requirement can be modeled as a composite relation:

$$\text{Re q2} = \text{Re q3} \wedge \text{Re q4} \wedge \text{Re q5} \quad (8)$$

The use of this construct is not limited to refining properties as sets of additional properties but it can also be used to compose requirements that apply the same domain on the same type of property to multiple objects of a system. An example of such a requirement could be "any component of the satellite shall be space rad-hard."

Automated Identification of Dependencies. In some cases, some requirements (or better said, their compliance) are related or depend on other requirements (or their compliance), all at the same level of encapsulation. That is, this kind of dependency does not refer to that which may exist across levels of encapsulation as the result of a requirements decomposition or flow down effort. Instead, it refers to dependencies that exist between requirements applicable to the same SOI. These dependencies emerge from the way in which different system properties may relate through established rules or laws, and they can be leveraged to assess how changes of requirements can propagate to other requirements [28] or to identify conflicting requirements early in the system development [29]. PMM enables the automated identification of these dependencies through its formal modeling of properties of objects.

For example, consider the requirements for a solar panel (denoted by o_2) on its output power capability at the beginning of life (BOL), its output power capability at the end-of-life (EOL), and its lifetime. Using notional values, they can be modeled using PMM as follows:

$$\text{Re q6 : In orbit} \Rightarrow val(o_2.P_{OUT,BOL}) \geq 1{,}200 \ W$$
$$\text{Re q7 : In orbit} \Rightarrow val(o_2.P_{OUT,EOL}) \geq 1{,}000 \ W \quad (9)$$
$$\text{Re q8 : In orbit} \Rightarrow val(o_2.LIFE) \geq 10 \ \text{years}$$

It is known that the output power capability of a solar panel at EOL is necessarily related to its output power capability at BOL and its lifetime, such that:

$$P_{EOL} = f(LIFE) \cdot P_{BOL} \quad (10)$$

where the function f is left undefined for simplicity. This relationship implies that the requirement Req7, or more precisely its attainability, depends on the requirements Req6 and Req8, as shown in Fig. 6. These dependencies are captured in PMM through the definition of the relationship in Eq. (10) to the properties of the objects it affects.

Hierarchical Nesting of Requirements. A hierarchical nesting of requirements occurs as the result of decomposing requirements at a given level of encapsulation into requirements for components at a lower level of encapsulation following an architecting effort. For example, a requirement on the maximum mass of a satellite during launch may be decomposed into several requirements that set limits on the maximum mass that each of its building components may have. This nesting is modeled in PMM through the nesting of the different object-property pairs.

For example, consider the notional requirement in Eq. (5) and assume that the satellite is decomposed into two main subsystems, the payload (denoted by o_3) and the bus (denoted by o_4). In PMM, it is said that o_1 is the composition of $\{o_3, o_4\}$. It is then possible to define, using notional values, the following requirements:

$$\text{Re q1.1 : } C_1 \Rightarrow val(o_3.p_1) <= 500 \ \text{kg}$$
$$\text{Re q1.2 : } C_1 \Rightarrow val(o_4.p_1) <= 400 \ \text{kg} \quad (11)$$

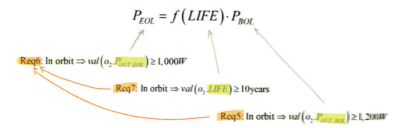

Fig. 6 Automated identification of requirement dependencies

11 Model-Based Requirements

Nesting occurs naturally given the composition relation between o_1 and $\{o_3, o_4\}$, and a relation between the requirements $Req1 \leq Req1.1 \wedge Req1.2$, where $x \leq y$ means that x is less than or equally restrictive to y.

As introduced earlier, the basic construct in PMM can be used to model behavioral requirements (e.g., those related to a function that the system must perform) and structural requirements (e.g., those that limit static attributes of the system) [30]. This is achieved by not restricting the *domain* element of the modeling construct, which enables it to get the form of a function, a state, or similar, and define the conditions as defining states or triggers (inputs). For example, a requirement for a satellite to deploy its solar panels upon command may be modeled as:

$$Req9 : \text{Upon Comand A} \Rightarrow val(o_1.\text{state}) = \{\text{Deployed}\} \quad (12)$$

Similarly, there is no distinction made in PMM between stakeholder needs and system requirements. In fact, due the general definition of the modeling construct, it can be applied to both closed and open systems. As a result, PMM is supposedly able to model both stakeholder needs and system requirements by representing conditions, objects, properties, and domains in line with the theoretical framework presented in section "Theoretical Framework."

While PMM is founded on a mathematical construct, its usage is not limited to using formulas. In fact, the different PMM constructs have been used to extend three modeling languages: SysML [24], Modelica [31], and VHSIC Hardware Description Language (VHDL) [32]. Because of its widespread use in the systems engineering community, only the SysML extension is shown in this chapter as an example.

The basic requirement formulation refines the requirement element with a new stereotype such that, instead of defining the requirement by a text string, it does so through the four main elements of the PMM construct: condition, object, property, and domain, as shown in Fig. 7 for the requirement formulated in Eq. (5).

Composition of requirements is captured by linking the different requirements with the symbol $\oplus$, as shown in Fig. 8 for the requirements formulated in Eqs. (6) and (7). The standard ≪constraint block≫ and ≪block≫ elements are used to establish requirement relationships that lead to dependencies.

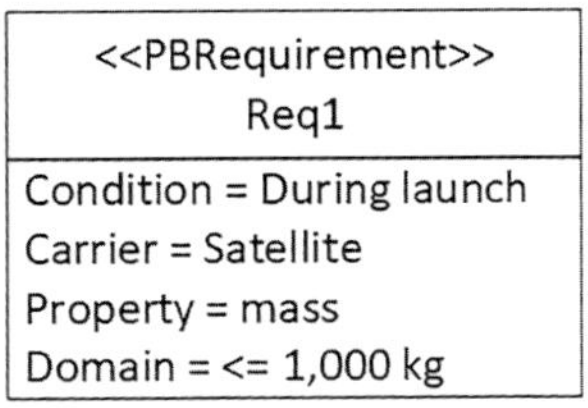

Fig. 7 SysML extension to use PMM constructs. Basic requirement construct

```
<<PBRequirement>>
Req1
Condition = During launch
Carrier = Satellite
Property = mass
Domain = <= 1,000 kg
```

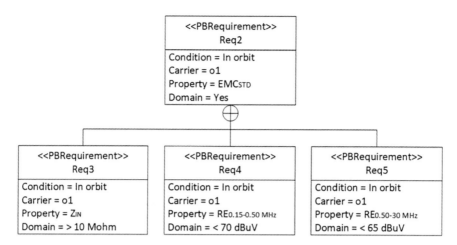

Fig. 8 SysML extension to use PMM constructs. Composite requirements

Semantic Extensions to Model the Problem Space

A different alternative to formally capture the problem space consists of defining an ontological framework, where the modeling constructs specify dedicated semantics for the problem space that are different from those employed to model systems. In this chapter, the True Model-Based Requirements (TMBR) [33] are presented as a formal modeling paradigm to model system requirements toward this vision.

TMBR's modeling constructs derive from the systems-theoretic modeling constructs of the Wymorian mathematical framework presented in section "Wymorian Models of Requirements" and leverage formal taxonomies of requirements [34] and interfaces [35]. These foundations guarantee avoiding formal flaws in problem formulation, such as enforcing design solutions or leaving the requirement unbounded [33].

It is important to note, however, that TMBR is only intended to model system requirements and not stakeholder needs. Following the theoretical framework presented in section "Theoretical Framework," TMBR attempts to model any type of requirement (i.e., functional, performance, resource, and environmental [34]) as a set (or sets) of required input/output transformations [33]. This is achieved by capturing any requirement as a construct of the type *The system shall accept/provide something under certain conditions through an interface* [33], where *something* is an item in the form of information, energy, or material [35], but in the form of a model without the use of textual statements.

In the context of SysML, TMBR can be implemented as an extension of behavioral and structural model elements. The usage of the different model elements relies on semantics that differ from those corresponding to the original model elements in SysML. Specifically, the models presented in this chapter capture solution spaces

11 Model-Based Requirements

(sets of solutions), not systems (single solutions). While SysML models are used for diagrammatic purposes, their *meaning* differs from the traditional SysML specification. In particular, TMBR's implementation in SysML is architected as follows:

- An extended sequence diagram captures the required logical transformation required to the system.
- Signals capture required logical inputs and outputs with their required attributes.
- Ports in block elements capture the required physical interfaces and their required properties through which inputs and outputs are conveyed.
- An extended state machine diagram is used to capture mode requirements, which capture the simultaneity aspects of requirements applicability.

The most basic modeling construct consists of using *Blocks* to represent the SOI for which requirements are being defined and extending *sequence diagrams* to capture the required flow of inputs (and outputs) to (and from) the system. In this way, the system remains a *solid line*, preventing the modeler from defining design-dependent requirements or inner aspect of the system; only the required system's behavior at its boundary in the form of external inputs and outputs is allowed. In (and out)-flows to (and from) the system are defined as items (i.e., energy, information, or material) not as actions, hence guaranteeing consistency with systems theoretic principles for system requirements. An example of how this construct can be applied to model the requirements associated to observing an area and providing data presented in previous sections is shown in Fig. 9. Note how the different constructs of the sequence diagram can be extended to capture the required input and output trajectories at once, including temporal dependencies, should they exist. In this particular case, the model indicates that the system must observe the required area and provide the imaged data (within a given time period of observation) while potentially be in the presence of clouds, rain, or snow.

Attributes of the *signal* element in SysML are extended to capture the required characteristics of the different inputs and outputs that the system must accept and provide, respectively. That is, whereas the standard SysML signal attribute captures the characteristics of the system input or output, this extended model (albeit using the same diagrammatic specification) captures the set of inputs the system must be able to accept and the minimum conditions that its output must fulfill. So, while the diagrams may look the same, the are read differently. As examples, models for two of the items used in Fig. 9 are shown in Fig. 10 using notional values. Note that, while each specific attribute might be defined as a dedicated requirement when using textual statements, they are directly associated to the model object they refer to in TMBR.

The required interfaces through which the required inputs and outputs must be exchanged are captured using *ports*, by allocating each signal element (representing an item) to the corresponding port (representing the interface that transfers or carries the item). While the relationship exists within the model, visualization can be done through an internal block diagram, as shown in Fig. 11. Similarly, the required properties of the interfaces are captured using *InterfaceBlocks*. Properties and values are used to capture requirements on the

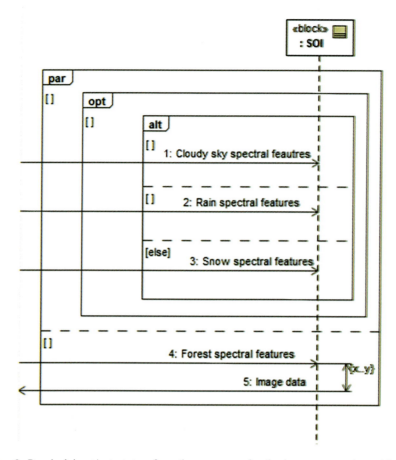

Fig. 9 Required input/output transformation sequence for the image area and provide data requirement

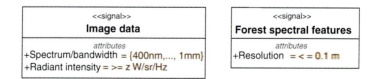

Fig. 10 Example of required characteristics of required input and output items

physical and transport (data) layers of the interface. An example of a modeled interface using notional values is shown in Fig. 12.

Finally, simultaneity of requirements applicability is captured by extending the use of SysML state machine diagrams rather than capturing all requirement in one large sequence diagram. This is defined as a *mode requirement*, which captures all requirements that "do not have conflicting requirements and that must be fulfilled

11 Model-Based Requirements

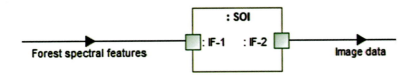

Fig. 11 Example of required allocation of items to interfaces

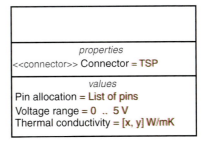

Fig. 12 Example of the use of an InterfaceBlock to capture the characteristics of an interface

simultaneously" [33]. Continuing with the fire detection example, the model in Fig. 13 captures the idea that the system must fulfill two different sets of requirements (*Clear Sky requirements* and *Nominal requirements*) depending on the presence of different operational conditions, *although not necessarily simultaneously.* The two boxes capture this aspect: Each box indicates the existence of a unique set of requirements that need to be fulfilled; they do not represent system states, as a state machine diagram would do in SysML. (Remember, while the same diagramatic construct is used here, the meaning of the diagram differs.) Sequence diagrams are used to model the specific requirements that apply under the different operational conditions. The transitions between the boxes, mapped to another sequence diagram, capture the conditions that make each set of requirements applicable. As shown in Fig. 14, Clear Sky Requirements only become applicable if there are no clouds, rain, or snow. In such a case, all requirements modelled in *Image area and provide data* (shown in Fig. 9 and relevant element models, such as those shown in Figs. 10 through 12) and in *Image fire and provide alert* must be fulfilled simultaneously. Otherwise, only those requirements modelled in *Image area and provide data* must be fulfilled.

It is critical to reinforce that the state objects in the state diagram do not represent *states* in the traditional sense of SysML. They are only used in this implementation of TMBR to capture operational scenarios under which different requirements are expected to be fulfilled at the same time. The model does not impose any design constraint for the system, such as what states it will have; such design decision is left open. Eventually, a *real* state machine diagram that captures the actual behavior of a potential system (not the required one) may have a completely different set of state elements (in the diagram) than the mode requirement diagram contains for indicating

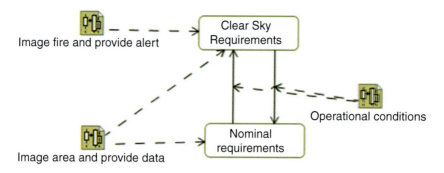

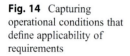

Fig. 13 Example of a mode requirement used to capture requirement simul-taneity

Fig. 14 Capturing operational conditions that define applicability of requirements

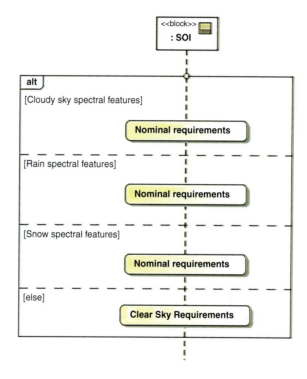

requirements applicability (i.e., as used in this chapter). This is because in a solution model, state elements capture system states, whereas, as indicated, state elements in TMBR capture operational conditions that differ in the requirements that need be fulfilled simultaneously. This choice avoids unnecessarily over constraining the solution space [21].

Elicitation, Derivation, and Trade-Off Analysis

Most techniques, if not all, that are employed to support requirements elicitation, derivation, and trade-off analysis when using textual requirements (e.g., those in [6, 36]) can be (and many should be) employed when using model-based requirements. However, using a model-based approach enables certain capabilities to support such requirements engineering activities in unprecedented ways. While many of these are currently under development and have not been shown at scale, they are presented here to hint at their potential. Note that the different capabilities are described with varying degrees of depth, depending on the availability of examples in the literature and/or simplicity of the presentation.

Better Problem Formulation Due to Automated Enforcement of Syntactic Rules

The usefulness to automate validation rules to check the syntactic correctness of systems model has been shown by several authors [37, 38]. The requirements modeling constructs presented in this chapter enable the utilization of similar kinds of rules. Some examples are described below.

Good requirements modeling approaches enforce staying at the right level of encapsulation. This means that the modeling constructs enforce the modeler to (1) avoid defining requirements for components that form *(are inside of)* the system (a.k.a. design-dependent requirements), and (2) enforce bounding the requirement to what is expected from the system (that is, the requirement can be attained by the SOI, in principle). Wymorian constructs achieve so by modeling requirements in the form of input and output trajectories. TMBR achieves so similarly, by defining requirements as input/output transformations and modeling the SOI as solid lines or empty boxes. PMM constructs achieve the same capability by defining unique objects as carriers of the requirements and assigning requirements to such objects. In this way, every requirement is always necessarily related to one object that has a precise definition. When requirements are consolidated for a given object, only those linked to it are applicable.

While these rules could also be defined for textual requirements, the diversity and flexibility of the natural language makes the effort quite challenging. For example, consider requirements for a system named "Telescope." Every time the requirement is written, there is a risk that the name is misspelled, or a dictionary has to be created and the name taken from such a list, which delays the writing process significantly. Similarly, this requires the name "Telescope" to be employed in every statement to be considered a telescope requirement, even though there may be a need to compose certain requirement, as indicated earlier. While this is again possible, it necessitates a heavy backbone structure to work. In contrast, in an MBSE environment, it suffices to select the object or the identifier. Similarly, one could control the verbs that are utilized in textual requirements. In contrast, the modeling constructs embed all sufficient possibilities.

The modeling constructs also support completeness of each required condition. As examples, Wymore enforces the specification of a mapping between sets of output trajectories and input trajectories, PMM requires four elements to be defined for every requirement, and TMBR requires that every exchanged item is transferred through a physical interface. Again, these constraints can also be defined for textual requirements. However, defining rules to check their compliance is hard given the flexibility of the natural language. Because those conditions are implemented through objects and relations in the model, automated checking consists of defining just a few rules.

More Comprehensive Requirements Derivation Due to Increased Semantic Precision

Based on anecdotal observation, there is a general consensus in the MBSE community that model-based artifacts capture information more precisely than document-based ones. Although this still needs to be formally confirmed [39, 40], the following example shows that, at least, model-based requirements promote higher precision in deriving system requirements from stakeholder needs [33].

Consider the need to illuminate something. A typical textual requirement that derives from this need could read something along the lines of *The system shall exhibit ABC color*. According to common guidelines to write requirements [5], such a statement would be considered a well-formulated requirement. However, when trying to capture such a requirement in the form of a model using, for example, TMBR, different competing models that capture different problem spaces are consistent with the textual statement. They are shown in Fig. 15.

The model in (a) indicates that the system must exhibit such a color inherently, without using any external source. The model in (b) indicates that the system has electrical power available for use when generating the color. The model in (c) indicates that the system can use the sunlight to provide the required color. Note that the system remains a solid line, so the three models do not capture different system solutions, but different operational concepts that are explored during requirements derivation. While such insights could definitely be gained when using textual requirements as well, they do not become as apparent because the textual statement was well formed, as earlier stated. The modeling constructs, however, enforce the modeler to explore, in the case of TMBR as in Wymorian models, different output trajectories for different input trajectories.

Early Verification and Validation

Requirement models have the potential to enable the automation of some early verification and validation activities carried out during need elicitation and requirements derivation, as well as during early development. Because requirement models can be easily parsed into computer-readable language, it is in theory possible to

11 Model-Based Requirements

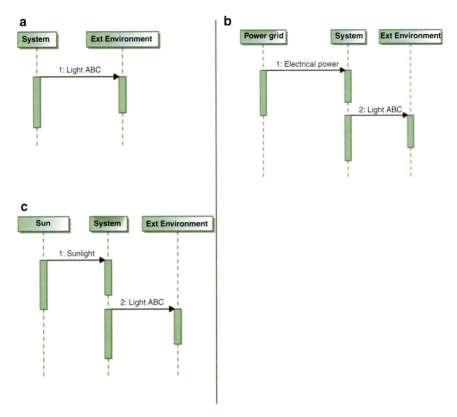

Fig. 15 Three different models of a color requirement, capturing in fact different requirements [from [33]]

easily map and contrast system models with requirement models. Some work in this direction has been performed, although maturation is still necessary.

Wymore, for example, provides a comprehensive description of formal functions that relate different kinds of system models to the requirement models. However, the use of these functions has not been shown in a tool environment yet to the best of the author's knowledge.

PMM uses its condition-object-property-domain construct as a benchmark that can be compared to a system model that may be constructed as condition-object-property-value, where *value* represents the actual value exhibited by the SOI and *domain* represents the set of values that are acceptable, as discussed earlier. Some aspects of this capability have been shown in implementations in VHDL [32] and Modelica [31].

In addition, such kind of construct and the dependencies that PMM captures are the bases of methods that can assess the existence of conflicting requirements before initiating developmental activities [29, 41]. Therefore, requirement models such as those constructed using PMM can be leveraged to assess their consistency in ways that textual requirements cannot.

Cognitive Assistance to Increase Completeness

Requirement models can be easily parsed into computer-readable language. By doing so, computers can be programmed to aid in requirements engineering (not just management) tasks.

Houston is an example of a concept for a virtual engineer/cognitive assistant that supports the requirements engineer during need elicitation and requirement derivation [42]. Houston, which leverages TMBR, *reads* the requirement models and evaluates them against a knowledge repository to identify requirement gaps. If Houston identifies potential gaps in the set of requirements, then it presents them to the (human) engineer, who then decides how to address the gaps. Here, gaps refer to the potential need to explore a missing scenario, situation, or condition. This concept is sketched in Fig. 16.

Once presented with the potential gaps, some actions presented to the (human) engineer may be to:

- Do not address the suggested scenario or conditions, that is, do not formulate/create requirement models for the identified gap.
- Expand a given set of required conditions to include those in the identified gap.
- Create a new requirement model that specifically addresses the identified gap.

The modeling constructs serve as the basis to build the knowledge repository from which evaluation rules are obtained. Table 3 provides some examples of the type of rules that can be defined. These rules can be obtained from experts in an organization or as projects progress and new organizational knowledge is created. In any case, the way in which the models are used significantly expands the benefits of syntactic checking (section "Better Problem Formulation Due to Automated Enforcement of Syntactic Rules"). In contrast to evaluate the completeness of a specific statement, this evaluation evaluates the potential completeness of an elicitation and derivation effort.

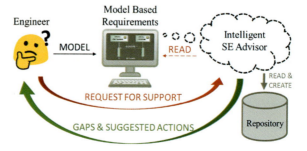

Fig. 16 Houston's concept [from ref. [42]]

11 Model-Based Requirements

Table 3 Examples of structural rules to support identification of requirements gaps [from Ref. [42]]

Rule	Description	Parameters

1	An input signal is removed (i.e., not being received) for a specific amount of time in a given scenario
	Examples:
	  A system loses its electrical power input while executing certain operation. The loss of power lasts 1 h
	  A system loses its connection to the Internet while uploading a file
		Signal (the input that becomes missing)
		Duration of signal not being received
		Starting point in time in which the signal stops being received

2	Two or more inputs with a strictly defined sequence of reception, including sequence with respect to one or more outputs, are received in a different sequence
	Examples:
	  A safety-critical system that expects to receive first an arming command and then a firing command receives first a firing command and then an arming command
	  A system that expects to be powered on when receiving data receives data before it has been powered on
	  A system that expects to receive user input after the system has provided a notification to the user receives the user input before the system has provided the notification
		Signal characteristic
		Excursion range
		Scenario where excursion occurs

Chapter Summary

Several approaches to model requirements have been presented in this chapter. They have been categorized in three groups, depending on the type of modeling construct they are based on.

The first group of approaches included those that are widely used in practice and/or implemented in MBSE tools. One of them specifies a dedicated class of elements in the modeling language to capture requirements. However, the requirement object is defined as a text string. Therefore, such an approach, while beneficial for traceability and requirements management purposes, cannot really be considered a model-based approach for requirements; requirements remain textual, not models. The other approach consists of flagging system models as requirements. It has been shown in this chapter that such an approach goes against the guidelines to formulate good requirements, leading to an unnecessary reduction of the solution space.

The second group of approaches included those that leverage mathematical constructs to specify requirements. Two have been presented. One of them, the Wymorian framework, models the problem space through two kinds of requirements, those that define what solutions belong to the solution space and which ones do not, and those that set an order on that space to capture the preferability of one solution over the other one. The modeling constructs for the requirements that define the content of the solution space are based on the premise of defining required output trajectories for expected input trajectories, and of defining technological restrictions imposed to the solution. The other approach, the PMM, models the problem space using a semilattice that specifies four conditions, based on the premise of constraining a property of an object given a condition. Implementations of this approach as extensions to several modeling languages, particularly SysML, VHDL, and Modelica, have been shown.

The third group of approaches included those that are based on defining dedicated semantics to capture the problem space. The TMBR approach has been presented as an example. The main premise of the modeling construct, as in Wymore, is that every requirement can be defined as an input/output transformation but includes specifications for physical interfaces and does not account for imposed technological limitations. A TMBR implementation has been shown as a semantic extension of SysML, where it leverages some of SysML's modeling constructs and graphical representations albeit having a different meaning.

Examples have been provided throughout the chapter to showcase the different approaches to modeling requirements.

The chapter has concluded with a discussion about how model-based approaches to capture requirements can affect the elicitation, derivation, and trade-off of needs and requirements. While existing techniques are applicable, using requirements models can enable analysis capabilities, automation, and cognitive assistance in unprecedented ways.

Several advances in model-based requirements are anticipated in the future. Most work in modeling the problem space has focused on system requirements, with no comprehensive and/or tested approach to model stakeholder needs. Advances in this area may necessitate the development of formal semantics through an ontological framework specifically dedicated to problem formulation, which is lacking today. Finally, automation in the areas of syntactic evaluation, semantic evaluation, and early verification and validation and the development of cognitive assistants to support elicitation, derivation, and trade-off analysis have only been hinted at. These are expected to significantly mature, scale, and evolve.

Cross-References

▶ MBSE for Acquisition
▶ Problem Framing: Identifying the Right Models for the Job
▶ Semantics, Metamodels, and Ontologies

References

1. INCOSE, *Systems Engineering Handbook: A Guide for System Life Cycle Processes and Activities*. version 4.0 ed. 2015, Hoboken, NJ, USA: John Wiley and Sons, Inc.
2. Buede, D.M. and W.D. Miller, *The Engineering Design of Systems: Models and Methods*. Third edition ed. 2016, Hoboken, NJ, USA: John Wiley & Sons, Inc.
3. Blanchard, B.S. and W.J. Fabrycky, *Systems engineering and analysis*. Vol. 4. 1990: Prentice Hall New Jersey.
4. Salado, A., R. Nilchiani, and D. Verma, *A contribution to the scientific foundations of systems engineering: Solution spaces and requirements*. Journal of Systems Science and Systems Engineering, 2017. **26**(5): p. 549-589.
5. INCOSE, *Guide for writing requirements*. 2012, The International Council of Systems Engineering.
6. Robertson, S. and J. Robertson, *Mastering the requirements engineering process. Getting requirements right*. 2012: Addison-Wesley.
7. Schneidera, F., H. Naughtona, and B. Berenbach, *New Challenges in Systems Engineering and Architecting*, in *Conference on Systems Engineering Research (CSER) 2012*. 2012, Elsevier B.V.: St. Louis, MO.
8. Helming, J., et al. *Towards a unified Requirements Modeling Language*. in *2010 Fifth International Workshop on Requirements Engineering Visualization*. 2010. Sydney, NSW.
9. Mordecai, Y. and D. Dori, *Model-based requirements engineering: Architecting for system requirements with stakeholders in mind*, in *2017 IEEE International Systems Engineering Symposium (ISSE)*. 2017: Vienna, Austria. p. 1–8.
10. Fockel, M. and J. Holtmann, *A requirements engineering methodology combining models and controlled natural language*, in *2014 IEEE 4th International Model-Driven Requirements Engineering Workshop (MoDRE)*. 2014: Karlskrona. p. 67–76.
11. Soares, M.D.S. and J. Vrancken, *Model-driven user requirements specification using SysML*. J. Softw, 2008. **3**(6): p. 57–68.
12. Adedjouma, M., H. Dubois, and F. Terrier, *Requirements Exchange: From Specification Documents to Models*, in *16th IEEE International Conference on Engineering of Complex Computer Systems*. 2011: Las Vegas, NV. p. 350–354.
13. Salado, A., *A systems-theoretic articulation of stakeholder needs and system requirements*. Systems Engineering, 2021. 24: p. 83–99.
14. von Bertalanffy, L., *General Systems Theory – Foundations, development, applications*. 1969, New York, NY: George Braziller, Inc.
15. Friedenthal, S., A. Moore, and R. Steiner, *A Practical Guide to SysML – The Systems Modeling Language*. 3rd ed. 2015, Waltham, MA, USA: Morgan Kaufman.
16. Long, D. and Z. Scott, *A Primer For Model-Based Systems Engineering*. 2nd ed. 2011, USA: Vitech Corporation.
17. Committee, L.S., *Lifecycle Modeling Language (LML) Specification*. Lifecycle Modeling Language website, 2015.
18. Friedenthal, S. and C. Oster, *Architecting Spacecraft with SysML: A Model-based Systems Engineering Approach*. 2017: CreateSpace Independent Publishing Platform.
19. Holt, J., S.A. Perry, and M. Brownsword, *Model-Based Requirements Engineering*. 2011: IET.
20. Pandian, M.K.S., et al., *Towards Industry 4.0: Gap Analysis between Current Automotive MES and Industry Standards Using Model-Based Requirement Engineering*, in *2017 IEEE International Conference on Software Architecture Workshops (ICSAW)*. 2017: Gothenburg, Sweden. p. 29–35.
21. Wach, P. and A. Salado, The Need for Semantic Extension of SysML to Model the Problem Space, in *Conference on Systems Engineering Research (CSER)*. 2020: Redondo Beach, CA, USA.
22. Bellagamba, L., *Systems Engineering and Architecting. Creating Formal Requirements*. 2012, Boca Raton, FL, USA: CRC Press.

23. Wymore, A.W., *Model-based systems engineering*. 1993, Boca Raton, FL: CRC Press.
24. Micouin, P., *Toward a property based requirements theory: System requirements structured as a semilattice*. Systems Engineering, 2008. 11(2).
25. Zeigler, B.P., A. Muzy, and E. Kofman, *Theory of modeling and simulation: discrete event & iterative system computational foundations*. 2018: Academic press.
26. Salado, A. and R. Nilchiani, *On the Evolution of Solution Spaces Triggered by Emerging Technologies*. Procedia Computer Science, 2015. **44**: p. 155–163.
27. Abbas, A.E., *Foundations of Multiattribute Utility*. 2018, Cambridge: Cambridge University Press.
28. Salado, A. and R. Nilchiani. *Assessing the Impacts of Uncertainty Propagation to System Requirements by Evaluating Requirement Connectivity*. in *23rd INCOSE International Symposium*. 2013. Philadelphia, PA (USA).
29. Salado, A. and R. Nilchiani, *The Tension Matrix and the Concept of Elemental Decomposition: Improving Identification of Conflicting Requirements*. IEEE Systems Journal, 2017. 11(4): p. 2128–2139.
30. Micouin, P., et al., *Property Model Methodology: A Landing Gear Operational Use Case*. INCOSE International Symposium, 2018. 28(1): p. 321–336.
31. Pinquié, R., et al. *Property Model Methodology: A case study with Modelica*. in *Tools and Methods of Competitive Engineering (TMCE)*. 2016. Aix-en-Provence, France.
32. Micouin, P., *Property-Model Methodology: A Model-Based Systems Engineering Approach Using VHDL-AMS*. Systems Engineering, 2014. **17**(3): p. 249–263.
33. Salado, A. and P. Wach, *Constructing True Model-Based Requirements in SysML*. Systems, 2019. **7**(2): p. 19.
34. Salado, A. and R. Nilchiani, *A Categorization Model of Requirements Based on Max-Neef's Model of Human Needs*. Systems Engineering, 2014. **17**(3): p. 348–360.
35. Kossiakoff, A., et al., *Systems Engineering Principles and Practice*. 2nd ed. 2011, Hoboken, NJ, USA: John Wiley & Sons, Inc.
36. Hull, E., K. Jackson, and J. Dick, *Requirements engineering*. 2005: Springer.
37. Giammarco, K., *A Formal Method for Assessing Architecture Model and Design Maturity Using Domain-independent Patterns*. Procedia Computer Science, 2014. **28**: p. 555–564.
38. Vinarcik, M., *Treadstone: A Process for Improving Modeling Prowess Using Validation Rules*, in *ASEE Virtual Annual Aconference*. 2020: Virtual.
39. Henderson, K. and A. Salado, *Value and benefits of model-based systems engineering (MBSE): Evidence from the literature*. Systems Engineering, 2021. **24**(1): p. 51–66.
40. Cratsley, B., et al., *Interpretation Dscrepancies of SysML State Machine: An Initial Investigation*, in *Conference on Systems Engineering Research*. 2020: Virtual.
41. Kannan, H., *Formal reasoning of knowledge in systems engineering through epistemic modal logic*. Systems Engineering, 2021. **24**(1): p. 3–16.
42. Salado, A. and R. Tan. *Structural Rules for an Intelligent Advisor to Identify Requirements Gaps using Model-Based Requirements*. in *IEEE International Conference on Systems, Man and Cybernetics (SMC)*. 2020. Toronto, Canada.

Dr. Alejandro Salado is an associate professor of systems engineering with the Department of Systems and Industrial Engineering at the University of Arizona. He conducts research in problem formulation, design of verification and validation strategies, model-based systems engineering, and engineering education. Before joining academia, Dr. Salado spent over 10 years in the space industry, where he held positions as systems engineer, chief architect, and chief systems engineer in manned and unmanned space systems of up to $1B in development cost. He has published over 100 technical papers, and his research has received federal funding from the National Science Foundation (NSF), the Naval Surface Warfare Command (NSWC), the Naval Air System

Command (NAVAIR), and the Office of Naval Research (ONR), among others. He is a recipient of the NSF CAREER Award, the International Fulbright Science and Technology Award, the Omega Alpha Association's Exemplary Dissertation Award, and several best paper awards. Dr. Salado holds a BS/MS in electrical and computer engineering from the Polytechnic University of Valencia, an MS in project management, an MS in electronics engineering from the Polytechnic University of Catalonia, the SpaceTech MEng in space systems engineering from the Technical University of Delft, and a PhD in systems engineering from the Stevens Institute of Technology. Alejandro is a member of INCOSE and a senior member of IEEE and AIAA.

Modeling Hardware and Software Integration by an Advanced Digital Twin for Cyber-physical Systems: Applied to the Automotive Domain

12

Applied to the Automotive Domain

S. Kriebel, M. Markthaler, C. Granrath, J. Richenhagen, and B. Rumpe

Contents

Introduction	380
State of the Art	382
Systems Engineering: Overcoming the Differences?	382
Model-Based: Unite Engineering Solutions?	384
Related Work	385
Best Practice Approach	387
Systems Engineering: Merging Engineering Disciplines	387
The Advanced System Model	393
The Advanced Digital Twin	398
Model-Based: More than a Description	400
Illustrative Examples	402
Level A: Customer Value	402
Level *B*: Operating Principle	403
Level *C*: Technical Solution	405
Level D: Realization	407

S. Kriebel (✉)
FEV.io GmbH, Aachen, Germany

BMW Group, Munich, Germany
e-mail: kriebel@fev.io

M. Markthaler
BMW Group, Munich, Germany

Software Engineering, RWTH Aachen University, Aachen, Germany

C. Granrath
FEV.io GmbH, Aachen, Germany

Mechatronics in Mobile Propulsion, RWTH Aachen University, Aachen, Germany

J. Richenhagen
FEV.io GmbH, Aachen, Germany

B. Rumpe
Software Engineering, RWTH Aachen University, Aachen, Germany

© Springer Nature Switzerland AG 2023
A. M. Madni et al. (eds.), *Handbook of Model-Based Systems Engineering*,
https://doi.org/10.1007/978-3-030-93582-5_21

Chapter Summary and Expected Advances 409
Cross-References 413
References 413

Abstract

Systems engineering deals with the development of complex systems where complexity is usually product domain-specific. Thus, it often fails to a large extent to integrate the mechanical and electrical engineering disciplines with the computer science discipline, including cultural issues. However, the success of future cyber-physical systems depends not only on hardware functionality but also progressively on its integration into distributed software functionality. The presented Advanced Digital Twin combines the prerequisites for efficient hardware and software integration, particularly for large and complex systems, like cyber-physical systems. Therefore, it is based on an Advanced System Model which comprises the respective architectural designs needed by the domains involved. Based on this resilient integrated architecture, the emerging artifacts can be reused which makes the application of model-based techniques economically reasonable. This enables automated quality checks, simulations, application of artificial intelligence, and big data analysis and serves as a thread through necessary cultural changes to set up a cross-functional and t-shaped collaboration.

Keywords

Digital twin · Model-based systems engineering · Cyber-physical system · Hardware-software integration · System architecture · Functional architecture · Logical architecture · Technical architecture · Automation · Reuse · Diversity · Cultural change

Introduction

It is common sense that for at least the last 30 years, mechanical systems are becoming more and more complex. Some say they follow Moore's law which is the observation that the number of transistors in dense integrated circuits doubles at least every 2 years. However, for mechanical systems, this doesn't mean that they double parts every 2 years. For cost reasons, it is even vice versa. Actually, it means the number of provided functions increases exponentially. These functions are provided jointly by contributions from the mechanic, the electrical/electronic (e/e), and the computer science domains. In the following, such combined systems are called cyber-physical.

Cyber-physical systems (CPS) (Broy et al. 2012; Kirchhof et al. 2020) comprise the most fascinating innovative technical products existing, like cars, planes, trains, ships, spacecrafts, satellites, etc., including cyber functions like traffic control, system-to-system communication, and customer information. This means they are pervasive and contribute to a huge extent to our economic power and advancement.

Unfortunately, mechanical systems engineering today still seems not well prepared to cope with the volatile, uncertain, complex, and ambiguous ("VUCA") product requirements of today's CPS.

In the automotive industry, for example, the VUCA gap addressed in this article is not only caused by modern functions, e.g., driver assistance, communication, and connectivity like network functionality (car2car) or near environment and backend communication (car2x). But it also addresses the real new challenge of autonomous driving with its huge bunch of necessary safe near-field and far-field environmental cognition and communication functions which shall replace the driver one day.

Additionally, a lot of former pure mechanical functions are today controlled by software as well, e.g., suspension, steering, braking, and combustion. Furthermore, these functions are progressively protected by safety and diagnostic functions. For increased customer comfort and product quality, additional functions are required like data analytics, big data evaluation, and artificial intelligence. How to cope with all these functions? How to reliably manage their interactions and dependencies?

Well-known and trained product development cycles for mechanical systems are not entirely matching the product and market needs any longer. Task forces are implemented more and more often as a tried and tested but cost-intensive survival technique of traditional players to achieve the market-driven delivery date. Collateral new market players are coming up with new product and process visions and are changing the game. Market pressure increases. The applied survival techniques are usually matching the schedules but hardly the cost and quality targets.

Economic Darwinism (Read 2010) forces mechanical industries to check their current product development processes for their applicability to CPS development, i.e., intensive and reliable integration of hardware and software. In order not to become a Dinosaur in economic Darwinism, the approach presented in this chapter introduces a way to continuously adapt the respective product development process and the required collaboration culture to the current state of the art in CPS Engineering (CPSE) and the new market conditions.

A comprehensive integrated modeling approach is introduced which facilitates hardware and software integration by an Advanced System Model. It also enables the optimization of time and budget required for the industrialization of a CPS by a specifically branched out Advanced Digital Twin. In addition, it increases product quality by arrogating preventive measures and facilitates the efficient handling of possibly necessary retrospective measures. The approach is based on function-oriented systems engineering combined with a solution-oriented model-based appendage similar to (Kriebel et al. 2017, 2018; Drave et al. 2019).

The integrated modeling approach presented in this chapter is named model-based cyber-physical systems engineering (MBCPSE). It merges the respective advantages of the underlying disciplines of mechanical and electrical engineering as well as computer science to overcome the technical and cultural gaps in cooperation and the increasing complexity of CPS. It shall be applied as a domain overlapping integrated development approach as neither a CPS nor the application of MBCPSE can be regarded as domain-specific any longer.

The current challenges of product development processes as well as existing approaches are presented in section "State of the Art" referring to the automotive domain as a well-known example. The details of MBCPSE based on the Advanced System Model and the resulting Advanced Digital Twin are shown in section "Best Practice Approach." The required modeling is illustrated in section "Illustrative Examples" by using an exemplary use case. Finally, in section "Chapter Summary and Expected Advances," the content is summarized, and expected future advances are outlined.

State of the Art

MBCPSE comprises the entire system life cycle starting from the concept phase and structuring the development phase to system design and analysis to the final disposal process (ISO International Organization for Standardization 2015; Walden et al. 2015). Furthermore, MBCPSE is an approach for accomplishing cross-disciplinary systems development. For this purpose, MBCPSE combines a set of cross-disciplinary architectures with separated development steps and artifacts. Consequently, MBCPSE enables a new approach to systems engineering for complex systems like CPS.

Systems Engineering: Overcoming the Differences?

As mentioned before, MBCPSE focuses on complex systems like CPS, in this article the automotive vehicle as a well-known example. A closer look at the term CPS shows that it addresses the areas "cyber," "physical," and "system." A system is defined as a "combination of interacting elements organized to achieve one or more stated purposes" (ISO International Organization for Standardization 2017).

Cyber relates to software systems and is characterized by the culture of computers and information technology, hence the discipline of computer science. Physical relates to the operation of natural forces generally, hence the disciplines of mechanical and electrical engineering. A CPS unites all three disciplines and therefore results in a high degree of system complexity.

Each discipline involved has its different approach to problem-solving and decision-making, i.e., to culture and psychology. In a CPS, these differences often lead to misunderstandings, frustration, delays, and consequently increasing development time and high costs (Kriebel 2018). To avoid these issues, it is necessary to identify potential impediments in order to avoid them but also to focus on the advantages of each discipline in order to keep their benefits. Therefore, the differences in these disciplines between skills and handling complexity are outlined in the following section.

Different Skills and Knowledge

To relate to and to differentiate these three disciplines from one another, it is essential to understand what mechanical engineers, electrical engineers, and computer scientists learn throughout their studies and particularly what they do professionally. The disciplines share many of the same intentions and problem-solving processes but in different dimensions. To identify and solve problems, all of them use many of the same mathematical and scientific principles.

The main difference is that in computer science, the focus is less on physical elements, but on dealing with a huge variety of functional solutions and innovations. Complexity handling techniques such as modeling are used to define suitable boundary conditions. Thus, computer science developers learn to continuously optimize boundary conditions and to efficiently manage the evolution of versions and variants. On the other hand, in mechanical and electrical engineering, the physical products with their physical laws define the boundary conditions. Once developed, they are fixed, and the products are continuously optimized within constant boundary conditions. Engineering developers learn to optimize within given boundary conditions and to manage efficiently the evolution of parameters.

However, the increasing proportion of software in physical products demands a product development process integrating the needs of cyber systems development. To cope with these innovative needs, the development of CPS must merge the advantages of each discipline. MBCPSE facilitates the separated application of the different skills and knowledge as outlined in section "Best Practice Approach."

Different Handling in Complexity

Besides different skills and knowledge, the disciplines use different methods to handle complexity based on what developers learned. Electrical and mechanical engineering usually use physical systems decomposition, which is based on (geometrical) components. For this, the decomposition starts with the entire system that is iteratively decomposed into subsystems until the emerging components are manageable to be developed independently. For subsequent technical modeling, CATIA (Dassault Systemes 2020) and MATLAB/Simulink (MathWorks 2019) are common and often optimized by simulations.

In contrast, computer scientists usually decompose the system according to its functional behavior. For this, computer scientists recursively break down the system at different levels of abstraction, until these become simple enough to be solved independently. For the subsequent structural and behavioral modeling, the Unified Modeling Language (UML) (OMG Object Management Group 2017) or Systems Modeling Language (SysML) (OMG Systems Modeling Language (OMG SysML) 2017) is common and often optimized by model-checking and code generation tools.

Figure 1 shows a selection of strengths of the product development techniques of computer science and mechanical engineering. They are assigned roughly to the well-known basic V-model (Boehm 1979; Scheithauer and Forsberg 2013).

However, an applicable solution for any systems engineering approach for complex systems must combine all three disciplines with all its characteristics

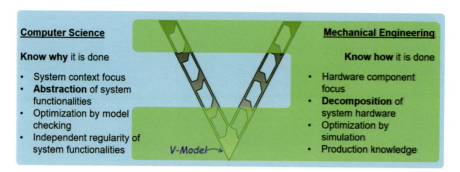

Fig. 1 Strengths of computer science and mechanical engineering in the context of the basic V-model

mentioned above. Mechanical and electrical engineers focus on hardware components, deal with complexity by decomposition of system hardware, optimize the components by simulations, and have particular knowledge about hardware production systems.

In contrast, computer scientists focus on the system context, deal with complexity by the abstraction of system functionalities optimized by model-checking, and have particular knowledge about the reuse of independent system functionalities. MBCPSE incorporates the shown different handling of complexity as outlined in section "Best Practice Approach."

Model-Based: Unite Engineering Solutions?

Models are found in every named discipline, and modeling takes place even when the experts are not aware of it. According to (Stachowiak 1973), a model represents an original, includes only relevant ("seeming") properties, and always fulfills a purpose with respect to the original. For example, in mechanical engineering, the finite element method is model-based and aims on the simulation of the behavior of a physical system.

Electrical engineering, for example, uses circuit models for a representation of the electrical system and differential models to describe the aspects of electromagnetism. Compared to these physical models, the models in computer science mainly serve the abstraction of these technical details to concentrate on the functional behavior and describe the logical system, i.e., the operating principle.

Regardless of its purpose, the role of modeling and the quality of models are extremely important not only within the disciplines but for the development of a CPS in general. The integration of the different models is difficult, respectively not possible, because a common "semantic linkage" is missing. The term semantic linkage encompasses linguistic aspects, such as expressions and the understanding of words, and also models, architectures, and their elements. For a successful model-based project, a common spoken language is not enough. In addition to a glossary for the precise definition of often-used terms (Walden et al. 2015), the models, architectures, and their elements such as interfaces, information, etc. must also be defined and their relationships precisely specified. Otherwise, the misunderstandings and frustration mentioned above are likely to occur and jeopardize the project success. Further, if there might be a semantic linkage available, integration of the different models fails technically due to different modeling languages, modeling guidelines, modeling focuses, and not aligned architectural principles and interfaces. The effort in the particular modeling is cost-intensive, while the models are not usable for an integrated modeling approach. Additionally, the reuse of the models is limited, and therefore the return of investment is little.

To meet this challenge, MBCPSE integrates all models in a comprehensive development model, the Advanced System Model introduced in section "Best Practice Approach." The applied model integration is according to well-known methods of computer science. The embeddedness with methods of computer science was chosen because the consistent use of complex models has improved the quality and efficiency of software development in computer science over the last decades (Rumpe 2017). This provides a consistent model-based methodology and aims at a significant raise of the return of investment.

Related Work

This section lists the relevant existing approaches and provides a brief evaluation of their applicability to CPSE. It also provides an opportunity to acknowledge the influence of existing research in the field. Even though the automotive industry has recognized mastering the complexity of CPS, new approaches are still required in the future (Broy 2006; Whittle et al. 2014; Giusto, and R. S., and S. M 2016). Current literature reflects that text-based systems engineering is still the status quo (Giusto, and R. S., and S. M 2016; Liebel et al. 2019). Nevertheless, several approaches exist for the model-based specification of CPS but are not applied as a standard methodology today.

One of the best-known standards in the field of systems engineering is part of the ISO standard 15,288 systems engineering-system life cycle processes (ISO International Organization for Standardization 2015). The standard can be understood as a collection of activities that are necessary to achieve the desired result. The philosophy is based on the systems engineer's ability to divide the given descriptions into a sequence of activities that are applicable to the problem. At the same time, logical sequences for determining the sequences of necessary activities are not specified.

Besides the ISO standard 15,288, well-known examples in the automotive industry include OOSEM (Walden et al. 2015), EAST-ADL (EAST-ADL Association 2013), REMsES (Braun et al. 2014), and SPES 2020 (Pohl et al. 2012). These approaches combine abstraction and decomposition steps which imply for their applicability that either all individuals or teams have to provide profound integrated knowledge in all involved disciplines. This is, of course, possible in smaller professional expert environments but does not usually scale to the needs of the industrial product development of large and complex systems. In fact, for applicability reasons, these approaches seem to focus on a maximum of two disciplines. In contrast, MBCPSE uses separated steps for abstraction and decomposition and documents them thoroughly. This allows exploiting the experts' knowledge of all three involved disciplines, mechanical engineering, electrical engineering, and computer science in separate steps as outlined in section "Best Practice Approach." Of course, there is the need for mutual understanding when defining interfaces and interferences of the CPS. However, this applies not to all individuals and teams involved in the project but to a smaller number of necessarily experienced architects.

The object-oriented systems engineering method (OOSEM) (ISO International Organization for Standardization 2015) is a stepwise approach to system design and specification. In the first step, the needs of the stakeholders are analyzed. In a second step, the system requirements are analyzed, followed by the definition of a logical architecture. In the last step, possible physical system architectures are considered. All the mentioned steps can be applied to different system hierarchies and run through iteratively. Overall, OOSEM represents a collection of activities for each phase of the four-pronged process, which can be used to specify the desired system. Procedures for the execution of the individual steps must be defined by the user in an individual application-specific way. Hence, there is no detailed description of the resulting views. Another difference is that the four steps combine abstraction and decomposition and the approach is not domain-specific.

EAST-ADL (Electronics Architecture and Software Technology-Architecture Description Language) is a domain-specific language in the automotive industry for software and system modeling (EAST-ADL Association 2013). The focus is on the specification of electrical and electronic vehicle architectures and software (EAST-ADL Association 2013). The four abstraction levels of vehicle, analysis, design, and implementation, system descriptions describe the tasks and communication of the system (EAST-ADL Association 2013). In EAST-ADL, the refinement mixes abstraction and decomposition, making it difficult to revise the specification/architecture in the event of a functional or technical change. In addition, the focus is strongly on e/e and software, which limits the integration of the mechanical approach.

Another MBSE approach is REMsES (Requirements Engineering and Management for software-intensive Embedded Systems) (Braun et al. 2014). REMsES uses concepts of the UML for the specification of software-intensive embedded systems. The three abstract and system decomposition layers are divided into a system level, function group level, and hardware and software level and separate the solution and problem level (Braun et al. 2014). Orthogonal to this abstraction is a categorization with views to the categories context, requirements, and draft (Braun et al. 2014). REMsES is similar to the approach presented here, especially in prioritizing the functional basis before the technical solution. However, REMsES mainly focuses on the computer science and e/e disciplines and combines abstraction and decomposition steps.

A similar methodology is SPES 2020 (Pohl et al. 2012). Similar to the approach of REMsES, the "decomposition layers" are arranged orthogonally to the "viewpoints" requirements, functions, logic, and technology. The decomposition layers are not predefined besides by a structural characteristic from the layer above. SPES 2020 is the only approach that combines all three disciplines, but the mechanical, electrical, and technical parts are separated at an early stage, and no order of the steps is given.

Summed up there are a number of model-based approaches available. However, they seem to be useful in academics or in an expert environment but are not meeting the entire needs for an industrial scalable CPS development as discussed above. Therefore, MBCPSE was developed and applied in the last couple of years as a best of breed approach within the disciplines involved.

Best Practice Approach

There are plenty of differences in engineering and management between mechanical engineering, electrical engineering, and computer science. However, the combination of these three disciplines is fundamental for the interdisciplinary development of CPS as outlined in the previous sections. In the following, MBCPSE is presented as an integrated best practice approach in industrial development which makes it different from the approaches presented in section "State of the Art."

Systems Engineering: Merging Engineering Disciplines

A generally known, widely spread, and well-proven approach to systems engineering is the basic V-model mentioned in section "Different Handling in Complexity," especially in the automotive industry. It starts with a concept phase, from which requirements are derived, followed by design and realization of the system. In these phases, the system is iteratively decomposed and abstracted in each step. Often the V-model is used to structure the development process by product decomposition layers but unfortunately neglects documentation at least in daily routines. This includes requirements and architectural views as well as linked test cases.

The V-model is usually augmented with plenty of domain-specific standards focusing in more details on relevant procedures. The ISO 26262 (ISO International Organization for Standardization 2018), for example, is the automotive functional safety standard and therefore embedded in established development methods for automotive vehicles. The aim of the ISO 26262 is to minimize risks and therefore classifies the risks of failure probabilities. To calculate the failure probabilities, the ISO 26262 proposes to use models describing the failure aspects. It consequently demands intense documentation as well as systematic linking between requirements, architecture, and test cases. The ISO 26262 is derived from the IEC 61508 (ISO International Organization for Standardization 2018), well-known in electrical engineering since 1998.

MBCPSE provides various model-based views as presented in section "The Advanced System Model." Thus, it facilitates the integration of specific procedures like the failure aspects of the ISO 26262 in the modeling and uses parts of the V-model for structuring the decomposition layer and the respective artifacts. Additionally, it enables systematic derivation of test cases out of requirements and architectural models. It shall be emphasized that the V-model is not used in MBCPSE as a process model. The applied principal architectural methods are decomposition and abstraction as introduced in the following. Every step of decomposition and abstraction is executed separately, inspired by the well-known OSI model (Zimmermann 1980) where application and communication are consequently separated. The separated decomposition and abstraction steps greatly simplify documentation of different architectural views on the CPS.

Starting with the system architecture as the prevailing geometric structure in a mechanical environment, the functional and logical architectures, prevailing in a computer science environment, are developed in separate steps but with a clear plan for integration. This enables to assign the necessary tasks to specialists of the respective disciplines. The technical architecture integrates the results to a joint product view and specifies the (hardware and software) component architecture which prepares for industrialization.

This defined self-contained granularity facilitates particularly the application of agile frameworks like the Scaled Agile Framework (SAFe) (Knaster and Leffingwell 2019) and its suitable agile methods as a process model. In this way, a modern and sustainable systems engineering culture can be progressively built which is attractive to young and experienced specialists as they can evolve continuously within the disciplines during product development.

Decomposition

The purpose of decomposition as an architectural tool is to divide the complexity of a system into less complex subsystems. This is usually done starting from the root system element, going through one or more intermediate layers depending on the system's complexity until a leaf system element is reached. The appropriate number of decomposition layers is dependent on the complexity of the product. Thus, it is possible that a vehicle is decomposed into three to five layers but an aircraft or spacecraft into more than eight layers. The number of layers is variable and can be unsymmetrical, if necessary. The resulting view is the **system architecture** of the product which is structured by **system elements** of types root, node, and leaf as presented subsequently.

Root: In MBCPSE, the decomposition of the system starts with layer 1 which represents the entire product. It contains one system element of type "root" as it is the starting point for the entire product decomposition. In the following, the term CPS is used for the product which is intended to be handed over to the customer. Hence, the boundaries of the CPS are specified by the root. External functionalities like backend support are regarded as part of the CPS environment but not of the CPS itself. Consequently, the CPS can operate in a higher system environment in a system of systems manner. However, it is perceived as a product on its own by the customer. The backend part of the root specifies the interface at product level and is specified in more details by the subsequent layers ending at component level. The customer relevant part of the root comprises the necessary content for the product manual.

Node: Usually, several connected intermediate layers are required to enable functionalities in a CPS. For example, the acceleration of an electric vehicle is a complex interaction of the chassis, the drive train, the power electronics, the electric machine, and the high-voltage storage. In MBCPSE, the class of these intermediate decomposition layers is called "nodes." Depending on the complexity of a node, additional decomposition according to the system architecture may lead to subordinate nodes.

Leaf: The system elements of the last decomposition layer are of type "leaf." They specify the lowest entities of the system architecture of the CPS under concern, the components. Leaves are usually subject for subcontracting or internal manufacturing when the CPS is industrialized. Figure 2 illustrates a cutout of a vehicle system architecture derived from the decomposition focusing on the torque functionality as an example.

The expression "function list" in the blocks shown in Fig. 2 may surprise in the context of a mechanical decomposition. However, it was chosen intentionally to emphasize that each single element of the system architecture has to provide a specific functionality regardless of its type, i.e., root, node, or leaf. The complete but concise specification of the system element functionality is one of the crucial points in systems engineering and, of course, even more so when dealing with CPS.

Each system element in a layer n can be assigned requirements from layer $n-1$. Consequently, not all requirements have to be modeled or implemented in one specific layer. As an example, the vehicle functions can be experienced by the clients, i.e., the final customers, whereas timing relevant security and safety functions are usually occurring only in successional layers. In the same way, mechanical

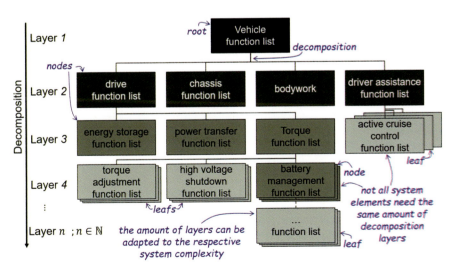

Fig. 2 Cutout of a vehicle system architecture focusing on torque functionality

requirements can be dealt with. For example, a power train momentum is specified in layer n and is further specified in layer $n + 1$ to decide whether the momentum is generated electrically or conventionally by a combustion engine. Of course, they can use the same gearbox specified as well on layer $n + 1$ as contributing to the power train in level n. Using the hierarchical decomposition of system elements persistently in a function-oriented manner facilitates an efficient handling of variants.

Out of experience, system inconsistencies usually occur already when specifying the system elements and their dependencies. The documentation seems to be well established for root and leaf elements but not for the nodes. This may also be due to the compulsory approval and homologation procedures which focus basically on the entire product or the relevant components as the specification of node system elements is often regarded as internal knowledge. Therefore, it occurs that specification of nodes is documented coarsely as this provides apparent efficiencies. However, if particularly the node system elements of the system architecture are not subject to a thorough change management, the connection between root and leaf is lost. Possible side effects through changing cannot be evaluated and tested as traceability is not established. Knowledge heroes who know why the system is designed in a certain way and task forces are necessary for the inevitable reengineering but, unfortunately, more budget and project time as well. Even worse are the effects when system inconsistencies are not detected at all and the CPS operation ends up with failure and loss, e.g., the maiden Flight 501 of the European Launcher Ariane 5 on June 4, 1996 (European Space Agency 2020).

At this point, MBCPSE emphasizes the function-oriented approach according to the advantages of the engineering disciplines involved and pointed out in Fig. 1. The knowledge of the heroes mentioned above shall be documented by means well established in computer science to make CPS organizations sustainably learn their know-whys to be independent from individuals and well prepared for necessary changes. The architectural principle for the anchoring of the know-why is the abstraction presented in the following section.

Abstraction

Based on the system architecture derived from the decomposition presented in the previous section, the functionality of each system element outlines its requirements. In particular, the task, the behavior, the interfaces, and the dependencies are specified. Referring to Fig. 1 and the previous section, the mechanical engineering know-how of the CPS domain was used to derive the system architecture. The abstraction introduced in this section applies computer science methods to document the different views of functionality, as it is not enough to specify the know-why in a sole collection of textual requirements. Too many aspects of the desired functionality would not be specified in this way.

Computer science has developed in the last decades several (model-based) methods and procedures to solve this issue, in response to the software crisis in the late 1960s and 1970s (Randell 2020; Dahl et al. 1972) and when focusing on new development processes and process improvement methods in the 1980s (Humphrey 2002) and 1990s (Booch et al. 2005). These fundamental contributions to today's

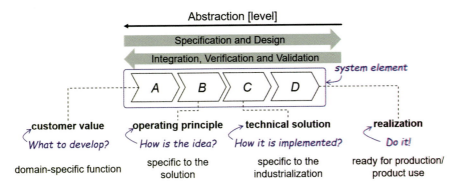

Fig. 3 Specification of a system element: abstraction levels and their purpose

state-of-the art software engineering were also used for the development of agile methods, e.g., SAFe and the agile manifesto in 2001 (Manifesto 2001).

Within MBCPSE, the functionality is used to integrate decomposition and abstraction. In contrast to the variable number of decomposition layers, the number of abstraction levels is defined to four, as shown in Fig. 3. These four abstraction levels are worked out for each system element as follows.

Customer Value: The purpose of level A is to specify the customer value of a system element by the required functions, resulting in a **function list**. The environmental conditions and the context conditions are determined for each function. It is important to point out that these functions are independent of the disciplines required to develop them. In other words, those functions can be mechanical, mechatronics, and software functions. For this purpose, level A specification elements are realized by.

- SysML use case diagrams (OMG Systems Modeling Language (OMG SysML) 2017) to provide an overview of the entire system at a high level of abstraction with services provided from the user's point of view.
- Textual requirements to supplement the use case diagrams.

The model includes both the diagrams and the textual requirements. The textual requirements supplement the diagrams with information that cannot or are not intended to be modeled. Consequently, there must be no redundancy between the diagrams and the textual requirements. If requirements are required outside the model, they are exported from the model and are not adapted since the model is the single source of truth.

Operating Principle: Each function of the function list created at level A is further specified by its operating principle at level B (see Fig. 3). The required functions are broken down into subordinated functions which are either local, i.e., within the system element, or functions of a child system element allocated in the subsequent decomposition layer. The resulting functions are then logically integrated at the function level. To achieve functional consistency at the system element level, all logically integrated functions of the function list have to be integrated into the **logical architecture**.

If possible, the operating principle is not intended to anticipate a technical solution, although certain product conditions may apply. Such product conditions can be rather simple but may have a major impact, e.g., a car has four wheels and is powered by an (electric) engine and not by a turbine. If necessary, they are added as (textual) requirements. Level B specification elements are realized by.

- SysML activity diagrams (OMG Systems Modeling Language (OMG SysML) 2017) to model sequences of actions.
- SysML state charts (OMG Systems Modeling Language (OMG SysML) 2017) to model states of the system under consideration and SysML sequence diagrams (OMG Systems Modeling Language (OMG SysML) 2017) to model the flow of a use case with the focus on interactions.
- Textual requirements to supplement the SysML diagrams.

Technical Solution: At level C, the function-oriented view of the operating principle is mapped to the mechanical view of the system, i.e., the functions of the function list are assigned to the provided **technical elements**. In other words, it is identified how the function and its operating principle are technically realized.

Technical elements are the abstract representation of one possible industrialization. This implies that technical elements correspond to system elements. However, it is possible that several system elements are bundled in one technical element or that one system element is distributed over several technical elements in one variant.

For example, a torque adjustment includes the functionality, electronics for regulation, mechanical parts such as the electrical machine, and software for intelligent control. All these technical elements can be mapped onto one system element, e.g., a component. Or the electronics and software could be outsourced to another system element for electromagnetic compatibility reasons. Since the maintenance of architectures involves effort, it is recommended to match the technical elements 1:1 with the system elements. Otherwise, both the technical and the system architectures have to be maintained separately, which leads to a significant additional effort and is prone to inconsistencies.

The required technical elements are further specified locally, i.e., within the decomposition layer, or using the functionality of a child technical element allocated in the subsequent decomposition layer. Furthermore, the technical elements determine which elements are realized in hardware or implemented in software, if the operating principle in the respective layer is already detailed enough. The resulting **technical architecture** integrates all technical elements to achieve technical consistency. Level C specification elements are realized by.

- SysML block definition diagrams to describe the relationships between the technical elements.
- SysML internal block diagrams to describe the interfaces and information flows between the technical elements.
- Textual requirements to supplement the SysML diagrams.

Realization: Based on the technical solution, the realization of the technical elements is specified at level *D*. The main focus of this level is to specify and document the necessary details for hardware and software in order to enable their efficient industrialization. The structure is called **component architecture** and remains local within the decomposition layer of the technical element.

The required components are further specified locally, i.e., within the technical element, or using the functionality of a child technical element allocated in the subsequent decomposition layer. The component architecture is consolidated by the resulting technical solution which integrates all technical elements at the same decomposition layer to achieve technical consistency.

The transparent and documented refinement and decision-making process from level *A* to level *D* result in the integrated specification of the system element. This helps to react quickly to change requests in any circumstance and is reusable. In the case of new technical findings, another type of realization can be quickly chosen without having to change the basic functionality of the technical elements. By this, local changes can be tested and approved locally when not changing the interfaces. This effect has a huge impact on possible cost reductions when industrializing the product.

The Advanced System Model

The Advanced System Model is the essential systems engineering element within MBCPSE. It provides all necessary architectural views and information of the CPS. It comprises the system architecture, the functional architecture, the logical architecture, the technical architecture, and the component architecture and parts of the e/e architecture, as shown in Fig. 4. Thus, the Advanced System Model provides the **semantic link** for the integration of all the disciplines involved as mentioned in section "Model-Based: Unite Engineering Solutions?".

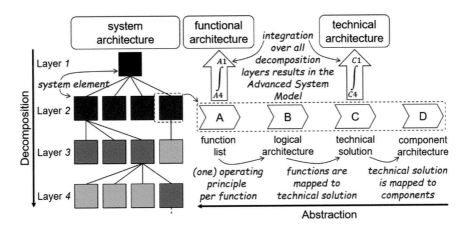

Fig. 4 The Advanced System Model

Each system element is defined by the decomposition presented in section "Decomposition" and specified by the abstraction introduced in section "Abstraction." Figure 4 illustrates the orthogonality of abstraction and decomposition of the Advanced System Model. The development of the advanced system architecture is outlined in this section.

In order to initially apply MBCPSE within an organization which has already developed a prevailing **system architecture**, this structure can be modeled as the system architecture of MBCPSE by an initial decomposition of the CPS as presented in section "Decomposition." The system architecture represents a primary abstract mechanical decomposition of the CPS and provides the **system elements** of types root, node, and leaf. The requirements for each system element are clustered according to its functionality. As an example, the decomposition of the root vehicle function list in Fig. 2 is partitioned into the nodes drive function list, chassis function list, body function list, and driver assistance function list. The further decomposed node (sub-)functions are then partitioned into further subfunctions until reaching the leaf functions in the leaf system element which provide pure hardware and software functions. Hence, a **function list** structures and specifies a system element. Consequently, the mechanical hardware focus is integrated with the functional focus of computer science, which provides the necessary **semantic linkage** between the two disciplines.

In this way, the **customer value** (level A) of the CPS is specified by several function lists, one for each system element. However, this would lead to system inconsistency as the specified function lists are worked out independently and are still not consolidated, i.e., there is no integration plan or architecture applied yet. To derive the **functional architecture** from the set of function lists, they have to be made consistent. Within MBCPSE, the principle of **parental communication** is applied. Each function of a function list shall be fully specified either internally within the system element or using the functions of a child system element only. In other words, each function serves the customer, the parent system element (see section "Decomposition"), or other functions of the same system element.

It is essential that the functions of different system elements of the same layer do not communicate directly but are controlled by the parent function of the layer above. The objective is again system consistency. The functions of the system element in layer n would not notice the information exchange of the functions of different system elements in layer $n + 1$ which may cause a state change of the CPS, and the entire system would risk inconsistent behavior. Needless to emphasize, this is one of the main sources for task forces, additional cost for rework, and system failure. In addition, applying the parental communication principle provides a design criterion for the functional architecture. The number of indirect communications between functions of the same decomposition layer via their parent functions shall be as small as possible.

The functional architecture may give the impression of over-specification or being too complex at first glimpse. Still, it provides system consistency and facilitates (re-)structuring and (re-)integration (particularly after changing) of the respective functions. This means individual parts of the system can be flexibly separated and reused without having to detach them from the entire network. Depending on the quality of the functional architecture, reuse can be deployed to a large extent at any layer. This provides a quite large leverage for cost savings at any stage of the product life cycle. Particularly in the automotive domain, this feature additionally contributes to customer satisfaction and risk minimization when applied to necessary changes in operative use.

Furthermore, the quality of the functional architecture can be measured by the parental communication principle. If, for example, an active cruise control function in Fig. 2 directly communicates with the torque adjustment function, this communication must communicate over five system elements. Consequently, it must be evaluated whether these functions are optimally deployed in the functional architecture or the system architecture, respectively.

For the specification of the customer value, it is also necessary to determine the customer. There are different kinds of customers possible. Most important is the end user, of course, when specifying the root system element. Specifying the other system elements, the functions of the function list of the parent system element are regarded as a customer. The third kind of customer is external stakeholders like legal regulations, standard specifications, and company standards which can occur in all decomposition layers. This also means that requirements can either be propagated through the Advanced System Model or be directly assigned to one system element. Requirement's consistency has to be checked continuously throughout the sequence of development phases, e.g., sprints or milestones.

Corresponding to the functional architecture, the underlying **operating principle** of each function (level *B*) shall be made consistent. This leads to the **logical architecture** of each function list, respectively the system element. The logical architecture also clusters and aligns various model elements of the different functions specified. Required data are to be aligned in resolution, timing, accuracy, repetition frequency, etc., e.g., speed information and engine state. This facilitates identifying commonalities and redundancies as well as reducing the development effort. Thus, information and data consistency are provided already at an early phase of the product development to avoid later cost-intensive troubleshooting when integrating the components. The operating principle shall be highly independent of technical influences to facilitate reuse.

Therefore, the principle of the **lowest technical definition** is introduced. It means that a detailed technical solution of a system element shall be defined by the child system elements in the lowest decomposition layer possible. The system elements of higher decomposition layers can, of course, limit the spectrum of technical solutions when technically necessary. As an example, the principle of the brake system of a vehicle has to be decided already at the root level as it has a major impact on all other parts of the system when using a brake parachute instead of an (electromagnetic) friction brake. It might be different, for example, for the power train when using an integrated combustion engine and/or an electric power train with or without fuel cells. If the different types of power trains fit into the same geometric and electrical interfaces, they can be gathered together as variants of a system element in the Advanced System Model.

It appears that in organizations developing complex systems, the hierarchical structure of the organization is mirrored to the system architecture in order to have assigned a one-on-one responsibility (Manifesto 2001). Of course, this effect can facilitate efficiencies by easier decision-making and correspondingly adapted processes. Nevertheless, it impedes overlapping innovation and technology changes as well as entraps to neglect thorough documentation. Particularly the structured and integrated modeling of the functional and logical architecture is a prerequisite for sustainable and efficient systems engineering.

At this stage of MBCPSE, the system architecture, the functional architecture, and the logical architecture of the CPS are consistently modeled and well documented. However, the system architecture is still a primary abstract mechanical decomposition as mentioned at the beginning of this section and has to be further technically detailed to be applicable for industrialization. The system architecture is therefore mapped to the **technical architecture**, i.e., the system elements are mapped to **technical elements** in all decomposition layers as outlined in the following. This implies that there are technical elements of type root, node, and leaf in analogy to the system elements and the mapping remains in the same decomposition layer.

There are three possibilities for how the mapping can take place. Firstly, the system elements can be mapped 1:1 to the technical elements. This means the technical solution for all system elements which is the entire CPS is defined from the beginning. This is suitable if the product doesn't change or innovate but in the given system elements. Secondly, the system elements can be mapped n:1 to the technical elements. This means a technical element integrates a number of system elements. This contributes particularly to savings of material costs. However, the additional cost for the more complex integration may not be planned. Thirdly, the system elements are mapped 1:m to the technical elements. This means the system is further decomposed and additional technical interfaces are introduced. It has to be particularly made sure that these interfaces are not affecting the functional and technical consistency of the CPS.

It is important to mention that all three mapping possibilities can be applied to different parts of the system architecture as this is a major asset of MBCPSE. There may be less innovative parts within the CPS that are already well established for industrialization where modeling efficiencies can support cost savings. For these parts, the details of the technical elements are integrated into the technical solution of the system elements. The principle of the lowest technical definition is broken intentionally for cost savings. For more innovative parts of the system architecture, the functional and logical architecture can be used to check whether the technical innovation under concern can be consistently integrated. However, the principle of the lowest technical definition shall be applied in general, allowing the mentioned exceptions. For CPS consistency reasons, the mapping of system elements to technical elements has to be checked when changing interfaces of the technical architecture.

After the mapping of the system elements to the technical elements, the technical elements are further specified locally and by the technical elements of the child technical element of the next lower decomposition layer. In this way, a consistent and complete technical architecture is created. If the technical element is of type leaf, the technical solution at level C is further detailed to the **component architecture**, which comprises the realization at level D for each technical element. The component architecture contains the requirements for subcontracting the components production and assembly.

According to the technical architecture and the component architecture, the industrialization of all parts of the CPS, inclusively their assembly, shall be made consistent. For all mechanical technical elements, the overlapping of constructed space is avoided by the thorough model-based specification of the technical architecture and the subsequent computer-aided geometric integration. For the mechatronics components, including pure computational components, additionally, the **e/e architecture** has to be developed.

In any case, the e/e architecture is a topic in CPS development on its own. However, the thorough specification of the functional and logical architecture at levels A and B and the technical and geometric architecture at levels C and D provide all necessary specifications to measure and evaluate possible e/e solutions. It is worth to be mentioned again that without a complete and consistent functional and logical architecture, the performance of an e/e architecture can only be evaluated after industrialization when finally approving the CPS itself, accompanied with additional time and budget needs for the required rework.

The abstraction and decomposition steps are performed and repeated multiple times until the CPS is fully specified and designed. If the requirements of any abstraction level in any decomposition layer cannot be met, the higher-level or higher-layer solution shall be reworked iteratively. This accelerates the development and helps to achieve an early cross-system consensus for the entire system, with all its system elements because the specification is consolidated before industrialization. Like this, the consistency and completeness of the Advanced System Model can be evaluated in early development phases. This early evaluation of the entire CPS can be done by continuously testing the CPS against the model views of the Advanced System Model.

Summed up, the Advanced System Model is the overarching element of MBCPSE. It comprises the required architectural views for the disciplines involved with an integrated semantic linkage of all models. Different technological solutions and variants for a deliberate technical element of the technical architecture can be easily added to the Advanced System Model as long as its interface is not changed.

Furthermore, the Advanced System Model extends the analysis, organization, checking, and derivation of elements for automated procedures. This includes simulations, test case generation, and model-checking as well as failure mode and effects analysis (FMEA). As the CPS is entirely and consistently structured by the technical architecture, all engineering procedures can be applied to a specific technical element. This facilitates the assignment of specialized teams and increases the quality of the results. Hence, it connects people through common interfaces and languages. In addition, the unified data sets enable big data approaches, e.g., digital shadowing (Bibow et al. 2020). With digital shadowing, the system and functional behavior can be checked according to its data without a predefined use case and requirement such as defined in the customer value. The behavior is tested on stored data and use cases from the field which is not feasible in this number of use cases by individuals. For this and further applications, structured data as presented in MBCPSE is necessary.

This consistent model view of MBCPSE facilitates a so-called 150% parts list. Of course, not all variants of the Advanced System Model are industrialized in one specific CPS version. Functional variants can be efficiently managed due to the system view of the functional architecture. Mechanical and electrical variants can be efficiently managed due to the technical architecture at level C and the component architecture at level D. Hardware and software solutions within the component architecture are to be realized specifically only. For this purpose, the realization uses different tools to (generate) code, design hardware, and integrate software and hardware.

The decision which variants shall be deployed is made for the industrialization of a specific version of the CPS for which the Advanced Digital Twin is derived from the Advanced System Model as outlined in the following section.

The Advanced Digital Twin

"A digital twin of a system consists of a set of models of the system, a set of digital shadows, and provides a set of services to use the data and models purposefully with respect to the original system" (Bibow et al. 2020).

The digital twin introduced in this section is the outcome of MBCPSE and comprises the complete, consolidated, and comprehensive model-based documentation of the specified version of the CPS to be industrialized. The set of models of the CPS is structured by semantically linked architectures which specify the relevant characteristics and functions of the CPS, as outlined in section "The Advanced System Model." Further, it provides a set of services to use data and models purposefully. For this, the Advanced Digital Twin of MBCPSE is derived from the Advanced System Model, presented in the previous section. As an enormous advantage, MBCPSE integrates the different architectures explicitly. Thus, each discipline integrates its architectural and technical assets, contributing the respective added value.

The functional and logical architectures are structured according to the system architecture, i.e., to the system elements, as this is ensured by the semantic linkage (see section "Systems Engineering: Merging Engineering Disciplines"). They provide the architectural aspects respectively the behavioral models representing the computer science point of view. The functional architecture models the consolidated and consistent customer value. It enables deriving specific development artifacts such as test cases for system and acceptance testing. The logical architecture specifies, consolidates, and integrates the respective operating principles. It facilitates, for example, the development of interface testing.

The technical and component architecture are initially structured according to the system architecture, i.e., to the system elements, as well. The system elements are then further mapped to the technical elements to decouple the mechanical and electrical point of view from the functional point of view. With this structured and reversible decoupling, MBCPSE provides a semantic linkage between the computer science and engineering disciplines. The technical elements provide the architectural aspects, respectively, the mechanical and electrical models, representing the engineering point of view, including the e/e architecture for the design of the vehicle electrical system. The technical architecture models the consolidated and consistent technical solution. It enables deriving specific development artifacts such as test cases for system test, e.g., crash test and onboard communication as well as component testing. The component architecture further specifies, consolidates, and integrates the respective requirements for subcontracting and industrialization.

Nonetheless, as already mentioned in section "Introduction," modeling has no economic purpose on its own. The nontechnical objectives of modeling a digital twin are to minimize the required time and budget for industrialization and operative use of the CPS. Thus, generating an **Advanced Digital Twin** is the phase of MBCPSE where the not neglectable additional effort for modeling the Advanced System Model shall pay off. Particularly, multiple generations of different Advanced Digital Twins, i.e., the reuse of the Advanced System Model, lead to a major increase in cost savings and customer satisfaction.

Based on a complete, consistent, and tested Advanced System Model, an Advanced Digital Twin can be efficiently branched out. For each technical element, the required variants are selected to set up the Advanced Digital Twin.

The maturity and quality of the technical elements may vary depending on the number of previous usages in earlier industrializations. However, adequate risk management covers possible maturity and quality issues by introducing effective measures for risk mitigation. The technical architecture shall not be changed at branching out or afterward. With such basic conditions, the respective industrialization shall remain in time and budget.

Since all product-relevant data is proven to be complete and coherent as well as available in a common pattern, i.e., semantically linked, the Advanced Digital Twin intrinsically provides consistency for integrated engineering, production, and operational approaches, as shown in Fig. 5.

Figure 5 shows how all elements of MBCPSE are connected, providing the systems engineering framework. It shows how decomposition and abstraction span the documentation of the CPS. Based on the structure of the basic V-model, the system elements are well defined by the Advanced System Model. The diagonal in Fig. 5 shows the apparent cost-saving mentioned in section "The Advanced System Model," i.e., the decomposition and abstraction steps are worked out and documented in one step. This doesn't enable iterative development due to the complexity of each step. Hence, the "diagonal development" skips at least 75% of the development steps and therefore seems to be efficient, but it is not sustainable. The similarity to the so-called waterfall model is obvious. With such coarse

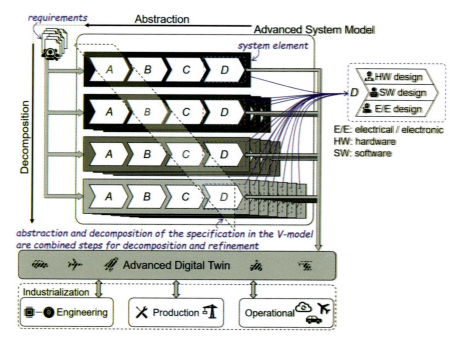

Fig. 5 The Advanced Digital Twin is branched out from the reusable Advanced System Model and provides all information and documentation for an efficient industrialization of one specific version of the CPS

documentation, the reusability of the development artifacts is not feasible with respect to robust testing and approval.

In contrast to the diagonal path, the separated steps of abstraction and decomposition in MBCPSE is not only about refining but also about being able to pull the elements apart again for modifications or technical replacements. Therefore, it is necessary to specify the entire abstraction and decomposition matrix of Fig. 5 with all its system elements.

Furthermore, the different system elements of the Advanced System Model, e.g., layer *2*/level *A* and layer *3*/level *C*, may be subject to different development styles, e.g., agile or conventional, as long as its interfaces and documentation results are not affected. A further crucial point is, of course, that all structural changes are to be done within the Advanced System Model and not within the Advanced Digital Twin, i.e., after having branched it out.

Model-Based: More than a Description

The sustainable management of any applied systems engineering and development methodology needs a cross-disciplinary base for communication. This cross-disciplinary base is preferably model-based to have a reliable basis for agile and fast processes. Reliable means that the model is the single source of truth and the necessary information for the system is included in the model. Since the costs and benefits of modeling must always be weighed up, the additional information has to be limited. This counts, of course, also for textual requirements.

Furthermore, it is important that once the modeling has started, the information from the model remains model-based. This means that a model is never converted into textual requirements and vice versa. This rule is applied because the conversion from model to text and back to model bears high risks of redundancies as well as the loss of information and must be avoided. Nevertheless, this procedure is still a common practice in model-based approaches. The requirements are adopted and satisfied depending on the depth of detail of the layer and level. Consequently, the model is not overloaded with unnecessary information and can be reused. Such a model has to support the top-down approach to reduce complexity and to enable an intellectual cross-disciplinary understanding of complex correlations. For this reason, the model needs a language that is

- Unambiguous.
- Domain-specific.
- Machine-readable.
- Understandable for people working in systems development.

An unambiguous and common language is necessary for the cross-disciplinary approach since the operating principles and technical solutions need to be understandable and comparable for every discipline. Otherwise, the interdisciplinary cooperation with its interaction between the disciplines will be lost within single models. The practical application of modeling shows that previously separated tasks elaborated separated teams building silos inside an organization are solved together in a cross-functional team in order to clarify all open questions, which are triggered by needs of explicit modeling. Hence, modeling especially fosters a cultural change from the strict division of labor into more agile organizations where each team member establishes its t-shaped profile through collaboration with others (Johnston 1978). A useful starting point for an unambiguous and common language is general-purpose modeling languages like SysML (OMG Systems Modeling Language (OMG SysML) 2017).

Based on a systems engineering method and an established modeling method, efficiencies can be improved with a domain-specific language (DSL). A DSL is less expressive than general-purpose modeling languages and limits the modeling options, but it is expressive enough to represent and address the problems and solutions for the specific domain (Brambilla et al. 2012). However, if the specification of a DSL is considered, experience with MBSE is required. This is because the scope and requirements of the desired DSL have to be communicated to experienced DSL developers. With the DSL, fewer redundancies, declarative descriptions, easier readability, and easier learnability due to the limited language set increases efficiency.

Additionally, the DSL has to be machine-readable to support all parties involved in the project with automated steps to monetize the development more efficiently. An illustrative example is the checking of functional dependencies using model-checking, automated functional safety checks, or automated test case creation (Drave et al. 2019; Drave et al. 2018; Hölldobler et al. 2019).

The unambiguous, domain-specific, and machine-readable language shall also be legible and comprehensible for people working actively in systems development. The project participants include modelers, system engineers, testers, functional safety engineers, and developers from various disciplines. All these groups must be able to rely on the information and automatically generated elements from the model. Furthermore, the model represents the status and process of the current project situation that affects various involved groups. A change in the operating principle possibly affects the subsequent development steps as well as integrating, verifying, and validating steps.

Illustrative Examples

The following section illustrates a best practice modeling approach for specifying CPS as presented in section "Best Practice Approach." As an example, the development of a Lane Keep Assist (LKA) function is shown. The LKA is a (product) function that is developed across domains and therefore has interdisciplinary requirements. For this purpose, the levels of abstraction according to section "Abstraction" respectively different views within the chosen SysML framework are exemplified.

The decomposition layer can either be of type node which means that the LKA function is part of a CPS or of type root which means the LKA function is the CPS by itself. It cannot be of type leaf as the components are not specified sufficiently. In any case, as an example, the following modeling represents an arbitrary decomposition layer representing the LKA perspective which would have to be integrated into an Advanced System Model of a CPS at the adequate decomposition layer. In other words, the function LKA is integrated firstly into the functional architecture and then secondly in the system architecture.

The content represents a sample version of the development artifacts and aims to illustrate the modeling methodology of MBCPSE without claiming completeness of the function information.

Level A: Customer Value

The customer value of abstraction is expressed as a black box view and represents the most abstract description of the function. A consideration of the inner operation of the function to fulfill its requirements is not part of this level. The focus is on the definition of clear function boundaries, the identification of all interacting functions and stakeholders, and the specification of black box requirements for the function.

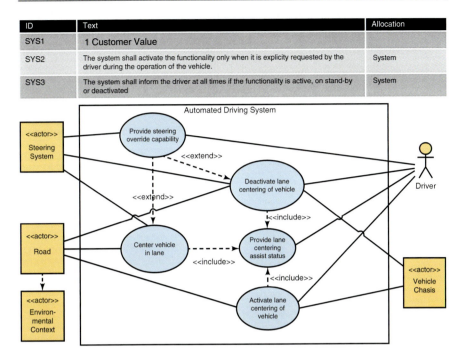

Fig. 6 The customer value of the Lane Keep Assist is modeled by use case diagrams augmented with textual requirements

These requirements form the central basis for the further development of the function, since the fulfillment of these customer (refer to section 3.1.3) requirements contributes decisively to the success and acceptance of the function to be developed. Further, the requirements of the customer value provide a founded and reusable test basis for system and acceptance tests (Fig. 6).

The LKA function interacts with the driver, the environment (e.g., the road), and the components of the vehicle, such as the steering system, to center the vehicle within the lane. In addition, the actors impose requirements on the function for activation and deactivation, display of the current status, and manual override of the function. The model-based specification is done with a SysML use case diagram and allows good visualization of the interrelations and dependencies of actuators and use cases. The specification of requirements has to be done as part of the diagrams as well as textual. As an optional aspect, there are advantages for machine post-processing if textual requirements are specified in a formalized manner.

Level *B*: Operating Principle

The next step of abstraction is the specification of the operating principle at level *B*. It is expressed in a first phase as a **white box view** and provides a solution-neutral

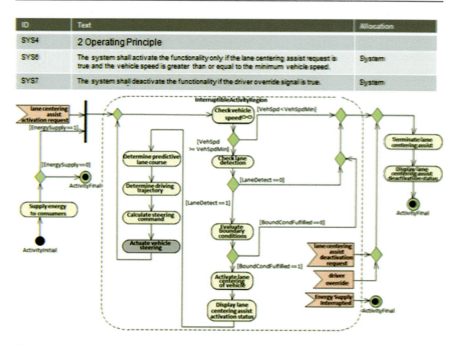

Fig. 7 The operating principle of the Lane Keep Assist is modeled by activity diagrams augmented with textual requirements

description of the function, the LKA in this case. Consequently, the inner operation of the function is considered without predefining its technical implementation. Especially functional dependencies, decision paths, function states, and information flows are focused by this form of specification. For the considered example, the operating principle is modeled in the form of an activity diagram and again described by additional textual requirements (see Fig. 7).

It is especially important to note that there shall be only one master of the specification. Even if additional textual requirements exist, they have to be a part of the function model and shall be automatically extractable from the function model. This avoids version conflicts and reduces consolidation and consistency issues, i.e., the well-known but inefficient additional rework effort which comes along with additive cost can be avoided. The actions included in the activity diagram are formulated solution-neutral to enable its reuse independently of the technical implementation.

For example, the determination and verification of the vehicle speed can be performed by GPS measurement or by means of built-in sensors but is not further specified in detail within the LKA. Figure 7 illustrates actions according to the places of execution in different colors. The highlighting allows identifying those actions that are executed by different components on leaf-level but are necessary for the modeled functionality. Overall, this model-based description represents an ideal functional behavior without considering any kind of functional misbehavior.

The essential substitute behavior in case of **functional misbehavior** is specified in the second phase of the operating principle. In order to consider different kinds of functional misbehavior, a stepwise extension (in accordance to (Granrath et al. 2019)) of the previously modeled desired behavior is added to the model of the first phase (see Fig. 7).

The first aspect is the function **diagnostics**. To ensure that the LKA works fine, all sensors shall be clean. Therefore, an example of the diagnostic aspect is the detection of a not properly cleaned sensor, which requires cleaning to ensure impeccable operation.

The second aspect considered is **degradation**. Here, the ideal behavior with maximum availability of the function is maintained by suitable substitute reactions. An example is the prevention of termination of the LKA due to a not detectable lane, for which purpose the pursuit of another vehicle can be used to stay in line.

The third aspect is the consideration of **safety**-relevant topics. For example, the plausibility check of the current position of the vehicle must be performed for the LKA in order to ensure the safety of the occupants and the environment.

The fourth aspect is **security**. This factor is of particular importance in the field of automotive driving, as there is a high risk of external control of the own or other vehicles, e.g., the LKA must ensure that the leading vehicle is trustworthy; otherwise, the function shall be terminated. If this aspect is not given, there would be a high risk in the vehicle pursuit mode that a risky and unwanted behavior occurs.

Besides the presented supplementary aspects of the operating principle, this level shall also be used to specify additional functionality necessary for the application of **big data** analyses and incorporated **artificial intelligence** algorithms.

The presented stepwise model-based system specification allows especially good complexity handling by continuously adding new aspects during development. In addition, this procedure allows for good cross-functional, i.e., interdisciplinary, collaboration, because depending on the development aspects under consideration, the necessary experts can be involved in a targeted manner, thus reducing high communication efforts caused by excessively large teams.

Level C: Technical Solution

The third level is the technical solution. This level is again divided into two phases. The first phase addresses the assignment of the system elements specified by the functional architecture to the technical elements of the technical architecture within the respective decomposition layer. For this purpose, the functional actions with the corresponding functional requirements are allocated to the full extent. This can be realized, for example, by partitions in the activity diagram of the operating principle. Consequently, in this phase, the system architecture in the form of interacting technical elements is further specified. In other words, the desired functional behavior is assigned to each technical element, and, even more important, the behavior of all technical elements is consistent with the specified operating principle.

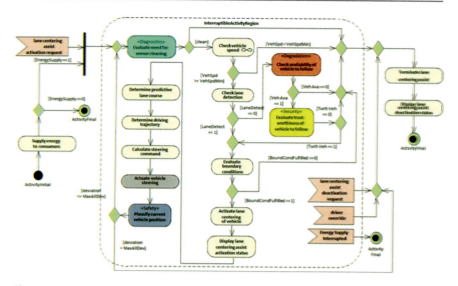

Fig. 8 The operating principle model of the Lane Keep Assist extended with functional misbehavior

However, the technical implementation of the technical elements is still not necessarily defined in this phase, following the rule of the lowest technical definition set up in section "Abstraction." This facilitates improved reuse of specification artifacts since the detailed specification of the technical elements in this phase is still independent of the realization. It allows an efficient mastering of the management of system variants and versions.

In the case of the LKA, the functional actions for the determination of the vehicle speed and for the digitalization of the measured raw values are assigned to a technical element "environment sensing" (see Fig. 8). For the realization of interactions between these technical elements, control flows and object flows are used, whereas the latter only represent information flows (Fig. 9).

In the second phase, the system architecture for the considered decomposition layer is defined if necessary, for further system design, again following the rule of the lowest technical definition. Each defined technical element is assigned either a technical element of the next decomposition layer or a unique type of realization intended for the product, i.e., a component. Thus, in the case of the LKA, the decision is made that the functional architecture element environment sensing is executed by physical components (e.g., two wheel speed sensors). More detailed consideration and specification of the wheel speed sensor component are given as part of the following decomposition layer of type "leaf." However, the functional and electrical interfaces have to be specified here.

To illustrate the system architecture, a representation in an internal block diagram is used. Functional and nonfunctional information requirements (e.g., resolution of information, timing) are additionally specified in textual form. After completion of the architecture definition, which contains decisions about the components to be technically realized and their interfaces, the details of the implementation and

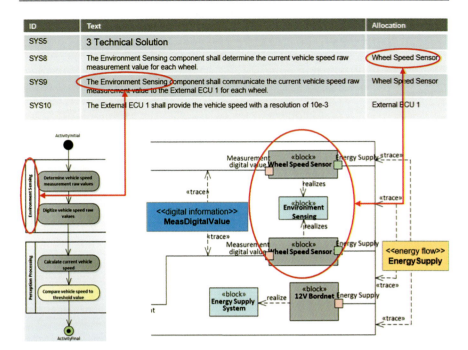

Fig. 9 The technical solution of the Lane Keep Assist is modeled by activity diagrams and internal block diagrams augmented with textual requirements with the objective to provide the functional behavior specified by the functional architecture and mapped to the technical elements

manufacturing of the individual components are defined. This includes all relevant functional and technical details of software and hardware components. For the realization of a software component, for example, the integration target to be used must be defined. For hardware components, details about materials like dimensions, resistance, and stability or manufacturing processes to be used have to be specified, for example.

Before the transition to the realization and the concluding industrialization of the system, the technical elements of the technical architecture defined here are mutually consolidated first and then considered and specified individually. When working on the decomposition layer of type leaf, a unique realization must be defined for each (hardware or software) component. After completion of the final specification, level D follows.

Level D: Realization

The last level of the abstraction is the detailed specification of each component of the system architecture assigned to the decomposition layer under concern, independent from whether it is realized in hardware or in software. The detailed specification is then used as the base for industrialization, i.e., as a specification for internal production or for subcontracting.

For this purpose, software models, e.g., with SysML or UML for discrete functionalities in combination with MATLAB/Simulink for continuous functionalities, and hardware models with computer-aided design are created. Based on the software models, software code can be generated automatically or completed by handwritten code. Once the code is generated, it is linked to the technical elements of the technical solution at level C as a valid realization and can be verified and validated virtually using the SysML specification.

It is important that the models used in this development step are completely different from the models made with SysML in the previous abstraction levels. No shortcuts are possible as the purpose of the models is different!

In close conjunction with the software realization, the hardware is designed and virtually tested. Based on the virtual designs, first prototypes are produced and evaluated. The easily adaptable solutions fit into the previously specified system architecture due to the clearly defined interfaces at all (abstraction) levels and (decomposition) layers. As a result, integration, variation, and validation are facilitated and reduce the possibility of unpleasant surprises due to unpredictable interfaces and communication problems. Each software and hardware prototype is verified and validated with test cases derived from the specification on the corresponding levels and layers within the Advanced System Model. The integration of the components is similarly verified and validated with test cases from preceding specified elements, e.g., the integration of the electric drive system is tested against the test cases of the operating principle to confirm the maturity of functions (Kriebel et al. 2018; Zimmermann 1980) (Fig. 10).

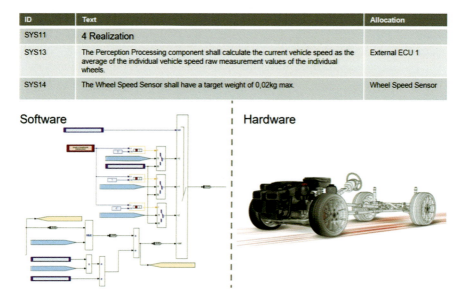

Fig. 10 The realization models the implementation and production results of a technical element augmented with textual requirements and specifies the requirements for industrialization

Chapter Summary and Expected Advances

The approach presented in this chapter is called MBCPSE. This is a rather bulky abbreviation for something rather smooth and intelligent. However, each of the three master issues integrated, i.e., model-based (MB), cyber-physical (CP), and systems engineering (SE), demands particular prerequisites, comes along with different requirements, and provides different possibilities. The core is an integrated systems engineering approach, particularly when developing and industrializing a cyber-physical system (CPS). As shown in section "State of the Art," there are multiple systems engineering approaches available in the disciplines of mechanical and electrical engineering as well as in computer science.

Each discipline focuses mainly on its specific requirements, product maturity steps, and culture. The necessary partner disciplines are hardly integrated with respect to intermediate results, milestones, reuse, error tolerance, and decision-making – not to mention the different ways of supplier management. This issue may be made more tangible by checking how and by whom decisions are made in the respective professional environment. It is an easy way to determine the core discipline and may lead the attention to other disciplines, as "diversity" is not meant to be only gender-specific or referring to nationalities. Diversity is indeed also focusing on disciplines as pointed out in the description of t-shaped profiles and cross-functional teams (Johnston 1978).

Before starting to create complex systems, it has to be made sure the people involved in the project are well trained on the same systems engineering approach and have the same understanding, the documentation is well defined, and most importantly they use the same language with respect to the systems engineering approach applied. This can be evaluated by asking a team to set up a glossary with common terms and definitions starting with "function," "model," and "system."

To handle very complex systems like today's CPS, MBCPSE provides an adjustable approach applicable for existing and completely new CPS developments. The adaptability is achieved through system elements derived from mechanical decomposition, their functional abstraction, their mapping onto technical elements, and the respective properly organized architecture, i.e., the architectures are semantically linked.

With the decomposition introduced in section "Decomposition," a reduction in complexity is achieved by dividing the system into system elements in subordinated layers. Starting from the root system element, the system is decomposed layer by layer into different nodes until the leaves are assigned to component elements. This approach ensures that all system-relevant relationships of the CPS are consistently and completely specified and documented and comprehensively modeled. Thus, side effects can be understood early in the development process or when changes are to be made within the system elements.

The root, each node, and each leaf demand specific functional requirements of the system, i.e., behavior and interfaces. To meet these functional requirements, they are clustered in functions for each system element which are specified by abstraction, as pointed out in section "Abstraction." The carried-out abstraction steps are separated from the decomposition steps to provide the semantic linkage, which integrates the engineering point of view with the computer science point of view. Like this, the specific design steps of each discipline can be performed independently and distinguished in terms of modeling and comprehensibility. Further, it provides the possibility to avoid side effects when modifying a function of a system element. Hence, the semantic linkage provides the mechanisms for an efficient local adaptation reinforced by the principles of model-based systems engineering.

The abstraction is divided into four levels and starts with level A the customer value. At level A, the environment and context conditions for each function are specified, resulting in the function list mentioned above. These functions are further broken down into internal functions and functions provided by a child system element resulting in the operating principle at level B. The operating principle provides a solution-neutral description of the function. At the next level, i.e., level C (technical solution), the functions of the operating principle are further technically specified and mapped to the technical elements which are used for industrialization. Subsequently, at level D (realization), this technical solution is further refined and documented for component realization in order to enable efficient industrialization. The results of all decomposition and abstraction steps span the Advanced System Model presented in section "The Advanced System Model."

By this, the Advanced System Model comprises the system architecture, the functional and logical architecture, as well as the technical and component architecture. As these architectures are semantically linked, the Advanced System Model provides all necessary documentation consistently for its augmentation with further relevant integrated system parts, e.g., the vehicle electric system or the integrated cooling system.

It may be criticized that MBCPSE leads to a large number of documents to be managed. However, based on a sound systems engineering approach, this helps to cope with the high complexity of a CPS as its architecture comprises the function-oriented view of a computer scientist (abstraction levels) and the engineering views of a mechanical and electrical engineer (decomposition layers, abstraction levels C and D). Particularly it replaces the rather unstructured discussion during requirement elicitation which causes many industry projects to be delayed already in early project phases. All developers are guided by a systematic approach in an integrated architectural framework which is semantically linked, the Advanced System Model. There is an additional positive effect caused by the MBCPSE documentation. As all model interfaces of a CPS are well defined, the different technical elements are behaving like black boxes to each other. Thus, they can be internally changed without causing side effects in other parts of the CPS as long as their interface is not changed and behaves as before. All interface tests and other parts of the CPS can be (re)used as outlined before.

A further advantage is the complete and consistent Advanced Digital Twin obtained from the Advanced System Model as detailed in section "The Advanced

Digital Twin." Since the Advanced Digital Twin contains all implemented architectures of the Advanced System Model, it creates a consistency for integrated engineering, production, and operational approaches. This facilitates, for example, industrialization, appropriate, flexible, and efficient responses to product changes and development of interface testing. Nevertheless, modeling a digital twin is always cost-intensive and needs special knowledge and, as a matter of fact, a lot of experiences as discussed in section "Model-Based: More than a Description." Domain experts in the mechanical industry, e.g., the automotive industry, often have little knowledge about standardized modeling languages for discrete systems like SysML and UML but excellent knowledge in modeling dynamical systems, e.g., with MATLAB/Simulink. Vice versa, computer scientists are usually not prepared to specify CPS beyond the functional architecture. Since experts who master mechanical and electrical engineering as well as computer science are rare, large CPS projects usually fail to meet their objectives as pointed out in section "Introduction."

Nonetheless, due to the high dependency of the functional and the technical architecture, MBCPSE also requires a few experts who master all three disciplines, the CPS architects. These CPS architects are the key players for a successful system architecture. However, the CPS architects shall not have to deal with the specific technical details but with the interfaces and the interferences of the parts of the Advanced System Model only, as outlined in section 3.1.3. The details of the functional architecture are assigned to a cross-functional team with a focus on computer science knowledge and the development of the technical architecture to a cross-functional team with a focus on engineering expertise. The realization of hardware and software is assigned to the respective expert teams. This leads to an easier exchange of expert knowledge. Furthermore, the independence of all development results within the CPS architecture enables simultaneous engineering and continuous integration which are the prerequisites for agile development, e.g., according to the SAFe framework. These modern working methods appeal to young people and emerging young disciplines, and these will again further develop the system's method.

In order to achieve such an integrated development environment, the system architecture of the Advanced Digital Twin enables implementing MBCPSE in every organization, by modeling the prevailing mechanical structure of the CPS. Development processes can be optimized in accordance with the system architecture as long as product changes stay within the underlying framework, e.g., technology updates or innovations. Next to the process, often the hierarchical organization is also following the system architecture. This may impede flexibility for necessary product advancements and innovations which can be taken further when applying MBCPSE, as it enables the adaptation of product, organization, and culture within an existing system architecture.

What at first sight seems like an anachronism, because modern approaches start on a greenfield with the functional architecture and leave the system architecture open at the beginning (see section "Related Work"), is at second sight a suitable approach to address existing organizations (brownfield systems) (Hopkins and Jenkins 2008). However, the adoption of a prevailing system architecture, which often reflects the organizational structure (Colfer and Baldwin 2016), limits the functional and technical architecture. This circumstance can be overcome by consistently applying MBCPSE starting with the prevailing system architecture, semantically linking it to the functional architecture, and resulting in a technical architecture for industrialization (see Fig. 4).

The function-oriented view will additionally provide new potentials for efficiencies which can lead to a completely different approach to structure processes and organization. In conclusion, it can be observed that since the last decades the product decomposition in engineering domains is usually structured by geometric interfaces. Due to countless optimization steps, this led to mirrored structures in organization and responsibilities. All other "minor" disciplines involved had to follow this prevailing structure and to bow their architectural needs and requirements to the still ongoing economic success story. This subordination led to relatively weak results in the product contributions of the electric engineering and computer science domain.

If the entire development organization shall be changed to follow an effective function-oriented approach, MBCPSE supports this objective by smoothly transferring the system architecture to the technical architecture. The iterative and agile application of MBCPSE development cycles leads to a function-oriented product development process. At this stage, the functions are developed at level A and level B, and level C and level D take continuously over for the technical solution. Hence, the technical architecture substitutes the not any longer needed initial system architecture. The functional architecture is free of initial technical limitations, as required by the mentioned greenfield approaches. It is a question of economic efficiency to adapt the organization to the function-oriented systems engineering approach.

In any case, such an extensive technical change implies an extensive cultural change that takes a serious amount of time and budget. In addition, it also requires a huge amount of discipline and management commitment to stay with the applied architecture and to incorporate upcoming changes properly. However, the application of MBCPSE definitely pays off when the Advanced System Model is reused in order to develop a couple of Advanced Digital Twins, by which the time to market is reduced significantly and task forces are avoided. Most importantly, it prevents an organization from becoming a Dinosaur. Modern agile processes are implemented, and the continuous achievement of reachable targets is mutually recognized and appreciated. The efficiency and effectiveness of each individual work are increasing as personnel continuously evolve within a cooperative agile culture.

Last but not least, the customer will appreciate the consistent and reliable behavior of the CPS which can be functionally continuously maintained and extended over the air avoiding cost-intensive major defects and failure.

Cross-References

▶ Exploiting Digital Twins in MBSE to Enhance System Modeling and Life Cycle Coverage
▶ Model-Based Hardware-Software Integration

References

P. Bibow, M. Dalibor, C. Hopmann, B. Mainz, B. Rumpe, D. Schmalzing, M. Schmitz, A. Wortmann Model-Driven Development of a Digital Twin for Injection Molding, International Conference on Advanced Information Systems Engineering (CAiSE'20), ser. Lecture Notes in Computer Science, S. Dustdar, E. Yu, C. Salinesi, D. Rieu, V. Pant, 12127. Springer International Publishing, 2020. 85–100.

B. W. Boehm, "Guidelines for Verifying and Validating Software Requirements and Design Specifications, Euro IFIP 79, P. A. Samet North Holland, 1979, 711–719.

G. Booch, J. Rumbaugh, and I. Jacobson, The unified modeling language user guide: Covers UML 2.0 ; thoroughly updated, the ultimate tutorial to the UML from the original designers, 2nd ed., ser. Safari Books Online. Upper Saddle River, NJ: Addison-Wesley, 2005. [Online]. Available: http://proquest.tech.safaribooksonline.de/032126797426

M. Brambilla, M. Wimmer, and J. Cabot, Model-driven software engineering in practice, ser. Synthesis lectures on software engineering. San Rafael, Calif.Morgan & Claypool, 2012, 1.

P. Braun, M. Broy, F. Houdek, M. Kirchmayr, M. Müller, B. Penzenstadler, K. Pohl, and T. Weyer, "Guiding requirements engineering for software-intensive embedded systems in the automotive industry: The REMsES approach," Computer Science - Research and Development, vol. 29, no. 1, pp. 21–43, 2014.

M. Broy, "Challenges in Automotive Software EngineeringProceedings of the 28th International Conference on Software Engineering, ser. ICSE '06. New York, NY, USA: ACM, 2006, pp. 33–42.

M. Broy, M. V. Cengarle, and E. Geisberger, "Cyber-Physical Systems: Imminent ChallengesLarge-scale complex IT systems, ser. Lecture Notes in Computer Science, R. Calinescu D. Garlan, Berlin: Springer, 2012,. 7539, 1–28.

L. J. Colfer and C. Y. Baldwin, "The mirroring hypothesis: Theory, evidence, and exceptions," Industrial and Corporate Change, vol. 25, no. 5, pp. 709–738, 2016.

O. J. Dahl, E. W. Dijkstra, and C. A. R. Hoare, Structured programming. GBR: Academic Press Ltd, 1972.

Dassault Systemes. 03.11.2020. Catia - Computer Aided Three-Dimensional Interactive Application: https://www.3ds.com/de/produkte-und-services/catia/.

I. Drave, T. Greifenberg, S. Hillemacher, S. Kriebel, E. Kusmenko, M. Markthaler, P. Orth, K. S. Salman, J. Richenhagen, B. Rumpe, C. Schulze, M. Wenckstern, and A. Wortmann, "SMArDT modeling for automotive software testing," Software: Practice and Experience, vol. 49, no. 2, pp. 301–328, 2019.

I. Drave, T. Greifenberg, S. Hillemacher, S. Kriebel, M. Markthaler, B. Rumpe, and A. Wortmann,"Model-Based Testing of Software-Based System Functions," Conference on Software Engineering and Advanced Applications (SEAA'18), 2018, 146–153.

EAST-ADL Association. 2013. EAST-ADL Domain Model Specification version V2.1.12. [Online]. Available: http://www.east-adl.info/Specification.html

European Space Agency. 03.11.2020. Ariane-5: Learning from Flight 501 and Preparing for 502. [Online]. Available: www.esa.int/esapub/bulletin/bullet89/dalma89.htm

P. Giusto, R. S., and S. M., "Modeling and Analysis of Automotive Systems: Current Approaches and Future Trends, Proceedings of the 4th International Conference on Model-Driven

Engineering and Software Development. SCITEPRESS - Science and Technology Publications, 2016, pp. 704–710.

Granrath, C., et al. The next generation of electrified powertrains: Smart digital systems engineering for safe and reliable products,SIA PARIS 2019 - Power Train & Electronics, 2019.

K. Hölldobler, J. Michael, J. O. Ringert, B. Rumpe, and A. Wortmann, "Innovations in model-based software and systems engineering," The Journal of Object Technology, vol. 18, no. 1, pp. 1–60, 2019.

R. Hopkins K. Jenkins, Eating the IT elephant: Moving from greenfield development to brownfield, ser. Safari Books Online. Upper Saddle River, N.J: IBM Press/Pearson plc, 2008.

W. S. Humphrey, Managing the software process, 28th ed., ser. The SEI series in software engineering. Boston: Addison-Wesley, 2002.

ISO International Organization for Standardization, "ISO/IEC/IEEE 15288–1:2015–05 Systems and software engineering: System life cycle processes," Berlin, 2015.

ISO International Organization for Standardization, "ISO/IEC/IEEE 24765:2017–09: Systems and software engineering | Vocabulary," Berlin, 2017.

ISO International Organization for Standardization. 2018. ISO 26262-10:2018: Road vehicles - Functional safety - Part 10: Guidelines on ISO 26262 Berlin

D. L. Johnston, "Scientists become managers-the 't'-shaped man," IEEE Engineering Management Review, 6, 3, 67–68, 1978.

J. C. Kirchhof, J. Michael, B. Rumpe, S. Varga, A. Wortmann,"Model-driven Digital Twin Construction: Synthesizing the Integration of Cyber-Physical Systems with Their Information Systems," Proceedings of the 23rd ACM/IEEE International Conference on Model Driven Engineering Languages and Systems. ACM, 2020, 90–101.

R. Knaster and D. Leffingwell, SAFe 4.5 distilled: Applying the scaled agile framework for lean enterprises. Boston: Addison-Wesley, 2019.

S. Kriebel. 2018. Pains in Modeling: SysML-based Deployment in an Engineering Domain: Invited Talk at 1st Workshop on Pains in Model-Driven Engineering Practice, Conference on Model Driven Engineering Languages and Systems (MODELS' 18): https://sites.google.com/view/pains-2018/home. Copenhagen, Denmark.

S. Kriebel, V. Moyses, G. Strobl, J. Richenhagen, P. Orth, S. Pischinger, C. Schulze, T. Greifenberg, and B. Rumpe, "The next generation of BMW's electrified powertrains: Providing software features quickly by model-based system design,"26th Aachen colloquium automobile and engine technology, 2017.

S. Kriebel, J. Richenhagen, C. Granrath, and C. Kugler, "Systems engineering with SysML the path to the future?" MTZ worldwide, 79, 5, 44–47, 2018.

G. Liebel, M. Tichy, and E. Knauss, "Use, potential, and showstoppers of models in automotive requirements engineering," Software & Systems Modeling, vol. 18, no. 4, pp. 2587–2607, 2019.

A. Manifesto,Agile manifesto,Haettu, 14, 2012, 2001.

MathWorks. 02.07.2019. Simulink - Simulation und Model-Based Design: https://de.mathworks.com/products/simulink.html.

OMG Object Management Group. OMG Unified Modeling Language, v2.5.1: Version 2.5.1 2017. [Online]. Available: http://www.omg.org/spec/UML/2.5.1

OMG Systems Modeling Language (OMG SysML): Version 1.5. 2017. [Online]. Available: http://www.omg.org/spec/SysML/1.5/

K. Pohl, H. Hönninger, R. Achatz, and M. Broy, Model-based engineering of embedded systems: The SPES 2020 methodology. Berlin and Heidelberg: Springer, 2012.

B. Randell. 03.11.2020. NATO Software Engineering Conference 1968. Schloss Dagstuhl. [Online]. Available: http://homepages.cs.ncl.ac.uk/brian.randell/NATO/NATOReports/

C. Read, "Dinosaurs and economic Darwinism," in The rise and fall of an economic empire: With lessons for aspiring economies. London: Palgrave Macmillan UK, 2010, pp. 256–270.

B. Rumpe, Agile Modeling with UML: Code generation, Testing, Refactoring. Springer International, 2017.

D. Scheithauer K. Forsberg, "4.5.3 V-Model Views," INCOSE International Symposium, 23, 1,. 502–516, 2013.

H. Stachowiak, "Allgemeine Modelltheorie," Wien: Springer, 1973.

D. D. Walden, G. J. Roedler, K. Forsberg, R. D. Hamelin, and T. M. Shortell, Systems engineering handbook: A guide for system life cycle processes and activities, 4th ed. Hoboken, NJ: Wiley, 2015.

J. Whittle, J. Hutchinson, and M. Rouncefield, "The state of practice in model-driven engineering," IEEE Software, vol. 31, no. 3, pp. 79–85, 2014.

H. Zimmermann, "OSI reference model-the ISO model of architecture for open systems interconnection," IEEE Transactions on Communications, vol. 28, no. 4, pp. 425–432, 1980.

Stefan Kriebel is technical director in the Business Unit Intelligent Mobility & Software and responsible for Future Mobility SW & EE Platforms at FEV Europe GmbH, Aachen, Germany. He studied Computer Science at TU Munich and received his Ph.D. from TU Munich with his thesis done on behalf of the Joint Research Centre (JRC) of the European Commission in Ispra, Italy, on time series analysis with artificial neural networks. He was with the BMW Group in various innovative projects assigned as responsible manager to several fields like electric power train, driving dynamics software, driver assistance software, and systems engineering. Earlier he was responsible manager for the software test of smart cards at Giesecke & Devrient GmbH, Munich, and with the Software Quality Department of the European Space Agency (ESA) at the European Space Research and Technology Centre (ESTEC), Noordwijk, Netherlands.

His main interest is state-of-the-art systems engineering integrating engineering and computer science disciplines, enhanced by digital transformation, agile transformation, and cyber-physical systems engineering. He is the author and co-author of several articles on these topics.

Matthias Markthaler is a function specialist at the BMW Group responsible for the test and integration of the vehicle energy system and the e/e architecture. He is currently pursuing a Ph.D. degree with his thesis on "model-based method for automated test case creation in the automotive industry based on a systems engineering approach" on behalf of the BMW Group as a Ph.D. fellow and the Department of Software Engineering at the RWTH Aachen University.

Since 2016, he has been working in different departments of integration and testing at the BMW Group. During this time, he was jointly responsible for the development and rollout of a model-based systems engineering method in the area of the electric drive system. He studied at the Federal University of São João del-Rei, Brazil, and at the Munich University of Applied Sciences, Germany, where he received his B.Sc. and M.Sc. degree in electrical engineering and information technology. His current research interests include the cooperation of the different disciplines in model-based systems engineering, the transformation to model-driven systems engineering, and cyber-physical systems engineering. He is the author and co-author of several articles on these topics.

Christian Granrath received his B.Sc. degree in mechanical engineering in 2014 and his M.Sc. degree in energy engineering in 2016 from RWTH Aachen University, Aachen, Germany. He is currently pursuing the Ph.D. degree in software and systems engineering at the Junior Professorship for Mechatronic Systems for Combustion Engines, RWTH Aachen University. As group leader at RWTH Aachen University, he is supporting the lecture "Software Development for Combustion Engines." In 2019, as a research associate at the University of Applied Sciences Aachen, he conducted a scientific training in systems engineering and agile development within the project "ERASMUS+ UNITED" to realize a knowledge transfer between European and Indonesian universities. His research interests include the fields of model-based and feature-driven systems engineering, agile software engineering, software architecture development and evaluation, as well as simulation model development for XiL applications in the automotive domain.

Johannes Richenhagen is vice president of Intelligent Mobility & Software at FEV Europe GmbH, where he previously held a number of responsible management positions. He studied mechanical engineering at the RWTH Aachen University, Germany, where he also received his Ph.D. with a thesis "Control Software Development for Automotive Powertrains with Agile Methods." He is assigned as associate lecturer on Software Development at the Junior Professorship for Mechatronic Systems for Combustion Engines, RWTH Aachen University.

His main interest is state-of-the-art systems engineering integrating engineering and computer science disciplines, enhanced by digital transformation, agile transformation, and cyber-physical systems engineering. He is the author and coauthor of several articles on these topics.

Bernhard Rumpe is heading the Software Engineering Department at the RWTH Aachen University, Germany. Earlier he had positions at INRIA/IRISA, Rennes, Colorado State University, TU Braunschweig, Vanderbilt University, Nashville, and TU Munich.

His main interests are rigorous and practical software and systems development methods based on adequate modeling techniques. This includes agile development methods like XP and SCRUM as well as model engineering based on UML-like notations and domain-specific languages. He has contributed to many modeling techniques, including the UML standardization. He also applies modeling, for example, to autonomous cars, human brain simulation, BIM energy management, juristical contract digitalization, production automation, cloud, and many more. In his projects, he intensively collaborates with all large German car manufacturers, energy companies, insurance and banking companies, a major aircraft company, a space company, as well as innovative start-ups in the IT-related domains.

He is author and editor of 36 books and editor in chief of the Springer international journal *Software and Systems Modeling* (www.sosym.org). His newest books *Agile Modeling with UML: Code Generation, Testing, Refactoring* and *Engineering Modeling Languages: Turning Domain Knowledge into Tools* were published in 2016 and 2017, respectively.

Integrating Heterogenous Models

13

Michael J. Pennock

Contents

Introduction	418
Key Concepts and Definitions	419
Ontologies and Models	419
Preserving Model Correspondence	420
Causes of Ontological Differences among Models	422
The Levels of Conceptual Interoperability Model	424
Best Practice Approaches	426
Model Integration Standards and Tools – Levels 1 and 2	426
Addressing Levels 3 Through 6	428
Illustrative Example	433
Expected Advances in the Future	435
Chapter Summary	437
Disclaimer	437
References	438

Abstract

A common problem in Model-Based Systems Engineering is that models may come from a variety of different sources and domains. When it is necessary to integrate these diverse models to answer engineering questions, it is sometimes referred to as integrating heterogenous models. Accomplishing this integration has been a persistently challenging problem. In this chapter, the problem is characterized as one of addressing critical differences in the ontologies that underlay the models. The Levels of Conceptual Interoperability Model (LCIM) is used to categorize these differences, and approaches to address the levels are described including model integration standards, semantic ontologies, and

M. J. Pennock (✉)
The MITRE Corporation, McLean, VA, USA
e-mail: mpennock@mitre.org

© Springer Nature Switzerland AG 2023
A. M. Madni et al. (eds.), *Handbook of Model-Based Systems Engineering*,
https://doi.org/10.1007/978-3-030-93582-5_23

bridging mechanisms. Finally, a multi-scale modeling example is used to illustrate how bridging mechanisms work.

Keywords

Heterogenous models · Ontology · Composability · Syntactic interoperability · Semantic interoperability · Multi-scale modeling

Introduction

A central concept in Model-Based Systems Engineering (MBSE) is the use of models, often digital models, to describe a system. When that system is large and complicated, it is likely that multiple models will be required to achieve an adequate description. These models may represent different system components, different perspectives of the same system, or both. Furthermore, these models may have been developed by different teams with different types of expertise and for their own purposes. These circumstances create a complication for the systems engineer because performing typical engineering tasks such as exploring the performance of design alternatives may require the integration of some subset of these models. Accomplishing this integration is sometimes referred to as composing or integrating heterogenous models.

Unfortunately, integrating heterogenous models has proven challenging in practice. As Taylor, et al. [1, p. 652] note "Composability is still our biggest simulation challenge." Furthermore, a National Science Foundation (NSF) workshop report on modeling and simulation noted that model reuse is "peculiarly fragile" [2]. The implication for MBSE is that model integration difficulties can substantially slow the implementation of MBSE approaches as well as reduce potential cost savings.

There is no one approach to integrating heterogenous models. Integrating heterogenous models is a vague term that is typically invoked when one attempts to integrate one or more models that do not seem to be designed to work together. Presumably, if the models were homogenous, they could be integrated easily. Instead, there must be some material differences among the models that interfere with their integration. Hence, they are heterogenous. This characterization is not particularly enlightening as there are many ways that models may differ, but not all differences are relevant to a particular integration effort. Integrating heterogenous models requires determining which differences are relevant and addressing them. This is often a case-by-case assessment.

The literature on integrating heterogenous models is vast, and the problem is common enough that it spans multiple scientific and engineering disciplines. Certain classes of model differences occur often enough that they have received their own designation and associated body of literature. Examples include multi-method, multi-resolution, multi-level, multi-domain, and multi-scale modeling. Given the breadth of literature on this topic, it is not possible to cover it comprehensively in this chapter. Instead, this chapter focuses on a subset of issues and methods that are likely to be relevant to systems engineers. Anyone attempting to integrate heterogenous models is advised to explore the literature relevant to the specific aspects of their model integration problem.

Key Concepts and Definitions

As noted previously, the term heterogenous models is not sufficiently descriptive to diagnose a particular model integration problem. In this chapter, one or more models are considered heterogenous if they employ different ontologies. The nature of these ontological differences can point to how they can be addressed via bridging mechanisms. To be successful, the selected bridging mechanisms must preserve the correspondence of the integrated model with the referent. Each of these concepts will be explained in turn.

Ontologies and Models

While ontology is a branch of philosophy, computer scientists have adapted the term ontologies to refer to "formal specifications of concepts representing entities of a specific knowledge domain and the relationships that can hold between these entities." [3, p. 60] When someone constructs a model of a system, they are explicitly or implicitly characterizing that system using a particular ontology. Consider the ideal gas law, $PV = nRT$, as a simple model of a gas. At minimum, the ontology consists of entities in the equation including P, pressure, and T, temperature. It also includes the equation itself which defines the allowable relationships among those entities. Of course, for the model to be useful, one would need to define concepts such as pressure and acceptable means to measure that pressure. There is also the underlying concept of a gas as a collection of point particles that do not interact. This suggests that, even for this simple model, the ontology is larger and more complicated than it seems at first.

Under different circumstances a different ontology may be required to model a gas accurately. For example, if particle size and particle interactions cannot be ignored, then the van der Waals equation, $(P + a/V_m^2)(V_m - b) = RT$, may be a more appropriate definition of the allowable relationships among the entities. However, new entities that depend on the specific gas being modeled have now been introduced to the ontology.

Depending on the question of interest one may choose to model the same system using different ontologies. For example, one could model a piece of silicon as a continuous solid, a lattice of atoms, or even as a collection of interacting quantum particles [4]. Each of these modeling approaches entails a different model ontology.

To complicate matters further, the construction of any given model may rely on more than one type of ontology. Hofmann [3] highlights two categories of ontology that are particularly relevant to model development: referential ontologies and methodological ontologies. The referential is the ontology used to describe the portion of the real world that one wants to model. An example would be the ontology above that describes a gas as a non-interacting collection of point particles. The methodological ontology refers to the ontology used by a particular modeling methodology. An example would be the set of objects and relations used to implement a discrete event simulation including queues, events, interarrival times, etc. When one attempts to model a system using a particular modeling approach, it requires mapping portions of the referential ontology to the methodological ontology.

Preserving Model Correspondence

For a model to be useful it must preserve correspondence with a reference system over a set of properties of interest. When this condition is satisfied, the model is said to be valid. An adaptation of Rosen's [5] analysis of models of natural systems is employed to explain this concept. The reference system that one is attempting to model is called the referent. Fig. 1 depicts a simple example of a model preserving correspondence with its referent. In this figure, the referent is R. The goal is to measure R's temperature and use the model to predict the pressure of R. The measured temperature is input into the model as T and used to predict the pressure, P. If the model is in correspondence with the referent, when one measures the pressure of R, it will be the same as the model's prediction for any given value of T. In mathematical terms, the graph in Fig. 1 commutes.

Of course, real world models do not exhibit this level of perfect correspondence. Instead, they only need to stay within acceptable error tolerances over a broad enough set of circumstances. Still, this highly simplified schematic is sufficient for explanatory purposes here.

Next, define a simple integrated model as one where two or more models are composed to make the prediction. Fig. 2 modifies Fig. 1 by requiring two models to go from temperature to pressure. First, Model 1 uses temperature to predict the value of an intermediate variable X. Model 2 uses the value of X to predict the value of pressure P. Again, if the measured pressure of R matches the predicted pressure P

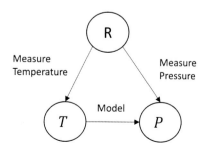

Fig. 1 A simple example of a model remaining in correspondence with its referent

13 Integrating Heterogenous Models

Fig. 2 A simple example of an integrated model remaining in correspondence with its referent

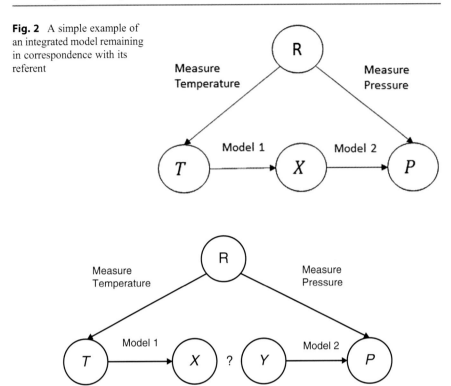

Fig. 3 A simple example of an attempt to construct an integrated model with inconsistent ontologies

over a range of values for T, the integrated model is in correspondence with its referent.

In an integrated model, the models must exchange data to preserve correspondence with the referent. In this simple example, Model 1 must pass the value of X to model 2. When heterogenous models are integrated, there is a problem because the models may use different ontologies. To illustrate this, the example must be modified one more time.

In Fig. 3, Model 2 does not accept variable X as an input. Instead, it only takes variable Y. This means the modeler must develop a mechanism that can translate from one ontology to another. The modeler must find a way to translate from X to Y while still preserving the correspondence relationship. In this chapter, this translation will be referred to as a bridging mechanism. The difficulty of constructing a bridging mechanism can vary greatly depending on the nature of the differences between the models. A difference in units such as meters versus feet may be addressed with a simple linear function. However, a more complicated difference such as the example of representing the piece of silicon as both quantum particles and a continuous solid may not be as easy to address. Unfortunately, the combination of two valid models is not necessarily valid [6].

Now consider that most engineering models are far more complex than the simple functions used in the example, and instead, the correspondence must be preserved over many different variables with many different relationships. Even worse, many of these relationships will not even be documented in the model itself but are unstated assumptions made by the person that created the model. The bottom line is that integrating heterogenous models is often far more challenging than simply translating outputs of one model to match the inputs of another.

A useful analogy is to consider the case of two people who speak different languages that are trying to communicate with each other. For the sake of the example, assume that one speaks English and the other speaks German. They find that some concepts are easily communicated because there is an almost one-to-one mapping between English and German. For everyday objects, the two only need to determine the corresponding word to exchange information. For example, "table" in English translates to "tisch" in German. "Car" in English translates to "auto" in German. But for more abstract concepts, the situation becomes more problematic. The German word "zeitgeist" does not have a corresponding word in English. If the German speaker would like to communicate this concept to the English speaker, it is going to take a lot more work. In fact, it might require constructing a lengthy explanation using many words, and even that may not be sufficient. The English speaker may need to construct their own internal representation of "zeitgeist" using concepts they know from English in order to make use of the data they were passed. Doing so is a mechanism for preserving correspondence with the reference concept when the two are communicating. It is this need to preserve correspondence that is so challenging when integrating heterogenous models.

Causes of Ontological Differences among Models

To understand why preserving correspondence can be difficult, it is necessary to discuss the reasons why models may use different ontologies. The first reason is that selection of a model ontology is typically driven by the questions of interest to the modeler. Models are a simplification of reality, so the modeler must choose which aspects of a system to keep and which to ignore. Often these choices are driven by economy as there is no reason to waste time and resources modeling phenomena that are not relevant to the question at hand. In fact, adding too many unnecessary concepts and details to a model may make it unusable. Revisiting the gas models discussed above, why use van der Waal's law of a gas when the ideal gas law is sufficient? Van der Waal's requires additional concepts and measurements that that may be unnecessary under many circumstances. This simplification is acceptable when it is applied appropriately. However, when one reuses models, particularly complicated models with many underlying, perhaps latent, assumptions, one can inadvertently apply the model outside its zone of validity. When one attempts to integrate two reused models, there is the added complication that models may have made inconsistent assumptions about the same underlying entities or relationships. In other words, the two models use different ontologies to describe the same referent.

Overcoming these types of ontological inconsistencies not only requires developing bridging strategies, but also finding a way to deconflict overlapping representations of the referent. For example, one approach is to introduce a "ghosted" object that can be used by multiple models, but the state of the object can only be updated by a single designated model [7].

The second reason that model ontologies can differ is due to limitations in the state of human knowledge. To use a deliberately extreme example, each scientific theory comes with its own ontology that describes relationships among a subset of phenomena. Classical mechanics defines concepts such as mass, energy, velocity, and momentum. In contrast, a common modeling approach in epidemiology defines the concepts susceptible, exposed, infectious, and recovered. While it is possible that these concepts from physics and epidemiology may be connected on some deeper level, it is not currently known how that should be done. While the gap in this example is obvious, in practical situations, the gaps can be more subtle. For example, there may be gaps in the current understanding of the relationships among certain micro and macro-economic phenomena.

In some cases, addressing incompatibilities among specific ontologies can be a major scientific activity. For example, a major goal of modern physics is to unify two of its major theories: quantum mechanics and general relativity. The two theories are based on two different ontologies, yet both are valid when used appropriately. On small scales where the strong, weak, and electromagnetic forces dominate, quantum mechanics is extremely accurate. On large scales, where gravity dominates, general relativity is extremely accurate. This situation is not problematic until one needs to model a system like a black hole that involves both small scales and high gravity. How these two ontologies should be integrated is one of the foremost unsolved questions in physics. While most applications of Model-Based Systems Engineering will not encounter this level of incompatibility among models, it has been argued that goal-dependence in ontologies is not merely a theoretical problem. It arises practice and may be fundamental [8].

Given the difficulties, why attempt to integrate heterogenous models at all? The first reason is that for many engineering problems, it is simply unavoidable. Like the quantum mechanics – general relatively example above, human knowledge is organized into ontologies, but the behavior of real systems often spans these ontologies. An aeronautical engineer is concerned with the structural, thermal, and aerodynamic aspects of an aircraft, and these may be modeled using separate ontologies. Unfortunately, they interact. A change in the aerodynamics may induce additional heating of aircraft components which can affect structural integrity.

The second reason is model reuse. As systems become larger and more complex, it becomes quite expensive to construct a system model from the ground up. Instead, it may be more cost effective to integrate existing models of systems components. But given that those models may have been developed with different questions and applications in mind, a certain amount of heterogeneity in the underlying ontologies is likely present and cannot be ignored.

To summarize:

- Models are considered heterogenous when there are substantial differences in the ontologies that underlay them.
- The combination of two valid models is not necessarily valid. The integrated model must preserve correspondence with the reference system.
- Integrating heterogenous models is dependent upon the questions of interest. An integration approach that is valid to answer one question may not be valid to answer another.
- Differences in underlying ontologies can arise from both limits in human knowledge and choices made to optimize models for specific applications.

The Levels of Conceptual Interoperability Model

To address ontological heterogeneity, it is useful to decompose the types of heterogeneity into a more refined set of categories. Diagnosis of the type of modeling heterogeneity encountered can point the modeler toward potential approaches to address it. A well-known model for categorizing model integration problems is the Levels of Conceptual Interoperability Model (LCIM) [9, 10].

The most recent version of the LCIM contains seven levels of interoperability that must be satisfied to successfully integrate simulation models [10]. While there is not a one-to-one correspondence between the layers of the model and types of ontologies, the layers describe concepts that must be addressed by the ontologies for the integration to be successful. The layers from the model are listed in Table 1. It is important to note that, strictly speaking, only the middle layers deal with model interoperability. While labeled interoperability, the upper layers actually address model composability. However, to remain consistent with the LCIM, all references to the layers will use the term interoperability for the remainder of the chapter.

Levels 1 and 2 address the requirements to exchange data between models. Level 1, Technical Interoperability, refers to the ability to use a physical connection to exchange data between models. Examples include networking protocols such as TCP/IP. Level 2, Syntactic Interoperability, means that there is an agreed upon grammar for data exchange among the models. Examples would be defined XML or JSON schema. As will be discussed below, approaches to deal with the first two layers are the most common and well developed. Those new to model integration

Table 1 The Levels of Conceptual Interoperability. (Adapted from [10])

Level 6	Conceptual interoperability
Level 5	Dynamic interoperability
Level 4	Pragmatic interoperability
Level 3	Semantic interoperability
Level 2	Syntactic interoperability
Level 1	Technical interoperability
Level 0	No interoperability

may mistakenly believe that Levels 1 and 2 describe the whole model integration problem. While addressing these levels is necessary, it is not sufficient.

Levels 3 and above impose additional ontological requirements on the integration. Level 3, Semantic Interoperability, requires that the there is agreement on what the exchanged data items mean. For example, a data exchange schema may include a value called "Pressure" that is to be exchanged between models. But are both models using the same definition of pressure? This is not as easy as it seems. For example, ISO 8625-1:2018, a standard for aerospace fluid systems defines 55 different types of pressure [11]. If the two models are using different definitions for pressure, the integrated model is likely to be invalid. Issues can also arise due to differences in methodological ontologies. For example, the SISO Standard for COTS Simulation Package Interoperability Reference Models identifies typical interoperability problems that can occur when integrating independent discrete event simulations [12]. Standards and reference models are examples of approaches to addressing Level 3.

Level 4, Pragmatic Interoperability, deals with the application context of the models. As noted previously, the question of interest and associated context has a major impact on how a model is constructed. Returning to the simple ideal gas law example, if one is asking about the relationship between pressure and temperature, is the gas in a rigid container of fixed volume or can the volume change? If one has an off-the-shelf gas model they wish to integrate, which context did the model creator assume? A failure to understand the context and questions a model was designed to answer can easily result in an invalid integrated model. Models that were constructed for similar contexts and questions are less likely to exhibit these issues. Approaches to addressing Level 4 involve referential ontologies and bridging mechanisms.

Level 5, Dynamic Interoperability, requires a consistent understanding of how the system states evolve over time. Continuing with the gas example, assume that one wants to integrate this simple model into a larger system model to track how the temperature of a system component will change as the system increases the pressure of the gas in the component. Pragmatic interoperability is achieved because the selected model appropriately assumes that the volume is fixed. However, this component will fail once a pressure threshold is exceeded. Consequently, the integrated model is initially valid, but as it increases the pressure past the failure point, it becomes invalid. It will show the component temperature continuing to rise. In the real system, the component would have long since failed. Thus, the integrated model lacks dynamic interoperability because it could not maintain correspondence with the reference system as the model state evolved. Such challenges are discussed in greater detail in Pennock and Gaffney [13]. Examples of approaches to address Level 5 involve referential ontologies and descriptive models that document dynamic state changes such as SysML™ [14] or Object-Process Methodology (ISO 19450) [15].

Level 6, Conceptual Interoperability, requires an understanding of the conceptualization used by each model. In the gas example, the ideal gas law implicitly conceptualizes a gas as a collection of point particles that do not interact. Van der Waal's equation, on the other hand, implicitly conceptualizes a gas as a collection of particles with volume. These particles can interact through collisions. Depending on the circumstances, these two conceptualizations can yield different predictions. Differences in conceptualizations used by models do not automatically preclude their integration, but any inconsistencies that adversely affect the correspondence of the integrated model with the reference system must be addressed. Conceptual interoperability issues can be the most challenging to resolve, and example approaches include partitioning and the introduction of fictional entities to enable bridging mechanisms. These will be discussed further in subsequent sections.

For an integrated model to be valid, the component models must be interoperable at all levels. When an effort to integrate models only addresses Levels 1 and 2 but is still successful, it only means that the upper levels were still satisfied implicitly. When models are built within the same domain, using standard methods, and for similar purposes, they are much more likely to use the same underlying implicit referential and methodological ontologies. Consequently, they are less likely to exhibit issues such as conceptual interoperability. However, as the degree of heterogeneity increases, the likelihood of these issues increases.

Broadly speaking, resolving interoperability issues at the upper levels is more challenging. While not a perfect correspondence, the upper levels more closely relate to referential ontologies and the lower levels more closely relate to methodological ontologies. Since addressing the upper levels is generally harder, most mature model integration approaches only address the lower levels. Addressing the upper levels is an area of active research, and the section "Bridging Mechanisms" will describe some of that research.

Best Practice Approaches

In order to integrate heterogenous models, it is necessary to identify relevant ontological differences among the models and address them. Approaches for doing so are dependent on the nature of these differences. As noted above, there are often multiple layers of ontologies that underlay any given model. Different ontological layers require different approaches to rectify differences.

This section will use the Levels of Conceptual Interoperability Model (LCIM) described in the section "The Levels of Conceptual Interoperability Model" to organize the approaches to integrate heterogenous models.

Model Integration Standards and Tools – Levels 1 and 2

There are several mature standards and tools that enable technical and syntactic interoperability to integrate heterogenous models. These standards and tools provide mechanisms to exchange data among models and coordinate execution. Ensuring interoperability for levels 3 and above is left to the model developer. Still these approaches provide crucial infrastructure for integrating computational models. This section describes representative examples of several common model integration frameworks that are used to construct integrated engineering models, but it is not intended to be a comprehensive list.

High Level Architecture (HLA) is an IEEE standard for building federated simulations and is documented in IEEE 1516–2010 [16]. As noted in the abstract for the standard, "The High Level Architecture (HLA)—Object Model Template (OMT) specification defines the format and syntax (but not content) of HLA object models" [17]. Federates are described via XML-based object models and can exchange data via a publish-subscribe mechanism. The HLA Run Time Infrastructure (RTI) provides a set of services to be used by the federated simulation including time management services to ensure that the correct time ordering of messages and interactions is maintained. While HLA was originally developed for federating military simulations, its application has expanded to a wide variety of domains [18].

The *Modelica Language*™ is an open access, object-oriented language used to integrate equation-based component models [19]. It is used by a number of different free and commercial modeling tools to implement models of physical systems where each component model may be from a different domain such as electrical, thermal, or mechanical. Consequently, it has seen widespread use in engineering applications. It is important to note that the Modelica Language focuses on specifying equations and is fundamentally acausal. While this limits its application to certain classes of model integration, it also avoids some of the upper-level interoperability issues.

Functional Mock-up Interface (FMI)™ is an open access standard that packages dynamic physical models for exchange and integration into simulations [20]. It is intended to support Model-Based Engineering through the integration of dynamic models of system components including both physical systems and control systems. Component models describe subsystems using differential or discrete time equations or algebraic equations. This enables an engineer to explore system design alternatives by integrating the component models. It is important to note that FMI only provides syntactic interoperability. Level 3 and above must be ensured by the model developer.

Simulink® is a widely used, commercial, MATLAB-based tool that can be used to integrate multi-domain simulations to support Model-Based Engineering [21]. It can also integrate models of physical and control systems that were developed using modeling tools other than MATLAB. As with FMI, it enables an engineer to explore system design alternatives by experimenting with models that describe different system components, but it also relies on the modeler to ensure Level 3 and above interoperability.

OpenMETA is an open-source model integration infrastructure that is specifically designed to support Model-Based Systems Engineering and Model-Based Engineering [22]. It can support design optimization, set-based design, and continuous integration, and it was designed to integrate existing engineering tools to achieve a "correct-by-construction" design that reduces design cycles and rework [23]. OpenMETA assists the modeler with establishing semantic (Level 3) interoperability by introducing the CyPhyML language to define model interactions and is combined with a semantic backplane.

ModelCenter™ is commercial tool for enabling data exchange among engineering tools and simulations [24]. It is vendor neutral and provides a drag and drop graphical interface for assembling integrated models into a workflow. It also provides visualization tools and decision support to enable Model-Based Engineering. ModelCenter can also integrate data from SysML modeling tools, which is of particular relevance to Model-Based Systems Engineering.

The *Discrete Event System Specification* (DEVS) is a methodological ontology for discrete event simulation [25, 26]. It formalizes the entities of a discrete event simulation as well as the relations among them. Zeigler, et al. [26] also introduce the Differential Equation Specified System (DESS) as a methodological ontology for differential equation-based models and use this in combination with DEVS to lay out the rules for integrating discrete event and differential-equation based simulation models. DEVS and DESS are the premier examples of using methodological ontologies to facilitate the integration of models that use different formalisms. This is sometimes called multi-formalism or multi-method modeling.

AnyLogic™ is a commercial, multi-method simulation tool that allows for the construction of models that combine the discrete event, agent-based, and system dynamics modeling paradigms [27]. Unlike most of the other Level 2 standards and infrastructures described in this section, AnyLogic is not intended to integrate independently developed models. Instead, it allows the model developer to construct a model where different objects in the model can be simulated using different formalisms. For example, one could construct a model that simulates individual humans using the agent-based formalism, and these agents make decisions based on macroeconomic variables that are simulated using the system dynamics formalism.

In general, the standards and tools described in this section facilitate data exchange among models, though each may emphasize addressing different aspects of the problem. Methodological ontologies such as DEVS provide some coverage of Level 3, semantic interoperability, as they standardize the way certain simulation objects are interpreted. However, they do not address all of Level 3 because they do not define and standardize domain specific entities such as temperature.

Historically, these tools and standards have yielded the most success with multi-physics modeling and integrating closely related engineering models. This may in part be because approaches to modeling these types of systems have fairly standardized domain ontologies and well understood relations among them. As one moves out of multi-physics toward biology, economics, and the social sciences, the domain ontologies become more variable and less standardized. Still as will be discussed in the section "Illustrative Example," even multi-physics modeling can encounter challenging interoperability problems.

Addressing Levels 3 Through 6

One could argue that ensuring interoperability for levels 3 through 6 is a fundamentally empirical problem because achieving it depends on the modeler's knowledge of the systems being represented as well as which representations and simplifications of those systems were chosen to implement any given model. Continuing with the simple gas model example, assume a modeler wants to integrate the previously described component model that assumes a fixed component volume under all circumstances. To detect the dynamic interoperability (Level 5) issue in the larger system model, they would have to know that 1) volume is a relevant variable, 2) the model assumes constant volume, and 3) that a sufficient increase of pressure could violate the constant volume assumption because the component ruptured in the real system. These pieces of knowledge are acquired empirically.

Use of a standard syntactic data exchange protocol as described in the previous section would not allow the modeler to deduce the pertinent pieces of knowledge. Even if the model has a published interface that only includes pressure as an input, the modeler cannot safely assume that volume is constant in the model. It may be that the volume is variable but determined endogenously through some other unknown mechanism. Consequently, it was never exposed in the interface.

To acquire the relevant pieces of knowledge to integrate the system model, the modeler must leverage some combination of internal knowledge, documentation, and experimentation. For example, the modeler may know through their own education in physics that volume is a relevant variable in the real system. Consequently, they check the model documentation to see if volume is assumed to be constant. If the documentation is absent or insufficient, they may run experiments using different combinations of inputs to try to infer what the model assumes about volume. While running experiments on the simple gas model might be feasible, running experiments on an extremely complicated model to infer internal assumptions can be a difficult and time-consuming activity. Once that is accomplished, additional experiments with the reference system may be required to establish the validity of the integrated system model.

Given the empirical nature of achieving the upper levels of interoperability, integrating heterogenous models requires leveraging two broad sources of knowledge: 1) previously acquired empirical knowledge captured in model documentation, standards, and ontologies; and 2) experimentation. Obviously, previously captured knowledge is preferred to experimentation. As Hofmann [3] observes, referential ontologies could, in principle address levels 3 through 6 simultaneously. However, this turns out to be challenging in practice. First, as one attempts to integrate models from a greater array of disciplines and application areas, the quantity of knowledge required becomes daunting. Second, referential ontologies cannot contain knowledge that has not been discovered yet. This is particularly problematic for complex systems where the knowledge of the component systems has been gained through isolating those systems. As a result, there may be unknown interaction effects in the integrated system. Third, as one moves away from physics toward the social sciences, the mapping between referential and methodological ontologies becomes more difficult [3]. This means that choosing the ontology for the referent depends on the question of interest.

Given these challenges, the state of practice for resolving upper-level interoperability issues is largely trial and error. This means upper-level interoperability issues are discovered over time through repeated testing and application. As experience is gained, an organization can create a stable set of well documented composable models that work well as long they continue to apply them in a similar manner. Still, there is always the risk that even a slight change in context (Level 4 – Pragmatic Interoperability) could lead to a loss of validity. In a survey of industries that use modeling and simulation extensively to support engineering design, it was found that these industries were cautious when reusing models for fear of inducing design flaws in the end product [28].

Keeping these caveats in mind, the following sections will discuss two common approaches for addressing upper-level interoperability issues, semantic ontologies and bridging mechanisms. Semantic ontologies capture accumulated knowledge to reduce the risks of Level 3 (Semantic Interoperability) problems. Bridging mechanisms are used to force the correspondence of the integrated model with the referent. They are often developed through trial and error and may be tuned to match experimental results from the reference system.

Semantic Ontologies

Semantic ontologies have a wide range of applications, and their development and use occupy an entire field of research. There are even specialized languages for constructing ontologies such as the Web Ontology Language (OWL) [29]. Consequently, it is not possible to provide a comprehensive review here. Instead, this section will focus on their relevance to modeling and simulation.

The role of a semantic ontology for facilitating model integration is to describe a reference system or domain as a standard set of objects along with standard definitions of those objects and the allowable relationships among them. They can also be used to document the provenance of models and datasets that can be used by those models. When models are developed using a shared semantic ontology, it can substantially reduce the risk of Level 3 interoperability problems.

For example, assume that a modeler has two models that they want to integrate. One has an output called atmospheric pressure, and the other has an input called atmospheric pressure. The modeler needs to feed the output atmospheric pressure from the first model into the associated input in the second model. If both models were developed using a shared semantic ontology and the model provenance verifies that, then it becomes much more likely that the two models are using the same definition of atmospheric pressure.

A real-world example of such an ontology is that used by the Global Change Information System (GCIS), which is maintained by the U.S. Global Change Research Program [30]. It organizes and links data sets, models, and research results that can be used to model and study climate change. Climate modeling, in general, is an example of an area where integrating heterogenous models has been successful. Winsberg [4] discusses the integration challenges faced by climate modeling at length. Successful climate research and associated integrated climate models have been accomplished through a great deal of trial and error. Ontology-based information systems such as the GCIS help to store and maintain some of that hard-won knowledge in a way that it can be used to develop future models and studies.

Another common type of semantic ontology is the domain specific modeling framework. These frameworks involve decomposing and standardizing modeling components to meet the needs of a particular domain or problem space. Examples include infrastructure modeling [31], logistics systems [32], healthcare [33], and space science missions [34]. What these frameworks are effectively doing is creating domain/problem specific reference ontologies. The hard-won empirical knowledge of what is important to integrating models for a given problem domain is encoded within the framework. As long as one integrates models that are consistent with a particular domain specific framework and applies those integrated models to the types of problems considered by the framework, the chances of successful integration should be much higher. However, this also means that these frameworks implicitly assume that the modeler ensures pragmatic interoperability by constraining the questions of interest and avoiding cross-domain integration.

Ontologies are so important to integrating heterogenous models, that their use to improve model composition has been examined extensively [3, 35–38]. Even so, there are a number of outstanding research questions about how they should be used when modeling complex systems [2]. As has been repeatedly emphasized throughout this chapter, ontologies can help address certain challenges to integrating heterogenous models, but they should not be viewed as a panacea.

Bridging Mechanisms

Once semantic differences among component models are understood, there remains the problem of resolving them. Furthermore, successful resolution of semantic differences would not address any outstanding Level 4 and above interoperability issues. One approach to accomplishing this is to refactor the component models using new, fully compatible ontologies. However, one could argue that doing so is tantamount to creating a whole new model rather than integrating heterogenous models. Furthermore, refactoring cannot address fundamental gaps in human knowledge or insurmountable computational issues. Some of these issues are discussed further in Pennock and Gaffney [13]. When refactoring is not a feasible or cost-effective way to proceed, the only option left to the modeler is to develop bridging mechanisms.

The concept of a bridging mechanisms is depicted in Fig. 4. The bridging mechanism addresses the incompatibility of the output of Model 1, X, with the required input for Model 2, Y, as depicted in Fig. 3. However, not just any bridging mechanism will do. The selected bridging mechanism must preserve correspondence with the referent, R. This is often where the difficulty of integrating heterogenous models lies, and the problem is compounded when one is attempting to bridge multiple models as the correspondence must be preserved when all the bridging mechanisms are imposed simultaneously. As an extension of the prior point that the combination of two valid models is not necessarily valid, the combination of multiple, valid, pair-wise bridging mechanisms is not necessarily valid. As a result, empirically based tuning of the bridging mechanisms may be required.

The difficulty of establishing bridging mechanisms can vary substantially depending on nature of the interoperability problem. At one extreme, a simple

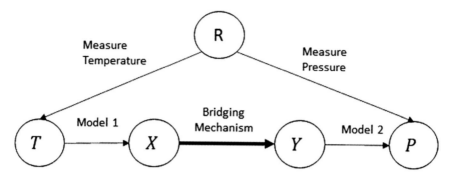

Fig. 4 The purpose of a bridging mechanism is to address the incompatibility of output X with input Y while still preserving correspondence with referent R

semantic (Level 3) issue such as a difference in units can be bridged through a simple unit conversion function. At the other extreme, bridging a conceptual (Level 6) interoperability issue may require the introduction of experimentally tuned "fictions" that have no basis in reality [4]. For example, Abraham, et al. [39] introduced fictious "silogen" atoms that have a mixture of properties of hydrogen and silicon to bridge a conceptual gap between quantum and molecular component models. Developing these types of bridging mechanisms can require a great deal of trial and error. This example will be discussed in greater detail in the section "Illustrative Example."

When attempting to develop bridging mechanisms, it is typically the most problematic difference among the ontologies that drives the categorization of the problem in the academic literature. For example, multi-method modeling involves integrating two or more models that employ different methodological ontologies such as integrating a discrete event simulation with a differential equation-based simulation. Multi-scale modeling involves integrating models with referential ontologies at different scales such as quantum and molecular, and multi-discipline modeling involves integrating models with referential ontologies from different disciplines such as structures and aerodynamics. Multi-scale modeling, in particular, has seen application across many domains. Examples include cancer biology [40], in-stent restenosis [41], chemical reactors [42], and chemical process equipment [43].

Several of these categories of bridging problems are considered research domains in and of themselves, each with its own body of literature. Consequently, if a modeler faces a bridging problem that falls into one of these areas, it is advisable to review the associated literature as there may be applicable bridging techniques that can be reused or adapted. Some representative examples of key areas include Multidisciplinary Optimization [44, 45], multi-scale modeling [46, 47], and multi-resolution modeling [48, 49, 50]. Within each of these areas, there can be many different techniques for bridging. A sampling of representative examples includes:

- Partitioned-domain methods [51]
- Heterogenous multi-scale method [52]

- Equation-free multi-scale method [53]
- Multiscale Modeling and Simulation Framework [54]

As Hoekstra, et al. [47] note regarding multi-scale modeling, bridging methods often depend on the application domain. Whether bridging mechanisms can be grouped into classes of generic methods is an open question. Consequently, concrete techniques for bridging mechanisms are often found described in the specific study or effort that developed them.

Illustrative Example

In this chapter, it has been asserted that ensuring conceptual interoperability is the most challenging aspect of integrating heterogenous models. This is because when two models rely on incompatible conceptual models, it is not always clear how to create bridging mechanisms that preserve correspondence with the referent. To illustrate the challenges as well as how they can be addressed, this section summarizes a successful effort by Abraham et al. [39] to build a multi-scale simulation to model fracture formation in silicon. This work was examined in depth by Winsberg [4], and any reader interested in addressing these types of conceptual interoperability issues is advised to review it.

It has long been recognized in physics that systems will exhibit qualitative differences in behavior at different spatial and temporal scales. As a result, different sets of abstractions are applicable at different scales. Thus, one may model a solid object as either a continuum or a discrete set of particles depending on the circumstances and question of interest. Each of these views leverages a different conceptual ontology of the solid. What makes the case of fracture formation in silicon interesting is that it cannot be successfully modeled using any one standard conceptualization.

The traditional way to model fractures in silicon is to use continuum mechanics. Thus, the underlying conceptual ontology treats the piece of silicon as continuous solid. The problem with this view is that it cannot capture the dynamic behavior of a fracture because the propagation of the fracture involves the interactions of atoms, which are not modeled. Taken even further, understanding the initial formation of the fracture involves quantum mechanical factors that would not be captured using a model of molecular interactions.

These circumstances might lead one to think that the whole problem can be modeled using quantum mechanics. However, this is impossible as modeling anything more than a small number of atoms using quantum mechanics is computationally intractable. Instead, to model the phenomenon, one must simultaneously consider linear-elastic theory, molecular dynamics, and quantum mechanics. The problem is that these three theories are inconsistent and incompatible. To make the simulation work, Abraham, et al. [39] had to develop bridging mechanisms to translate parameter values back and forth among the three views.

The first issue is that one cannot model the same thing using three different, incompatible conceptualizations. To address that issue, the piece of silicon must be partitioned into three different regions as depicted notionally in Fig. 5. The area where the fracture forms is modeled using a quantum tight binding (TB) approach. As the fracture gets larger, it moves into the area modeled using molecular dynamics (MD). Finally, when the fracture is large enough, it is modeled using a finite element (FE) approach to capture the continuum dynamic aspects. Thus, the reference system has been decomposed into three subsystems, each described by a different conceptual ontology.

This situation leads to the second issue. To model the fracture propagation, the interactions of these three regions must be modeled. In other words, one needs to model how a change in one region affects the adjacent region. This means that one must integrate and exchange data among the three conceptually incompatible models. A simple exchange of parameter values will not suffice because the three models do not use the same types of parameters nor do they even represent the same types of objects. In other words, the three models lack conceptual interoperability (Level 6). To resolve this situation, Abraham, et al. developed what they called "hand-shaking algorithms," but in this chapter would be termed bridging mechanisms.

To implement the bridging mechanisms, it was necessary to introduce two new partitions beyond those depicted in Fig. 5, one for each inter-model boundary. These new regions are depicted notionally in Fig. 6. The FE-MD region facilitates bridging the finite element and molecular dynamic regions, and MD-TB region facilitates bridging the molecular dynamic and quantum tight binding regions. An overarching Hamiltonian was defined for the combined model which is made up of separate Hamiltonians for each of the five regions. For the specifics, see [39, p. 541], but the idea is to ensure conservation of energy across the combined model.

Fig. 5 Notional view of Abraham, et al.'s approach to partitioning the silicon into regions modeled using different conceptual ontologies. The regions are not to scale

Fig. 6 Notional view of Abraham, et al.'s introduction of handshake regions. The regions are not to scale

To implement the FE-MD bridge, the FE mesh size was reduced to molecular scales in the handshake region, and each atom on the MD side can be associated with a displacement on the FE side. Displacements on the FE side and forces from the MD side were each given half weights in the FE-MD Hamiltonian. It is also important to note that the FE-MD region must be far from the TB region. Otherwise, it would not be possible to perform a one-to-one assignment of atom to displacement in the boundary region.

To implement the MD-TB bridge, it was necessary to deal with the "dangling" molecular bonds on the edge of the TB region. To accomplish this, fictitious "silogen" atoms were introduced on the boundary of the handshake region that have some properties of silicon and some properties of hydrogen. These tie off the dangling bonds and allow the Hamiltonians to update properly. Winsberg [4] makes the important observation that this type of bridging has a "semi-empirical" aspect. There is no such thing as a silogen atom, so it is not derived from or consistent with theory. However, it serves the purpose of passing state information between the incompatible views in a manner that makes the state transitions for both the MD and TB regions stay consistent with the real-world system. In other words, the bridging mechanism is semi-empirically formulated to preserve correspondence with the referent.

In this example, partitioning the block into three regions, each modeled with separate, conceptually inconsistent models means that relationships on the regional boundaries are lost. If not properly addressed, the combined model could achieve states that are not achievable in the real system. This means that the state restrictions must be built back in somehow. That is the role that these bridging mechanisms and "fictions" play. However, since they are not always derived from theory, they must be developed via trial and error and can be application specific. For example, how far is far enough to put the FE-MD boundary from the TB region? This must be resolved with experimentation. Furthermore, the use of silogen atoms on the MD-TB boundary would only work for this silicon application. Modeling fractures in another material will likely require a different fiction.

When all of the pieces are integrated together, this model can simulate dynamic fracture behavior in silicon more accurately than traditional approaches that only use the continuum dynamics paradigm. It is a prime example of how conceptual interoperability issues can be overcome to integrate heterogenous models to simulate phenomena that could not otherwise be modeled. However, the key takeaway is that overcoming upper-level interoperability issues can require solutions that are both empirical and application specific.

Expected Advances in the Future

In recent years, there have been substantial improvements in tools that enable technical and syntactic interoperability among models, and this trend is expected to continue as multiple industries push for enhanced digital engineering capabilities. At this point, advances in syntactic integration are less about new research and more about tool vendors implementing and adhering to standards for models and data exchanges.

To improve semantic and pragmatic interoperability, advances will depend more on communities of practice, industries, professional organizations, and standards organization than tool vendors. It is expected that such organizations will continue to develop and promulgate domain ontologies and domain specific languages to facilitate the integration of models that address typical problems within their respective domains.

Beyond standards, improvements in modeling practice should also facilitate the integration of heterogenous models. Key among these improvements is designing models for reusability. Increasingly, engineering model developers will adopt and adapt practices from software development where reuse has long been a central concern. While academic research and instruction will aid in the spread of design for reusability, communities of practice and professional organizations will likely play a significant role.

Another important practice is model curation. Advances in model curation will not only require engineering modelers to document and curate models but will also require approaches to store, manage, and discover models. Academic and industry research in this area is expected to improve these practices, and proper curation will enable modelers to better identify potential integration issues. While publicly curated repositories of engineering models would be ideal to facilitate model-based systems engineering, it seems likely that intellectual property concerns will limit sharing by private companies. Still, it is expected that the international scientific community will encourage and populate publicly available curated repositories that may be useful to practicing engineers.

For the foreseeable future, advances in establishing dynamic and conceptual interoperability will remain problematic and subject to significant trial and error. Much research remains to be done in this area. Possible advances include:

- The identification of useful classes of bridging mechanisms and associated rules for applying them
- Algorithms or heuristics that could be used to identify potentially conflicting portions models targeted for integration
- Model testing approaches to identify conceptual conflicts among models
- Application of machine learning techniques to develop bridging mechanism in high data environments

Overall, the engineering modeling community should continue to see improvements in the integration of physics-based models. This should be particularly true for industries and application domains where there is a favorable benefit/cost ratio to incentivize working through integration issues. Application domains with relatively stable modeling environments that are used frequently would be prime candidates.

However, as systems engineering becomes increasingly concerned with large-scale complex systems, the need to integrate concepts and models from human, social, and organization modeling will grow. Integrating such concepts with engineering models is expected to remain challenging for the foreseeable future.

Chapter Summary

Integrating heterogenous models is a persistently challenging problem for Model-Based Systems Engineering. When designing large-scale or complex systems, it is likely that system components will need to be modeled using many concepts and techniques from different domains and modeling paradigms. Certain engineering tasks such as performing design trades may require integration of these diverse models.

Heterogeneity among models arises from differences in the ontologies that underlay the models. Differences in ontologies can arise from both limits in human knowledge and choices made to optimize models for specific applications. While not all differences are relevant, some can inhibit model integration by preventing the correspondence of the integrated model with the reference system. The key point is that just because two models are valid when used independently, it does not automatically follow that the integration of those models is valid.

To successfully integrate heterogenous models, it is necessary to identify and address critical ontological differences. The nature of the differences determines the appropriate approach to do so. In this chapter, the LCIM was used to categorize ontological differences.

To address technical and syntactic interoperability issues, there are a wide range of model integration standards implemented in commercial and open-source tools. These enable the exchange of data among heterogenous models but are not always sufficient to ensure the validity of the integrated model.

Sematic interoperability issues can be mitigated using semantic ontologies. These ontologies define standardized entities and relationships. While in principle they can address interoperability levels above semantic, they are often domain specific.

Any interoperability issues that cannot be addressed through integration standards or semantic ontologies, must be addressed through bridging mechanisms. Bridging mechanisms enable the exchange of data among models while preserving correspondence with the reference system. However, they are often custom crafted for each integration, which limits their reusability.

Generally speaking, it is more challenging to address the upper levels of the LCIM than the lower levels. While there are well established methods for ensuring technical and syntactic interoperability, there are no standardized methods for ensuring semantic, pragmatic, dynamic, or conceptual interoperability among heterogenous models. Addressing these areas remains an area of open research.

Disclaimer

Approved for Public Release; Distribution Unlimited. Public Release Case Number 22–1519. The author's affiliation with The MITRE Corporation is provided for identification purposes only and is not intended to convey or imply MITRE's concurrence with, or support for, the positions, opinions, or viewpoints expressed by the author.

References

1. S. J. Taylor, A. Khan, K. L. Morse, A. Tolk, L. Yilmaz, J. Zander and P. J. Mosterman, "Grand challenges for modeling and simulation: simulation everywhere – from cyberinfrastructure to clouds to citizens," *Simulation,* vol. 91, no. 7, pp. 648–655, 2015.
2. R. Fujimoto, C. Bock, W. Chen, E. Page and J. H. Panchal, Eds., Research challenges in modeling and simulation for engineering complex systems., Berlin: Springer, 2017.
3. M. Hofmann, "Ontologies in Modeling and Simulation: An Espistemological Perspective," in *Ontology, Epistemology, and Teleology for Modeling and Simulationj,* A. Tolk, Ed., Heidelberg, Springer, 2013, pp. 59–87.
4. E. Winsberg, Science in the age of computer simulation, University of Chicago Press, 2010.
5. R. Rosen, Fundamentals of measurement and representation of natural systems, North Holland, 1978.
6. E. W. Weisel, R. R. Mielke and M. D. Petty, "Validity of Models and Classes of Models in Semantic Composability," in *Proceedings of the Fall 2003 Simulation Interoperability Workshop,* Orlando FL, 2003.
7. A. Tolk, "The elusiveness of simulation interoperability: what is different from other interoperability domains," in *2018 Winter Simulation Conference (WSC),* 2018.
8. D. Danks, "Goal-dependence in (scientific) ontology," *Synthese,* vol. 192, no. 11, pp. 3601–3616, 2015.
9. A. Tolk and J. A. Muguira, "The levels of conceptual interoperability model," in *Proceedings of the 2003 fall simulation interoperability workshop,* 2003.
10. W. Wang, A. Tolk and W. Wang, "The levels of conceptual interoperability model: applying systems engineering principles to M&S," in *Proceedings of the 2009 Spring Simulation Multiconference,* 2009.
11. ISO, *ISO 8625-1:2018 Aerospace – Fluid systems – Vocabulary – Part 1: General terms and definitions related to pressure,* Geneva, 2018.
12. SISO, *Standard for COTS Simulation Package Interoperability Reference Models (SISO-STD-006-2010),* SISO, 2010.
13. M. J. Pennock and C. Gaffney, "Managing epistemic uncertainty for multimodels of sociotechnical systems for decision support," *IEEE Systems Journal,* vol. 12, no. 1, pp. 184–195, 2016.
14. OMG, *OMG System Modeling Language (OMG SysML), version 1.6,* OMG, 2019.
15. D. Dori, "Object-process methodology," in *Encyclopedia of Knowledge Management,* 2nd ed., D. Schwartz and D. Te'eni, Eds., IGI Global, 2011, pp. 1208–1220.
16. IEEE Standards Association, *IEEE 1516-2010 – IEEE Standard for Modeling and Simulation (M&S) High Level Architecture (HLA)– Framework and Rules,* IEEE, 2010.
17. IEEE Standards Association, "IEEE 1516.2-2010 – IEEE Standard for Modeling and Simulation (M&S) High Level Architecture (HLA)– Object Model Template (OMT) Specification," 2010. [Online]. Available: https://standards.ieee.org/standard/1516_2-2010.html. [Accessed 7 December 2020].
18. A. Tolk, A. Harper and N. Mustafee, "Hybrid models as transdisciplinary research enablers," *European Journal of Operational Research,* 2020.
19. Modelica Association, "Modelica Language," 2020. [Online]. Available: https://www.modelica.org/modelicalanguage. [Accessed 7 December 2020].
20. Modelica Association, "Functional Mock-up Interface," 2020. [Online]. Available: https://fmi-standard.org/. [Accessed 7 December 2020].
21. MathWorks, "Simulink," 2020. [Online]. Available: https://www.mathworks.com/products/simulink.html. [Accessed 9 December 2020].
22. MetaMorph, "OpenMETA," 2020. [Online]. Available: https://openmeta.metamorphsoftware.com/. [Accessed 9 December 2020].
23. J. Sztipanovits, T. Bapty, S. Neema, L. Howard and E. Jackson, "OpenMETA: A model-and component-based design tool chain for cyber-physical systems," in *From programs to systems.*

The systems perspective in computing, S. Bensalem, Y. Lakhnech and A. Legay, Eds., Berlin, Springer, 2014, pp. 235–248.
24. Phoenix Integration, 2020. [Online]. Available: https://www.phoenix-int.com/. [Accessed 9 December 2020].
25. A. I. Concepcion and B. P. Zeigler, "DEVS formalism: A framework for hierarchical model development," *IEEE Transactions on Software Engineering,* vol. 14, no. 2, pp. 228–241, 1988.
26. B. P. Zeigler, H. Praehofer and T. G. Kim, Theory of Modeling and Simulation, 2nd ed., Amsterdam: Academic Press, 2000.
27. AnyLogic, 2020. [Online]. Available: https://www.anylogic.com/. [Accessed 10 December 2020].
28. F. E. Mullen, "Dynamic multilevel modeling framework phase – Feasibility," U.S. Department of Defense, Modeling and Simulation Coordination Office, Washington, DC, 2013.
29. W3C, "OWL," 2012. [Online]. Available: https://www.w3.org/OWL/. [Accessed 12 December 2020].
30. USGCRP, 2020. [Online]. Available: https://data.globalchange.gov/. [Accessed 12 December 2020].
31. P. T. Grogan and O. L. de Weck, "The ISoS modeling framework for infrastructure systems simulation," *IEEE Systems Journal,* vol. 9, no. 4, pp. 1139–1150, 2015.
32. T. Sprock and L. F. McGinnis, "Simulation model generation of discrete event logistics systems (DELS) using software design patterns," in *Proceedings of the Winter Simulation Conference 2014,* 2014.
33. B. P. Zeigler, "Discrete event system specification framework for self-improving healthcare service systems," *IEEE Systems Journal,* vol. 12, no. 1, pp. 196–207, 2016.
34. T. J. Bayer, M. Bennett, C. L. Delp, D. Dvorak, J. S. Jenkins and S. Mandutianu, "Update-concept of operations for Integrated Model-Centric Engineering at JPL," in *2011 Aerospace Conference,* 2011.
35. M. Hofmann, J. Palii and G. Mihelcic, "Epistemic and normative aspects of ontologies in modelling and simulation," *Journal of Simulation,* vol. 5, no. 3, pp. 135–146, 2011.
36. L. McGinnis, E. Huang, K. S. Kwon and V. Ustun, "Ontologies and simulation: a practical approach," *Journal of Simulation,* vol. 5, no. 3, pp. 190–201, 2011.
37. C. Partridge, A. Mitchell and S. de Cesare, "Guidelines for developing ontological architectures in modelling and simulation," in *Ontology, Epistemology, and Teleology for Modeling and Simulation,* A. Tolk, Ed., Berlin, Springer, 2013, pp. 27–57.
38. A. Tolk and J. A. Miller, "Enhancing simulation composability and interoperability using conceptual/semantic/ontological models," *Journal of Simulation,* vol. 5, no. 3, pp. 133–134, 2011.
39. F. F. Abraham, J. Q. Broughton, N. Bernstein and E. Kaxiras, "Spanning the length scales in dynamic simulation," *Computers in Physics,* vol. 12, no. 6, pp. 538–546, 1998.
40. Y. Liu, J. Purvis, A. Shih, J. Weinstein, N. Agrawal and R. Radhakrishnan, "A multiscale computational approach to dissect early events in the Erb family receptor mediated activation, differential signaling, and relevance to oncogenic transformations," *Annals of biomedical engineering,* vol. 35, no. 6, pp. 1012–1025, 2007.
41. A. Caiazzo, D. Evans, J. L. Falcone, J. Hegewald, E. Lorenz, B. Stahl, D. Wang, J. Bernsdorf, B. Chopard, J. Gunn and R. Hose, "A complex automata approach for in-stent restenosis: two-dimensional multiscale modelling and simulations," *Journal of Computational Science,* vol. 2, no. 1, pp. 9–17, 2011.
42. D. G. Vlachos, "Multiscale integration hybrid algorithms for homogeneous–heterogeneous reactors," *AIChE Journal,* vol. 43, no. 11, pp. 3031–3041, 1997.
43. A. Yang and W. Marquardt, "An ontological conceptualization of multiscale models," *Computers & Chemical Engineering,* vol. 33, no. 4, pp. 822–837, 2009.
44. J. Agte, O. De Weck, J. Sobieszczanski-Sobieski, P. Arendsen, A. Morris and M. Spieck, "MDO: assessment and direction for advancement—an opinion of one international group," *Structural and Multidisciplinary Optimization,* vol. 40, p. 17–33, 2010.

45. J. R. Martins and A. B. Lambe, "Multidisciplinary design optimization: a survey of architectures," *AIAA Journal,* vol. 51, no. 9, pp. 2049–2075, 2013.
46. E. Weinan, Principles of multiscale modeling, Cambridge University Press, 2011.
47. A. Hoekstra, B. Chopard and P. Coveney, "Multiscale modelling and simulation: a position paper," *Philosophical Transactions of the Royal Society A: Mathematical, Physical and Engineering Sciences,* vol. 372, 2014.
48. P. K. Davis and J. H. Bigelow, "Experiments in multiresolution modeling (MRM)," RAND Corporation, Santa Monica, 1998.
49. P. K. Davis and A. Tolk, "Observations on new developments in composability and multi-resolution modeling," in *2007 Winter Simulation Conference,* 2007.
50. A. Kunoth, "Multiresolution Methods," in *Encyclopedia of Applied and Computational Mathematics,* B. Engquist, Ed., Berlin, Springer, 2015.
51. W. A. Curtin and R. E. Miller, "Atomistic/continuum coupling in computational materials science," *Modelling and simulation in materials science and engineering,* vol. 11, no. 3, 2003.
52. E. Weinan, B. Engquist, X. Li, W. Ren and E. Vanden-Eijnden, "The heterogeneous multiscale method: A review," *Communications in Computational Physics,* vol. 2, no. 3, pp. 367–450, 2007.
53. J. Dada, O and P. Mendes, "Multi-scale modelling and simulation in systems biology," *Integrative Biology,* vol. 3, no. 2, pp. 86–96, 2011.
54. B. Chopard, J. Borgdorff and A. G. Hoekstra, "A framework for multi-scale modelling," *Philosophical Transactions of the Royal Society A: Mathematical, Physical and Engineering Sciences,* vol. 372, 2014.

Michael Pennock is a Principal Systems Engineer at the MITRE Corporation where he specializes in digital engineering. Previously, Michael was a faculty member in the School of Systems and Enterprises at the Stevens Institute of Technology where he worked to create new approaches to design and evolve large-scale systems that consist of interacting engineered and social components. His research interests focus on understanding the issues associated with the computational modeling of socio-technical systems and systems of systems in the national security and health care domains. Michael has also worked as a senior systems engineer in various lead technical roles for the Northrop Grumman Corporation. He holds a Ph.D. in Industrial Engineering from the Georgia Institute of Technology and Bachelor's and Master's degrees in Systems Engineering from the University of Virginia.

Improving System Architecture Decisions by Integrating Human System Integration Extensions into Model-Based Systems Engineering

14

D. W. Orellana

Contents

Introduction	442
State of the Art	443
Human System Integration Ontology	447
HSI Ontology – Mechanisms	447
Best Practice Approach	452
Develop or Extend a Human System Ontology	454
Stand Up a Model-Based Environment	454
Create System Diagrams and Export Task Allocation and Workflow for Analysis	455
Evaluate Results, Build Alternatives, and Evaluate Alternatives	456
Illustrative Examples	456
Case Study System Overview	456
Ontology Extension or Modifications	456
Architecture and Analysis	456
The Human System Integration Analysis	459
Results of Analysis	460
Chapter Summary	466
Cross-References	467
References	467

Abstract

Model-based systems engineering (MBSE) plays an increasingly important role in the development of complex systems. Currently, systems architecture models (e.g., descriptive SysML models) have focused more on depicting machine interactions with little consideration for human characteristics that are needed to make holistic architectural decisions. This chapter describes a human system integration (HSI) extension which facilitates integration of system architecture models with human task models. This integration allows tighter coupling between

---

D. W. Orellana (✉)
ManTech International Corporation, Los Angeles, CA, USA
e-mail: douglas.orellana@mantech.com

© Springer Nature Switzerland AG 2023
A. M. Madni et al. (eds.), *Handbook of Model-Based Systems Engineering*,
https://doi.org/10.1007/978-3-030-93582-5_27

system architecture and analysis with a human agent. It also presents an ontology broker for tool integration. The ontology broker supports information scalability captured in the modeling ecosystem when making architectural decisions. A case study of an unmanned aerial system and an image analyst assesses whether architectural decisions resulting from tighter integration can improve the human-system performance. The results of the study show that architectural changes made and subsequent analysis of the human-system performance produce superior analysis by reducing analyst workload, eliminating bottlenecks, and achieving overall improvement in how the human analyst interacts with the system.

Keywords

Systems engineering · Human system integration · Model-based systems engineering · Model-based engineering · Human machine interface · Human performance models · Workload analysis · System functional analysis

Introduction

Model-based systems engineering (MBSE) plays an increasingly important role in the development of complex systems. Although engineers have used models for centuries and humans use mental models on a daily basis, using descriptive and analytical models in systems architecting and systems engineering has only been in the forefront of conversation within the last decade.

As systems evolved into more complex entities, the number and types of tools needed for understanding them have also increased. As a result, each tool has its own set of rules and lexical semantics. Lexical semantics define the domain-specific meaning of terminology and relations. As more tools are needed for understanding system complexity, so has the need for increasing and formalizing modeling semantics. Since its inception in 2007, the System Modeling Language (SysML) has become the de-facto descriptive modeling language used by system architects and system engineers. As more descriptive models continue to integrate with analytical models, the opportunity to integrate other viewpoints into the system model increases. Viewpoints specify rules for building and using views based on a stakeholder. SysML has no formal rules on how to model a human element and human considerations. This chapter adds a human viewpoint by extending current modeling semantics.

Today, with the changing role of the human from system operator to human agent within the system [1], and with the need for greater system adaptability, greater importance is placed on designing human-machine interfaces that facilitate human-machine interactions. A human agent carries authority within the system to act on the system's behalf to decide appropriate responses in accordance with the input. To decide the appropriate response, the human has to have a level of trust that is dependent on the consistency of the machine response to the human [2]. As such, architects must be able to build a trust through the transparency and consistency of the human system interface [3, 4]. Accomplishing these objectives requires explicit accounting of the human element needs accompanied by appropriate models that support human-machine trade-offs analysis. In sharp contrast, current systems engineering practices address human system integration as an afterthought (i.e., after architectures have been specified). In this situation, as changes to the system accumulate, redesign costs can spiral out of control. Unfortunately, people not trained in human system integration (HSI) cannot communicate with HSI engineers. In large part this is due to differences in understanding human considerations in the system design. Even within the HSI and human factor communities, there is no general agreement with respect to common terminology for human considerations. Human-system collaborations necessitate extending current systems engineering and integration of tools necessary for evaluating human-system dynamics. Coupled system and human analysis allows for analyzing the human element holistically within the system, especially during the architecting phase of system development. According to the International Council of Systems Engineering (INCOSE), the architecting phase is where up to 70% of systems costs get allocated [5].

State of the Art

Modeling is a standard practice in engineering. System architecting and engineering employ models for representing complex systems to assist understanding, reduce inconsistencies among model elements, enable reasoning, and enable trade space evaluation for system design synthesis.

The system model is developed with various viewpoints (lenses). Each viewpoint is expressed by model views that address stakeholder concerns. The integration of views, viewpoints, and analysis results enables the creation of a system picture that addresses all the concerns expressed in the viewpoints [6]. [6] looks at providing an effective way to express the viewpoints of the stakeholders by building views in SysML that represent stakeholder stories and are executed through a 3D virtual world where the various concerns can be played out through the game engine. Playing out the stakeholder lens allows the larger stakeholder community to have transparency and enable collaboration and understanding of the concerns.

Even without using a standard tool or pen and paper, engineers depend on cognitive models for better understanding their domains. Cognitive models are descriptive accounts of concepts, typically focused on understanding the processes of a concern. Cognitive models are powerful tools for understanding an approach to solving analytical problems and building the knowledge base of the problem [7, 8]. [7] uses cognitive models to better understand human considerations underlying the design and acceptance of expert systems. Through a system user taxonomy, based on knowledge of the system, [7] uses an inside-out/outside-in approach for exploring the human consideration through the processes users would execute. [8] builds on the cognitive model approach and uses a network-based environment to analyze task allocation between humans and machine. Network-based task simulation provides a more rigorous way to allocate functions, translating cognitive models to more formal and standardized models. Formal models are necessary as system complexity increases to better capture the different facets of complexity [9]: scale, diversity, connectivity, and optimization. In particular the architect can track elegant system characteristics and metrics [10] for making better trade system alternatives [11–13].

[10] describes a heuristic-driven process for elegant system design. As part of this process, it identifies 12 elegant system characteristics that could be used to assess system for elegance: purposivity, parsimony, transparency, scalability, bonding, efficiency, evolvability, affordability, usability, utility/impact, and predictability. The 12 characteristics are compiled from different viewpoints that present stakeholder concerns. Some of these concerns that apply to elegant system such as adaptability, affordability, and utility are also characteristics that are concern when engineering resilient systems. [11] discusses engineering resilient systems processes and tools and how to best integrate viewpoints together to inform your trades. These trades are dependent on being able to assemble a library of models together to assess for resilience. In a similar fashion, as you look to assess for human consideration, assembling models together will allow to analyze overall system performance through the combination of machine and human models.

Human-centered model-based systems engineering focuses on explicit representations of human actions from multiple perspectives. Given there is no explicit way to represent these perspectives in current SysML-based architecture modeling, there is a need to increase modeling capability by extending semantics and syntax, semantics being an agreed meaning for terms and syntax being a set of rules by which the terms relate. Extending semantic and syntax for new domains has been the cornerstone of the modeling community. SysML itself is an extension of the Unified Modeling Language (UML) [14]. Before SysML was developed, system architects and engineers attempted to use UML for system definition [15]. SysML made it easier for system architect and engineers to model systems. These diagrams allow the system architect and engineer to capture and allocate functions and tasks to the system elements. SysML is a good foundation for the systems engineering community to build upon and extend [16]. By considering the human actions in new semantics and syntax, you increase the ability to assess machine and human behavior through a systematic approach, relying on understanding of human characteristics [17]. [18] discusses how various aspects of HSI are applied across system design life cycle from requirement definition through decommission of systems. During requirement definition, [18] describes the need to interview the end user for expectation and concerns. Through design phases, various psychological and physiological considerations need to be taken into account (e.g., surrounding environment) as well as the interfaces being used between the machine and the human. Then there is training where the goal is to have the human have a level of understanding of the machine in which they can build trust and familiarity of use. Although [18] explains the importance of the human consideration across the life cycle, there is no single holistic approach that considers HSI as an integral system element. Importantly, modeling and simulation can provide the basis of creating a convenient workspace for conducting HSI trade-offs necessary during system development [19]. Unlike current approaches to couple SysML with Matlab and Simulink [20] or SysML to Modelica [21] to understand mechanical dynamics and physical-based phenomena, current HSI modeling tools are independent of the architecture process and the decision-making in the conceptual design of the system. HSI modeling tools include Micro Saint Sharp and IMPRINT for task analysis, ACT-R and Cognitecture™ for agent-based simulation focused on cognitive models to understand human behavior, Jack for ergonomic considerations, and many others. Consequently, any major design decision after the architecture has been laid out and started to be implemented can potentially have significant monetary implications.

Until recently, building common semantics and syntax for HSI has not had much traction within the HSI community or the systems engineering community. There have been two attempts [22, 23] of building partial constructs to be used in the architecting and engineering of systems, but neither has come to fruition or been fully implemented or used.

In an attempt to fulfill the gap in architecting humans into system, the Integrated Definition (IDEF) methods community developed IDEF 8, Human-System Interaction Design Method. IDEF 8 was created to look at three different levels of human system interaction: (1) overall system operation, (2) role-centered scenario of system use, and (3) design objectives implemented through a library of metaphors used as best practices for detailed design [22]. The overall system operation in IDEF 8 consists of representing the operational concept (OPSCON) through a set of models and text that focus on the use of the machine by the user in a series of tasks done by the user and machine. The role-centered scenarios go into more detail identifying inputs and outputs from the machine and user as well as specific sequences that lead to corner case usage of the system. Typically, the inputs and outputs identified in the scenarios are type of information being conveyed by the user to the machine or vice versa. The metaphor library is a common set of templates that define objects, operations, and human gestures for an interaction. For example, a button would have an operation select, and typically a human using a computer might point, click, hold, and release. These templates are meant to begin to lay out design decisions for input-output devices to be used by the human.

Recognizing the need for human viewpoint of the system, the North Atlantic Treaty Organization (NATO) undertook an effort that examines ways for better evaluating human system compatibility and created human views that document the unique implications of humans in system design [23]. Both attempts of human views did not focus on system architecting and system development but rather focused on lower-level design detail and acquisition, respectively. Adding human views does present an opportunity to include HSI constructs, providing a suitable mechanism for decision-making to ensure that this significant cost driver is addressed upfront. Although the NATO human views are used for acquisition, upfront understanding manning and skills needed are showing value enough for the Department of Defense to transition its use of human views [24].

More recently, systems engineers create SysML diagrams that explicitly represent user interfaces and humans as subsystems. User interfaces and human behavior are decomposed into model components based on the senses: visual, auditory, and touch

[25]. Decomposition and allocation provide a useful mechanism for communication among engineers; however, there can be substantial differences in model completeness when there are loose metamodels for guidance. The stricter the semantics and syntax can be implemented, the easier it is to provide consistency from one model to another. Semantics and syntax need to evolve over time to address new challenges so as you enforce semantic and syntax there still needs to be some inherent balance of adaptability to be able to extend to new problem sets. These efforts though provide a path toward analyzing and identifying functional allocation and opportunities for automation.

Through architectural definition and design, machine and human performances are analyzed to optimize the system, and trade-off must be made in how functions are allocated across the system. Functional allocation includes making decisions on which function may be automated where humans do not wish to perform or cannot perform the function [26]. As technology keeps improving, automated systems are more responsive to the humans they interact with [27]. Managing these interactions requires determining the correct level of automation that minimizes human performance costs [26]. [26] proposes four types of functions that an architect could choose to automate: (1) information acquisition, (2) information analysis, (3) decision and action selection, and (4) action implementation. These four types of function are correlated to the cognitive four-stage model of human processing [26] identifies: sensory processing, perception, decision-making, and response selection. Choosing the right level of automation for desired human functions during requirement analysis and functional analysis is important, and architects must survey the functions and capabilities the machine must do and begin allocating those functions between the machine and the human. [26] analyzes the functions at each stage of the human processing model and lays out heuristics to analyze and choose the level of automation from the machine does everything to the machine does nothing providing an automation framework that can be used as one architects. Although [26] also proposes to tie these decisions through more concrete criteria such as consequences, reliability, and cost, [28] uses a computational simulation framework that provides objective measures for allocating functions between human and machines. The computational simulation framework, Work Models that Compute (WMC), integrates three types of models: resource models that capture both physical and information resources, action models that capture tasking to perform a scenario successfully, and agent models that represent machines and humans that can do the tasks laid out in the action models. The agent models contain constraints attached to both machine and human such as task saturation limits, delays, etc. Through assembly of these models and running through a simulation engine, WMC can assess the functional allocation given system performance measures. The use of WMC highlights the need for ensuring that architecture models are integrated with dynamic models to provide the necessary quantitative measures for making architectural trades.

Human System Integration Ontology

An absolute and integral part of any design is proper communication among all stakeholders in the development process [17]. The HSI ontology becomes an integral part of human and machine consideration integration so that architect's human specialty engineers can understand each other's models. An ontology is a structure of concepts (entities) and logical relationships. It captures the vocabulary being used by given consistent definitions to entities (semantics), how to express the relationship between entities (syntax), and the usage of the entities given different contexts (context sensitivity), where context is the scope in which a concept is understood. Due to their data-driven focus, ontologies help model-based environments with an inherent focus on entities and data instances captured in the modeling tools. Models are driven by strict axioms to be useful, and ontologies provide a method for those axioms to be explored and defined, whether it is the ontology of mathematical concepts like differential equations (which are driven by known physical rule sets) or modeling languages such as SysML (that provide a very simplistic view for meta-object relationships within systems engineering (i.e., a block is composed of parts)).

There are several areas that affect HSI; the HSI ontology spans various areas within the framework of the system modeling pillars and other considerations that will give a more holistic system view with the perspective of the human. Collectively, these factors will provide the semantic underpinnings for defining and managing the human element within the mission and system context. A unified view of these factors is presented in the top-level HSI ontology (Fig. 1). The HSI ontology tries to best communicate the unified view through the use of the UML profile diagram. A profile diagram is typically used to extend UML and SysML with domain-specific notation. The profile diagram is a structural diagram that describes extension where ontology classes can be represented as stereotypes and identify its relationships to one another.

The HSI ontology offers a unifying means of concerns and expectations of the human element. Considering the various factors in these areas can proceed to increase communication between system architects, engineers, and human factors/human system specialists. Specifically, the HSI ontology provides the building blocks to bring human element considerations upfront versus relegating it to the detailed design phase. The HSI ontology informs us that HSI is composed of mechanisms, requirements, human agents, behavior, structure, and parametric constructs. The HSI ontology can guide the HSI processes and facilitate communication among stakeholders by allowing a better representation of the man-machine system into a descriptive model that will drive analytical models. The key concepts represented in Fig. 1 are summarized in the following subsections.

HSI Ontology – Mechanisms

The mechanism portion of the HSI ontology is focused on the human system integration processes, procedures, tests, and verification required. This view of the

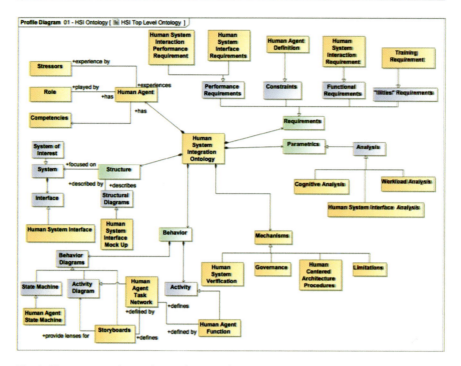

Fig. 1 Human system integration top layer ontology

ontology focuses on assuring appropriate precautions have been taken for integrating the human element into the overall system. The integration procedures attempt to circumvent adverse effects and failures between components [29] with a specific focus on effects produced by the human agent as well as effects produced by that impact of the human agent.

HSI Ontology – Requirements

During requirement exploration one of the key steps in understanding human requirements is to interview potential human agents to better understand their roles: determine requirements for skills of roles that allow better selection of personnel [17]. The roles played by human agents are captured during the development of concept of operations where scenarios are defined through a series of tasks the system must accomplish to achieve its objective, including environmental effects it must operate in and external interfaces it is constrained to. As the scenarios are built, interviews of human agents with skills sets that match the tasks described in the scenarios can be conducted to begin to understand the constraints and limitations a human agent can have through the scenarios (Fig. 2).

In order to integrate the human agent into the system, more attention in the specification of the human agent and HSI must be specified at all levels of abstraction. The ontology attempts to explicitly highlight these human-centered

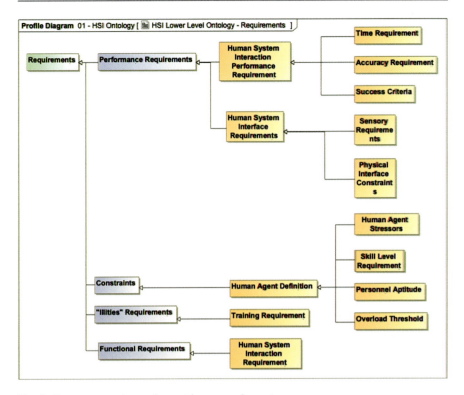

Fig. 2 Human system integration ontology – requirements

requirements in the modeling environment as you would highlight any other system functional and performance requirement. By explicitly highlighting these requirements, it will be easier to trace the requirements to the other aspects of the modeling pillars: behavior, structure, and parametric. The written requirements complement the system model, as to overcome limitations on inferring what is not explicitly modeled in the system model [30].

HSI Ontology – Human Agent

In order to optimize the system, the machine should complement the human agent and match human characteristics to the agent functions and performance needs [31]. It is equally important to specify human agent characteristics as well as other system agents. The human agent characteristics should include, but not be limited to, physical traits, cognitive limitations, sensory performance, and social factors (Fig. 3).

The human agent extends the block and actor types in SysML to better specify human agent as a component in the system under development. By using the block and actor base types, the human agent type inherits the ability to be described in a use case diagram but at the same time be further defined and characteristics and ownership of operations (functions). In this area, the human agent will have played

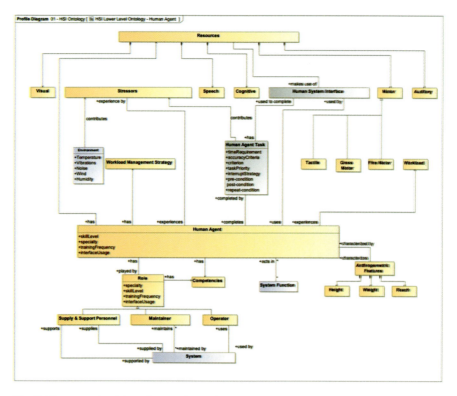

Fig. 3 Human system integration ontology – human agent

a certain role in the system operations and system capabilities. Along with this role, a set of constraints specify the strengths and weaknesses of the human agent using a minimum skill set required for the role played by the human agent. The definition of roles and skill sets should help the system architects better match the human agent to the role the human plays in overall system performance. At the organizational level, it will also facilitate skill gaps analysis using design science research methods (DSRM) [32] which quantitively determines skill gaps and in turn encourages organization upskilling or further training for better system optimization. DSRM looks at the skill gap by analyzing what a role requires to what a human agent brings. It runs through a six-step process that identifies the problem skill gap, defines the objective for solution, develops the solution, demonstrates the solutions, evaluates solution, and then results out to the organization. During the evaluation phase, [32] runs the solution with a group of human agents that play a role to get accurate measures on whether or not solutions to the skill gaps required work.

HSI Ontology – Behavior

Modeling behavior using SysML and other object-oriented modeling languages is based on use cases and use case scenarios. A use case represents a set of actions

14 Improving System Architecture Decisions by Integrating Human System... 451

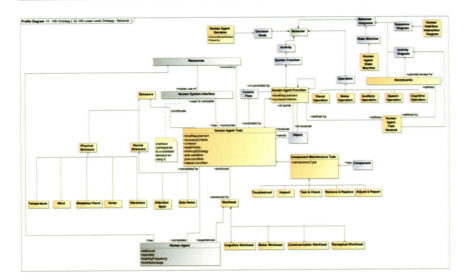

Fig. 4 Human system integration ontology – behavior

defining the interaction between roles and the system under interest. Use case scenarios are one path through actions represented by the use cases. Both concepts attempt to capture system usage through high-level interactions of system stakeholders and actors with the system. These use cases and use case scenarios enhance written requirements by refining the requirements to create a descriptive model. Not only does the requirement refinement describe the interactions, but it also shows external visible exchanges, explores user expectations, and defines intended purpose of system usage (Fig. 4).

The use cases are further refined through activity diagrams and sequence diagrams. Activity diagrams are well suited for explaining task flow, while the sequence diagram can depict a stronger step by step sequence as it relates to the interactions between machine and human (e.g., using fine motor skills for a controller). In both of these artifacts, it is important to explicitly detail which functions, operations, and accompanying attributes can potentially enhance the analysis of the human element. When a decision cannot be made functions, operations and attributes can be left unallocated and treated as area of concern within the model for further analysis. The functions, operations, and attributes need to allow for ease of transition from conceptual architecture to detailed design, allowing for these same identified functional definitions to be used in the HSI tools for further analysis. By having this level of integration, aspects that are analyzed in detailed design are considered upfront to account for the human element impact on the overall architecture, not just the performance of the system.

State machines are used to describe behavior in event-driven form. The events identified in this artifact can occur in one of the states or can drive a transition from one state to another. As in the case with machines, aspects of humans can be described using state machines, for example, depicting cognitive states for when a human should use short-term memory versus long-term memory. By creating specific state machines for the human element, the architect specifies certain behavior, which the human element must exhibit in response to certain events. This level of detail will inform the training necessary for the human agent to recall the actions needed to take given events.

HSI Ontology – Structure

The structural diagrams describe the system structure through blocks and parts. Within the framework, any system object can be defined using the block object. In a similar manner, the HSI ontology will be able to extend the semantics and syntax used in the structural diagrams to describe the human agent as well as human system interfaces. These extended semantics and syntax allow these two concepts to be considered upfront and closely tied to the top-level requirements. The ontology also extends the attributes and parameters looked at in this context (Fig. 5).

HSI Ontology – Parametrics

The parametric diagrams support engineering analysis of critical system parameters (often the measure of effectiveness and measure of performance). The evaluation of these metrics pertains to performance, physical characteristics, and "illities." The parametric pillar has not been used much until recently. One of the biggest problems with parametrics has been the limited options native to SysML tools. Fortunately, that problem has been reduced through the integration with external tools such as Model Center, Matlab, Simulink, and others as seen in [33, 34], where they traced and link mission and performance metrics of cube satellites to analytical models. Similarly, workload analysis, task analysis, and other HSI analysis could be traced and linked to the human aspects of the system model as will be shown in section "Illustrative Examples" (Fig. 6).

Best Practice Approach

We take a four-step approach for investigating human-system interaction based on the Shadow UAV system [35, 36]. The process comprises of the following:

1. Analyzing the overall system architecting process and human analysis to develop and extend a human system integration ontology.
2. Develop a model-based environment that includes the tools needed to architect, integrate, and analyze for human system performance. The environment includes profiles or schemas to extend tools' ability to build system artifacts representation of human system semantics and syntax.
3. Once the foundational pieces are in place, SysML diagrams are created with new semantics and syntax described in the ontology to show how pre-existing system-centered architecture can benefit from tight coupling to task analysis performed in IMPRINT. SysML was chosen to build out the descriptive model due to its

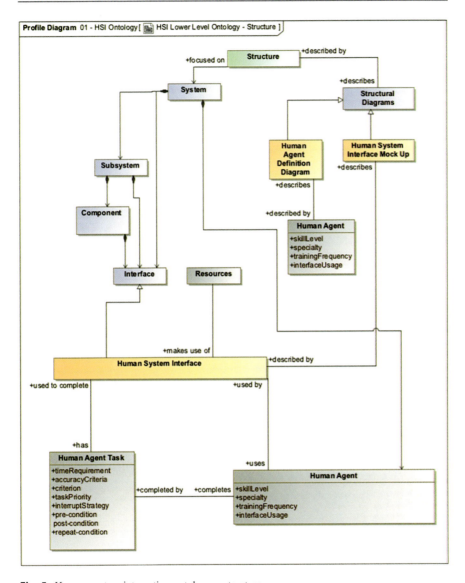

Fig. 5 Human system integration ontology – structure

widespread use throughout the DoD, its wide support in the tools, and using any other language would have incurred other integration challenges.

4. After the architectures are laid out, alternatives are evaluated to assess their impact to human and machine. Adjustments are made to the architecture and assess for impact.

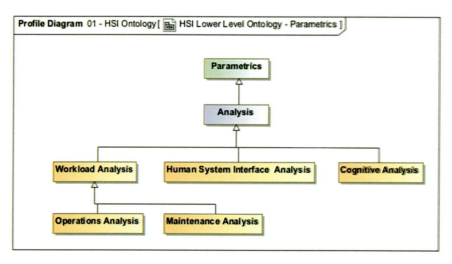

Fig. 6 Human system integration ontology – parametrics

Develop or Extend a Human System Ontology

Section "State of the Art" describes a human system ontology that was built by analyzing current system architecting process and task-based analysis. The HSI ontology is meant to evolve over time and is expected to be extended and modified beyond its current use.

Stand Up a Model-Based Environment

Establishing a model-based environment is crucial to creating a seamless thread between architecture tools and HSI tools. The human-centered model-based environment used in section "Illustrative Examples" consists of Cameo EA, an ontology broker, and IMPRINT. Cameo EA, a systems architecture tool, was used to develop the descriptive models using multiple language profiles: SysML and Unified Profile for DODAF and MODAF (UPDM). Descriptive models are graphical-based models describing real-world events and the relationships between factors responsible for them. In order to bring in the human considerations into the descriptive modeling efforts, the HSI ontology is implemented within a Cameo EA domain-specific language profile that is used in conjunction with the main language profiles (Fig. 7).

Once the HSI profile is implemented and the system model is developed using the semantics and syntax the ontology defines, the modeling efforts from traditional systems engineering and human system integration efforts are integrated. In order to integrate two traditional tools, the HSI ontology is leveraged to develop the ontology broker. The ontology broker is one of many options that could be used to connect

Fig. 7 Model-based environment implementation

tools, but the reason this approach was chosen was due to future scalability as point-to-point integrations would become widely to manage as you integrate more tools into the environment. For the illustrative example, the ontology broker was built using Cameo Workbench, but since then startup vendor SBE Vision has built an ontology broker that is more scalable and flexible to the ontologies being used and the tools within the environment, creating a hub spoke model of integration.

The ontology broker serves as the master translator and keeper of the axioms defined in the ontology. The adapters built to interface the traditional tools to the ontology broker map the data being captured in the tools to the ontology, allowing the right information captured in Cameo EA to seed the human system integration analysis (an analytical model). Due to its central role in the tool chain, the ontology broker allows the tools to be ignorant of each other and only know that the data being subscribed align to the semantics and syntax needed within its own environment. The ontology broker translates not only the semantics but also the syntax that each tool uses.

Create System Diagrams and Export Task Allocation and Workflow for Analysis

Using the tools in the model-based environment, the system is represented by behavioral, structural, and parametric diagrams in SysML. The architectural process begins by developing a Concept of Operations (CONOPS) of the system that focuses on the mission tasks. Once the CONOPS is well understood, usually an operational concept (OPSCON) is created for better understanding the system-to-system interactions or, in this case, the machine-to-human interaction. The OPSCON is then decomposed into lower level of behavioral diagrams and then allocated to the structural components of the system: machine or human. The human allocated behavior will then be used to continue with the appropriate human integration analysis.

Evaluate Results, Build Alternatives, and Evaluate Alternatives

Once the initial behavioral allocation analysis of the system is complete, the analysis will show whether or not you meet human agent thresholds and how critical it might be to re-allocate behavior from a human to a machine. This re-allocation will inherently change the machine and quite possibly require new analysis to see how the performance changes from one alternative to another.

Illustrative Examples

Case Study System Overview

The US Department of Defense has been moving forward using more unmanned capabilities due to more hazardous missions, smaller numbers in personnel, and more demands on resources. As such, they are slowly integrating unmanned platforms to cooperatively operate within standard forces to assist in mission objectives.

For this case study, a geospatial mission was chosen to study the interactions between the unmanned aerial vehicle (UAV) and the image analyst. All the data used for this case study was gathered from publicly available literature from the Army Research Laboratory (ARL) in Aberdeen, Maryland [35, 36]. The models built were focused on the lenses of the human agent, in particular the image analyst and its interface with the UAV.

The case study partially replicates the geospatial UAV study done at ARL [36] due to data availability, but appropriate level of details and fidelity was extracted to drive the architecture and analysis from the system architecture to the detailed human system design.

Ontology Extension or Modifications

Since the ontology described in section "State of the Art" was created with this case study in mind, there was no need for modification or extensions.

Architecture and Analysis

To replicate the systems engineering process as closely as possible, the geospatial UAV system model started with developing the overarching CONOPS through adaptation of the UAV studies presented by ARL [35, 36]. Figure 8 shows that the UAV and its supporting personnel receive the orders and plan the mission. Once the mission is planned, the UAV goes through its launch procedures and eventually is launched and travels to the area of interest, while in the area of interest, it performs counter-improvised explosive device operations. Once the mission is complete, the UAV goes back to base and performs landing and post-mission procedures.

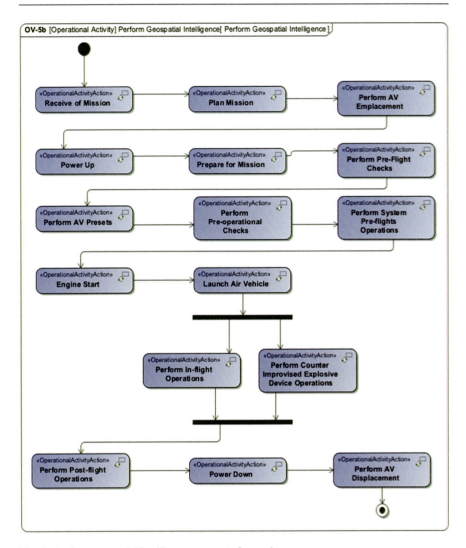

Fig. 8 Perform geospatial intelligence concept of operations

For the purposes of this case study, the perform counter-improvised explosive device operations operational activity was decomposed to better understand the operations functions needed to complete the operations use case. This lower-level operational level detail was needed as the system views were created, and the model focuses in on system functions that support the task.

Figure 9 describes the mission scenario which is focused on the image analyst and its interactions with the systems to coordinate and execute the mission. In this OPSCON, there are five agents: the UAV, UAV operator, UAV mission control station, geospatial intelligence (GEOINT) analyst control station, and the image

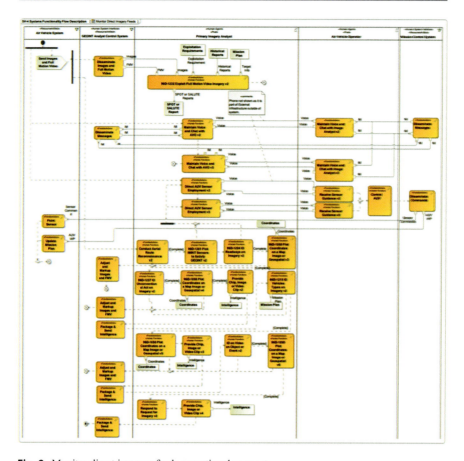

Fig. 9 Monitor direct imagery feeds operational concept

analyst. The scenario modeled starts by geospatial information from the UAV being sent down to the GEOINT control station, where the image analyst can exploit video and images being captured. As the feed comes in, the analyst must process features on the feed of potential areas of interest. Once features are identified, intelligence is forwarded to other analysts for further processing and data fusion with other intelligence sources. If the analyst decides that he or she needs more information in a certain geographical location, he or she can cue the UAV operators with new mission requirements.

Typically, at this point in the systems engineering process, the system architects focus in on decomposing the functions relative to the system of interest and rarely focus on human agent tasking decomposition. Due to the HSI extension, not only was a mock-up of a human system interface captured in the system model, but also human functions were decomposed using task networks in the same system model that decomposed the system functions down to further components. The following figures are examples of some of the human functions that were decomposed.

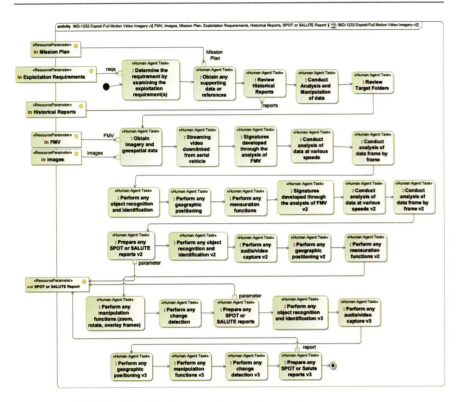

Fig. 10 Exploit full motion video imagery task network example

Typically, this may or may not happen at this stage, depending on the level of information available to the architects and the understanding of the human system integration engineer. On some occasions, the high-level human function will be identified and be left for further definition in the human system analysis and brought back up into integration with the rest of the system model when better understood (Figs. 10 and 11).

The Human System Integration Analysis

Whether parts or all of the human functions are defined in the system architecture, starting the human system integration analysis upfront in the architecting process not only strengthens the architecture but also allows for architecture to be adapted and optimized to the human agents.

For the case study, the human agent functions and tasks were identified in the system architecture. Once the functions and tasks necessary to define the GEOINT scenario were completed, the system architecture model was converted into the inputs of an IMPRINT model to begin the human system integration analysis. The

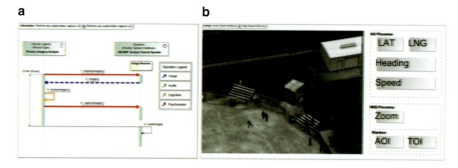

Fig. 11 (a) Perform any audio/visual capture VACP sequence diagram; (b) primary image analyst GUI mock-up

elements created in the IMPRINT model mirrored the human elements of the system architecture elements in Cameo EA with all attributes that are defined according to the HSI profile. As seen in Fig. 12, the converted human elements are constructed with the same relationships defined in the system architecture.

Results of Analysis

Evaluate Architecture and Its Impact

As discussed previously, the human functions and tasks were extracted from an ARL study [36]. In this study, the tasking for the image analyst was surveyed for task time and workload in accordance with visual (V), auditory (A), cognitive (C), and psychomotor (P) resources. In the ARL study, they ran the model and analyzed for overall workload (Ow). Ow in IMPRINT is defined as Ow = V + A + C + P. Overloads occur when any single resource goes above 7 or the Ow goes above a 40. Figure 13 shows the value scales for each resource that is attached to each task described in the model.

In order to replicate the ARL study, the case study ensured that the workload analysis results resulted in similar results as the original study. Figure 14a shows the original workload analysis results. Similar to the pattern of the original results, the replicated study had similar shape and workload spikes, as seen in Fig. 14b.

Unlike the previous study that explained how these results integrated into the system architecture, the replicated study allowed for deeper analysis and highlights which human functions need attention and how they integrate to system functions. As described by [35, 36], the analysis was done by having the image analyst conduct all functions manually and without the use of any assistive aids or automation. Unfortunately, [35, 36]'s study was not tightly integrated to the system architecture and was conducted separately from the architecting effort. With the human-centered model-based systems engineering approach that has been developed in this research, these results can be tied back to a very tightly coupled relationship between the human system integration analysis and the overall system architecture.

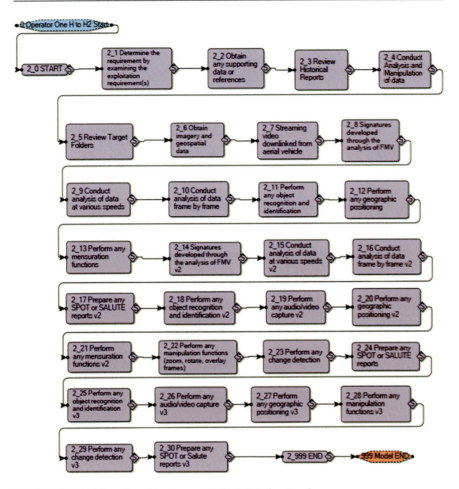

Fig. 12 Exploit full motion video imagery IMPRINT task network

Evaluate Alternative Architecture and Its Impact

The three areas of workload overload identified in Fig. 14 were analyzed for evaluating aspects of the human tasking attributed to the overload. Then, each human function was analyzed for how they integrated to the system functions. As the functional picture was explored, system functions that could aid the image analyst reduce his or her workload were explored. The workload analysis identified functions of opportunity where we could reduce human agent workload and trace directly to the system architecture due to the tight integration with the human system integration profile.

With the workload results in hand, it was obvious that the human functions that could be improved would be any identification function. By allocating automated identification functions to the GEOINT analyst control system to assist the human

Fig. 13 IMPRINT VACP scale values [35]

Value	VISUAL SCALE
0.00	No Visual Activity
1.00	Visually Register/Detect (detect image)
3.70	Visually Discriminate
4.00	Visually Inspect/Check (static inspection)
5.00	Visually Locate/Align (selective orientation)
5.40	Visually Track/Follow (maintain orientation)
5.90	Visually Read (symbol)
7.00	Visually Scan/Search/Monitor (continuous)
	AUDITORY SCALE
0.00	No Auditory Activity
1.00	Detect/Register Sound
2.00	Orient to Sound (general orientation)
4.20	Orient to Sound (selective orientation)
4.30	Verify Auditory Feedback
4.90	Interpret Semantic Content (speech)
6.60	Discriminate Sound Characteristics
7.00	Interpret Sound Patterns (pulse rate, etc.)
	COGNITIVE SCALE
0.00	No Cognitive Activity
1.00	Automatic (simple association)
1.20	Alternative Selection
3.70	Sign/Signal Recognition
4.60	Evaluation/Judgment (consider single aspect)
5.30	Encoding/Decoding, Recall
6.80	Evaluation/Judgment (consider several aspects)
7.00	Estimation, Calculation, Conversion
	PSYCHOMOTOR SCALE
0.00	No Psychomotor Activity
1.00	Speech
2.20	Discrete Actuation (button, toggle, trigger)
2.60	Continuous Adjustive (flight or sensor control)
4.60	Manipulate
5.80	Discrete Adjustive (rotary, thumbwheel, lever)
6.50	Symbolic Production (writing)
7.00	Serial Discrete Manipulation (keyboard entries)

agent, the human agent VACP resource workload would reduce, in accordance with the scale values used to measure those resource values in IMPRINT.

As seen in Fig. 15, the light-yellow color functions in the GEOINT analyst control system are the added system functions that assist the human agent to reduce workload. As part of these added features, instead of having the raw feed, the system functions provide cues that allow the human agent to filter out areas that are not of concern. The new semantics and syntax used in the HSI profile allow to understand the exact data needs for the human agent's situational awareness allowing the architect to increase pick alternative human-system interfaces that increase system performance.

Once the changes to the functional architecture had been made, the new VACP values were entered into the IMPRINT model to rerun the discrete event VACP analysis over a 100 run Monte Carlo simulation. Figure 16 shows the results of the new architecture with the automated assist functions in place.

Even with the new automated assistance, the image analyst workload still reaches two peaks that are over the workload threshold of 40. The first peak is very early on

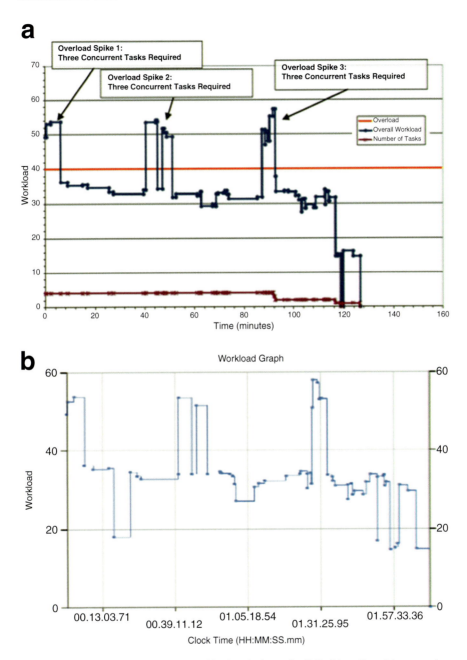

Fig. 14 (**a**) Original primary analyst workload analysis results [35]; (**b**) replicated image analyst workload analysis results

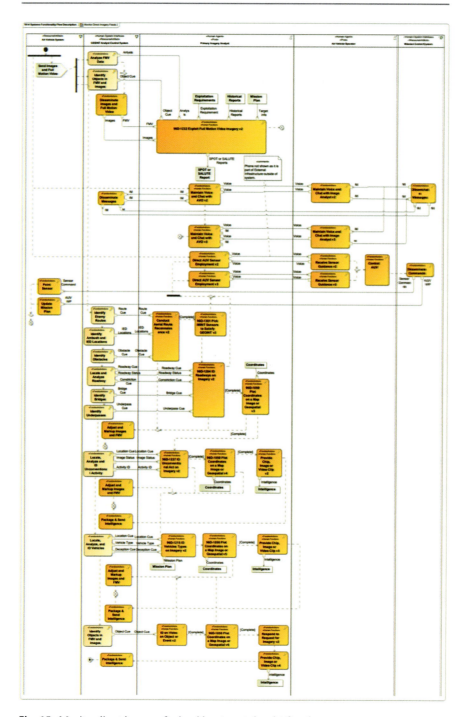

Fig. 15 Monitor direct imagery feeds with automated assist functions

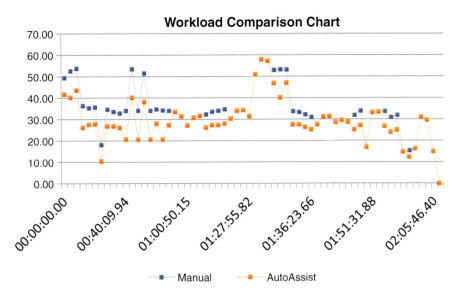

Fig. 16 Architecture workload comparison

and is slightly over the threshold at 43.6. Due to the slight overload for a short period of time (about 2 min), the architecture would be able to stand. The second peak, though, lasts for a little over 3 min with a peak overload at 58. In Fig. 16, the image analyst workload demands of both functional architectures are compared. Overall, with the new architecture, there was a 13% average reduction in workload throughout the operational concept, with a maximum workload reduction of 43% in some moments in time.

Since automated aids did not reduce the workload, the architecture was re-evaluated to see if other architectural changes could be made. Revisiting the original study, [36] studied other human agents as part of a larger concept of operations that could affect architecture results through team cooperation. Under the right conditions in a team environment, an analysis can be looked at in which other human agents are assigned tasks and are allowed to be shed and taken on by other human agents. Due to the focus of this case study being on individual impact to a system, this is a shortfall of the human system integration ontology and will need to be researched and added at a future date. The ontology used will however facilitate said future research, since the core items are already part of the ontology. Further, it will allow architects to identify key items needed for human agent situational awareness that are critical for human-to-human task distribution or if a human is a secondary option to the machine.

Chapter Summary

This chapter presented a human-centered model-based systems engineering approach comprising methods, processes, and tools that transform how architectures are employed for exploring human interactions, interfaces, and integration. An illustrative case study of the introduction of HSI methods into MBSE is presented for a UAV image analyst. This case study, along with the methods, processes, and tools, allows evaluating the impact to allocation of collaborative functions on human performance by system architects and human system integration engineers. The previous study [35, 36] allocated functions related to image analysis completely on the image analyst to manually assess. The image analyst had to receive the direct full motion video or still images from the UAV and evaluate for various factors that would get cued for further processing and data fusion to other systems and personnel. The described methods, processes, and tools presented in this chapter modified the baseline architecture that included image identification and cueing automation. With these changes the image analyst was still responsible for reviewing the full motion video and still images but had the luxury of the full motion video and still imagery being prepopulated with areas of interest that would then be corrected as needed by the image analyst. The image analyst moved from a worker role to a supervisor role, which reduced overall workload. In fact, workload analysis showed that the architecture with automation reduced workload on average 13% throughout for the scenario analyzed.

Along with direct comparison of architecture optimization for the human agent, the new methods, processes, and tools enable sensitivity analysis, exploring the sensitivity of changes to the machine on human performance. Another analysis that could have been run using the processes, methods, and tools was how each addition of an automation function changed the human performance until reaching full automation of the identification functions. Although the research did not cover human error analysis as a primary research focus, a by-product of the integrations presented in this research is to include human error analysis as another factor. The sensitivity analysis could explore the level of failure or errors due to changes on how the machine handles human interaction, interfaces, and integration.

Another aspect shown in the case study was a new set of tools, which made it easier to ascertain what information the human needs for situational awareness purposes. Situational awareness becomes a key constraint when the architecture moves the human agent from a worker role to a supervisor role. In the case study, these considerations were not as critical, since the system was to designate areas of interest that could be used for intelligence through data fusion. For other systems, where the supervisory role is critical to success of the mission or can affect life, having better knowledge of what the human agent must have to react to could mean the difference between success and failure, where the worst consequence could lead to death. For example, in the UAV market today, there is a push to put more artificial intelligence onboard the UAVs in order to minimize control of the human agent and allow the growth of multiple planes to be controlled by one agent. The problem with this scenario is that if a failure occurs, in order for the human agent to react, first the agent must acknowledge that there is a failure and then figure out the status, as well as what actions he or she can take to correct the situation. The level of situational awareness on the ground is usually not as high as in the UAV, since UAVs tend to have multiple sensors and, in some cases, functional duplication on measurements. All this data tends not to get to the ground due to bandwidth limitations on communication systems or mission constraints; however, the new methods, processes, and tools give a better picture to the information a human agent needs in order to make decisions. By understanding more explicitly what the vehicle could be using to make decisions and what data is actually getting to the human agent, the architect can analyze gaps in situational awareness early on the development life cycle versus in test and evaluation or, worse yet, in operations and sustainment.

As a result of these modeling advances, system architects and system engineers are able to analyze for the human agent using a model-based systems engineering approach. This extension serves as a framework for future work in extending model-based systems engineering to other disciplines and to the system "ilities." In the HSI community, it advances current practices by expanding the HSI consideration out of the system design phases to throughout the system life cycle by enabling HSI considerations during system architecting and systems engineering methods and processes.

Although this research was focused on increasing human performance by extending MBSE for human considerations, one of the outcomes of this research has been to expand the use of ontologies to drive model integration by becoming the foundation for integrating MBSE methods, processes, and tools through a more data-centric approach. The human model-based environment built as part of this research was not based on point-to-point solutions, but semantic integration through an ontology broker. This approach changes how models are integrated and allows MBSE environments to be scalable – the driven force being the ontology versus the tools – keeping methods and processes tool agnostic. The focus is taken away from the tool-to-tool integration and focused on the ontology broker that talks to each tool for the correct data in accordance with the ontology axioms.

Cross-References

▶ Model-Based Human Systems Integration
▶ Semantics, Metamodels, and Ontologies

Acknowledgment I would like to thank Dr. Azad Madni for his countless hours of mentoring during this research.

References

1. A. M. Madni, "Integrating Humans With and Within Complex Systems," CrossTalk, vol. May/June, pp. 4–8, 2011.
2. Madni, A. M. (2017). Mutual Adaptation in Human-Machine Teams. ISTI-WP-02-012017.

3. Madni, A. M., & Freedy, A. (1986). Intelligent Interfaces for Human Control of Advanced Automation and Smart Systems in Human Productivity Enhancement. Training and Human Factors in Systems Design, 1(J. Zeidner), 318–331.
4. Madni, A. M. (2015). Expanding Stakeholder Participation in Upfront System Engineering Through Storytelling in Virtual Worlds. Systems Engineering (Vol. 18).
5. International Council on Systems Engineering, Systems Engineering Handbook, 3.2.2. San Diego, CA, 2011.
6. A. M. Madni, "Expanding Stakeholder Participation in Upfront System Engineering Through Storytelling in Virtual Worlds," 2015.
7. A. M. Madni, "The Role of Human Factors in Expert Systems Design and Acceptance," Hum. Factors J., vol. 30, no. 4, pp. 395–414, 1988.
8. A. M. Madni, "HUMANE: A Designer's Assistant for Modeling and Evaluating Function Allocation Options," in Proceedings of Ergonomics of Advanced Manufacturing and Automated Systems Conference, 1988, pp. 291–302.
9. J. Axelsson, "Towards an Improved Understanding of Humans as the Components that Implement Systems Engineering," in Proceedings 12th Symposium of the International Council on System Engineering, 2002, pp. 1–6.
10. A. M. Madni, "Elegant Systems Design: Creative Fusion of Simplicity and Power," Syst. Eng., vol. 15, no. 3, pp. 347–354, 2012.
11. R. Neches and A. M. Madni, "Towards Affordably Adaptable and Effective Systems," Syst. Eng., vol. 16, no. 2, pp. 224–234, 2012.
12. T. Bahill and A. M. Madni, Trade-off Decisions in System Design. Springer, 2017.
13. A. M. Madni, Transdisciplinary Systems Engineering: Exploiting Convergence in a Hyperconnected World. Springer, 2018.
14. Object Management Group, "Unified Modeling Language, Infrastructure v2.4.1," no. August. 2011.
15. H.-P. Hoffmann, "UML 2.0-Based Systems Engineering Using a Model-Driven Development Approach," CrossTalk J. Def. Softw. Eng., pp. 1–18, 2005.
16. E. Herzog and A. Pandikow, "SysML – An Assessment," in Syntell AB, SE 100, 2005.
17. P. K. Balakrishnan, "Analysis of Human Factors in Specific Aspects of System Design," in INCOSE International Symposium, 2002, pp. 1–9.
18. Landsburg, A. C., Avery, L., Beaton, R., Bost, J. R., Comperatore, C., Khandpur, R., ... Sheridan, T. B. (2008). The Art of Successfully Applying Human Systems Integration. *American Society of Naval Engineers Journal*, *120*(1), 77–107. https://doi.org/10.1111/j.1559-3584.2008.00113.x
19. J. M. Narkevicius, "Human Factors and Systems Engineering – Integrating for Successful Systems Development," in Proceedings of the Human Factors and Ergonomics Society Annual Meeting, 2008, vol. 52, no. 24, pp. 1961–1963.
20. Vanderperren, Y., & Dehaene, W. (2006). From UML/SysML to Matlab/Simulink: Current State and Future Perspectives. In *Proceedings of the Design Automation & Test in Europe Conference*. Ieee. https://doi.org/10.1109/DATE.2006.244002
21. M. Bajaj, D. Zwemer, R. Peak, A. Phung, A. G. Scott, and M. W. Wilson, "SLIM: Collaborative Model-Based Systems Engineering Workspace for Next-Generation Complex Systems," in IEEE Aerospace Conference, 2011, pp. 1–15.
22. Mayer, R. J., Crump, J. W., Fernandes, R., Keen, A., & Painter, M. K. (1995). *Information Integration for Concurrent Engineering (IICE) Compendium of Methods Report*. Wright-Patterson Air Force Base, Ohio. Retrieved from http://www.idef.com/pdf/idef3_fn.pdf
23. H. A. H. Handley and R. J. Smillie, "Architecture Framework Human View: The NATO Approach," J. Syst. Eng., vol. 11, no. 2, pp. 156–164, 2008.
24. K. Baker, A. Stewart, C. Pogue, and R. Ramotar, "Human Views: Extensions to the Department of Defense Architecture Framework," 2008.

25. Watson, Michael & Rusnock, Christina & Miller, Michael & Colombi, John. (2017). Informing System Design Using Human Performance Modeling. Systems Engineering. 20. https://doi.org/10.1002/sys.21388.
26. Parasuraman, R., Sheridan, T. B., & Wickens, C. D. (2000). A Model for Types and Levels of Human Interaction with Automation. IEEE Transactions on Systems, Man, and Cybernetics. Part A, Systems and Humans : A Publication of the IEEE Systems, Man, and Cybernetics Society, 30(3), 286–97. Retrieved from http://www.ncbi.nlm.nih.gov/pubmed/11760769
27. Dolan, N., & Narkevicius, J. M. (2005). Systems Engineering, Acquisition and Personnel Integration (SEAPRINT): Achieving the Promise of Human Systems Integration. In Meeting Proceedings RTO-MP-HFM-124 (pp. 1–6). Neuilly-sur-Seine, France.
28. Ijtsma, Martijn & Pritchett, Amy & Ma, Lanssie & Feigh, Karen. (2017). Modeling Human-Robot Interaction to Inform Function Allocation in Manned Spaceflight Operations.
29. A. M. Madni and M. Sievers, "Systems Integration: Key Perspectives, Experiences, and Challenge," Syst. Eng., vol. 16, no. 4, pp. 1–23, 2013.
30. N. G. Leveson, "Intent Specifications: An Approach to Building Human-Centered Specifications," IEEE Trans. Softw. Eng., vol. 26, no. 1, pp. 15–35, 2000.
31. N. L. Miller, J. J. Crowson Jr, and J. M. Narkevicius, "Human Characteristics and Measures in Systems Design," in Handbook of Human Systems Integration, H. R. Booher, Ed. John Wiley & Sons, Inc, 2003, pp. 699–742.
32. McKenney, Martin & McKenney, Martin & Handley, Holly. (2020). Using the Design Science Research Method (DSRM) to develop a Skills Gaps Analysis Model. IEEE Engineering Management Review. PP. 1-1. https://doi.org/10.1109/EMR.2020.3011704.
33. S. C. Spangelo et al., "Applying Model Based Systems Engineering (MBSE) to a Standard CubeSat," in 2012 IEEE Aerospace Conference, 2012, pp. 1–20.
34. S. C. Spangelo, "Model Based Systems Engineering (MBSE) Applied to Radio Aurora Explorer (RAX) CubeSat Mission Operational Scenarios," in IEEE Aerospace Conference, 2013, pp. 1–18.
35. B. P. Hunn and O. H. Heuckeroth, "A Shadow Unmanned Aerial Vehicle (UAV) Improved Performance Research Integration Tool (IMPRINT) Model Supporting Future Combat Systems," Aberdeen Proving Ground, MD, 2006.
36. B. P. Hunn, K. M. Schweitzer, J. A. Cahir, and M. M. Finch, "IMPRINT Analysis of an Unmanned Air System Geospatial Information Process," Aberdeen Proving Ground, MD, 2008.

Douglas Orellana is ManTech's VP of Intelligent Systems Engineering in the Innovation and Capability Office, focused on developing the next-generation solutions powered by computing and AI.

Dr. Orellana has been recognized for his academic excellence, professional work, and community involvement. In 2019, his paper on SE Ontology was awarded best forward-thinking paper at the Conference in Systems Engineer Research, and the Engineer's Council awarded him Outstanding Engineering Achievement. In 2012, INCOSE awarded him a Research Award. In 2010, Great Minds in STEM named him HENAAC's Most Promising Engineer, and SHPE selected him in 2009 as Promising Engineer.

Dr. Orellana earned a Doctorate in Astronautical Engineering from the University of Southern California; his research focused on integrating human considerations into the system architecting process though the use of Model-Based Systems Engineering. He earned a master's degree in Systems Engineering and his Bachelor's degree in Electrical Engineering from Johns Hopkins University.

Model-Based Human Systems Integration

15

Guy André Boy

Contents

Introduction	472
State of the Art: History and Evolution	474
Task and Activity	474
Evolution of Engineering and Associated Human Factors	475
System Knowledge Impacts Design Flexibility and Resource Management	477
Use of Digital Twins During System's Whole Life Cycle	478
Key Concepts and Definitions for a Human-Centered Systemic Approach	480
What Does "System" Really Mean?	480
Emergent Functions and Structures	482
Looking for Separability, Emergence, and Maturity	482
Domain Experience Integration and Artificial Intelligence Solutions	484
What Does "Experience" Mean?	484
Toward Model-Based Experience Integration: Human-AI-SE Cross-fertilization	485
Coordinating Technology, Organization, and People (TOP)	486
Concrete Chapter Contribution: The PRODEC Method	488
Procedural and Declarative Knowledge	489
An Instance of PRODEC	490
An Illustrative Example of PRODEC Use	491
Discussion: Challenges, Gaps, and Possible Futures	492
Departing from Technology-Centered MBSE	492
Human-Centered Modeling Limitations and Perspectives	493
HCD Based on Virtual Environments as Digital Twins	493
Summary	495
References	495

G. A. Boy (✉)
CentraleSupélec, Paris Saclay University, Gif-sur-Yvette, France

ESTIA Institute of Technology, Bidart, France
e-mail: guy-andre.boy@centralesupelec.fr; g.boy@estia.fr

© Springer Nature Switzerland AG 2023
A. M. Madni et al. (eds.), *Handbook of Model-Based Systems Engineering*,
https://doi.org/10.1007/978-3-030-93582-5_28

Abstract

Human systems integration (HSI) is an essential field of systems engineering (SE) that emerged, departs, and encompasses from its initial components that are human factors and ergonomics, human-computer interaction, engineering, and domain experience. Current capabilities and maturity of virtual prototyping and human-in-the-loop simulation (HITLS) enable virtual human-centered design (HCD) that can be combined with SE to realize HSI. HSI is almost necessarily model-based; it uses HITLS and requires a homogenized human and machine systemic representation. Virtual HCD enables us to take into account both human and organizational elements not only during the design process but also during the whole life cycle of a system. These new capabilities are made possible by digital tools that enable virtual environments that in turn should be made tangible. Digital twins can be solutions for supporting HSI, operations performance, and experience integration. Tangibility is therefore a crucial concept in model-based HSI (MBHSI), which should be both analytical and experimental, based on appropriate scenarios and performance metrics essentially supported by domain experience. An aeronautical example illustrates an instance of MBHSI.

Keywords

Human systems integration · Systems engineering · Human factors and ergonomics · Human-computer interaction · Human-in-the-loop simulation · Modeling · Virtual prototyping

Introduction

When the first draft of this chapter was started, the intention was to provide a model-based systems engineering (MBSE) contribution to human systems integration (HSI). Thinking about it, it became clear that HSI is necessarily model based. This is the reason why the title of this chapter is Model-Based Human Systems Integration (MB-HSI), which reinforces the intrinsic modeling need for HSI. In addition, modeling in HSI cannot be considered without simulation and even more importantly human-in-the-loop simulation (HITLS).

HSI emerged in the beginning of the 2000s as an approach to considering the human element in the design and management of a complex sociotechnical system during its whole life cycle. A sociotechnical system (also called human-machine system) is composed of humans and machines interacting with each other.

The progressive and exponential accumulation of software during the last three decades naturally induced the need for considering human-centered design (HCD) seriously [9, 11]. More specifically, HCD was made possible thanks to the development of computer-based prototyping that enables "virtual HCD" (VHCD). VHCD involves HITLS that enables carrying out activity analyses and further considers tangibility. Virtual HCD is further developed in section "State of the Art: History and

Evolution" of the chapter. Two tangibility meanings should be articulated in the design and management of complex sociotechnical systems: physical tangibility (e.g., grasping a physical object) and figurative tangibility (e.g., grasping a concept or an abstraction). Today, some tangibility problems can be anticipated through the use of 3D printing capabilities, for example.

Eugène Ionesco, a French avant-garde theater writer, depicted human existence in a tangible way. Claude Bonnefoy published, after Ionesco's death, a magnificent interview he had with him, where the following citation is an excellent perspective for VHCD: "In a dream, we're still in a situation. In short, I believe that dreaming is both a lucid thought, more lucid than in the waking state, a thought in images, and that it is already theatre, that it is always a drama because we are always in a situation (French citation: *"En rêve, on est toujours en situation. Bref, je crois que le rêve est à la fois une pensée lucide, plus lucide qu'à l'état de veille, une pensée en images et qu'il est déjà du théâtre, qu'il est toujours un drame puisqu'on y est toujours en situation."*)" [32]. Virtual prototyping and advanced visualization techniques and tools provide us with means that transform dreams into images, theater plays, and consequently tangible designs. Brenda Laurel introduced this concept in her book, *Computer as Theater*, in the beginning of the 1990s [37].

This evolution toward a human-centered systemic approach requires to define what the concept of "system" really means. Indeed, the concept of system is not only a matter of machines, but it is also a matter of people and organizations. This is the content of section "Key Concepts and Definitions for a Human-Centered Systemic Approach" of the chapter, where a system is defined as an articulation of structures and functions and issues of complexity and maturity of human-machine systems are developed.

Section "Domain Experience Integration and Artificial Intelligence Solutions" is an attempt of clarification of the concept of experience, as well as its utility in HSI. A systems engineering framework for model-based experience integration is provided. This framework could be based on appropriate artificial intelligence (AI) concepts, methods, and tools. Domain experience is thought in the phenomenology sense (i.e., meaning and subjectivity of people's experience from observation and compiled knowledge in a given domain).

Section "Concrete Chapter Contribution: The PRODEC Method" presents a new method, called PRODEC, based on the acquisition, analysis, and use of procedural (PRO) and declarative (DEC) knowledge for VHCD. Human-centered design of complex systems is a matter of identification of relevant human and machine entities, considered as systems, which can be physical and/or cognitive (cyber). Procedural knowledge is about operational experience that is often expressed in the form of stories by subject matter experts. Declarative knowledge is about objects and agents involved in the targeted human-machine system being designed. The PRODEC process involves human-in-the-loop simulation to incrementally create and maintain appropriate performance models. An aeronautical application is provided to illustrate the use of the PRODEC method.

A discussion is started in section "Discussion: Challenges, Gaps, and Possible Futures" where the approach departs from technology-centered MBSE [31] toward human-centered design intimately fused into systems engineering (SE) that naturally leads to human systems integration. Human-centered modeling limitations are explored leading to an epistemological endeavor.

The summary provides perspectives in model-based human systems integration where possible directions of research are projected not only on human modeling but also on human-centered digital twins.

State of the Art: History and Evolution

Task and Activity

It is useful to introduce the distinction between task and activity. A task is what is prescribed to be performed. For example, preparing breakfast in the morning is a task. It can be described in the form of subtasks, including "taking fruits out of the fridge," "putting cereals and soymilk in a bowl," "preparing coffee," and so on. It becomes clear that a task is linguistically described by a verb and a direct object complement. A task can also be described by a procedure, which can be a list of subtasks or a more sophisticated algorithm of subtasks, like a computer program that includes sequences, choices, loops, and other things of that type. Flying an aircraft is based on procedure following, for example. If a task is usually prescribed before actual performance using a procedure, this procedure can be occasionally adapted. For example, when breakfast becomes a brunch, other ingredients could be added to the everyday routine, and you can make another more elaborated procedure for that occasion. In terms of task, you are always brought back to follow a procedure, whether already made or made on the spot.

However, at operations time, things may not happen exactly as expected. There are events or circumstances that may disturb the execution of the task. In these cases, procedures do not apply as prescribed, and human operators need to make something up. Consequently, their actual activity departs from the task (i.e., what you do effectively is different from what was prescribed). For example, you realize that you forgot to buy coffee, and the subtask "preparing coffee" cannot be executed as prescribed. At this point, you have to solve a problem. You may decide to choose the task "preparing tea," or "going to the shop, if open, and buy coffee." Activity could be called effective task that is an adaptation of the task to the circumstances in real time. The expression, "deviations between task and activity," is typically used.

In life-critical environments, procedures greatly support operations (i.e., performance), but problem-solving skills and knowledge are also necessary to handle unanticipated circumstances or unexpected events. Reasons of task-activity deviations can be intrinsic (e.g., a human operator suddenly becomes sick) or extrinsic (e.g., weather conditions cause an electric shutdown). In both cases, adaptation is required and effectively happens. This is the reason why it is essential to anticipate possible activities in human systems integration. This has been a major difficulty for a long time. The next section presents an evolution of engineering and associated human factors disciplines that contributed to bring solutions to this task-activity issue.

Evolution of Engineering and Associated Human Factors

The evolution of human-related engineering approaches can be decomposed into three eras (Fig. 1). Prior to the 1980s, the most important disciplines in engineering were mathematics, physics, mechanical engineering, and industrial engineering. This era was dominated by hardware and human physical issues at work. Human factors and ergonomics (HFE) started to develop after World War 2 as an evaluation discipline, populated with occupational medicine doctors, caring about workers' health and safety. Health at work dealt with physical injuries, humidity, noise, heavy physical workload, and so on. Safety was in terms of physical danger. When new systems were built, people's activity had to be observed and analyzed before design and after the complete development of the system. This was fine, except for the fact that human factors specialists were always fighting with engineers to convince them that significant sociotechnical changes should be done on the system developed. They were always too late, after design and development processes were over and budget was almost all spent.

The second period started with the development of personal computers during the 1980s and later on with the Internet. Universal access to computing grew exponentially. Software engineering, computer graphics, AI, and cognitive engineering developed very rapidly. Software and computing were implemented in the form of

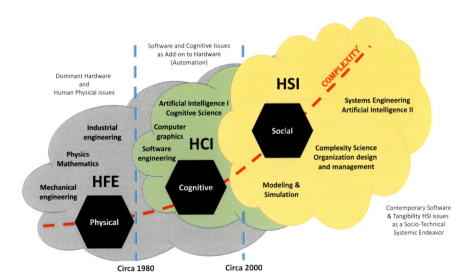

Fig. 1 Evolution engineering and associated human factors

add-ons to hardware. This is what automation is about (e.g., automation of airplanes, office automation, and so on). People incrementally discovered new cognitive issues that were unknown up to then. The shift was from doing to thinking. Human-computer interaction (HCI) developed as a discipline, making graphical user interfaces (GUIs) a necessity. Usability engineering became a dominant practice in HCI, because it became possible to test a system with end users in the loop. HCI developed because both fundamental and practical supports for interaction design and task analysis were needed as a major technique in HCD. Interaction design is very contextual (i.e., user experience is crucial to HCD, and tests should be performed in a large number of contexts). HCI provided what is available today in terms of GUIs, data and information visualization, computer-supported cooperative work (CSCW), and smartphones, for example. Commercial aircraft flight decks that increasingly include layers and layers of software and computer-based pointing devices were called "interactive cockpits." Pilots started to interact with computers more than with mechanical devices of their aircraft.

HCD first focused on usefulness and usability engineering [46] and the development of ISO 9241-210:2010(E) standards, for example. At the same time, even if participatory design was developed within the HCI community [45], it should be recognized that aeronautical engineering design was for a long time, and still is, based on participatory work with experimental test pilots. Participatory design requires everyone to understand each other and more specifically the definition of a common language, based on the various concepts of the domain. It should be evolutionary because it is impossible to get a definitive ontology but a stabilized explicit ontology. This language should support analysis, design, and evaluation of complex systems.

The importance of complexity science in SE, virtual modeling and simulation (M&S), and organization design and management became effectively clear in the beginning of the 2000s. Human systems integration (HSI) was born as a combination of HCD and SE. The problem became less about automation since any technological project is starting nowadays on a computer, using PowerPoint, for example, to present how the future system will look like and it could be used. The next step is typically a more detailed virtual M&S of that system, even human-in-the-loop simulation (HITLS) where people, potentially end users, have the possibility to test the future system being developed, in a virtual way. However, even if this is great news (i.e., activity can be tested at design and development time), the problem of tangibility remains.

Current sociotechnical systems are complex because they are extremely interconnected, and what people's roles are or should be within these systems should be figured out. Interconnectivity is not only between us and technology but also between us through technology. It is therefore important and timely to improve our understanding on what these systems are really about. In addition, since AI is back, how AI and SE can be harmonized and work together should be better understood. Why? This is because systems are becoming more autonomous and therefore have to be more appropriately coordinated. Automation has been developed using knowledge of very well-known situations. Autonomy is currently seen as the next technological step. It consists in developing systems that are equipped with enhanced situation awareness, decision-making, planning, and action capabilities. However, it would be better to focus more on human autonomy as well as organizational autonomy by developing technology and organizational setups that provide flexibility, instead of current automation rigidity. Therefore, working on human and machine autonomy is a must to better understand how to handle often-forgotten non-linearities.

System Knowledge Impacts Design Flexibility and Resource Management

System knowledge, design flexibility, and resource commitments are three parameters that should be followed carefully during the whole life cycle of a system. HSI aims to increase the following sufficiently early [6]:

- System knowledge, that is, knowing about systems at design, development, operations, and disposal times, how the overall system, including people and machines, works and behaves.
- Design flexibility, that is, keeping enough flexibility for systems changes later in development and usages.
- Resource commitments, that is, committing as late as possible on expensive resources (e.g., hardware) during the life cycle of the overall system.

When a technology-centered approach is used (typically what has been done up to now), system knowledge increases slowly in the beginning, growing faster toward the end of the life cycle (i.e., from early design to disposal). Design flexibility drops very rapidly, leaving very few alternatives for changes, because resource commitments were too drastic and too early during design and development processes.

Observing the way systems are being designed today helps realize the shift from the twentieth century, where the engineering design process was from hardware to software with constant automation during the last three decades, to the twenty-first century, where the engineering design process is from software to hardware since all designs start on a computer and eventually end up being 3D printed. This shift is good news because there is no need to commit too early on hard-to-modify resources. At the same time, the engineering design team can learn about the system being designed from its digital twin that can be used and then tested (i.e., using HITLS). This means that activity analysis can be done at design time, and results can be injected into the requirements of the system before it is physically built. Even better, enough design flexibility can be kept for a longer period of time.

This drastic revolution makes emerge issues of tangibility that need to be addressed seriously. Tangibility is not only a matter of physics; it is also a matter of intersubjectivity (i.e., mutual understanding) between end users and designers and ultimately between end users and developed technology. Note that in human-machine systems, machines are the concrete realization of designers; this is the reason why user-designer intersubjectivity is at stake. In other words, end users should be able to understand what machines are doing at appropriate levels of granularity and why. Consequently, complexity analysis has become tremendously important in our increasingly interconnected world. This is even more significant when machines include AI software that provides behaviors difficult to understand and therefore requires explainability mechanisms to keep trust and collaboration with the people involved.

At this point, the need for virtual M&S for HCD and consequently HSI becomes obvious. Virtual M&S is mandatory to develop and use HITLS for making virtual HCD possible (i.e., enabling activity observation and analysis in a virtual world that resembles reality and therefore getting the right system for the right task in the right context). Tangibility tests should be done.

Use of Digital Twins During System's Whole Life Cycle

The digital twin concept was presented to the industry in 2002 under the term of "Conceptual Ideal for Product Lifecycle Management (PLM)" at the University of Michigan (Grieves, 2016). NASA adopted the term "digital twin" [17, 25, 47, 62]. A digital twin is "a set of virtual information constructs that fully describes a potential or actual physical manufactured product from the micro atomic level to the macro geometrical level" [26]. This definition usually refers to product's structure. It should be extended to product's function. Figure 2 presents a concept map of the digital twin concept.

A digital twin (DT) is a virtual instance of a physical/cognitive system that enables to simulate dynamic phenomena of both structures and functions of a system. Madni et al. (2019) claimed that DT extends MBSE. According to Madni and his colleagues [40], a digital twin can be used as a model of a real-world system to represent and simulate its structure, performance (i.e., function), health status, and mission-specific characteristics during the whole life cycle of the system and incrementally update it from experience (e.g., malfunctions experienced, maintenance, and repair history). In other words, a digital twin can be used as a recipient of experience feedback information and support for system performance (e.g., preventive and timely maintenance based on knowledge of the system's maintenance history and observed system behavior). A digital twin is therefore a great support to improve understanding of the various relationships between system design and usages. In addition, a digital twin enables to support traceability and logistics along the whole life cycle of a system.

Considering a digital twin as a system digital model, in the SE sense, a distinction should be made between predictive digital DT and explanatory DT. A predictive DT is typically a very well-tested digital analog that produces similar outputs as the system would produce in response to the same inputs. It is usually simple and defined in a limited context. It is consequently short-term, rigid, and focused on a specific process or phenomenon. It can be used for marketing and, in some specific cases, when the domain is sufficiently mastered to operationally predict crucial

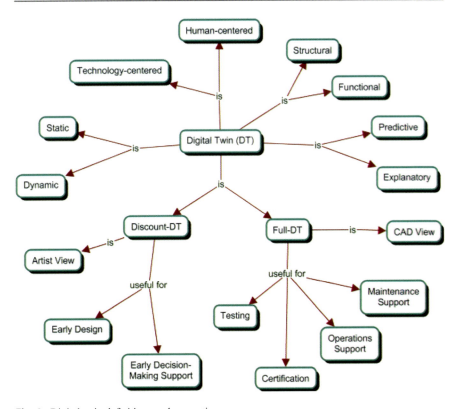

Fig. 2 Digital twin definitions and properties

system's states. In contrast, an explanatory DT is defined by an ontology of the domain. It is longer-term, flexible, and generic within the domain being considered. It can be used for analysis, design, and evaluation of a complex system, as well as for documenting its design and development process and its evolutionary solutions. Therefore, a digital twin is a digital model that enables running simulations to predict behavior and performance of a real-world system and/or explain why this system behaves the way it behaves.

A digital twin could be considered as a sophisticated interactive notebook that provides a vivid representation of the system being considered. Validation and certification of a DT are never finished. A DT is constantly modified by integrating new features, as well as modifying and/or removing old features. Considering system's life cycle, digital twins can be used in a variety of industrial activities including product design; engineering optimization; smart manufacturing; job-shop; scheduling; human-machine collaboration; operations diagnostic and decision-making; prognostics and health management; maintenance management; and, more generally, product lifecycle data management.

A DT could be used as a mediating tool to collaboratively test a very early concept within a design team that includes a group of experts with different backgrounds.

Note that the design team includes targeted users or human operators of the system to be developed. Each member of the design team should understand the same thing as the others. This is the reason why team members should have the same objectives and share the same situation awareness (SA) of what is being designed and further developed [22]. In this case, the DT represents this shared SA (SSA). Each member of the design team can see the same thing and eventually manipulate it. It is therefore a great support for participatory design. A DT for SSA starts with a discount DT (DDT). This could be done with the help of an artist, capable of producing a DT in the form of a cartoon or animation of the targeted system to be developed. DDT increases design team intersubjectivity through incremental modifications using collective critical thinking and experience feedback.

Once the DT is fully completed and approved by all members of the design team, it can be used to develop computer-aided design (CAD) of the system in depth. Using real numbers determining system's structure and function contributes to tangibilize the whole thing. The DT then becomes more rational and tangible. Once a "final" version is approved, physical construction of the system can start. This is another stage of tangibilization. Once a satisfactory version of the physical prototype is constructed, it can be tested in the "real world." Nevertheless, handling this HCD participatory process toward HSI requires everyone to speak the same systemic language. Consequently, the concept of system should be clarified in light of HSI.

Key Concepts and Definitions for a Human-Centered Systemic Approach

What Does "System" Really Mean?

A system is a representation of a natural or artificial entity. For example, physicians talk about cardiovascular or neural systems; anthropologists talk about communities of people and social groups as organizational systems; and engineers talk about mechanical and civil engineering systems. Figure 3 presents a synthetic view of what

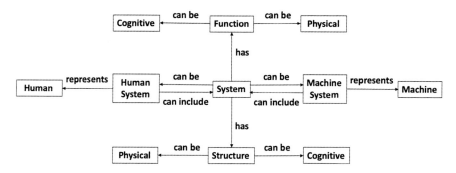

Fig. 3 Cognitive-physical structure-function system representation

a system is about. A system is recursively defined as including people, machines, and systems. A system can then be considered as a system of systems (SoS). In addition, a system has at least a structure and a function that can be physical and/or cognitive. In practice, a system has several structures and several functions articulated within structures of structures and functions of functions.

At this point, let's notice that an SoS is defined in the same way as Minsky [44] defined an agent as a society of agents. Russell and Norvig [55] defined an agent as an architecture (i.e., structure) and a program (i.e., function). For a long time, engineers considered a system as an isolated system or a quasi-isolated system. As for an agent in AI, which has sensors and actuators, a system in SE has sensors to acquire an input and actuators to produce an output. In an SoS, each system is interconnected to other systems either statically (in terms of systems' structures) or dynamically (in terms of systems' functions). Summarizing, an SoS is projected onto a structure of structures, usually called an infrastructure, where a network of functions could be allocated. It should be noted that the resulting network of functions is not necessarily a direct mapping on the related infrastructure.

The definition of a system is intrinsically recursive (Fig. 4), as an agent is defined as a society of agents in AI [6]. Therefore, in this chapter, "system" and "agent" are representations that have the same meaning. A system's function is defined by a role, a context of validity, and resources that can be physical and/or cognitive, human or machine systems, or agents. In addition, concepts of system and resource are very similar. A resource of a system is a system itself.

Complexity generated by several levels of recursion, shown on Fig. 4, has a direct impact on engineering design and validation of systems being developed. If resources are rigidly allocated to a system, when external context does not match the context of validity of the system, serious issues may arise at operations time. Alternatives are dynamic resource allocation either by adapting existing resources to new jobs or creating new resources. This will be further analyzed later on in this chapter. Resources and contexts are orthogonal and should be articulated. From a methodological point of view, a context-resource hyperspace can be a very useful

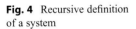

Fig. 4 Recursive definition of a system

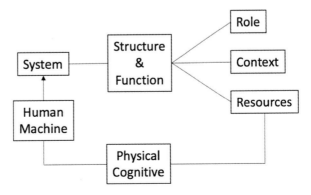

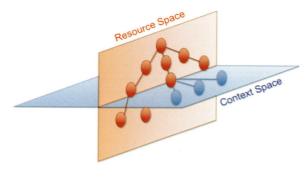

Fig. 5 A functional context-resource hyperspace (adapted from Boy, 2011)

topological support in engineering design and SE. Figure 5 presents a functional context-resource hyperspace that can be mapped onto a dual structural hyperspace.

Emergent Functions and Structures

Following up on the task-activity distinction, in addition to the topological and teleological definition of a function in terms of role, context, and resources, it is interesting to also consider a function in a logical sense. That is, a function transforms a task into an activity. In other words, when a human-machine system is at work, it produces various kinds of activity resulting from the execution of the various tasks that it is assigned to do. Bertalanffy [5, 58] said "a system is a set of elements in interaction." In addition, a system at work does not stay the same during its lifetime. It learns. Such learning is a matter of incorporation of emergent behaviors and properties. Emergence comes from activity.

Figure 6 shows emerging functions in yellow (i.e., functions coming from problem-solving of unanticipated issues and that should be compiled and incorporated in future practice) and potentially emergent structures in pink (i.e., structures that were previously ignored and need to be incorporated in the design of the overall system).

Looking for Separability, Emergence, and Maturity

There are three important factors that characterize sociotechnical systems: separability, emergence, and maturity. Physiologists are aware and use the separability concept for a long time to denote the possibility of separating momentarily an organ from the human body without irreversibly damaging the entire human body, considered as a system of systems. Some organs, such as the brain, cannot be separated because the human being could die from this separation. Those organs, as systems, have to be investigated and treated while connected to the rest of the body.

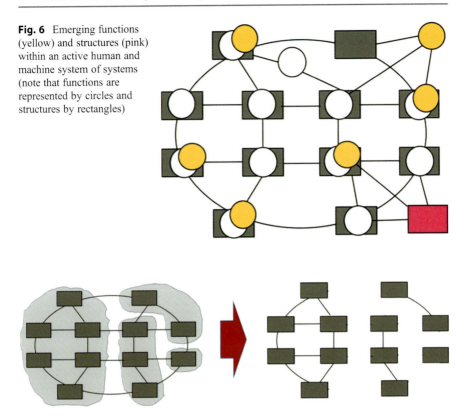

Fig. 6 Emerging functions (yellow) and structures (pink) within an active human and machine system of systems (note that functions are represented by circles and structures by rectangles)

Fig. 7 Example of three separable systems of a SoS (Boy, 2020)

It is therefore crucial to depart from the system design approach in silos and integration just before system delivery. This is not a problem when sub-systems are separable (Fig. 7). Clumsy integration, often done too late in the development process, is likely to cause surprises and even a few catastrophes at operations time. This is the reason why adjustments are always required, operationally either via adapted procedures or human-machine interfaces and in the worst case more drastic redesign of the whole system.

The lack of consideration for the separability issue in air traffic management (ATM) is a good example, where, for a long time, most air and ground technologies were designed and developed in isolation. Recent programs, such as SESAR (Single European Sky ATM Research), try to associate air and ground stakeholders. The "TOP model" (shown on Fig. 8 and developed in the next section of the chapter) supports design and development teams in the rationalization of interdependencies between technology, organizations, and people [9]. This requires observing and analyzing various activities using virtual M&S to identify emerging properties and functions of technologies under development, and not await to discover them at operations time.

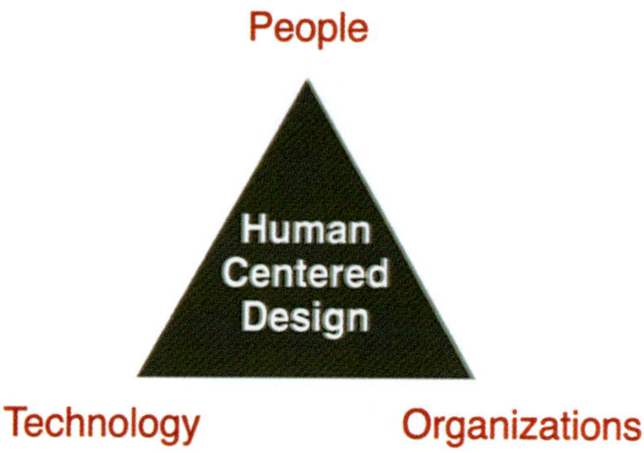

Fig. 8 The TOP model for HCD

Domain Experience Integration and Artificial Intelligence Solutions

There is a lot to say on experience feedback, also called in-service experience or user experience depending on the domain at stake (e.g., aeronautical, nuclear, or computer industry). In life-critical systems, experience feedback, often known as REX or RETEX, is cumulative and happens to be heavy duty but contributes to create and maintain a safety culture, for example. More specifically, ultra-safe industries [2], such as the nuclear industry, produce procedures and rules that have become numerous and can constrain operations [13]. This is mainly due to cumulative experience feedback mechanisms that contribute to pile up large numbers of regulatory-based requirements, which end up in such numerous procedures and rules. In human-computer interaction, user experience, often known as UX, is used to refine user interfaces with respect to activity observation and analysis. This section is an attempt to rationalize what experience integration means. But first, let us define the term "experience."

What Does "Experience" Mean?

The French philosopher and sociologist Auguste Comte introduced positivism that considers authentic knowledge as based on sense experience and positive verification [19]. The German philosopher Edmond Husserl introduced phenomenology in the beginning of the twentieth century as the study of consciousness and conscious experience [29, 30]. Among the most important processes studied by phenomenology are intentionality, intuition, evidence, empathy, and intersubjectivity. The positivism-phenomenology distinction opens the debate on objectivity and subjectivity. Our occidental world based most of our engineering approaches on positivism which led to developing a very precise and verifiable syntax, often leaving semantics somewhere behind, perhaps because semantics is full of subjectivity.

Winograd and Flores [65] provided a perspective for AI and HCI based on phenomenology. This chapter extends this perspective to engineering design and SE. Therefore, referring to phenomenology, the concept of experience that will be used in HSI is about meaning and subjectivity coming from people's experience in a given work environment. Gathered people's experiences will help the construction of typical episodes or scenarios. HSI considers that the classical positivist approach is no longer sufficient, often ineffective, and inappropriate, leaving aside crucial non-linearities that come back to us at operations time in the form of what is currently called "unexpected events" (e.g., COVID-19 pandemic). Experience is gathered and integrated incrementally during (sometimes long) periods of time. Typical episodes are assimilated and accommodated in the form of schemas, in Piaget sense [49, 50]. Piaget's schemas can be represented by cases in AI and lead to case-based reasoning.

Complex human-machine systems are living entities where "normal" events are experienced and/or observed, and emergent phenomena are incrementally discovered; they all are reported as experience. Good human-centered design should focus on the discovery of such emerging phenomena, in HITLS during design and development and in-service experience during the whole life cycle of a system.

Toward Model-Based Experience Integration: Human-AI-SE Cross-fertilization

Cross-fertilization of SE and AI has been already mentioned in this chapter when the analogy between system and agent was described and more specifically the analogy between a system as a system of systems in SE and an agent as a society of agents in AI [7]. The Research Council of the Systems Engineering Research Center (SERC) and a US Defense Department-sponsored University Affiliated Research Center (UARC) recently developed a roadmap structuring and guiding AI and autonomy research [43]. This roadmap outlines "digital engineering transformation aspects both enabling traditional SE practice automation (AI4SE) and encourage new SE practices supporting a new wave of automated, adaptive, and learning systems (SE4AI)." Even if this roadmap mentions that automation and human interaction research (denoted as manned/unmanned teaming) was "an essential part of systems engineering of these systems," nothing was said on neither organizational issues and problems to be solved nor how human-systems integration would be done. This is the reason why HSI takes even more importance in our growing digital world where autonomy and flexibility have become essential [6].

Why is the shift from automation to autonomy crucial in HSI? Control and management of life-critical systems are typically supported by operations procedures and automation. Automation is usually thought as automation of machines. Analogously, operation procedures can be thought as automation of people [9]. Problems come when unexpected situations occur, and rigid assistance (i.e., procedures and automation) may not work any longer, because system's activity is out of its context of validity. Outside system's context of validity, the rigidity of both procedures and

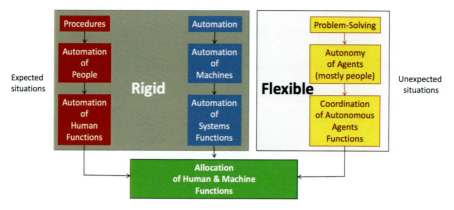

Fig. 9 Procedures, automation, and problem-solving leading to the allocation of human and machine functions [6]

automation rapidly leads to instability and sometimes radical breakdowns unfortunately. In these cases, instead of following procedures and monitoring automation, people need autonomy to solve problems. The more people have appropriate knowledge and experience, the more they are autonomous and have the capabilities to solve problems. They also need to have appropriate technological and/or organizational support. Therefore, autonomy is a matter of appropriate technological support enabling flexibility, coordinated organizational support, and people's knowledge and knowhow. Off-nominal situations management involves functions of autonomous human and machine agents that need to be coordinated. Figure 9 presents these three options, which lead to the difficult problem of function allocation.

In all cases, functions from automated or autonomous systems/agents need to be correctly allocated to the right systems/agents whether they are people or machines. Such allocation cannot be done only statically a priori but also dynamically with the evolution of context. Consequently, systems should be flexible enough to be able to be modified incrementally. Flexibility should result from such function allocation based on the TOP model.

Coordinating Technology, Organization, and People (TOP)

On the technology side, machine functions should be flexible enough to be modified if required and appropriately usable. HSI requires integration of experience and expertise, which is often available in the form of cases solved in the past and potentially reusable in similar situations. AI extensively developed knowledge-based systems and case-based reasoning, which can be very effective when associated with supervised machine learning. Handling cases requires appropriate situation awareness, and therefore AI can supply approaches such as intelligent visualization, which involves deep learning. Case-based reasoning (CBR), as presented by Aamodt and Plaza [1], is very useful in the context of experience feedback management. CBR is based on four main processes: retrieval of one (or several) similar case(s) similar to a current case; reuse of previously used cases to elaborate a working-in-progress solution; revision of the working-in-progress solution by modifying its structures and functions until a satisfactory solution is found; recording of the successful solution for later reuse; and potentially making it more generic. CBR could be developed within a statistical framework in order to perform probabilistic inference as opposed to deterministic inference [64].

Machine learning (ML) has become dominant in AI because it offers algorithms that enable to reduce huge data sets to "meaningful" information. Consequently, ML can be very interesting within our flexible autonomy endeavor. Indeed, flexible autonomy should be based on experience, which is acquired incrementally through a large number of try-and-error activities and therefore big data sets. As a matter of fact, positive experience is as important as negative experience. If incidents and accidents are very well documented and can serve as useful data for learning, the focus should be put even more on positive experience (i.e., things that went well). ML algorithms are developed to make sense of large amounts of data. They enable the elicitation of patterns, information organization, anomalies and relationships detection, as well as projections making. These algorithms will enable fine-tuning of increasingly autonomous systems performance in a safe, efficient, and comfortable manner. They can contribute to improve task execution precision. Most important here is the definition of "meaningful" information, which cannot be done without well-done human-centered design. AI algorithms are not neutral; they incorporate human-made requirements and technological constraints that determine all logical and mathematical formula that they use.

Intelligent visualization [24, 33, 61] is a growing field of investigation that attempts to develop visualization methods and tools that enable complex data to be better understood by people. In other words, it helps people to be more familiar with complex systems when they are appropriately visualized. Remember the adage, "A picture is worth a thousand words." In addition to static pictures, dynamic animations, simulations, and movies can be displayed to improve understanding of complex data and concepts. AI can support intelligent visualization as demonstrated in visual analytics [35]. More specifically, human visual exploration could be supported by data mining for knowledge creation and management.

On the organizational side, three systemic interaction models can be considered [6]:

- Supervision is a process that enables a system (i.e., a supervisor) to supervise interactions among other systems that interact among each other. Supervision is about coordination. This interaction model is used when systems do not know each other and/or do not have enough resources to properly interact with each other toward a satisfactory performance of their constituting systems.
- Mediation is a process that enables systems to interact with each other through a mediation space made of a set of mediating systems, such as ambassadors and

diplomats. This model is used when systems barely know each other but easily understand how to use the mediation space.
- Cooperation is when systems are able to have a socio-cognitive model of the SoS which they are part of. Each system uses a socio-cognitive model of its environment to interact with the other systems, maximizing some kinds of performance metrics. Note that this model is collective and democratic. This interaction model is used when systems know each other through their own socio-cognitive model, which is able to adapt through learning from positive and negative interactions. Other models could be used such as dominance of a system over the other systems (i.e., a dictatorial principle).

On the human side, people can be designers, engineers, developers, certifiers, maintainers, operators or end users, trainers and dismantlers (not an exhaustive list). People, in the TOP model, have activities and jobs. Anytime technology and/or organization changes, people may change their activities and/or jobs. Sometimes, new technology may lead to people losing their jobs, or conversely new jobs (i.e., functions) should be created and therefore a new set of people might be hired (i.e., a new structure should be created within the organization). People have their own human factors issues, such as fatigue, workload, physical and cognitive limitations, and creativity [6].

Sociotechnical SoS infrastructure can be hierarchical or heterarchical, for example. Evolution of digital organizations drastically changed people's jobs going from the hierarchical army model to the heterarchical orchestra model [9], with musicians, some of them being conductors and compositors. More formally, an orchestra playing a symphony (i.e., a product) requires five interconnected components:

- Music theory as the common language (i.e., a framework for collaborative work).
- Scores produced and coordinated by composers (i.e., coordinated tasks to be executed).
- Workflow coordinated by a conductor (i.e., system of systems activity).
- Musicians performing the actual symphony (i.e., the actual system of systems).
- Audience listening produced symphony (i.e., end users of the product).

Concrete Chapter Contribution: The PRODEC Method

Human-centered design of complex systems is a matter of identification of the multiple human and machine entities, considered as systems, which can be physical and/or cognitive (cyber). Systems can indeed be modeled by roles, contexts of validity and resources that are systems themselves. Therefore, these properties need to be properly identified. PRODEC is a method to this end [12]. It is based on the distinction used in cognitive science and computer science between procedural and declarative knowledge. Procedural knowledge is about operations experience that is often expressed in the form of stories by subject matter experts. Declarative knowledge is about objects and agents required in the design of a targeted human-machine system. The PRODEC method is articulated around the elicitation of procedural knowledge from subject matter experts and abduction of various human and machine systems properties and attributes. This abduction process is based on creativity and validation of systems being targeted. The PRODEC process may take several iterations to converge. It is highly recommended to run human-in-the-loop simulation for such validation and therefore incrementally create and maintain appropriate performance models and simulation capabilities.

Procedural and Declarative Knowledge

The distinction between procedural and declarative (or logical) knowledge is not new. In the early stages of AI, this distinction was used to denote procedural and declarative programming. Procedural programming languages are high-level abstraction of computer instructions that enables the programmer to express an algorithm in a line-by-line sequence of instructions. Procedural programing languages originated from FORTRAN [34, 42] and include Pascal [66], C [51], and Python [21]. Conversely, declarative programming languages enable programmers to declare a set of objects that have properties and capabilities. They originate from LISP [4] and includes Haskell, Caml, and SQL, for example. Object-oriented programming originated from Smalltalk and was inherited from both paradigms, including Java and C++, for instance. These object-oriented languages enable programmers to declare objects and their properties as well as specific procedures called methods. Declared objects are further processed as they are by a software inference engine.

Computer science, AI, and cognitive psychology cross-fertilized for a long time. Indeed, procedural knowledge and its distinction with declarative (or conceptual) knowledge have been developed in several fields related to cognition such as educational science [41] and development psychology [56], including mathematics education [16, 27, 60], user modeling [20], and experimental psychology [38, 54, 63].

If the theater metaphor is used, a theatrical play is usually available first in a procedural manner. A writer produces an essay that tells a story. Then, a director selects, in a declarative manner, actors who have to read the essay and learn their roles and scripts procedurally and coordinate them. PRODEC has been designed to be used in HCD to benefit from both operations experience (i.e., human operators will be asked to provide their salient operations stories) and definition of objects and agents involved in the targeted human-machine system to be designed (i.e., the design team will provide prototypes at various progressive levels of maturity).

With this method, how operations are performed prior to starting any design is explored first. A procedural scenario is developed with experienced people, as a timeline of events. PRODEC is based on the claim that stories told by subject matter experts can be easily translated into procedural scenarios. Once several procedural scenarios are elicited, human-centered designers are able to extract meaningful objects and agents, which are described both functionally and structurally. This constitutes declarative scenarios (i.e., organizational configurations). Of course, this articulation of procedural and declarative knowledge can, and should, be repeated as many times as necessary and possible to get a consistent and implementable human-machine system prototype, further developed and validated.

The production of procedural and declarative knowledge should be guided by a framework that supports the main factors that include artifacts to be designed and developed, users who will use them, the various tasks to be executed by these artifacts, the organizational environment where they will be deployed, and the various critical situations in which these tasks will be executed (see the AUTOS Pyramid in [10]).

An Instance of PRODEC

At this point, let's provide an instance of the Business Process Modeling Notation (BPMN) (BPMN is based on a flowcharting technique tailored for creating graphical models of business process operations, similar to UML (Unified Modeling Language) activity diagrams. BPMN is procedural (i.e., it enables the description of procedural information with different graphical elements in the form of scripts, episodes, sequences, and so on, which mixes the ways agents interact with each other – it is a program or a routine in the computer science sense).) associated with an extended version of the Cognitive Function Analysis (CFA) method. BPMN is a standard for business procedural process modeling [48, 57] and a language that supports procedural knowledge elicitation and graphical formalization. CFA has been developed for declaratively identifying human and machine cognitive (and physical) functions and their relationships to support HCD [11]. CFA has been upgraded as Cognitive and Physical Structure/Function Analysis (CPSFA) to handle systems as agents [6].

Note that alternative procedural and declarative methods and tools could be used to instantiate the PRODEC method. The BPMN-CPSFA PRODEC method is then the following:

1. Elicit and review all tasks involved in the achievement of various goals.
2. Describe them as BPMN graphs (procedural scenarios).
3. Elicit cognitive (and physical) functions in the form of roles (associated to tasks and goals), contexts, and associated resources (declarative scenarios).
4. Describe and refine elicited resources' structures and functions (using CPSFA formalism).
5. Iterate until a satisfactory solution is found.

Resources are typically human and/or machine agents, in the AI terminology, and systems, in the SE sense. Contexts express persistent situations that can be either normal, abnormal, or emergency. Contexts are usually defined in the form of combined spatiotemporal conditions. For example, a postman job in a normal context can be expressed in terms of time (i.e., every weekday from 8 am to 5 pm) and space (i.e., a well-defined neighborhood).

The PRODEC method has been used in an air combat system project called MOHICAN. This project aimed at deriving performance metrics to assess collaboration between pilots and cognitive systems, as well as trust in such cognitive systems. Before deriving such metrics, relevant human and machine functions should be elicited and further tested against such metrics.

First, task analyses were developed in the form of procedural scenarios (i.e., BPMN graphs). Then, function analyses were developed in the form of CPSFA declarative scenarios (agent-based configurations). Acquired air combat functions knowledge greatly determined the kinds of metrics that should be used for performance evaluation. For example, the "Acquire Information" function could be assessed from various viewpoints that include accuracy, time, workload, meaningfulness, and so on. This depends on context and available resources. Functions, either physical or cognitive, were declared in terms of role, context, and resources.

An Illustrative Example of PRODEC Use

In the MOHICAN study, the BPMN-CPSFA PRODEC process describes (1) tasks to be completed within the cockpit; (2) their distribution among agents involved (e.g., Pilot and Weapon System Officer, decision-making assistant system); (3) required resources to complete each subtask (e.g., time, weapon system, air-to-air picture for situation awareness, etc.); and (4) the various agents as well as the interdependency between them (e.g., the pilot needs navigation information processed by the Weapon System Officer to achieve a subtask). When a satisfactory solution is found, it is typically implemented and tested into a human-in-the-loop simulation (HITLS). Testing results are then used to re-instantiate a BPMN-CPSFA PRODEC process. As an example, an emerging task, "Remind the pilot with safety altitude and safety heading," was discovered. This task is highlighted in the blue box in Fig. 10.

More specifically, when the situation degrades (e.g., autopilot height interception does not perform as expected), simulations have revealed the pilot's need for information (e.g., safety altitude reminder and heading to remain on the planned route) that experts didn't plan in their procedural projections. The main emergent function involved is "collaboration," which can be expressed in the form of role, context of validity, and resources required to make it useful and usable. It is clear in this example that the function "collaboration," implementing the task "remind the pilot with safety altitude and safety heading," can be allocated to either a human being (i.e., the Weapon System Officer) or a machine (i.e., a virtual assistant) thanks to an algorithm based on system status, flight parameters, and minimum height monitoring.

This example shows how emergent functions are discovered from an initial task analysis providing procedural scenarios (left hand side of Fig. 10), themselves used in HITLS (middle of Fig. 10) with subject matter experts (e.g., pilots), leading to

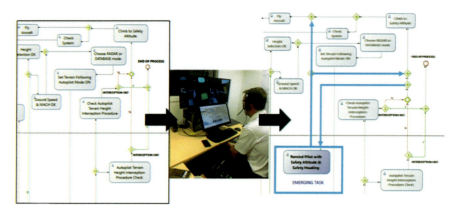

Fig. 10 A MOHICAN project's example of PRODEC use

activity observation and further analysis and finally discovering emergent behaviors, properties, and functions (left hand side of Fig. 10).

Discussion: Challenges, Gaps, and Possible Futures

Departing from Technology-Centered MBSE

Even if systems engineering (SE) is based on a holistic approach, it is often too much technology centered and not enough human centered. Consequently, it fails sometimes spectacularly [14]. Uncertainty management is one of the main reasons. World changes very fast. Therefore, requirements and solutions should be constantly adapted. SE definitely requires flexibility. This is the reason why HSI should be better developed based on domain experience, creativity, HITLS, activity analysis, human-centered complexity analysis, as well as organization design and management [9]. Unifying HCD and SE will shape appropriate HSI and facilitate the production of successful systems.

The recent SERC research roadmap listed seven requirements relevant to AI4SE [43]: tools and domain taxonomies and ontologies; semantic rules; inter-enterprise data integration; automated decision framework; authoritative data identification; digital assistance; and digital twin automation. This roadmap specifies that HSI "will no longer be a specialty area but will be front and center to system definition." This chapter brings solutions for the operationalization of this endeavor. In addition, AI is considered as data science, including machine learning. Case-based reasoning, multi-agent systems, and human-robot interaction appear to be useful AI approaches and methods. What is called "hybrid Human/AI systems" is based on automated reasoning machine agents helping humans understand complex data. The term "virtual assistant" was used in the MOHICAN project. Human-autonomy teaming between humans and increasingly autonomous machines is a crucial endeavor.

Human-Centered Modeling Limitations and Perspectives

What kinds of models will be useful and usable for HSI? Many human models for system analysis, design, and evaluation have been developed over the years. Let's cite a few. Originally, Simon and Newel's model [59] of information processing has been developed and extensively used in HFE and HCI. It was followed by Rasmussen's model [52] that provided a valuable and very much used framework for cognitive engineering. Besides these fundamental models, other more applied and targeted models were developed in the aerospace domain as support for simulation purposes, such as MIDAS [18] and MESSAGE [15]. These models attempted to mimic human behavior and cognitive processes (e.g., they were able to land an aircraft safely and mimic human errors). For example, MESSAGE has been used to figure out workload assessments during the certification of two-crewmen commercial aircraft. On another example, MIDAS supported exploration of computational representations of human-machine performance to aid designers of interactive complex systems by identifying and modeling human-automation interactions with flexible representations of human-machine functions. MIDAS helped producing guidelines in aeronautics and air traffic management [28].

Designers can work with computational representations of the human and machine performance, rather than relying solely on expensive hardware simulators and real flights with humans in the loop, to discover problems and ask "what-if" questions about the projected mission, equipment, or environment. The advantages of this approach are reduced development time and costs and early identification of human performance limits, plus support for training system requirements and development. This is achieved by providing designers accurate information early in the design process, so impact and cost of changes are minimal. After almost 40 years of experience of such simulated human models, it is obvious that there are several directions of investigation, such as virtual human-in-the-loop simulation (HITLS) where real humans interact with simulated machines and simple human models that support the development of metrics and scenarios useful for HITLS.

HCD Based on Virtual Environments as Digital Twins

The shift from twentieth-century engineering associated with corrective ergonomics to twenty-first-century human-centered design based on digital prototyping, HITLS, and tangibility assessment opens SE to a radical transformation where HSI becomes central. Engineering design enables now to involve humans in the loop within a control and management virtual space, which incrementally becomes more tangible [8]. Figure 11 presents the tangibilization process of virtual HCD.

The term "control & management space" in Fig. 11 is generic, referring to a control room or a vehicle cockpit. Since it is deliberately assumed that the environment is multi-agent, agents being people and/or machines, initial agents being designed are virtual. These agents do not include the real people who are interacting with the control and management space within which agents are incrementally

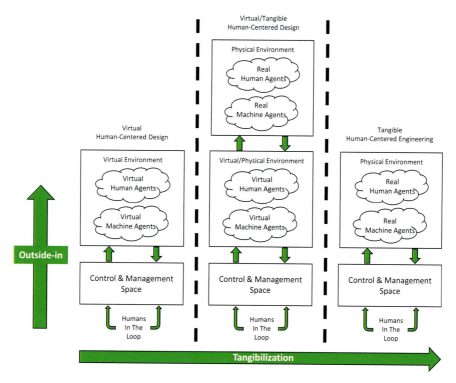

Fig. 11 Tangibilization process in three steps: from virtual to tangible

tangibilized in an incrementally more physical environment. For example, let's consider that our goal is the design and development of a fleet of robots replacing people on an oil and gas offshore platform. A control and management room (space) is developed in the first place where real people will have to deal with a simulator of a virtual fleet of robots both moving and interacting with a virtual oil and gas offshore platform. Activities of these people are observed and analyzed to produce modifications of structures and functions involved in the simulation. This virtual HCD (VHCD) process is further pursued until a satisfactory design is reached, in terms of safety, efficiency, and comfort, for example. Then, one or two robots, as well as the platform, can be started to be tangibilized in a physical playground. A mixed virtual/tangible HCD process can then be initiated as for the VHCD, same agile processes, and so on. The tangibilization process can then continue until everything is tangible.

In addition, once a human-machine system is fully developed, virtual environments that were used in VHCD can be used and refined as digital twins constantly evolving using experience feedback. In other words, system's interactive documentation is now available in the form of digital twins.

Summary

Summing up, model-based human systems integration (MBHSI) is a very promising field of investigation. Possible directions of research are not only on human modeling but also on HITLS and more specifically human-centered digital twins, with tangibility in mind. This chapter presented an approach that enables improvement of design flexibility, resource management, and system knowledge though HITLS. Therefore, virtual prototyping that supports virtual human-centered design needs to be continuously improved.

New approaches are currently developed using HCD in MBSE [36] and more specifically virtual HCD in increasingly autonomous complex systems [3]. From a more general standpoint, using human and social sciences in systems engineering (i.e., HSI) differs from using hard sciences that include science, technology, engineering, and mathematics (i.e., STEM disciplines) by the fact that they force to explore human systems activity. More specifically, the main difference between conventional MBSE [23, 39, 53], that is, typically technology centered, and MBHSI relies on constant search for emerging behaviors through activity observation and analysis and, consequently, agile incorporation of emergent functions and structures into human-centered systems engineering practice.

An epistemological endeavor is in front of us. HSI requires more fundamental research efforts on human-centered and organizational topologies and ontologies that should support MBHSI. A novel system science where technology, organizations, and people (i.e., The TOP model) can be studied together in a rational and consistent manner needs to be harmonized.

References

1. Aamodt A, Plaza E (1994) Case-Based Reasoning: Foundational Issues, Methodological Variations, and System Approaches. *Artificial Intelligence Communications,* 7(1):39–52
2. Amalberti R (2001) The paradoxes of almost totally safe transportation systems, *Safety Science*, 37:109–126
3. Bellet T, Deniel J, Bornard JC, Richard B (2019) Driver Modeling and Simulation to support the Virtual Human Centered Design of future Driving Aids. *Proceedings of INCOSE International Conference on Human Systems Integration (HSI2019),* Biarritz, France
4. Berkeley EC, Bobrow DG (eds) (1964) *The programming language LISP: Its operation and applications*. MIT Press, Cambridge, MA, USA
5. Bertalanffy L (1968) *General System Theory: Foundations, Development, Applications*, Revised ed., New York, NY, USA: Braziller
6. Boy GA (2020) *Human Systems Integration: From Virtual to Tangible*. CRC Press – Taylor and Francis Group, USA
7. Boy GA (2019) Cross-Fertilization of Human-Systems Integration and Artificial Intelligence: Looking for Systemic Flexibility. *AI4SE: Artificial Intelligence for Systems Engineering*. REUSE, Madrid, Spain
8. Boy GA (2016) *Tangible Interactive Systems*. Springer, UK
9. Boy GA (2013) *Orchestrating Human-Centered Design*. Springer, UK
10. Boy GA (ed) (2011) *Handbook of Human-Machine Interaction: A Human-Centered Design Approach*. Ashgate/CRC Press–Taylor and Francis Group, USA

11. Boy GA (1998) *Cognitive Function Analysis*. Praeger/Ablex, CT, USA
12. Boy GA, Dezemery J, Hein A, Lu Cong Sang R, Masson D, Morel C, Villeneuve E (2020) PRODEC: Combining Procedural and Declarative Knowledge for Human-Centered Design. FlexTech Technical Report, CentraleSupélec and ESTIA, France
13. Boy GA, Schmitt KA (2012) Design for Safety: A cognitive engineering approach to the control and management of nuclear power plants. *Annals of Nuclear Energy*. Elsevier. https://doi.org/10.1016/j.anucene.2012.08.027
14. Boy GA, Narkevicius J (2013) Unifying Human Centered Design and Systems Engineering for Human Systems Integration. In: Aiguier M, Boulanger F, Krob D, Marchal C (eds), *Complex Systems Design and Management*, Springer, U.K. 2014. ISBN-13: 978-3-319-02811-8
15. Boy GA, Tessier C (1985) Cockpit Analysis and Assessment by the MESSAGE Methodology. *Proceedings of the 2nd IFAC/IFIP/IFORS/IEA Conference on Analysis, Design and Evaluation of Man-Machine Systems*, Villa-Ponti, Italy, September 10–12. Pergamon Press, Oxford: 73–79
16. Carpenter TP (1986) Conceptual knowledge as a foundation for procedural knowledge. In: Hiebert J (ed), *Conceptual and procedural knowledge: The case of mathematics*. Lawrence Erlbaum Associates: 113–132
17. Caruso P, Dumbacher D, Grieves M (2010) Product Lifecycle Management and the Quest for Sustainable Space Explorations. *AIAA SPACE 2010 Conference and Exposition*. Anaheim, CA
18. Corker KM, Smith BR (1993) An architecture and model for cognitive engineering simulation analysis: Application to advanced aviation automation. *AIAA Computing in Aerospace 9 Conference*. San Diego, CA
19. Comte A (1865) *A General View of Positivism*. Translated by Bridges, J.H., Trubner and Co., 1865 (reissued by Cambridge University Press, 2009; ISBN 978-1-108-00064-2
20. Corbett AT, Anderson JR (1994) Knowledge tracing: Modeling the acquisition of procedural knowledge. *User modeling and user-adapted interaction*, 4(4):253-278
21. Deitel PJ, Deitel H (2019) *Python for Programmers: with Big Data and Artificial Intelligence Case Studies*. Pearson Higher Ed, ISBN-13: 978-0135224335
22. Endsley MR, Jones DG (2011) *Designing for Situation Awareness: An Approach to User-Centered Design*. CRC Press, 2nd edn. ISBN 978-1420063554
23. Estefan JA (2008) *Survey of Model-Based Systems Engineering (MBSE) Methodologies*. INCOSE MBSE Initiative. Rev. B. International Council on Systems Engineering. San Diego, CA
24. Garcia Belmonte N (2016) Engineering Intelligence Through Data Visualization at Uber. (retrieved on July 29, 2019: https://eng.uber.com/data-viz-intel/)
25. Glaessgen EH, Stargel D (2012) The digital twin paradigm for future NASA and US Air Force vehicles. *AAIA 53rd Structures, Structural Dynamics, and Materials Conference*. Honolulu, Hawaii
26. Grieves M (2016) *Origins of the Digital Twin*. Concept Working Paper. August. Florida Institute of Technology / NASA. https://doi.org/10.13140/RG.2.2.26367.61609
27. Hiebert J, Lefevre P (1986) Conceptual and Procedural Knowledge in Mathematics: An Introductory Analysis. In: Hiebert J (ed), *Conceptual and Procedural Knowledge: The Case of Mathematics*, 2: 1–27. Erlbaum, Hillsdale, NJ, USA
28. Hooey BL, Gore BF, Mahlstedt EA, Foyle DC (2013) *Evaluating NextGen Closely Spaced Parallel Operations concepts with validated human performance models: Flight deck guidelines*. NASA TM-2013-216506. Moffett Field, CA: NASA Ames Research Center
29. Husserl E (1989) *Ideas pertaining to a Pure Phenomenology and to a Phenomenological Philosophy*. Second Book. Trans. R. Rojcewicz and A, Schuwer. Dordrecht and Boston: Kluwer Academic Publishers. From the German original unpublished manuscript of 1912, revised 1915, 1928. Known as Ideas II
30. Husserl E (1963) Ideas: A General Introduction to Pure Phenomenology. Trans. W. R. Boyce Gibson. Collier Books, New York. From the German original of 1913, originally titled Ideas pertaining to a Pure Phenomenology and to a Phenomenological Philosophy, First Book. Newly

translated with the full title by Fred Kersten. Dordrecht and Kluwer Academic Publishers, Boston, 1983. Known as Ideas I
31. International Council on Systems Engineering (INCOSE), A World in Motion - Systems Engineering Vision 2025, July 2014
32. Ionesco E (1996) *Between life and dream (Entre la vie et le rêve)*. Interview with C. Bonnefoy. Collection Blanche, Gallimard, Paris, France
33. Kruchten N (2018) Data visualization for artificial intelligence, and vice versa. (retrieved on July 30, 2019: https://medium.com/@plotlygraphs/data-visualization-for-artificial-intelligence-and-vice-versa-a38869065d88)
34. Kupferschmid M (2002) *Classical Fortran: Programming for Engineering and Scientific Applications*. CRC Press. ISBN 978-0-8247-0802-3
35. Keim D, Andrienko G, Fekete JD, Carsten Görg C, Kohlhammer J, Melançon G (2008) Visual Analytics: Definition, Process and Challenges. In: Kerren A, Stasko JT, Fekete JD, North C., *Information Visualization - Human-Centered Issues and Perspectives*, Springer: 154–175
36. Kim SY, Wagner D, Jimenez A (2019) Challenges in Applying Model-Based Systems Engineering: Human-Centered Design Perspective. *Proceedings of INCOSE International Conference on Human Systems Integration (HSI2019)*, Biarritz, France
37. Laurel B (1991) *Computers as Theatre*. Addison-Wesley. ISBN 0201550601.
38. Lewicki P, Czyzewska M, Hoffman H (1987) Unconscious acquisition of complex procedural knowledge. *Journal of Experimental Psychology: Learning, Memory, and Cognition*, *13*(4), 523
39. Long D, Scott Z (2011) *A primer for Model-Based Systems Engineering*. Vitech Corporation. ISBN 978-1-105-58810-5
40. Madni AM, Madni CC, Lucero SD (2019) Leveraging Digital Twin Technology in Model-Based Systems Engineering. *Systems*. https://doi.org/10.3390/systems7010007
41. McCormick R (1997) Conceptual and procedural knowledge. *International journal of technology and design education*, 7(1–2):141-159
42. McCracken DD (1961) *A Guide to FORTRAN Programming*. Wiley, New York. LCCN 61016618
43. McDermott T, DeLaurentis D, Beling P, Blackburn M, Bone M (2020) AI4SE and SE4AI: A Research Roadmap. *InSight Special Feature*. Wiley Online Library. https://doi.org/10.1002/inst.12278
44. Minsky M (1986) *The Society of Mind*. Touchstone Book. Published by Simon and Schuster, New York
45. Muller MJ (2009) Participatory Design: The Third Space in HCI. *The Human-Computer Interaction Handbook*, Hillsdale: L. Erlbaum Assoc: 1061–1081
46. Nielsen J (1993) *Usability Engineering*. Academic Press, Boston, MA. Available on amazon.com
47. Piascik R, Vickers J, Lowry D, Scotti S, Stewart J, Calomino A (2010) *Technology Area 12: Materials, Structures, Mechanical Systems, and Manufacturing Road Map*. NASA Office of Chief Technologist
48. White, S.A. (2004). Business Process Modeling Notation. Retrieved on November 23: https://web.archive.org/web/20130818123649/http://www.omg.org/bpmn/Documents/BPMN_V1-0_May_3_2004.pdf
49. Piaget J (1952) The Origins of Intelligence in Children. New York: Norton
50. Piaget J (1954) *The Construction of Reality in the Child*. New York: Ballantine
51. Prinz P, Crawford T (2015) *C in a Nutshell: The Definitive Reference 2nd Edition*. Kindle Edition. O'Reilly Media, ASIN: B0197CH96O
52. Rasmussen J (1983) Skills, rules, knowledge; signals, signs and symbols, and other distinctions in human performance models. *IEEE Transactions on Systems, Man and Cybernetics*. Vol. 13: 257-266
53. Reichwein A, Paredis C (2011) Overview of architecture frameworks and modeling languages for model-based systems engineering. In *ASME Proceedings*: 1–9

54. Richard JF (1983) *Engineering logic versus use logic* (*Logique de fonctionnement et logique d'utilisation*). Research Report, # 202, INRIA, Le Chesnay, France
55. Russel S, Norvig P (2010) *Artificial Intelligence – A Modern Approach*. Third Edition. Prentice Hall, Boston, USA. ISBN 978-0-13-604259-4
56. Schneider M, Rittle-Johnson B, Star JR (2011) Relations among conceptual knowledge, procedural knowledge, and procedural flexibility in two samples differing in prior knowledge. *Developmental Psychology*, *47*(6), 1525
57. White, S.A. & Bock, C. (2011). *BPMN 2.0 Handbook Second Edition: Methods, Concepts, Case Studies and Standards in Business Process Management Notation*. Future Strategies Inc. ISBN 978-0-9849764-0-9.
58. SEBoK (2020) Guide to the Systems Engineering Book of Knowledge (retrieved on 5 May 2020: https://www.sebokwiki.org/wiki/What_is_a_System%3F)
59. Simon HA, Newell A (1971) Human problem solving: The state of the theory in 1970. *American Psychologist, 26*(2), pp. 145-159. https://doi.org/10.1037/h0030806
60. Star JR (2005) Reconceptualizing procedural knowledge. *Journal for research in mathematics education*: 404–411
61. St. Amant R, Healey CG, Riedl M, Kocherlakota S, Pegram DA, Torhola M (2001) Intelligent visualization in a planning simulation. Proceedings Intelligent User Interfaces IUI'01). Santa Fe, New Mexico. ACM. 1-58113-325-1/01/0001
62. Tuegel EJ, Ingraffea AR, Eason TG, Spottswood SM (2011) Reengineering Aircraft Structural Life Prediction Using a Digital Twin. *International Journal of Aerospace Engineering*
63. Willingham DB, Nissen MJ, Bullemer P (1989) On the development of procedural knowledge. *Journal of experimental psychology: learning, memory, and cognition*, *15*(6), 1047
64. Wilson RA, Keil FC (eds) (2001) *The MIT encyclopedia of the cognitive sciences*. MIT Press, Cambridge, MA. USA
65. Winograd T, Flores F (1986) *Understanding Computers and Cognition: a new foundation for design*. Initially published by Ablex Publishing Corporation, Norwood, NJ. Now Addison-Wesley, Boston, MA, USA. ISBN 978-0-201-11297-9
66. Wirth N (1971) *The Programming Language Pascal*. Acta Informatica, Volume 1:35–63

Guy André Boy, Ph.D., is FlexTech Chair Institute Professor at *Paris Saclay University* (CentraleSupélec) and Chair of *ESTIA* Science Board, Fellow of the *Air and Space Academy*, and Chair of the *Human-Systems Integration Working Group* of International Council on Systems Engineering (INCOSE). He was University Professor and Dean, *School of Human-Centered Design, Innovation and Art*, and HCD Ph.D. and Master's Programs at the *Florida Institute of Technology* (2009–2017), as well as a Senior Research Scientist at the *Florida Institute for Human and Machine Cognition* (IHMC). He was Chief Scientist for Human-Centered Design at *NASA Kennedy Space Center* (2010–2016). He was member of the Scientific Committee of the SESAR program (*Single European Sky for Air Traffic Management Research*) from 2013 to 2016. He was the Chair of the 2012 *International Space University* (ISU) *Space Studies Program* (SSP12) FIT/NASA-KSC local organizing committee. He was Adjunct Professor at the *École Polytechnique* in Paris. He was Board Member of the Complex Systems Engineering Master Pedagogic Committee at *Paris Saclay University*. He was the President and Chief Scientist of the *European Institute of Cognitive Sciences and Engineering* (EURISCO, a research institute of Airbus and Thales). He co-founded EURISCO in 1992 and managed it since its creation to its closing in 2008. Between 1980 and 1991, he worked in artificial intelligence and cognitive science for ONERA (the *French Aerospace Lab*) as a research scientist and *NASA Ames Research Center* as the *Advanced Interaction Media Group* Lead (1984–1991). Engineer and cognitive scientist, he received his masters and doctorate degrees from the French Aerospace Institute of Technology (ISAE-SUPAERO), his Professorship Habilitation (HDR) from *Pierre and Marie Curie Sorbonne University*, and his Full Professorship Qualifications in Computer Science and Psychology. Boy actively participated to the introduction of *cognitive engineering* in France and its development worldwide. He was the co-founder in 2004 of the *French Cognitive Engineering School* (ENSC) in Bordeaux. He co-founded the HCI-Aero Conference series, in cooperation with the *Association for Computing Machinery* (ACM). He is the author of more than 200 articles and two textbooks, *Intelligent Assistant Systems* (Academic Press, USA, 1991) and *Cognitive Function Analysis* (Praeger, USA, 1998), and the editor of the *French Handbook of Cognitive Engineering* (Lavoisier, France, 2003) and the *Handbook of Human-Machine Interaction* (Ashgate, UK) in 2011. His most recent books are *Orchestrating Human-Centered Design* (Springer, UK, 2013), *Tangible Interactive Systems* (Springer, UK, 2016), *Human Systems Integration: From Virtual to Tangible* (CRC Taylor and Francis, USA, 2020), and Design for Flexibility: A Human Systems Integration Approach (Springer Nature, Switzerland, 2021). He is a senior member of the ACM (Executive Vice-Chair of ACM-SIGCHI from 1995 to 1999), Corresponding Member of the *International Academy of Astronautics*, and Chair of the Aerospace Technical Committee of *International Ergonomics Association* (IEA).

Model-Based Hardware-Software Integration

16

Joe Cesena

Contents

Introduction (Problem Statement, Key Concepts, Terms, and Definitions)	502
State-of-the-Art (Review of the Literature)	503
Hardware-Software Integration Modeling Methodologies	506
Functional Allocation	506
Instance Specifications for Data Intensive Applications	509
Capturing Data as Properties	509
Typing Ports	510
Item Flows	511
Allocation Relationships	512
Illustrative Example	513
Chapter Summary	521
References	522

Abstract

This chapter discusses modeling hardware-software integration and examines different methodologies that may be used in capturing hardware-software (HW-SW) integration relationships. As such, this paper does not focus on definition or elaboration of the Systems Modeling Language (SysML) constructs themselves, so much as how those constructs may be applied when describing relationships between physical and functional (behavioral) elements.

Model-based systems engineering (MBSE) and SysML allows for the creation of relationships to other data elements such as behavior or commands and monitors that software uses. For example, interface blocks that capture electrical signal characteristics or protocol definitions may be typed to pins used by an internal block diagram (IBD), or activity diagram swim lanes that capture component functions may be allocated to some other system model element such as a block or part. Note that if the swim lanes are typed by physical elements, this can

J. Cesena (✉)
Lockheed Martin, Sunnyvale, CA, USA
e-mail: joseph.r.cesena@lmco.com

© Springer Nature Switzerland AG 2023
A. M. Madni et al. (eds.), *Handbook of Model-Based Systems Engineering*,
https://doi.org/10.1007/978-3-030-93582-5_82

also help in understanding the functional allocation of behavior to physical elements in the system. Leveraging capabilities such as this, in addition to capabilities such as defining low-level information in blocks or instances so they may be aggregated and typed to activity parameters and ports, provide powerful mechanisms to tie detailed hardware characteristics to behavior and thus enhance hardware-software integration.

Keywords

Model-based systems engineering · Hardware-software integration · Embedded systems · SysML

Introduction (Problem Statement, Key Concepts, Terms, and Definitions)

Hardware-software (HW-SW) integration plays a key role in creating deterministic outcomes and realizing the expected behaviors in a system. Without HW-SW integration, there is no guarantee that when a user (or some other element within the system) sends the system a command, the system will respond correctly to the command. Or in the case of receiving data back from the system (usually in the form of telemetry), there is no guarantee that the system would know what data it is receiving. As such, HW-SW integration is a key activity in developing systems and integrates disciplines such as software engineering, electrical integration, and fault management, into the overall behavior and command and data handling operation of a system.

To couple the desired or relevant physical elements of a system (such as processors, pins, or wires) with the functional behavior (often defined by software), it is often necessary to associate a representation of a physical element (such as a computer or processor) or interface (wire, connector, or pins) with the functional behavior invoked. For example, it may be desired to associate a software command that travels over an interface with specific electrical characteristics of the data, clock, and strobe pin definitions that are necessary to support a given communication protocol.

Model-based system engineering offers a very effective platform in performing HW-SW integration. In the modeling environment, a user is able to capture the data of interest and create relationships between the different data elements, at varying levels of abstraction. For example, blocks, parts, and ports could be part of block definition diagrams (BDDs) and internal block diagrams (IBDs), respectively, that capture structure. The structure could represent systems that have interfaces, connectors, or message protocols.

Additionally, the SysML model environment allows for the creation of relationships to other data elements such as behavior or commands and monitors that software uses. For example, interface blocks that capture electrical signal characteristics or protocol definitions may be typed to pins used by an IBD, or activity diagrams themselves may be allocated to some other system model element such as a block or part. In the case of activity diagrams, if the swim lanes themselves are typed by physical elements, this can also help in understanding the functional allocation of behavior to physical elements in the system. Note that the techniques described in this chapter focus on taking advantage of some of the capabilities provided with activity diagrams in the SysML 1.5 definitions of object flows, pins, and the ability to type swim lanes by either blocks or parts.

Finally, the data and relationships captured in the SysML model may be communicated in traditional representations such as interface control documents (ICDs) or other views and formats that are optimized to stakeholder needs.

The following sections describe MBSE approaches that help capture the HW-SW interface(s) and thus help provide early validation that the interface definition has been correctly captured, thus avoiding costly mistakes later that could result in either hardware or software changes to correct for the intended system behavior.

State-of-the-Art (Review of the Literature)

This section describes the existing approaches and methods for performing the HW-SW integration function using model-based systems engineering (MBSE). Additionally, an overview of model frameworks used to perform the HW-SW integration task will be provided, in addition to a short summary of which Systems Modeling Language (SysML) constructs are often used to capture HW/SW integration considerations. However, first it may make sense to first try to assess the state of hardware-software integration, before describing how to do it in a model-based way.

Hardware-software integration can span from very low-level, system on a chip (SoC) applications all the way up to aerospace applications in which multiple computers may be communicating to each other over digital interfaces. In either case, there is often hardware from which data must be extracted, a sensor from which a signal is read (note that this sensor may convert an analog signal to a digital signal), a connector such as a cable, or, in commercial applications, a wireless connector such as Wi-Fi or Bluetooth and finally software that interprets the data and may convert the interpretation into action vis-à-vis a command to a mechanism or another computer.

As such, the hardware-software integration task often then involves ensuring that the various hardware platforms are compatible. In other words, the interfaces between hardware, sensors, and connectors are all electrically compatible. On the software side, given the arrival and assessment of information, the proper behavior and functions are performed. Finally, hardware-software integration is ensuring that together, they satisfy the system needs and requirements.

Note that at the HW-SW integration level, nuances begin appearing with considerations such as at what rate can software issue commands to computers or mechanisms on the receiving end. Alternatively, how quickly can software read and evaluate data that it receives? Or finally, are there constraints that must be adhered to imposed by the hardware and/or the environment?

While hardware-software integration is recognized as part of the engineering development process, there is not much literature describing generalized best practices in HW-SW integration as a discipline unto itself. This may be due to the very application-specific nature that doing this kind of integration takes on, since it often cuts to the core of the specific hardware/software application and design of a given system.

If the HW-SW integration aperture is now expanded to include model-based systems engineering, three modeling standards emerge as the primary workhorses in capturing, describing, and doing the requisite analysis for hardware-software integration. The first is the Object Management Group (OMG), Unified Modeling Language (UML), and Modeling and Analysis of Real-Time and Embedded Systems (MARTE) profile.

The MARTE profile adds capabilities for model-driven development of Real Time and Embedded Systems (RTES), to provide support for specification, design, and verification/validation. MARTE focuses on performance and scheduling analysis; however it also defines a general framework for quantitative analysis which intends to refine/specialize other kinds of analysis. Many papers are available that explore in more detail the use of MARTE in capturing and analyzing real-time embedded systems and as such will not be covered here.

The second modeling framework is the Architecture Analysis and Design Language (AADL) framework, which is also referred to as the Society of Automotive Engineers (SAE) AADL AS5506. Like MARTE, AADL defines a language for describing both the software architecture and the execution platform architectures of embedded and real-time systems. AADL categorizes the hardware components into "execution platforms" and "applications" which represent the software module applications. AADL is extensible to UML, including a UML profile to assist in capturing and integrating a broad range of other real-time concurrent processing domains. Again, there is literature available that explores the use of AADL to capture, define, and analyze real-time embedded systems. As such, this chapter will defer to that literature to explore more deeply the application of AADL to embedded systems.

The third modeling framework, and the focus of this chapter, is use of the System Modeling Language (SysML) to capture, describe, and perform the requisite analysis for hardware-software integration. Like MARTE and AADL SysML provides the necessary constructs to capture both the hardware and software elements to analyze and perform the hardware-software integration task.

Looking across the literature to find previous work in describing the use of SysML for the hardware-software integration application, there is a paper by Shames, Sarrel, and Friedenthal presented at the 26th Annual INCOSE International Symposium (IS 2016) titled, "A Representative Application of a Layered Interface Modeling Pattern." This paper describes a SysML-based hardware-software integration approach that presents a layered methodology. This methodology begins with the necessary hardware abstractions (e.g., subsystem, component, and sub-component) and, at the protocol layer, uses behavioral elements (state machines, sequence diagrams, and activity diagrams) to capture functions at each side of the interface.

Additionally, when looking across the SysML literature, it is commonplace for SysML textbooks to define the behavioral and physical architecture constructs of the language that may in turn be used to represent or capture the behavioral or architectural items of interest to modeler and stakeholders. However, some constructs are particularly suited to the hardware-software integration task, and as such, some of the common SysML constructs used for modeling hardware-software integration examples used in this chapter are summarized in Table 1 as a quick reference. Note

Table 1 Common SysML constructs used for HW-SW integration

SysML construct	Description
Block	An element that describes the structure of a system or other element of interest. A block may include both structural and behavioral features that represent the state of the system and behavior that the system may exhibit
Part property	A property that specifies a part with strong ownership by its containing block. It describes a local usage or a role of the typing block in the context of the containing block
Internal block diagram (IBD)	An IBD is a static structural diagram owned by a block that shows its encapsulated structural contents: Parts, properties, connectors, ports, and interfaces
Block definition diagram (BDD)	A BDD is a static structural diagram that shows system components, their contents (properties, behaviors, constraints), interfaces, and relationships
Instance specification	An instance configuration can be used if a block simply consists of a set of values or properties
Interface block	A specialization of a block that do not have behaviors or internal parts. Interface blocks are used to type proxy ports and block properties
Activity diagram	A diagram that represents the sequence of actions that describe the behavior of a block or other structural element; the sequence is defined using control flows. Actions can contain input and output pins that act as buffers for items that flow from one action to another as the task carried out by the action either consumes or produces them. These items can be information or anything else that can be produced, conveyed, or consumed, depending on the modeling objectives and stakeholder needs
Reference association	A reference association between two blocks means that a connection can exist between instances of those blocks in an operational system. Additionally, those instances can access each other for some purpose across the connection
Stereotype	A stereotype defines how an existing metaclass may be extended, enabling the use of a platform, a domain-specific terminology, or a notation in place of, or in addition with, the ones used for the extended metaclass
Allocation	A mechanism for associating elements of different types, or in different hierarchies, at an abstract level
Item flow	An item flow describes the flow between items

that details of these constructs can be found in any number of SysML textbooks and as such will not be repeated here. Note that specific needs could also drive the use of elements that are not listed.

In addition to the core SysML constructs that are defined and described through various texts, professional organizations such as the International Council of Systems Engineering (INCOSE) and the Object Modeling group (OMG) also provide access to papers and presentations that speak to specific use cases and examples where a specific HW-SW integration task or embedded system description was performed.

This chapter intends to describe generalized approaches that may be applied and tailored to specific modeling objectives and stakeholder needs.

Hardware-Software Integration Modeling Methodologies

The following sections provide overviews of different modeling approaches that may serve as different tools that a modeler can put "in the toolbox" for hardware-software integration. The "correct" tool or approach really depends on what the objectives of the modeler and stakeholders are, and as such, the correct approach may be a single view or diagram or may be some combination of different approaches.

The approach below describes the mechanisms and lays the groundwork to not only provide the hooks to describe hardware-software integration at the protocol level (e.g., specific 1553 message protocols or specific TCP/IP message protocols) but also up at the "system software level" where different elements may come into play that introduce timing or sequencing constraints when interacting with other hardware components and/or software processes hosted on other platforms.

The System Modeling Language (SysML) offers specific diagrams to address behavior (activity diagrams, sequence diagrams, and state machines), as well as specific diagrams to address structure (block definition diagrams (BDDs), and internal block diagrams (IBDs)). However, to perform the hardware-software integration function, techniques that allow for the intersection of these SysML constructs will be used to gain insight into what elements of the hardware facilitate the desired behavior of a system. In other words, defining and capturing the pathways signals take to invoke the desired behavior.

Functional Allocation

Not to be confused with the SysML allocation (which will be discussed in more detail later), functional allocation is a straightforward way to allocate functions or behavior that software may perform. In complex systems, this can be an important activity to define which element or component is responsible for a given function, in addition to identification of an interface between two elements or components.

For example, in cases of fault management, this kind of functional allocation can be very helpful and insightful in understanding if requirements to maintain safety are necessary or performed within a particular component and/or identification of interfaces in which notification messages are sent to other elements as information and/or to take action upon.

Figure 1 describes how a block definition diagram (BDD) can have a block elaborated through an internal block diagram (IBD). The model elements define the structure of a system or assembly, and in particular, the IBD elaborates the interfaces between parts. Note that the IBD can also define ports on the interfaces that allow us to "type" those ports. The ability to type the port with a block or interface block allows us to augment the ports with additional data.

Additionally, the activity diagram swim lanes can then be typed by the relevant parts of the block (note that it would also be acceptable to type the swim lanes by the blocks as well). Activity diagrams are excellent mechanisms to capture the functions that the system or assembly performs. In turn, these functions or behaviors are often performed by software. In addition to being able to type the swim lanes with the relevant architectural elements, typing the activity diagram pins provides another tool or mechanism in which physical characteristics may be intertwined with behavioral characteristics and thus integrate hardware and software.

While there is no prescriptive sequence to begin the HW-SW integration effort, it is often helpful to capture the expected system behavior as the architecture is defined. Capturing the behavior in an activity diagram allows for the allocation of behaviors to the relevant architectural components and demonstrates how the components interact to achieve the objective behavior, such as a control computer issuing a command to a component as an example. Or alternatively, a closed-loop control behavior may be described in which data is read in by the control computer, and based on the value, a particular command may be issued (on or off, left or right, etc.)

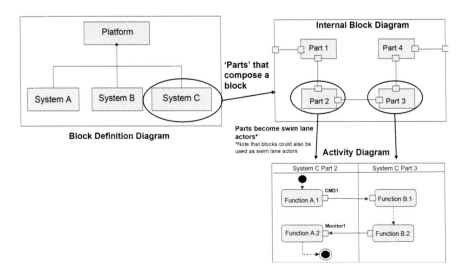

Fig. 1 Combination of HW and behavioral elements

Beyond just the definition of behavior, the activity diagrams possess key features that help with HW-SW integration. Once the swim lanes actors have been typed to identify the architectural elements, the object flows between the actions/functional behaviors illustrate the accepted inputs and generated outputs. Additionally, the pins on the activity diagram action may be typed with pertinent information such as the name of the command issued, the name of the monitor received, and/or the interface type or other information the modeler may find relevant that is captured in the block used to type the pin.

Additionally, a simple way to capture identification of interfaces is to utilize naming conventions to identify the signal types (or other identifier relevant to the modeler) when defining the pins. Doing so makes it readily apparent to a user what signal type (as an example) a given object flow represents (e.g., SpaceWire, Analog, 1553, etc.). As mentioned above, the pins also provide us an opportunity to type the interface with more detailed information. Figure 1 shows a pin named as CMD1 and another pin named Monitor1, but the selected name should be one that is meaningful to the user. Note as well that the object flow itself could be named.

Finally, when discussing activity diagrams in the context of HW-SW integration, an important aspect to capture with respect to HW-SW integration is timing considerations. Often different interface types have different bandwidth or data throughput capabilities which can drive the frequency at which either software can read in data or issue commands. SysML 1.5 provides a mechanism to capture constraints, which can be attached with notes, and with some modeling tools within the action. In either case, the constraint is contained within curly brackets. Figure 2 illustrates a simple case in which a 100-millisecond time duration has been defined between commands

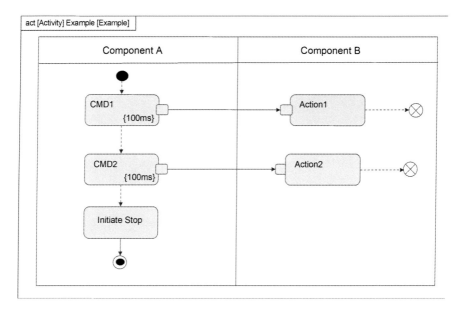

Fig. 2 Activity diagram with timing constraints that define the time interval between commands

that are issued. Capturing timing considerations such as these can be key drivers in communicating to software what constraints are in place or exist with respect to latency and bandwidth considerations as they apply to various hardware elements.

Instance Specifications for Data Intensive Applications

When modeling and capturing HW-SW Integration information, complex systems can often contain thousands, if not tens-of-thousands pin connections, registers, connectors, or wires that support the behavior invoked. As such, instance specifications may be used to capture data that share similar characteristics. (A detailed description of instance specifications can be found in the SysML 1.5 specification.)

While loading this data, it is also possible to use other element types (such as blocks) to represent the desired elements (pins, registers, etc.). However, instance specifications are recommended here since elements (pins or registers as an example) of the same data type do not require any relevant attributes to be re-defined/recreated every time a "new" element is created as they would with a block. Figure 3 illustrates how a block may be created that defines a pin of type RS422. Instances may then be created of this block that represent several individual pins, all of the same type. (Note that while instances are recommended in this case, it is recognized that the modeler may have reasons to represent this data as other model elements such as blocks.)

Capturing Data as Properties

Once the data has been captured (either as blocks or instances), the data can then be used to create flow properties. Flow properties can be useful to define for HW-SW

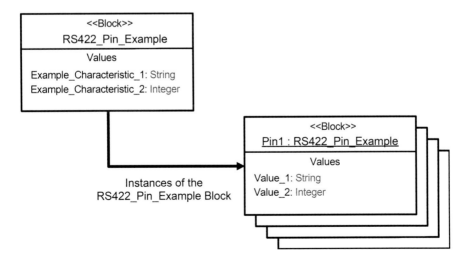

Fig. 3 Instance specifications may be made from a block that represents data of a similar type or grouping (RS422, SpaceWire, etc.)

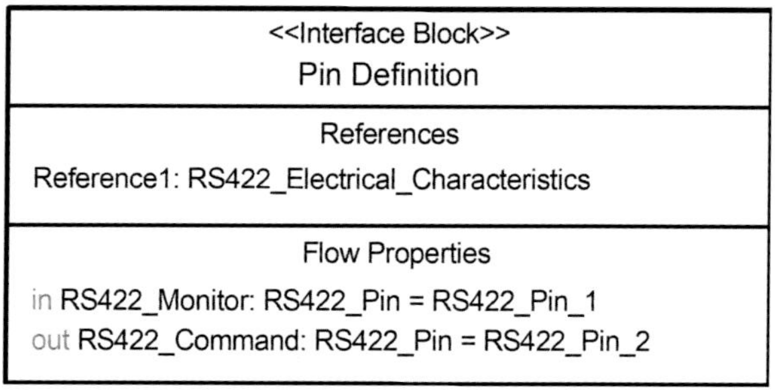

Fig. 4 Interface blocks may be used to capture flow properties or reference properties

<<Interface Block>>
Pin Definition
References
Reference1: RS422_Electrical_Characteristics
Flow Properties
in RS422_Monitor: RS422_Pin = RS422_Pin_1 out RS422_Command: RS422_Pin = RS422_Pin_2

integration purposes, since they allow us to not only capture the defined information (data such as commands or telemetry monitors), or pins, or whatever is expected to flow over the interface.

If on the other hand the data is more characteristic in nature, such as electrical signal characteristics, other properties such as a reference property (if the data has already been defined as a block or instance) or a value property may be used.

Figure 4 is an example of an interface block that contains both reference properties and flow properties. The reference property is typed by the "RS422_Electrical_Characteristics" block, which (as an example) could contain value properties that define the relevant electrical signal characteristics for that interface. Additionally, the flow properties specify the flow direction ("in" for monitors and "out" for commands). The RS422_Monitor flow property is typed by the "RS422_Pin" type. Since the RS422_Pin block has an instance that represents a specific pin, the "RS422_Pin" has a default value of RS422_Pin_1, which, for this example, is the pin that carries the monitor signal. As a compliment to the monitor example, the command has the flow property direction reversed, typed by the RS422_pin block (same signal type as the telemetry monitor), with a default value that represents a different pin.

With interface blocks created that capture and collect the relevant data, these blocks may be used as powerful mechanism to now type ports with detailed information that describes the interface.

Typing Ports

As discussed previously, once data of interest has been captured in an interface block as either reference properties or flow properties, this block may then be used to type ports on internal block diagrams (IBDs). The capability to type ports in this manner allows the modeler freedom to create abstracted diagrams that help improve readability and understanding while maintaining the detailed information as typed ports. Note also that as long as the data is captured in the typed port, the data may then be exported and included as a table that may accompany an interface control document or any other report or documentation that is exported out of the model.

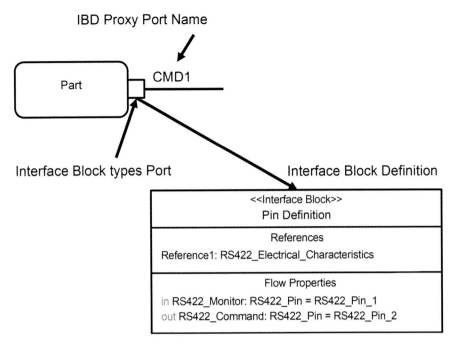

Fig. 5 An interface block may be used to type a port on an internal block diagram to augment the diagram with additional information

It should be noted as well that as a modeling consideration, there is some "level-setting" in terms of the scope and detail that a diagram contains, as well as the information and detail that is contained in the interface block that is used to type the pin or port. For example, if the IBD contains ports for every pin, an interface block that is used to type that port (which represents a pin) may only contain electrical characteristics for that pin. These kinds of considerations are left to the modeler, and the stakeholder objectives set forth when creating the model (Fig. 5).

Item Flows

Item flows are another method that may be used to capture information describing the hardware software interface. Item flows can be placed on internal block diagrams (IBDs), allowing for the direction in which data flows to be specified. Items flows can be typed by the blocks that are described in section "Typing Ports"; and as that section describes, those interface blocks can be defined with flow properties or reference properties as needed.

However, it should be noted that per the name, item flows are really intended to capture the data that flows between interfaces. As such, item flows are a perfectly suitable construct for things such as data, commands, or telemetry monitors that flow.

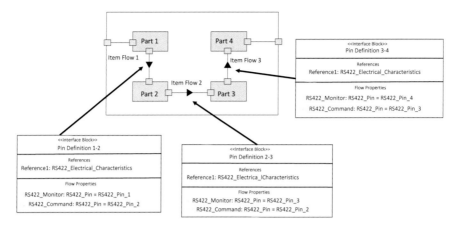

Fig. 6 This illustration shows how interface blocks can be used as item flow properties that capture the data (monitors and commands) that flow over the interface

Further, item flows can be allocated to the object flows defined in activity diagrams, creating an explicit linkage between the interface defined that supports the behavior defined in the activity diagram. Allocation relationships will be discussed in more detail in section "Allocation Relationships."

Figure 6 conceptually illustrates (and therefore not intended to be SysML compliant) how interface blocks may be used as an item flow property, that, in turn, captures the monitors and commands as flow properties.

Allocation Relationships

Finally, a SysML construct that can also be very useful in associating structure to behavior to perform hardware-software (HW-SW) integration is the allocation relationship. The allocation relationship is a mechanism that allows us to associate elements of different types, in this case behavior with structure.

As mentioned in section "Item Flows," allocation relationships may be used to associate object flows defined in activity diagrams with item flows used in internal block diagrams (IBDs). Relationships such as these may be made between any structural and behavioral elements that should be associated. For example, allocations may be made at the diagram level between the relevant activity diagrams and IBDs, or if data such as commands or monitors are captured as either instances or blocks, this data can be allocated to IBD ports, connectors, or item flows (if not already captured as part of an interface block). Additionally, allocations between functions, and the elements that host those functions can be a more explicit way to show functional allocation than just having the function "owned" by a swim lane actor.

Illustrative Example

To illustrate the methods to perform HW-SW integration as described in section "Hardware-Software Integration Modeling Methodologies," they will be applied to a simple example satellite component that will be referred to as the satellite electronics box (SEB). The scope of this example will be aligned to provide the kind of detail and data that is normally captured as part of an interface control document (ICD) that describes the external interfaces for the component under analysis (it is expected that in practice the SEB would have interfaces to other components that it communicates with). To be specific, this includes:

- High-level architecture description (e.g., cards that compose the component and contain any relevant external interfaces – as such, internal interfaces (bottom of card connections/backplane interfaces) of the component will be considered out of scope and not described as part of this example).
- Behavioral descriptions of box and card functions (this example will focus on power-on/start-up) including any timing constraints.
- Definition of electrical characteristics to include signal characteristics, protocols, and pins.

To begin, a block definition diagram is created that describes the high-level architecture of the component (satellite electronics box) for this example (Fig. 7):

With the architecture defined, an activity diagram can now be created that describes the power-on of the satellite electronics box (SEB) and contains the SEB parts as swim lanes. Note that the activity diagram has a pre-condition which assumes that operational voltage is present at the SEB when the power supply card closes the power relay. (This command presumably comes from the primary computer on the satellite) (Fig. 8).

When looking at the activity diagram, there are some things to note. First, the activity diagram is a key construct in terms of software integration. Just by

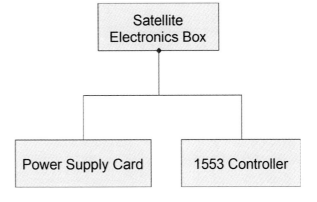

Fig. 7 This block definition diagram defines two part properties of the satellite electronics box (SEB). The power supply card and the 1553 controller

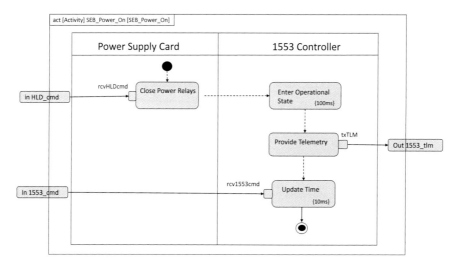

Fig. 8 This activity diagram defines the power-on sequence for the On-Board Processor

examination of the diagram, it can be seen that software must perform the following functions:

- Issue a discrete command to close the power relays for the satellite electronics box (SEB). (Note the pre-condition mentioned earlier in which it is assumed that operational voltage is present at the relay.)
- After application of power, the 1553 controller, which accepts command and provides telemetry from SEB, will begin providing telemetry after 100 ms (as captured in the timing constraint).
- The final step is to set the time, which comes over 1553 command. Also note the timing constraint here which allows the SEB 10 milliseconds to implement the time update.
- One other feature to note – by using a naming convention for the activity parameters and the action pins, the type of interface that the commands are issued over can be seen. High-Level Discrete (HLD) for closing the power relays and 1553 for setting the time. In addition, if specific commands and monitors are defined, the activity diagram parameters can by typed by interface blocks or allocated by the specific command or monitor.

This very simple example demonstrates the concepts and kind of information that the activity diagram conveys that helps assist with HW-SW integration. Some applications may benefit from additional state machine descriptions or sequence diagrams (inclusion of these additional diagrams is up to the modeler and/or stakeholder needs). However, now that the general behavior has been defined for this power-on activity, the supporting hardware configuration that enables the behavior can be explored in more depth.

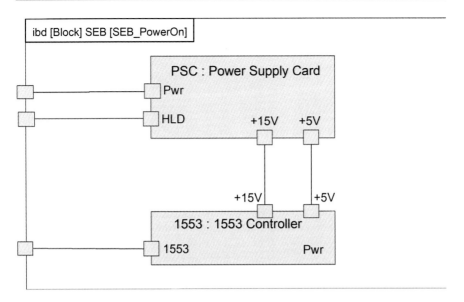

Fig. 9 This internal block diagram (IBD) defines the external interfaces for the power supply card (PSC) and 1553 controller parts of the satellite electronics box (SEB)

It is easiest to begin the electrical definition that supports the SEB power-on with an internal block diagram. The internal block diagram (IBD) defines the signal interfaces for the power supply card and the 1553 controller. As part of this definition, the IBD contains proxy ports, which may be used to type the interfaces with the electrical data of interest.

From the IBD in Fig. 9, it can be seen that there are three interface types captured – High-Level Discrete (HLD), 1553, and an internal power interface between the power supply card and the 1553 controller. Interface blocks can now be defined to capture the desired electrical characteristics.

However, before diving into the interface blocks, it is worth noting some of the data that will go into the interface block. First, blocks that capture the electrical characteristics for the 1553 interface, and the electrical characteristics for the High-Level Discrete interface, as well as capturing the power characteristics for the power interface for the power supply card are defined.

Note that for this example, the internal power interfaces between the power supply card and the 1553 controller are not expanded upon. Additionally, the MIL-STD-1553 specification for the 1553 electrical characteristics is simply referenced. However, should it be desired by the modeler to add more detail in these areas, the pattern remains the same. Internal interfaces (proxy ports) would be typed by blocks that contain the relevant data, and value properties within those blocks would have the desired information.

The block definitions that will serve as the basis for the electrical properties used/referenced by the interface blocks are defined below in Fig. 10. Note as well, that for a digital interface, such as the 1553 interface, it is also possible to define the

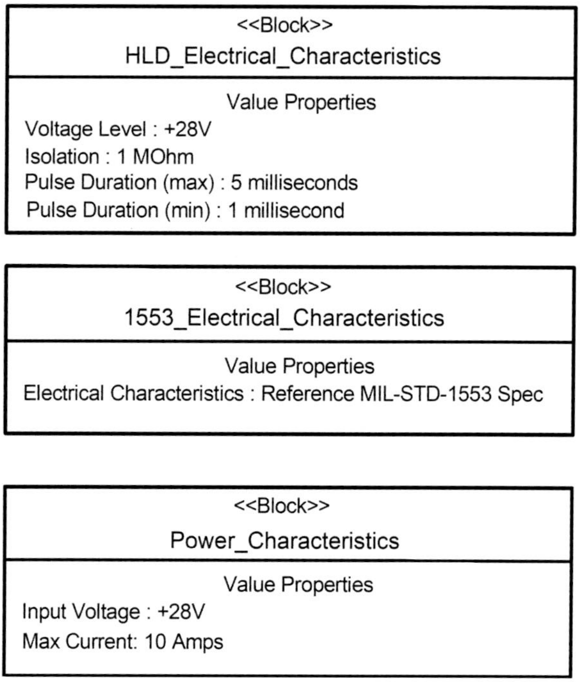

Fig. 10 Blocks define the electrical characteristics for their respective interfaces

```
                    <<Block>>
            HLD_Electrical_Characteristics
                   Value Properties
    Voltage Level : +28V
    Isolation : 1 MOhm
    Pulse Duration (max) : 5 milliseconds
    Pulse Duration (min) : 1 millisecond
```

```
                    <<Block>>
            1553_Electrical_Characteristics
                   Value Properties
    Electrical Characteristics : Reference MIL-STD-1553 Spec
```

```
                    <<Block>>
                Power_Characteristics
                   Value Properties
    Input Voltage : +28V
    Max Current: 10 Amps
```

commanding protocol (or other protocols as necessary) as well. Figure 11 illustrates a BDD that defines the 1553 Bus Controller to Remote Terminal protocol structure. If desired, this protocol could be further elaborated through sequence diagrams or other behavioral constructs belonging to the 1553 Bus Controller to RT Protocol Block.

An additional piece of "foundational" data to define before creating the interface blocks are the pin definitions that support the electrical characteristics and, in the case of the 1553, the command protocol. Figure 12 defines the two blocks. The first block defines the pin characteristics for the High-Level Discrete (HLD), and the second block defines the pin characteristics for the 1553 Interface. To the right of the blocks are instances of those blocks that represent specific pins. Note that as part of the instance definition, an included specific value is the other pin connection. In this case, it is just denoted as "OtherBox_J2_x." Again, the example here is representative, as when creating the pin blocks, the desired property values could be included as desired or necessary by the modeler and then filled in with the specific values when the instance is created.

The final pieces of "foundational" data that must be created are the command and monitor blocks that will also become instances. Figure 13 illustrates the command and monitor blocks that in turn become specific command and monitor instances. Two command instances were created, one for the HLD power-on command and the other for setting the time. Additionally, a specific instance was created for a voltage

16 Model-Based Hardware-Software Integration 517

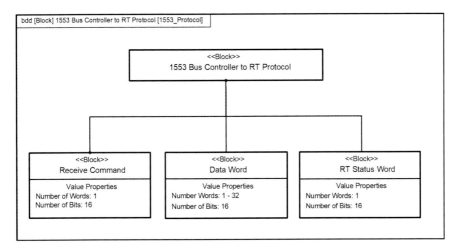

Fig. 11 The 1553_Protocol Block Definition Diagram defines the 1553 Bus Control to RT Protocol

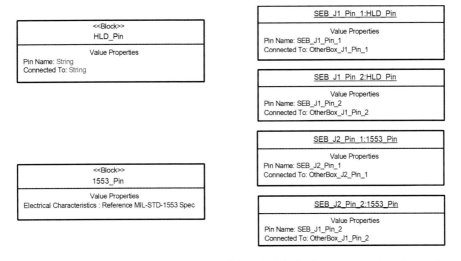

Fig. 12 Instances representing pins are created from the blocks that represent pins of a certain signal type, such as HLD or 1553

monitor that indicates the voltage value/reading after the HLD to power the example satellite electronics box (SEB) was received.

Now that the foundational data such as electrical signal characteristics, protocols, and pins have been defined, interface blocks that can "group" this data together and be used to type proxy ports on the IBD can now be created. In the case of the commands and monitors, blocks containing flow properties are created in order to type the activity parameters on the activity diagram.

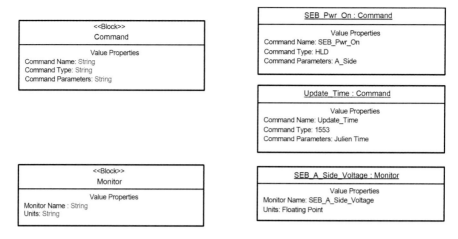

Fig. 13 Instances representing commands and monitors are created from the blocks that represent either a command or monitor

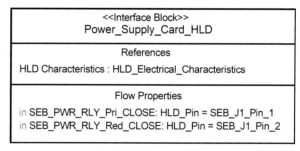

Fig. 14 Power supply card HLD interface block that captures the electrical characteristics and pins for the HLD interface

Figure 14 shows the interface block defined for the power supply card HLD interface. From Fig. 14, it can be seen that the interface block references the HLD electrical characteristics as well as the pin definitions as flow properties. This interface block will be used to type the HLD proxy port in the SEB internal block diagram.

Figure 15 shows the interface block defined for the 1553 controller interface. From Fig. 15, it can be seen that the interface block references the 1553 electrical characteristics as well as the 1553 protocols. Additionally, the pin definitions are captured as flow properties. This interface block will be used to type the 1553 proxy port in the SEB internal block diagram.

Finally, Fig. 16 shows the interface block defined for the power supply card power interface. From Fig. 16, it can be seen that the interface block references the power characteristics. This interface block will be used to type the PWR proxy port in the SEB internal block diagram.

With the interface blocks now defined, the proxy ports in the internal block diagram can now by typed by the respective interface blocks as shown in Fig. 17.

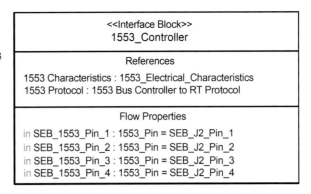

Fig. 15 1553 controller interface block that captures the electrical characteristics, protocol, and pins for the 1553 controller interface

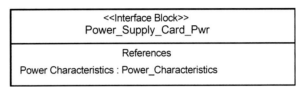

Fig. 16 Power supply card interface block that references the power characteristics for the power supply card interface

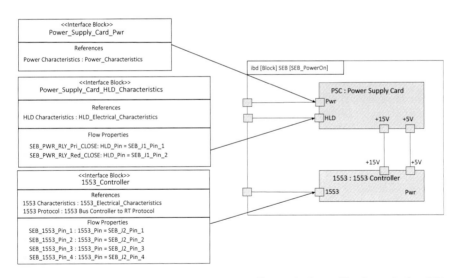

Fig. 17 Interface blocks are typed to the proxy ports of interest in the satellite electronics box IBD

(Note that Fig. 17 shows the interface block used to type the proxy port in the diagram itself. This is done for illustrative purposes only and is not intended to be compliant with what is actually allowed to be displayed on the diagram.)

In turn, the blocks that capture the commands and monitors as flow properties can be used to type the activity parameters in the SEB power-on activity diagram as

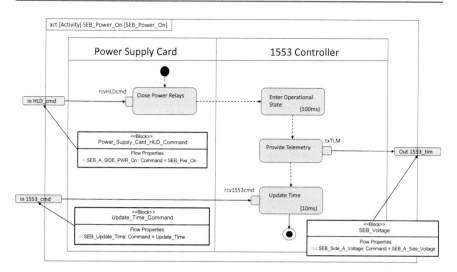

Fig. 18 Activity parameters are typed by the blocks that define commands and monitors in the satellite electronics box to capture the relevant commands and monitors of the activity

shown in Fig. 18. Note that it would be equally acceptable and effective to type the action pins with the command and monitor definition blocks. (Also note that Fig. 18 shows the block used to type the activity parameter in the diagram itself. Like Fig. 17, this is done for illustrative purposes only and is not intended to be SysML compliant with what is actually allowed to be displayed on the diagram.)

In reviewing the steps taken so far, a block definition diagram of the satellite electronics box (SEB) capturing the parts of the box that have external interfaces that will be elaborated has been created. Additionally, an activity diagram that captures the behavior that software will invoke or initiate, the timing constraints associated with that sequence, and activity parameters and pins that capture the command has also been created. Blocks and instances of those blocks have been created where appropriate to capture the foundational/low-level information such as pins and electrical characteristics. Finally, after the foundational data was created, the data was incorporated as reference properties and flow properties into blocks and interface blocks that are then typed to activity parameters and proxy ports in the activity diagram and internal block diagram, respectively. While specific modeling and stakeholder needs may result in variation to the specific approaches taken (e.g., typing pins instead of activity parameters), all the basic HW-SW integration information is now in place. Everything is now in place to examine ways in which the relationships between behavior (activity diagram in this example) and structure (e.g., IBD) are (or can be) explicitly tied together.

In terms of ownership and connectivity, it is important to understand that the satellite electronics box (SEB) owns the power-on activity diagram. The SEB block also owns the internal block diagram describing the physical description as well. Additionally, since each of these diagrams reference the low-level details, all the necessary HW-SW data is either owned by the SEB block or can be accessed from the diagrams that the block owns.

16 Model-Based Hardware-Software Integration

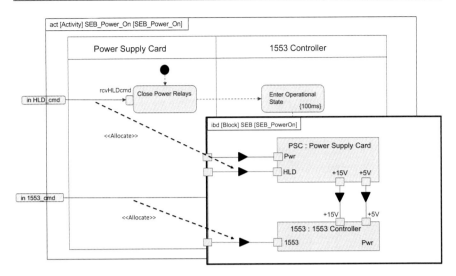

Fig. 19 Object flows may be allocated to IBD item flows to associate specific data elements that flow over representations of the same interface

There are additional methods and options available to create explicit relationships between individual behavioral artifacts and the low-level data. Primarily through allocation relationships. For example, if item flows are included in the internal block diagram connectors, and in turn, those item flows contain the command or monitor values that flow over those interfaces, allocations can be created between those item flows, and the object flows that represent those commands in the power on activity diagram. This relationship is conceptually illustrated in Fig. 19.

Since allocations are an easy way to explicitly represent cross-cutting relationships between structure and behavior, allocations can be created between many elements, spanning allocations at the diagram level between the diagrams themselves, or at much more granular levels at the individual data items.

This example has illustrated how some of the methods discussed in section "Hardware-Software Integration Modeling Methodologies" may be applied when performing a hardware-software integration task. While specific stakeholder needs and systems may drive unique or different modeling methodologies than presented here, it is hoped that these applications throw light on general approaches that may be adapted to specific needs.

Chapter Summary

This chapter presented different model-based system engineering SysML techniques and constructs for performing the hardware-software integration task. Since hardware-software integration fundamentally involves the definition of behavior and the supporting hardware elements that enable that behavior, it was identified how the SysML behavioral constructs can be elaborated to contain the data that helps with the hardware-software integration task.

This includes creating a block definition diagram to capture the high-level architecture (structure) and utilizing an activity diagram to capture a sequence (behavior). These two constructs would then serve as the cornerstones for adding additional detail necessary for HW-SW integration.

Based on the block definition diagram, parts were defined, which allowed for the creation of an IBD, and thus definition of the physical functional interfaces using proxy ports at the end of the connectors. The proxy ports could then be typed by interface blocks, which had properties that consisted of low-level electrical data such as the signal electrical characteristics, pins, and the communication protocol used if relevant. Alternatively, it was mentioned that item flows could also be used to capture the data that flows over the interface.

On the behavioral side, the activity diagram utilized the parts defined by the block definition diagram as the swim lanes, utilized naming conventions on the pins and activity parameters to help identify commands and monitors, and incorporated timing constraints so that software has a complete picture in terms of commands to issue, subsequent telemetry verification, and timing considerations.

Finally, after introducing these concepts, an example in applying them to a satellite electronics box (SEB) was provided, which was intended to represent a simple electronics box used in a satellite that is commanded by another on-board computer that can issue commands and receive telemetry from the SEB.

As mentioned in the example, specific system and stakeholder needs can drive different modeling approaches or techniques. Future revisions could explore the elaboration of additional behaviors through the addition of state machines or sequence diagrams that either add more insight into the primary behavior or supplement additional behaviors such as communication protocols. A demonstration of how requirements can be traced to the various interface characteristics could be included as well.

Additionally, the use of opaque behaviors to capture the software code/logic as part of the activity diagram could also be explored. This would not only allow for capture of the code and logic used when deployed but also potentially support simulation to verify behavior.

References

1. Object Management Group. OMG Systems Modeling Language (OMG SysML™). V1.5. Available at: http://www.omg.org/spec/SysML/
2. A Practical Guide to SysML, The Systems Modeling Language, Third Edition by Sanford Friedenthal, Alan Moore, and Rick Steiner, Morgan Kaufmann, 2014
3. Shames, Peter M, Sarrel, Marc A, A modeling pattern for layered system interfaces, 25th Annual INCOSE International Symposium (IS2015), Seattle, WA, July 13 – 16, 2015
4. Shames, Peter M, Sarrel, Marc A, Friedenthal, Sanford, A modeling pattern for layered system interfaces, 26th Annual INCOSE International Symposium (IS2016), Edinburgh, Scotland, UK, July 18–21, 2016

5. SysML Distilled: A Brief Guide to the Systems Modeling Language, First Edition, by Lenny Delligatti, Pearson Education, Inc. 2014
6. MIL-STD-1553B specification, Revision B, 1978. Available at https://quicksearch.dla.mil/Transient/FBAD0D30C74441A5AC56F712AAE521DF.pdf
7. Object Management Group. OMG MARTE Available at: https://www.omg.org/omgmarte/
8. Demathieu, Sebastion, MARTE tutorial: An OMG UML profile to develop Real-Time and Embedded Systems, 13th SDL Forum, Paris, France, September, 2007
9. Imran Rafiq Quadri, Abdoulaye Gamatié, Pierre Boulet, Samy Meftali, Jean-Luc Dekeyser, Expressing embedded systems configurations at high abstraction levels with UML MARTE profile: Advantages, limitations and alternatives, Journal of Systems Architecture, Volume 58, Issue 5, 2012, Pages 178–194
10. Society of Automotive Engineers. Architecture Analysis & Design Language (AADL) AS5506. Available at https://www.sae.org/standards/content/as5506/?utm_source=google&utm_campaign=TN_Mobilus-Core_MB
11. Software Engineering Institute (SEI). Architecture Analysis and Design Language. Available at https://www.sei.cmu.edu/our-work/projects/display.cfm?customel_datapageid_4050=191439&customel_datapageid_4050=191439
12. Model-Based Engineering with AADL: An Introduction to the SAE Architecture Analysis & Design Language by Peter H. Feiler, David P. Gulch, Addison-Wesly, 2013
13. M. Faugere, T. Bourbeau, R. d. Simone and S. Gerard, "MARTE: Also an UML Profile for Modeling AADL Applications," 12th IEEE International Conference on Engineering Complex Computer Systems (ICECCS 2007), Auckland, 2007, pp. 359–364.
14. UML 2 and the Unified Process: Practical Object-Oriented Analysis and Design, Second Addition by Jim Arlow, and Il Neustadt, Addison-Wesley, 2005
15. Embedded Systems: Analysis and Modeling with SysML, UML and AADL, First Edition by Fabrice Kordon, Jerome Hugues, Agusti Canals, Alain Dohet, ISTE and John Wiley & Sons, Inc., 2013

Mr. Cesena is a model-based systems engineering (MBSE) specialist, developing MBSE technology that includes application of model-based methods to various spacecraft system engineering products, as well as investigation into automation of system engineering spacecraft product generation through either extension of model tool capabilities and/or machine learning. Mr. Cesena helped pioneer the use of model-based system engineering at Lockheed Martin Space by leading one of the first successful applications to satellite design.

Part IV

Quality Attributes Tradeoffs in MBSE

Exploiting Digital Twins in MBSE to Enhance System Modeling and Life Cycle Coverage

17

Azad M. Madni, S. Purohit, and C. C. Madni

Contents

Introduction	528
MBSE and Digital Twin Technology: State of the Art	529
Model-Based Systems Engineering	529
Digital Twin Technology	530
Best Practice Approach	533
Methodology	533
Process	534
Ontology Use in Digital Twin Definition	535
Illustrative Example	535
Experiments	535
Implementation	538
Digital Twin and Machine Learning	539
Quantitative Analysis	540
Chapter Summary	545
Cross-References	545
References	545

Abstract

MBSE has made significant strides in the last decade. Today, its focus is on extending system life cycle coverage and improving system model accuracy by leveraging methods such as formal modeling, machine learning, and analytics in

A. M. Madni (✉)
Systems Architecting and Engineering, Astronautical Engineering Department, University of Southern California, Los Angeles, CA, USA
e-mail: azad.madni@usc.edu

S. Purohit
University of Southern California, Los Angeles, CA, USA
e-mail: shatadkp@usc.edu

C. C. Madni
Intelligent Systems Technology, Inc., Los Angeles, CA, USA
e-mail: cmadni@intelsystech.com

© Springer Nature Switzerland AG 2023
A. M. Madni et al. (eds.), *Handbook of Model-Based Systems Engineering*,
https://doi.org/10.1007/978-3-030-93582-5_33

conjunction with exploiting digital engineering technology. The growing convergence between digital engineering and MBSE and the advent of digital twins has made it possible to enhance system modeling and increase life cycle coverage. Digital twins are computational models of systems that have bidirectional communication with the real-world systems and evolve in synchronization with them. This chapter presents how digital twin technology can be leveraged in MBSE and offers an example of digital twin creation for an unmanned aerial vehicle (UAV) that can dynamically replan its path in response to disruptions such as inflight damage or changes in environmental conditions. This chapter also presents results of early experimentation with a digital twin and describes how operational analysis and modeling can be enhanced by leveraging the digital twin construct.

Keywords

Model-based systems engineering · Digital twin · Virtual prototype · Simulation

Introduction

This chapter describes the growing convergence between MBSE and digital engineering (DE) and the exploitation of digital twins in MBSE to improve system modeling and increase system life cycle coverage. It explicitly addresses how digital twin technology can be integrated into MBSE and how the resulting system model can be improved in accuracy and augmented with AI to enhance decision-making for the physical twin.

Digital twins are computational models of real-world systems that have bidirectional communication with the real-world counterparts and evolve synchronously with them. Recent advances in digital twin technology make it an ideal complement to and enabler of MBSE allowing MBSE to address the later stages of the system life cycle. In the following paragraphs, the key terms used in this chapter are defined.

Virtual Prototype: A generic virtual model of a physical system; represents a class of systems; and it can be used to explore certain functions before they exist in a physical system.

Digital Twin is a computational model of a particular physical system (i.e., it is an instance of a class of systems) *with bidirectional communication* with the physical counterpart. A digital twin coevolves with the physical system and reflects the state, status, and history of the system.

Digital Thread: The communication framework that allows connected data flow and integrated view of an asset's data throughout its life cycle across traditionally siloed functional perspectives [1, 2, 3].

Machine Learning: A computational method for automated data analysis that can detect patterns in data, build models, and predict future data or perform decision-making under uncertainty.

The remainder of this chapter is organized as follows. Section "MBSE and Digital Twin Technology: State of the Art" reviews the state of the art of MBSE and digital twin technology. Section "Best Practice Approach" presents a best practice approach to incorporate digital twins into MBSE. It presents the use of domain ontology to scope the model and facilitate digital twin-enabled model update. Section "Illustrative Example" describes exemplar experiments performed with physical and digital twins for an illustrative problem. Section "Chapter Summary" provides a summary of the chapter and potential advances.

MBSE and Digital Twin Technology: State of the Art

Model-Based Systems Engineering

While digital or electronic models have been in use in engineering since the 1960s, the early models were disparate and based on different assumptions, modeling methods, and semantics. With the advent of MBSE, models were placed at the center of systems engineering and became the authoritative source of truth (ASOT). Importantly, they replaced document-centric systems engineering. Models can include a variety of submodels such as M-CAD, EDA, SysML, and UML models, as well as physics models. The move to MBSE allows engineering teams to readily understand design change impacts, communicate design intent, and analyze a system design in terms of its properties before building the system. The primary focus of most MBSE efforts in industry is to: integrate data through models; realize an integrated end-to-end modeling environment that supports understanding of all elements that impact a design; and uncover and resolve undesirable outcomes. MBSE introduces and integrates diverse models and views to create a centralized model that facilitates greater understanding of an evolving system (i.e., a system under development). Importantly, MBSE provides a foundation for a Model-Based Engineering Enterprise, where multiple enterprise views and functions are facilitated using a centralized digital repository. As important, MBSE provides timely and early insights during systems engineering [4–6].

MBSE also supports automation and optimization, allowing systems engineers to focus on generating value and making effective trade-offs among competing objectives and system attributes. The key to a successful model-based initiative is scoping the problem and proactively managing the modeling process with the end in mind. MBSE goes well beyond traditional system specifications and Interface Control Documents by creating an integrated system model from which multiple views can be extracted [7, 8]. The system architecture provides the construct for data integration and transformation across the system's life cycle. The system/product dataset contains several MBSE artifacts: system architecture, system specification, 3D views, system attributes, design requirements, supplementary notes, and Bill of Material.

A critical activity in systems engineering is upfront trade-off analysis that ensures developing the best value system that satisfies mission/customer needs. As mission complexity increases, it becomes increasingly difficult to understand the factors that impact system performance [9, 10]. Integrating analytics within the rubric of a formal architecture can provide data-driven insights into system characteristics that typically do not surface in traditional model-driven analysis. Integrated tools allow systems engineers to analyze many more configurations in relation to mission scenarios, helping to identify key requirements "drivers" and lowest cost alternatives for system design.

System architecture models, represented in SysML, can serve as the focal point for integrated analytics comprising simulation tools, analysis results capture tools, context manager, requirements manager, and key trade-offs among architectural parameters. Analysis context specifies the scope of the analysis; parametric views inform the necessary sensitivity analysis, requirements diagrams capture stakeholder needs, thresholds, and "drivers" that circumscribe the tradespace. This type of model-centric framework provides a consistent and managed computational environment for analysis.

By placing analytics in the hands of system architects, insights into requirements and architectural features that drive performance and cost can be gleaned. Thereafter, multidimensional analysis can provide architects valuable perspectives that help identify the "knee in the curve" between cost and performance in an n-dimensional tradespace [9].

Integrated modeling and analysis can provide the basis for decision support to systems architects and engineers. Decision-makers will potentially have more information to draw on and more options to consider before reaching conclusions. Integrated analytics increase the amount of available information while also helping decision-makers make sense of the data. Finally, MBSE tools can help to explore, visualize, and understand a complex tradespace and can potentially provide early insights into the impact of decisions related to both technical solutions and public policies [10].

Today there is growing acceptance of MBSE in defense, aerospace, automotive, and consumer products industries. The expansion of MBSE to include analytics and trade-off analysis is dramatically increasing MBSE adoption rate within multiple industries. Finally, increased collaboration among MBSE practitioners and distributed research teams is helping to rapidly advance model-based approaches.

Digital Twin Technology

Digital Twins, a key concept in Digital Engineering today but with roots that go as far back as 2003, holds significant potential for enhancing MBSE. Digital Engineering focuses on communication and data sharing within an organization, whereas MBSE focuses on exploitation of data/models for decision-making [2, 11]. The term "digital twin" was initially coined in the context of product life cycle management (PLM) in 2003 [12–14]. At the time, the technology to implement digital twins was not sufficiently mature. Digital twins today are computational models that have bidirectional communication and evolve synchronously with their real-world counterparts. Therefore, digital twins reflect the state and status of the real-world counterpart. The real-world counterpart can be a system, a component, a process, or a human [15]. Digital twins can be employed in MBSE in different ways:

- Prototype and iteratively test and evolve the design of a real-world system/product.

- Monitor the performance of the physical counterpart and intervene in its operation, if needed.
- Collect data from a team of physical twins (aggregate twins) to approximate their behavior and use the resultant model to support predictive maintenance.

The concept of a digital twin is inexorably tied to modeling and simulation technology which has been around for decades. With digital twins, initial models created tend to be incomplete and potentially incorrect due to limited knowledge about the system at the beginning of the modeling activity. On the other hand, models created using traditional Modeling & Simulation (M&S) technology tend to be generic models of the system of interest. With the advent of Internet of Things (IoT), Industry 4.0, and applied analytics, the generic system model can be transformed into a digital twin, a computational model and virtual replica of a particular real-world system, process, or person [16]. This transformation becomes possible because the generic computational model can be updated to reflect the state, status, and maintenance history of a specific physical system, and then evolve in synchronization with the physical system. The digital twin can then be used to make better decisions about the real-world system than previously possible. Digital twins usually have onboard sensors that support two-way interaction between the physical world and the virtual environment of the digital twin.

The range of possibilities for employing a digital twin of various levels of sophistication is extensive. Digital twins can support a variety of use cases in different industries [17–19]. For example, in healthcare medical devices (e.g., insulin pump) can have a digital twin, patients can have a digital twin to monitor and record how a patient would respond to different treatments in healthcare, and process can have digital twins to analyze whether or not intervention is needed. More sophisticated twins are being employed to model cities while accounting for traffic transit, power consumption, and pollution considerations.

Until recently, hardware associated with a physical system and operational analysis were not an integral part of MBSE methodologies [20]. With the advent of digital twin technology, integrating physical system hardware into MBSE provides ample opportunities to learn from real-world data.

Taking the plunge into investing in digital twin technology has to be tempered with caution. Several questions need answering to conclude that pursuing a digital twin approach is worthwhile for an organization. It begins with identifying the problem being addressed and determining whether digital twin is suitable for that problem. Beyond that, it is necessary to answer whether the contemplated digital twin is sustainable, and if so, who will be assigned to ensure its sustainability. A fundamental and obvious question that needs to be answered is whether the digital twin is no more than a costly version of a solution that could have been readily satisfied with a digital document or electronic report. Is the digital twin a real-time system? The critical point here is that a sound business case needs to be made. This business case needs to be driven by use cases that help convey the value proposition of a digital twin for that business.

The real-world use case in this chapter is for a Formula One race car, and the business case that can be made. It begins with creating a simulation that reflects the real racetrack experience, i.e., it serves as a digital twin and enables extraction of maximum benefit from limited on-track testing time. The purpose of the digital twin simulator, in this case, is to:

- Help the team set up the car to run faster
- Rapidly advance car development
- Increase team's ability and speed to fine-tune car enhancements
- Determine where the team needs to improve
- Identify where its competition is the strongest
- Pinpoint where the team has weaknesses
- Identify performance areas to enhance
- Test new design features and understand their impact on performance before going on the track

The next step is the "data to decisions" mapping, i.e., selecting the data that needs to be captured from the physical twin to extract business value and/or operational success by enabling informed decision-making [19]. In such a data-rich environment, vast amounts of information can be gathered in a short period and run through machine learning algorithms to derive insights and uncover patterns that enable teams to make more informed decisions faster. Appropriately constructed digital twins can offer insights that help with operational optimization [3].

Much of the work with digital twins thus far has been confined to individual digital twin supporting a specific application. In 2021, researchers at MIT came up with a probabilistic model foundation for enabling predictive digital twins at scale [21]. These researchers developed a model that allows the deployment of digital twins at scale, e.g., creating digital twins for a fleet of drones. A mathematical representation that takes the form of a probabilistic graphical model provides the foundation for "predictive digital twins." This model was applied to an Unmannned Aerial Vehicle (UAV) in a mission scenario in which the UAV suffers minor wing damage in flight and has to decide whether to land, continue, or reroute to a new destination. In this case, the digital twin experiences the same damage in a virtual world and faced the same decision as the physical UAV experiences in the real world. These researchers discovered that custom implementations require significant resources, a barrier to real-world deployment. This problem is further compounded because digital twins are most useful in situations where multiple similar assets are being managed. To this end, these researchers created a mathematical representation of the relationship between a digital twin and the corresponding physical asset. However, this representation was not specific to a particular application or use. The resulting model mathematically defined a pair of physical and digital systems, coupled through two-way data streams that enable synchronized evolution over time. The digital twin parameters were first calibrated with data collected from the physical UAV so that the digital twin was an accurate reflection of the physical twin from the start. Then, as the physical UAV's state changed over time through wear and tear from flight time logged, these changes were observed and used to update the state of the digital twin, so it reflected the state of the physical UAV. The resultant

17 Exploiting Digital Twins in MBSE to Enhance System Modeling and Life... 533

digital twin then became capable of predicting how the UAV was likely to change and used that information to optimally direct the physical UAV in the future.

The graphical model allowed each digital twin to be based on the same underlying computational model. However, each physical asset maintained a unique "digital state" that defined a unique configuration of the model. This approach made it easier to create digital twins for a large collection of physically similar assets.

In their experiment, the UAV was the testbed platform for all activities – from collaboration experiments to simulated "light damage" event. The digital twin was able to analyze sensor data to extract damage information, predict the impact of UAV-structural health in future activities of the UAV, and recommend changes in maneuvers to accommodate the changes.

The key idea behind this advance is maintaining a persistent set of computational models that are constantly updated and evolved alongside the physical twin over its entire life cycle. The probabilistic graphical model approach helps "seamlessly" span the different phases of the physical twin's life cycle. In their problem, this property is seen in the graphical model which seamlessly extends from the calibration phase to the operational, in-flight phase. The latter is where the digital twin becomes a decision aid for the physical twin.

Best Practice Approach

Methodology

Model-Based Systems Engineering (MBSE) is the formalized application of modeling to support system requirements, design, analysis, verification, and validation activities beginning in the conceptual design phase and continuing throughout development and later life cycle phases. The models are based on a set of assumptions and have a common syntax, and compatible semantics. Such models can potentially answer questions of interest to stakeholders. Subsequently, experimentation with system models using simulations can be conducted to verify dynamic system behaviors with a variety of assumptions and what-if conditions.

At lower levels, state, time, and time synchronization need to be accommodated in the process. The Inference and Analysis block needs to be aware of these variables because this information is needed to determine what adjustments (if any) are needed. Also, since the Inference and Analysis block determines the needed changes, then this block needs to know what the digital twin did with the inputs it received. Since this closed-loop process can potentially produce oscillations if the update rate is too high, the update rate has to be set to prevent such oscillations from occurring.

It is important to recognize that the realizability of a digital twin is bounded by the complexity of the system. For a typical spacecraft, jetliner, or large factory, there can potentially be dozens of states allocated to multiple modes, multiple interacting parallel processes that react to potentially thousands of inputs. For such complex systems, developing a full-blown digital twin simulation is infeasible because of the sheer complexity of the system. Therefore, digital twin development of the system should focus on subsystem, component, and human agent levels to assure the problem is tractable.

Process

In MBSE, initial assumptions can be incorrect because initial models are created with limited data and incomplete knowledge of the system. Furthermore, assumptions may not be valid because they tend to be based on inadequate understanding and limited evidence [3]. Failure to revisit the initial assumption is one of the most significant issues that need to be addressed in MBSE. It is in this area that digital twin technology can play an important role.

Digital twin provides an effective means to replace assumptions with actual data or revise assumptions thereby increasing model fidelity. Specifically, new evidence from the physical system (twin) can be acquired and used to update the digital twin model [6, 22, 23]. The data captured in digital twins serves to enhance traceability between system requirements and real-world behaviors.

With digital twin-enabled MBSE [20], the model's accuracy increases with new data (Fig. 1). As shown in this figure, inference and analysis can complement MBSE, thereby producing data and insights that can be used to increase model accuracy leading to superior decisions and predictions. Also, as shown in Fig. 1, data acquired from the physical system (i.e., physical twin) in the real world can be used to update system models in the digital twin. Subsequently, analysis and inference can lead to new insights that can be used to update centralized digital models throughout the system's life cycle. Figure 2 depicts the cost curves for MBSE, traditional SE, and Digital Twin-Enabled MBSE across the system life cycle. Color codes are used to distinguish traditional systems engineering, MBSE, and digital twin-enabled MBSE curves. The costs and gains associated with Digital Twin-Enabled MBSE implementation are presented.

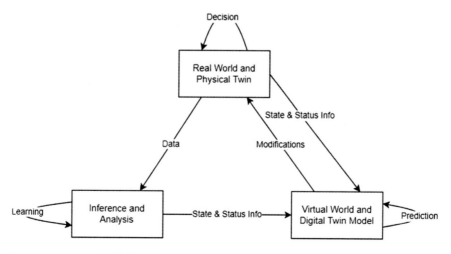

Fig. 1 Digital twin-enabled MBSE process overview

17 Exploiting Digital Twins in MBSE to Enhance System Modeling and Life...

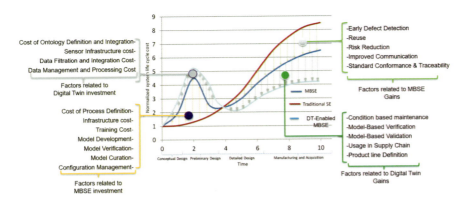

Fig. 2 Application of digital twin across MBSE life cycle with associated investments and gains [6]

As shown in Fig. 2, for Digital Twin-Enabled MBSE, additional costs are incurred in the initial phase of the system life cycle. These costs are related to ontology and metamodel definition, development, and integration, sensor infrastructure implementation, data processing, data management, and configuration management. Metamodel provides the structure for the ontology. Thus, defining the terms (i.e., ontology) that are entered in a database is usually associated with defining the database schema (metamodel). In general, defining the terms is easier than defining the metamodel. However, substantial gains (compared to incurred costs) can be expected during the later phases of the system life cycle, considering the time value of money. Consolidating information from the enterprise and organizational level and continuously updating the system model in the digital twin facilitates superior decision-making across the system life cycle.

Ontology Use in Digital Twin Definition

Figure 3 presents an exemplar partial ontology for a digital twin-enabled model update. The ontology view depicts a relevant aspect of closed-loop model-based systems engineering. The ontology answers questions, such as what are the different components of the digital twin, what are the various aspects of the digital twin that are updated during the system life cycle, and what kind of data are collected related to the system's temporality.

Illustrative Example

Experiments

The first step in integrating digital twin technology into MBSE is to create a digital twin and then incorporate facilities in the digital twin to update its model using data acquired from the physical twin operating in the real world. In this simple

Fig. 3 Ontology for digital twin-enabled model update

experiment, the contextual model of the system is used to demonstrate the value of digital twin technology (i.e., modifying the model during system operation in the real world). The contextual model captures the external entities present in the system's environment that influence system behavior. Physical properties such as size, shape, location, and velocity are all captured in the system model. Environment factors such as threat, visibility, and weather are captured in the contextual model.

The experimentation setup consists of 2 UAVs (Fig. 4). Each UAV system consists of a sensing subsystem, a communication subsystem, a battery subsystem, a propulsion subsystem, and a control subsystem. The UAVs operate in an indoor maze environment comprising static and dynamic entities (ex. static: table, dynamic: another UAV) and are subject to external forces. The mission of the UAV system is to search for a predefined object in the environment. Both UAVs communicate external visual and acoustic data, battery subsystem health, internal subsystem temperature, external temperature, altitude, location, communication bandwidth, and speeds to the digital twin model. These data are transferred to the database for preprocessing. The Inference and Analytics server performs data collection, integration, filtering,

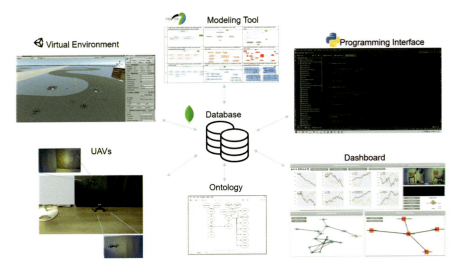

Fig. 4 Experimentation setup

feature extraction, transformation, and analysis. For example, the video stream coming from the UAV is converted into frames of fixed size, color threshold, and contrast values are adjusted. Frames are calibrated for distortion and passed through a trained Recurrent Convolutional Neural Network (RCNN) model, where the objects in the video feed are identified and tagged. Also, velocity vectors of objects (i.e., objects in the environment) are calculated based on their movement in the frames and UAV velocity vectors. Objects with similar velocities are clustered in the same group. In this case, the contextual model of the system is continuously updated based on results from analytics.

The planning and decision-making problem involves multiple levels of decision-making, with varying levels of difficulty. For example, rule-based, heuristic, and probabilistic models can be employed in response to the increasing complexity of the navigation problem and the availability of information about the system.

The experimentation scenarios comprise multiple phases of increasing difficulty. The first phase of the multi-UAV search mission has only static objects in the environment, and the system uses rule-based navigation. A dynamic object is introduced into the environment in the next phase. After that, the object is identified and clustered in a new group due to its dynamic nature. The generation of new clusters induces the system to switch to a more complex navigation model. In the next phase, external forces act on one of the UAVs, causing it to change speed, and location. The inference /analytics server then creates a correlation function between the sudden change in internal system attributes and the appearance of the external dynamic object.

The human-system interface (Fig. 5) allows the human to intervene and add inputs in the experimentation setup. Figure 5 is the bottom right corner pane of the dashboard shown in Fig. 4. This interface enables the creation of tags, editing the

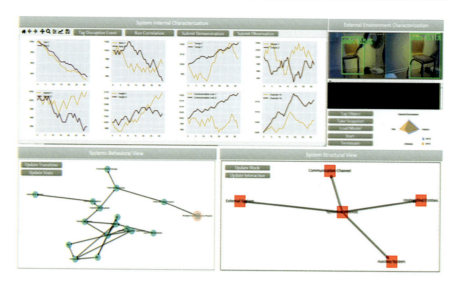

Fig. 5 Human system interface of experimentation setup

correlation function, control of the system, and updates to the model. In the experiment, the operational phase data of the system is used to update the model.

Implementation

Making model updates during simulated (or real-world) system operation requires the system representation to be sufficiently flexible to incorporate new information and changes. Furthermore, observations need to be captured and stored in digital repositories to support decision-making. This requires the system architecture to include a perception subsystem. Making models "closed-loop" [6] requires an investment in: (a) physical or virtual sensors during system prototyping, development, integration, testing, operation, and maintenance; and (b) testbed instrumentation for data collection.

Virtual simulations capable of interacting with physical systems require communication and computing infrastructure that enables the creation and automatic update of online models. A modest initial investment in the infrastructure that includes a perception subsystem (i.e., sensors) is required. However, significant benefits can be expected to accrue with repeated use of the infrastructure during simulation-based experiments. A primary technical risk in implementing a state-based model is the combinatorial explosion inherent in such models. Since data is collected continuously and the model once constructed is updated continually, dealing with a massive influx of data requires prior data filtering and processing. The research methodology proposes an approach to containing the combinatorial explosion is to employ heuristics and contextual filtering [4]. This approach, which separates data analysis and filtering from model building and communication can counter this problem.

Digital Twin and Machine Learning

For simple products, processes, and systems, digital twin technology offers value without having to employ sophisticated techniques such as machine learning. This is because simple applications can be parameterized using a limited number of variables and lend themselves to easily discoverable linear relationships between inputs and outputs. However, in the real world, most systems have multiple data streams with widely varying characteristics. For such systems, machine learning and data analytics can enable making sense of and discovering patterns in data [6]. Machine learning in this context implies any algorithms applied to data streams to reduce uncertainty in the knowledge of the system and the environment. Data analytics implies any statistical/AI algorithm that helps discover patterns and trends from data. Collectively, these methods can potentially contribute new insights that can be exploited in various ways.

In the context of digital twins, machine learning can be employed to:

- Evaluate data in real time
- Automate complex tasks
- Adjust behavior with minimal or no need for supervision
- Increase likelihood of desired outcomes
- Produce actionable insights that can potentially lead to cost savings

A digital twin model typically receives substantial data from multiple sources such as sensors mounted on the physical twin, environmental sensors, software functions, and human inputs. These data streams can be filtered and subsequently processed in real time using machine learning algorithms such as reinforcement learning (RL). RL can be used to incrementally learn previously unknown system states and state of the environment. Data analytics can be used to extract features and patterns using clustering techniques.

Digital twin technology also lends itself to the use of intelligent agents to improve *digital twin prediction* and *physical twin decision-making*. Intelligent agents can potentially augment data collected from the physical twin. Comprehensive datasets can be constructed by evaluating digital twin behavior using simulation. Digital twin behavior can be tested with use cases and disruptive scenarios. Appropriate tools can be used to generate event-driven or agent-based simulations to explore the behavior and interactions of the digital twin.

The accuracy of the digital twin model can be expected to improve gradually and automatically throughout the system life cycle [24, 25]. In this case, the model update function requires some intelligence. Initial assumptions and system constraints within the model can be progressively refined over time. Search, optimization, and constraint satisfaction agents can be used on data from the physical twin to gain insights into the state of the environment and physical twin in the real world. Knowledge-based and probabilistic methods within agents can be used to reason and update the digital twin. With each iteration of the digital twin model update, the accuracy of the model can be expected to increase. Comparing digital twin simulation results and physical twin behavior provides essential insights into improving the digital twin model and simulation. With increasing model accuracy, the digital twin model becomes a more accurate representation of the physical counterpart. Improved model accuracy implies more precise answers to questions, and an increase in the number of questions that can be answered.

Time-series data from the physical twin in the operational environment is used by the intelligent agents to identify unique patterns. These insights can be used during the system development life cycle to identify the most used and unused functions. System architects can then decide to enhance existing functions to improve customer satisfaction and potentially discard unused functions to reduce cost and increase ROI. Also, insights gleaned from digital twin data can be used to inform business decisions such as whether or not to develop certain functions as service. Operational data collected during disruptions can be analyzed and used to develop resilience capabilities in the system.

Quantitative Analysis

One of the MBSE experimentation testbed capabilities allows users to integrate and evaluate the performance of various machine learning algorithms for different system parameters. An exemplar digital twin simulation was created to evaluate both models and algorithms. The simulation environment consists of a UAV attempting to locate a predefined object in an indoor building setup. In this search operation, the UAV (agent) receives a negative reward for colliding with objects in the environment and a positive reward for touching the predefined goal object. Users may evaluate multiple reinforcement learning algorithms for the UAV agent. Multiple models were tested that acquire data (i.e., observations) from the environment

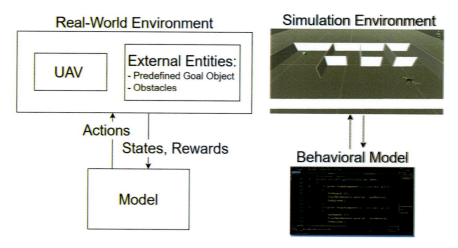

Fig. 6 Learning cycle

17 Exploiting Digital Twins in MBSE to Enhance System Modeling and Life... 541

and take appropriate actions. Five reinforcement learning algorithms were evaluated in this experiment: Proximal Policy Optimization (PPO); Soft Actor-Critic (SAC); PPO with Generative Adversarial Imitation Learning (GAIL); PPO with Behavioral Cloning (BC); and PPO combined with GAIL and BC [26–29]. Model parameters such as Cumulative Reward, Episode Length, Policy Loss, and Entropy are evaluated against Simulation Runs for each reinforcement-learning algorithm. Figure 6 shows the learning cycle, while Fig. 7 presents the quantitative analysis setup.

As shown in Fig. 7, the structural SysML model of the digital twin is mapped to the 3D virtual environment. To accomplish this mapping, a Python XMI translation tool was built, that automatically populates the asset container in the simulation

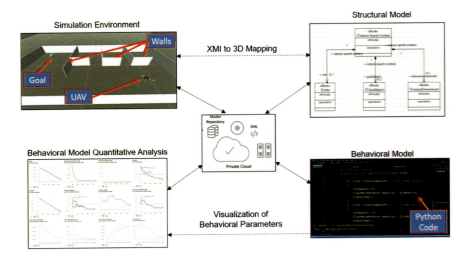

Fig. 7 Quantitative analysis setup

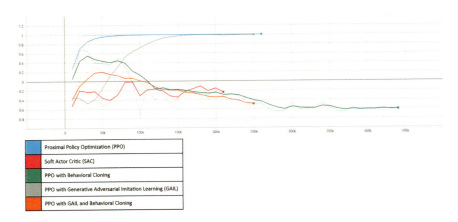

Fig. 8 Simulation runs versus cumulative reward

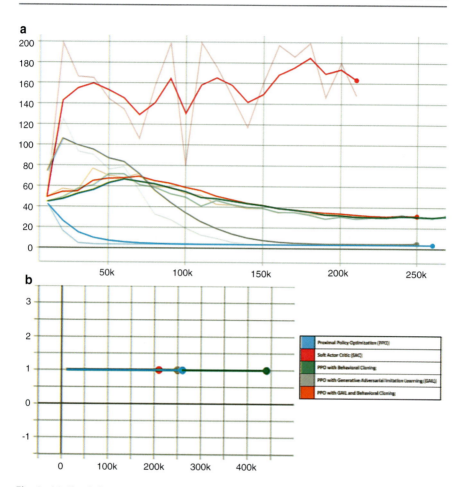

Fig. 9 (a) Simulation runs versus episode length. (b) Is training Boolean

environment from SysML models. Objects in the simulation environment are stored in an asset container. Figure 7 displays the SysML Block Definition Diagram for the "Indoor Search" scenario context. This diagram presents the entities that are part of the scenario. The blocks in the diagram are arranged under the "ScenarioEntities" package. The structured packaging of SysML model entities facilitates the extraction of model elements.

The UAV, Indoor Environment, and Goal Object blocks are extracted from the SysML scenario context model. These block names are then matched with existing 3D objects in the repository. When a match is found, the object is duplicated from the repository with the duplicate being placed in the 3D virtual environment's asset container. When a matching 3D object is not found in the repository, the translation program creates an empty object in the asset container for the user to further modify. The user can employ the asset container to populate objects in the simulation. The

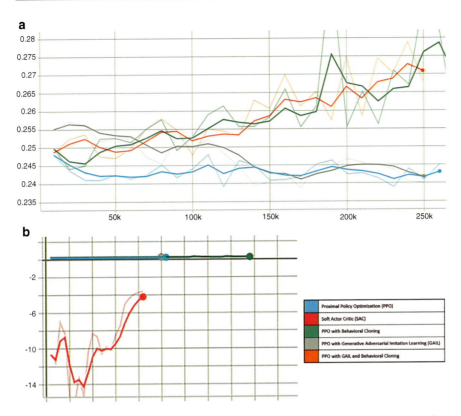

Fig. 10 (**a**) Simulation runs versus policy loss (excluding SAC). (**b**) Simulation runs versus policy loss for all the models

observations and actions are defined for the simulation setup, and the rewards mechanism is created in the experiment. Five models are evaluated in the experiment: Proximal Policy Optimization (PPO); Soft Actor Critic (SAC); PPO with Generative Adversarial Imitation Learning (GAIL); PPO with Behavioral Cloning (BC); and PPO combined with GAIL and BC [26–29]. The Unity ML agents kit was used to extend the testbed capabilities [30].

Figure 8 presents simulation runs on the horizontal axis and cumulative rewards gained by the agent for a particular model on the vertical axis. For PPO and PPO with GAIL, the mean cumulative episode reward increases as the training progresses. For the remaining agents, the reward decreases, indicating that the agent is not learning in the given simulation environment for the given simulation runs.

Figure 9a shows that the mean length of episodes goes down in the environment for successful agents as the training progresses. The "Is training" Boolean in Fig. 9b indicates whether the agent is updating its model or not.

The different models exhibit different behaviors for a given simulation setup. Also, the way the user constructs the training environment impacts model performance differently.

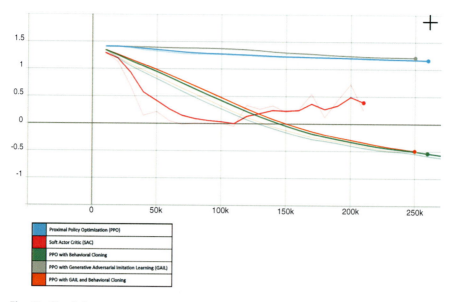

Fig. 11 Simulation runs versus entropy

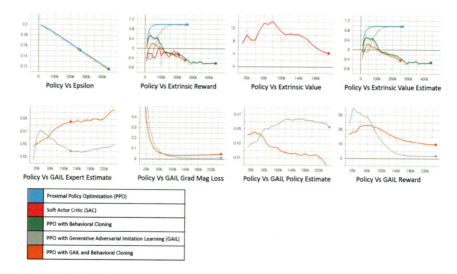

Fig. 12 Model-specific parameter evaluation

The policy loss parameter indicates the degree to which the policy is changing (Fig. 10a, b). Various models have different profiles, and for most models, the magnitude of policy loss is shown to be decreasing, which indicates successful training.

In Fig. 11, entropy indicates the randomness in model decisions. Slow decrease of this parameter is an indicator of successful training. As shown in Fig. 11, entropy profiles are different for the different agents.

Additionally, it is possible to analyze specific model parameters such as GAIL expert estimate, GAIL policy estimate, and GAIL rewards (Fig. 12).

In sum, the digital twin-enabled MBSE testbed enables quantitative analysis of behavioral models using a variety of reinforcement-learning algorithms. Users can manipulate various model parameters and run experiments with different assumptions, models, and learning algorithms. The testbed capabilities enable users to frame the decision, observation, and reward problem more holistically. Various use cases and scenarios can be considered before defining agent behaviors. The holistic approach and quantitative analysis allow users to determine effective strategies for intelligent agents.

Chapter Summary

This chapter has presented a "best practice" methodology, implementation, and experimentation with digital twin within an MBSE framework. The chapter presents the key elements of the methodology including the underlying ontology, illustrative example, implementation, experiments, results, and implications. The DT-enabled MBSE approach essentially transforms modeling into a closed loop process [6]. A multi-UAV scenario is used in the experiments to demonstrate the overall approach. Several machine-leaning algorithms to process incoming data are evaluated and compared in terms of their performance. An overarching ontology is used to integrate the major subsystems needed for experimentation, and to support digital twin model update based on data from the physical twin. The key contribution of this chapter is a systematic approach for incorporating and experimenting with digital twin technology in MBSE.

Cross-References

▶ Digital Twin: Key Enabler and Complement to Model-Based Systems Engineering
▶ Semantics, Metamodels, and Ontologies

References

1. Bone, M., Blackburn, M., Kruse, B., Dzielski, J., Hagedorn, T., Grosse, I., 2018. Toward an Interoperability and Integration Framework to Enable Digital Thread. Systems. 2018; 6(4):46.
2. Kinard, D., 2010. The Digital Thread–Key to F-35 Joint Strike Fighter Affordability, Aerospace Manufacturing and Design http://www.onlineamd.com/amd-080910-f-35-joint-strikefighter-digital-thread.aspx.

3. West, T.D., Pyster, A., 2015. Untangling the Digital Thread: The Challenge and Promise of Model-Based Engineering in Defense Acquisition, INSIGHT, 18(2), pp. 45-55, August.
4. Madni, A.M., 2018. Transdisciplinary Systems Engineering: Exploiting Convergence in a Hyper-connected World. New York, NY: Springer.
5. Madni, A.M., Erwin, D., 2018. Next Generation Adaptive Cyber Physical Human Systems, Year 1 Technical Report, Systems Engineering Research Center, September.
6. Madni, A.M., Madni, C.C., Lucero, D.S., 2019. Leveraging Digital Twin Technology in Model-Based Systems Engineering, MDPI Systems, special issue on Model-Based Systems Engineering, Publication, March.
7. Grieves, M., Vickers, J., 2017. Digital Twin: Mitigating Unpredictable, Undesirable Emergent Behavior in Complex Systems, F.-J. Kahlen et al. (eds.), Transdisciplinary Perspectives on Complex Systems, Springer International Publishing, Germany.
8. Hoffenson, S., Brouse, P., Gelosh, D.S., Pafford, M., Strawser, L. D., Wade, J., Sofer, A., 2019. Grand Challenges in Systems Engineering education, in S.C. Adams et al. (eds.), Systems Engineering in Context, Proceedings of the 16th Annual Conference on Systems Engineering Research.
9. Madni, A.M., Purohit, S. 2019. Economic Analysis of Model-Based Systems Engineering. MDPI Systems, special issue on Model-Based Systems Engineering, Publication, February.
10. Madni, A.M., Sievers, M., 2018. Model-Based Systems Engineering: Motivation, Current Status, and Research Opportunities, Systems Engineering, Special 20th Anniversary Issue, vol. 21, issue 3.
11. Kraft, E.M., 2015. HPCMP CREATE-AV and the Air Force Digital Threat, 53rd AIAA Aerospace Sciences Meeting, Kissimmee, FL.
12. Datta, S.P.A., 2016. Emergence of Digital Twins, arXiv e-print (arXiv:1610.06467).
13. Datta, S.P A., 2017. Emergence of Digital Twins – Is this the march of reason? Journal of Innovation Management, 5(3), 14–33. https://doi.org/10.24840/2183-0606_005.003_0003
14. Folds D. J. and McDermott T. A., 2019. "The Digital (Mission) Twin: an Integrating Concept for Future Adaptive Cyber-Physical-Human Systems," IEEE International Conference on Systems, Man and Cybernetics (SMC), Bari, Italy, 2019, pp. 748-754, https://doi.org/10.1109/SMC.2019.8914324.
15. Ghosh, A.K., Ullah, S., Kubo, A., 2019. Hidden Markov model-based digital twin construction for futuristic manufacturing systems. Artificial Intelligence for Engineering Design, Analysis and Manufacturing. 1-15. https://doi.org/10.1017/S089006041900012X.
16. Kritzinger, W., Karner, M., Traar, G., Henjes, J., Sihn, W., 2018. Digital Twin in manufacturing: A categorical literature review and classification. IFAC-PapersOnLine. 51. 1016-1022. https://doi.org/10.1016/j.ifacol.2018.08.474.
17. Marr, B., 2017. What Is Digital Twin Technology – And Why Is It So Important? Forbes, https://www.forbes.com/sites/bernardmarr/2017/03/06/what-is-digital-twin-technology-and-why-is-it-so-important/#78b97b8a2e2a.
18. Morton, S.A., McDaniel, D.R., Sears, D.R., Tillman, B., Tuckey, T.R., 2009. Kestrel—a fixed wing virtual aircraft product of the CREATE program, in Proceedings of the 47th AIAA Aerospace Sciences Meeting, Orlando, Fla, USA, January, AIAA 2009-338.
19. Ocampo, J., Millwater, H., Crosby, N., Gamble, B., Hurst, C., Reyer, M., Mottaghi, S., Nuss, M., 2020. An Ultrafast Crack Growth Lifing Model to Support Digital Twin, Virtual Testing, and Probabilistic Damage Tolerance Applications. https://doi.org/10.1007/978-3-030-21503-3_12.
20. Madni, A.M., Purohit, S., Madni, A., 2020. Digital Twin Technology-Enabled Research Testbed for Game-Based Learning and Assessment: Theoretical Issues of Using Simulations and Games in Educational Assessment, O'Neil, H. (Eds.), Taylor & Francis, Spring.
21. Kapteyn, M.G., Pretorius, J.V.R. & Willcox, K.E. A probabilistic graphical model foundation for enabling predictive digital twins at scale. Nat Comput Sci 1, 337–347 (2021). https://doi.org/10.1038/s43588-021-00069-0.

22. Chen, P.C., Baldelli, D.H., Zeng, J., 2008. Dynamic flight simulation (DFS) tool for nonlinear flight dynamic simulation including aeroelastic effects," in Proceedings of the AIAA Atmospheric Flight Mechanics Conference and Exhibit, Honolulu, Hawaii, USA, AIAA 2008-6376.
23. Glaessgen, E.H., and Stargel, D.S., 2012. The Digital Twin Paradigm for Future NASA and U.S. Air Force Vehicles.
24. Stojek, M., & Pietraszek, J., 2015. Simulation-Based Engineering Science Challenges of the 21st Century. Applied Mechanics and Materials, 712, 3–8. https://doi.org/10.4028/www.scientific.net/AMM.712.3
25. Tuegel, E.J., Ingraffea, A.R., Eason, T.J., Spottswood, S.M., 2011. Reengineering Aircraft Structural Life Prediction Using a Digital Twin, International Journal of Aerospace Engineering, https://doi.org/10.1155/2011/154798.
26. Schulman, J., Wolski, F., Dhariwal, P., Radford, A., & Klimov, O. (2017). Proximal policy optimization algorithms. arXiv preprint arXiv:1707.06347.
27. Haarnoja, T., Zhou, A., Hartikainen, K., Tucker, G., Ha, S., Tan, J., ... & Levine, S. (2018). Soft actor-critic algorithms and applications. arXiv preprint arXiv:1812.05905.
28. Torabi, F., Warnell, G., & Stone, P. (2018). Behavioral cloning from observation. arXiv preprint arXiv:1805.01954.
29. Ho, J., & Ermon, S. (2016). Generative adversarial imitation learning. arXiv preprint arXiv:1606.03476.
30. Juliani, A., Berges, V., Teng, E., Cohen, A., Harper, J., Elion, C., Goy, C., Gao, Y., Henry, H., Mattar, M., Lange, D. (2020). Unity: A General Platform for Intelligent Agents. arXiv preprint arXiv:1809.02627. https://github.com/Unity-Technologies/ml-agents.

Dr. Azad M. Madni is a University Professor of Astronautical Engineering, holder of the Northrop Grumman Fred O'Green Chair in Engineering, and the Executive Director of University of Southern California's Systems Architecting and Engineering Program. He is also the Founding Director of the Distributed Autonomy and Intelligent Systems Laboratory. He has joint appointments in the Department of Aerospace and Mechanical Engineering and Sonny Astani Department of Civil and Environmental Engineering. He is a member of the National Academy of Engineering and Life Fellow/Fellow of AAAS, IEEE, AIAA, INCOSE, IISE, IETE, AAIA, SDPS, and the Washington Academy of Sciences. He is the founder and CEO of Intelligent Systems Technology, Inc., a high-tech company which specializes in model-based and AI approaches to addressing scientific and societal problems of national and global significance. He cofounded and currently chairs the IEEE SMC award-winning technical committee on Model-Based Systems Engineering. He pioneered the field of transdisciplinary systems engineering to address problems that appear intractable when viewed solely through an engineering lens. He is the creator of the TRASEE™ educational paradigm which combines storytelling with the Science of Learning principles to make learning enjoyable while enhancing retention and recall. He is the author of the highly acclaimed book, *Transdisciplinary Systems Engineering: Exploiting Convergence in a Hyper-Connected World* (Springer 2018) and the coauthor of *Tradeoff Decisions in System Design* (Springer, 2016). He is the Coeditor-in-Chief of four systems engineering volumes. He has received more than 75 honors and awards from professional engineering societies, industry, government, and academia. His IEEE awards include *2021 IEEE Aerospace and Electronic Systems Judith A. Resnik Space Award*, *2019 AESS Pioneer Award*, and *2020 IEEE Systems, Man and Cybernetics Norbert Wiener Outstanding Research Award*. His INCOSE awards include the *2011 Pioneer Award, 2019 Founders Award*, and *2021 Benefactor Award*. He has also received prestigious educational awards including *2019 AIAA/ASEE John Leland Atwood Award* for excellence in engineering education and research and the *2021 Joint ASEE SED/INCOSE Outstanding Systems Engineering Educator* Inaugural Award. In 2020, he received the *NDIA Lt. Gen. Ferguson Award* for Excellence in Systems Engineering. In 2019, he received the *ASME CIE Leadership Award* and the *Society of Modeling and Simulation International Presidential Award*. In 2016, Boeing honored him with a *Lifetime Achievement Award* and a *Visionary Systems Engineering Leadership Award* for his "impact on Boeing, the aerospace industry, and the nation." He received his PhD, MS, and BS degrees in Engineering from University of California, Los Angeles. He is a graduate of AEA/Stanford Institute Program for Technology Executives.

Shatad Purohit, PhD Student, University of Southern California. Shatad is a doctoral student in the Astronautical Engineering Department specializing in Systems Architecting and Engineering. He is pursuing his doctorate under Professor Azad Madni. He has been involved in MBSE for several years as a practitioner and now as a researcher. His specific areas of interest are systems architecting, digital engineering, resilience engineering, machine learning, and applied analytics. He is a student member of IEEE, INCOSE, and AIAA. He has been actively involved in the Conference on Systems Engineering Research as a member of the organization team, paper contributor, and reviewer. He received his MS degree in Systems Architecting and Engineering from USC, and his BS degree from Swami Ramanand Teerth Marathwada University. He has published research papers in IEEE Systems Journal, IEEE SMC International Conference, AIAA SciTech, IEEE Aerospace Conference, and CSER.

Carla C. Madni, EVP for R&D and COO, Intelligent Systems Technology, Inc. Carla is a principal systems engineer specializing in the engineering of complex and intelligent human-machine systems for human performance enhancement. She has served as Program Manager and Coprincipal Investigator on numerous R&D projects sponsored by DARPA, AFRL, ONR, ARL, NIST, DOE, and DHS S&T. Her specific areas of interest are simulation-based training systems, intelligent decision-aiding systems, digital engineering, and uses of digital twin technology in Model-Based Systems Engineering research and education. She received her BS degree from Tulane University in Biomedical Engineering, and her MS degree from University of California, Los Angeles, in Engineering Systems with specialization in Distributed Problem Solving and Artificial Intelligence. She is a member IEEE and the Human Factors and Ergonomics Society. She has supported the Conference on Systems Engineering Research as a sponsor since 2008. She has led several successful Small Business Innovation Research projects sponsored by various government agencies. She is currently involved in supporting General Motors and Cruise LLC in conducting studies in the safety of autonomous vehicles.

Model-Based Mission Assurance/ Model-Based Reliability, Availability, Maintainability, and Safety (RAMS)

18

Luca Boggero, Marco Fioriti, Giuseppa Donelli, and Pier Davide Ciampa

Contents

Introduction	550
From Document-Based to Model-Based Reliability and Safety Assessment	551
Model-Based Reliability and Safety Assessment: A Literature Review	553
Functional Hazard Analysis (FHA)	555
Fault Tree Analysis (FTA)	557
Failure Model and Effects Analysis (FMEA)	560
Reliability Analysis	562
Summary Table	563
A Recommended Approach for Model-Based Reliability and Safety Assessment	563
Model-Based Functional Hazard Analysis	564
Model-Based Fault Tree Analysis	566
Model-Based Failure Modes and Effects Analysis	567
Model-Based Reliability Block Diagram	569
Advantages of the Model-Based Reliability and Safety Assessment	570
Examples of Application of the Recommended Model-Based Reliability and Safety Assessment	571
Model-Based Functional Hazard Analysis: Aileron Command	575
Model-Based Fault Tree Analysis: Aileron Command	576
Model-Based Failure Modes and Effects Analysis: Aileron Command	579
Model-Based Reliability Block Diagram: Aileron Command	579

L. Boggero (✉) · G. Donelli
German Aerospace Center (DLR), Hamburg, Germany
e-mail: Luca.Boggero@dlr.de; Giuseppa.Donelli@dlr.de

M. Fioriti
Politecnico di Torino, Turin, Italy
e-mail: marco.fioriti@polito.it

P. D. Ciampa
Institute of System Architectures in Aeronautics, MDO Group, German Aerospace Center (DLR), Hamburg, Germany
e-mail: Pier.Ciampa@dlr.de

© Springer Nature Switzerland AG 2023
A. M. Madni et al. (eds.), *Handbook of Model-Based Systems Engineering*, https://doi.org/10.1007/978-3-030-93582-5_34

Chapter Summary 583
Cross-References 584
References 585

Abstract

Among all the disciplinary analyses performed during the development of a new aircraft, reliability and safety play the most important role for the certification and operation of the aircraft. Traditionally, reliability and safety analyses are document-based, i.e., a vast quantity of tables and reports collect all the assumptions, inputs, decisions, and results. However, this approach entails several disadvantages hampering the quality and the effectiveness of the analyses, especially with the introduction of novel technologies that make the aircraft a system with ever-increasing complexity. Therefore, the research community proposes innovative model-based approaches to support the development of a new aircraft. This chapter specifically focuses on the model-based approaches proposed in literature dealing with reliability and safety analyses.

Keywords

Model-based safety analysis · System modeling language · Reliability analysis · Safety analysis · Flight control system · Electro-hydrostatic actuator

Introduction

The main prerequisite of a civil transport aircraft is to safely complete its mission without any damage to passengers and crew. However, considering the number of aircraft components and equipment, the heavy load they have to sustain and the operating environment, in-depth analyses of the aircraft safety must be carried out to satisfy regulatory requirements. These analyses begin from the first phases of the aircraft design, and they last until the end of the project constantly increasing the detail level. Moreover, several studies compose safety analyses that involve evaluations from the single system component to the entire system architecture using both bottom-up and top-down approaches. All the failure modes of each system component have to be identified and deeply analyzed to define the possible impacts onto the near equipment and the whole aircraft. Nevertheless, the whole aircraft and all its subsystems should be analyzed to prevent any failure that would compromise the flight safety. **RAMS** is the engineering discipline that aims at analyzing all the aircraft systems, equipment, and components in order to reduce at minimum any potential damage that would hamper the aircraft safety or the scheduled mission. The acronym stands for *Reliability, Availability, Maintainability* and *Safety*. These attributes play an important role in the design and production of aircraft. Their definitions are the following:

- **Reliability:** The probability of an item to perform a required function under stated conditions for a specified period of time [1]

- **Availability:** Measure of the degree to which an item is in an operable state and can be committed at the start of a mission when the mission is called for at an unknown (random) point in time [1]
- **Maintainability:** The ability of an item to be retained in, or restored to, a specified condition when maintenance is performed by personnel having specified skill levels, using prescribed procedures and resources, at each prescribed level of maintenance and repair [1]
- **Safety: The state of being free from harm or danger** [2]

The present chapter focuses on the *reliability* and *safety* of aircraft. Due to the high importance of these two system attributes, hundreds of standards (e.g., [3–5]) have been published in the years to prescribe guidelines on how to conduct reliability and safety analyses on several types of systems, not only aeronautical ones but including also other domains, for example, automotive, nuclear energy, rail transportation, and software. Generally, documents (e.g., reports, tables) report all the results that can be obtained from the analyses conducted by following the guidelines provided in these standards. However, this traditional approach has several limitations, as the following section outlines in more details. In brief, limitations encompass lack of clarity, miscommunication and misunderstandings among engineers, spread of information in multiple sources, difficult traceability between system design characteristics and results of reliability, and safety analyses. Therefore, new approaches based on models are being proposed since the last years. This chapter aims at illustrating and discussing the most promising and relevant studies available in literature. All these studies propose methods and tools for the model-based assessment of reliability and safety, in order to relieve the complexity of these analyses and to trace the design characteristics of the system with its reliability and safety requirements.

From Document-Based to Model-Based Reliability and Safety Assessment

In the last years, different techniques and methods have been developed to assess systems reliability and safety. Examples of these techniques are ([3, 6]) Reliability Block Diagram (RBD), Functional Hazard Analysis (FHA), Fault Tree Analysis (FTA), Failure Modes and Effects Analysis (FMEA), Failure Modes, Effects and Critical Analysis (FMECA), Zonal Safety Analysis (ZSA), Markov Analysis (MA), and Common-Cause Analysis (CCA). All the aircraft manufacturers use these techniques during the aircraft reliability and safety assessment processes [7]. In addition, there are also handbook-based methods that rely on documents as MIL-HDBK-217 [8] and Nonelectronic Parts Reliability Data (NPRD) publications to estimate reliability properties of several system components, from electronic to mechanical ones. Even if the US National Academy of Sciences strongly criticizes these methods due to their inaccuracies and deficiencies, different commercial and military avionic applications exploit them [9].

All already mentioned techniques and methods rely on a traditional document-based approach. However, a significant limitation affects this kind of approach. The entire system development information, including reliability and safety data, is in fact spread among several documents. This negatively impacts the quality of the system development process, since ambiguities, lacks of clarity, and misunderstandings may affect this process. Reliability and safety analyses supported by documents could be therefore prone to errors and inconsistencies, because it is difficult to accurately manage a huge quantity of data, results, assumptions, and numbers collected in thousands and thousands of pages of multiple reports. This widespread of information also makes almost impossible the complete tracking of all the decisions taken during the analyses and development process, therefore losing many relationships (e.g., cause-effect) among input data, assumptions and constraints, and derived solutions. This lack of traceability also inhibits the design and analysis of alternative solutions, which are identified when innovative system architectures having different technologies are generated from a conventional architecture. Therefore, all the analyses, including those addressing reliability and safety aspects, should be adapted to all the new configurations. However, this process is time-consuming and error prone, and a document-based approach certainly doesn't foster it.

All the limitations and disadvantages of the traditional document-based approach are motivating many organizations and engineering communities since the past years to shift to a more innovative approach based on models. More specifically, the most relevant organization that is promoting the shift from a document-based to a model-based approach is the International Council on Systems Engineering (INCOSE) [10]. According to INCOSE, models can be built to support activities that are traditionally performed by means of a document-based approach, e.g., development of system requirements, identification of functions that should be performed by the systems, determination of one ore multiple system architectures, validation, and verification tasks. These activities are part of a Systems Engineering process [11], and a model-based approach may enhance them, if adopted. This is the reason why it is expected that the so-called Model-Based Systems Engineering (MBSE) will play an increasing role in the practice of Systems Engineering in the next decades [12].

The main potential advantages that are pushing the model-based approach over the document-based one are the following [13]:

- **Enhanced collaboration and communication** within the design teams. Models are used as lingua franca between several developers.
- The system model is a **single source of truth**, containing the entire information in a unique place instead of having it spread in multiple documents.
- Models improve the **management of the complexity** that especially affects innovative systems, meanwhile assuring **consistency** and **completeness** of the information.
- **Traceability and tracking of design decisions** can be improved, hence quickly and easily adapting the models after variations (e.g., new system architectures) during the design process.
- Models can foster validation and verification activities, therefore entailing **reduced development risks and time**.
- **Models may be reused** and adapted for the development of new systems.

All these potential advantages make the use of modeling in several activities of the development process appealing. One of these activities regards the assessment of systems reliability and safety. This chapter indeed focuses on how a model-based approach, or more specifically an MBSE one, can drastically improve the reliability and safety evaluation of on-board systems. However, more than one challenge derives from this shift from document to model-based reliability and safety analyses. In particular, two main challenges are present:

1. The first challenge is determining how all relevant information traditionally collected in documents may be represented through MBSE technologies (including standards, formats, and tools).
2. The second challenge questions how to effectively exploit and take advantage of model-based reliability and safety analyses, considering that the shift from a document-based approach might represent a huge investment and burden.

In order to give a proper answer to both questions, the first part of the present chapter investigates and reports how the research community is tackling the model-based assessment of system reliability and safety so far. Section "Model-Based Reliability and Safety Assessment: A Literature Review" intends to cover the main research studies focusing on the model-based reliability and safety analyses, highlighting suggested guidelines and potential advantages. Section "A Recommended Approach for Model-Based Reliability and Safety Assessment" instead describes the best practice approach recommended in this chapter, addressing both the challenges previously exposed. Then, section "Examples of Application of the Recommended Model-Based Reliability and Safety Assessment" collects examples of application of this approach, strengthening the recommendation in adopting these suggested guidelines in the model-based reliability and safety evaluations. Finally, the conclusive section "Chapter Summary" ends the chapter by summarizing all the contents and mentioning future potential updates and improvements.

Model-Based Reliability and Safety Assessment: A Literature Review

Due to all the benefits of a model-based approach previously mentioned, many research studies investigate how modeling activities can effectively support safety and reliability analyses.

Joshi et al. propose in [14, 15] the employment of modeling and simulation tools (e.g., Simulink [16] and SCADE [17]) to support activities of system safety analysis as the standard ARP 4761 [3] prescribes. The authors introduce an approach called **model-based safety analysis** (**MBSA**), according to which engineers create formal models to support system safety analyses. Traditionally, models capture the *nominal behavior* of systems, i.e., they represent the correct functioning of the systems as intended by design. However, with an MBSA approach, formal modeling activities can address also the *fault behavior* of systems, including potential system failures.

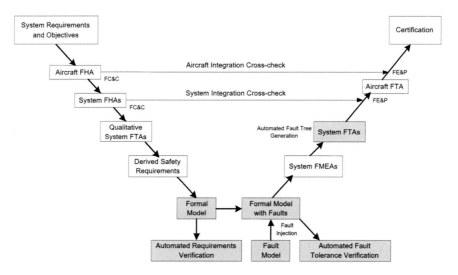

Fig. 1 "V" diagram representing the safety assessment process defined in the standard ARP 4761 and supported by models in a MBSA approach [14]

Figure 1 depicts this approach, where models support the traditional **safety assessment process** recommended by in ARP 4761 and represented in a "V" diagram.

The primary step of this MBSA approach is the definition of a formal specification of the system model. Formal specification languages supporting textual representations (e.g., textual languages like RSML^{-e} [18] and LUSTRE [19]) and graphical representations (e.g., Simulink and SCADE) model the behavior of the system under design. Architecture description languages instead model logical and physical architectures of the system (e.g., [20, 21]). Other than modeling the system nominal behavior, formal *fault models* are created. These models formalize common failure modes, but they may also encode other information about the faults, such as fault propagation and fault hierarchy. In this way, system design and verification activities may be both improved, since the formal models representing the *nominal* and *off-nominal* behavior of the system may reduce the effort required to the analysts and increase the quality of the analyses.

However, models can be built not only to support design and verification activities, but to support also other activities of a Systems Engineering product development process that are traditionally based on documents, hence including the modeling of system requirements, functions, and system architectures. Therefore, many researchers have conducted several studies in the past to integrate reliability and safety evaluation activities in a MBSE approach. Some studies (e.g., [22–24]) employ Unified Modeling Language (UML) [25] as standard modeling language. Other kinds of modeling language are also proposed in literature, for instance, the Architecture Analysis and Design Language (AADL) [26]. However, the INCOSE promotes the system modeling language (SysML) [27] as a standard modeling language for MBSE activities. SysML [28] is a general-purpose graphical modeling language, consisting of nine types of diagram that represent the functional behavior and the structure of systems. SysML is generally employed in the design of conventional and innovative systems, but several works propose the use of this modeling language for safety and reliability evaluation activities. The use of this standard language provides a consistent, well-defined, and well-understood means to communicate the requirements and the corresponding design among engineers.

The following sections describe in details methods that aim to generate safety and reliability analyses from SysML diagrams. The aim of the literature hereafter described is to integrate Systems Engineering with safety and reliability assessment. This integration is done to avoid the gap between the design process and the safety and reliability analysis and therefore to automatically update the analysis results after design changes. These studies mainly address the generation from SysML models of the following analyses: **Functional Hazard Analysis (FHA), Fault Tree Analysis (FTA)**, and **Failure Model and Effects Analysis (FMEA)**. Particularly, section "Functional Hazard Analysis (FHA)" reports two literature papers exploring how the FHA can be effectively associated with the MBSE using SysML language. Section "Fault Tree Analysis (FTA)" instead selects four papers that show how the extraction of the FTA from SysML diagrams is becoming a more and more automatic process over years. Furthermore, the application of the developed methods to several systems gives evidence of the easy adaptability of the SysML diagrams for the FTA of complex and multiple systems. In section "Failure Model and Effects Analysis (FMEA)," the description of three research works found in literature provides an overview on how to link the functional design phase using SysML with FMEA tables. Unlike the previously mentioned analyses, only a single study has been found in literature addressing reliability analyses supported by SysML diagrams. Section "Reliability Analysis" describes this study. Finally, section "Summary Table" reports a table summarizing all the previous mentioned research works, in order to give the reader a complete overview of the model-based safety and reliability analyses addressed in this chapter.

Functional Hazard Analysis (FHA)

Within hazard analysis, a hazard is a specific set of system conditions that leads to a unique potential system failure [29]. Hazard analyses are performed to identify these hazards, their effects, and causal factors. Hazard analyses define risk for every single hazard in the way that safety design measures can be adopted to eliminate or mitigate them. A typical hazard analysis is the **Functional Hazard Analysis (FHA)**. This analysis aims at identifying failure modes, severity, and risk associated with each system function [30]. Safety analysts generally perform the FHA during the initial phase of the aircraft design process when the focus is mainly on the functions that the system should be performed, instead on the components of the system architecture. This analysis is indeed carried out first at aircraft level, evaluating the functions of the whole aircraft. Afterward, when the aircraft functions are allocated to aircraft systems, the safety analysts perform FHA for each subsystem. The FHA might be easily performed by following the information modeled and described by the behavior and architectural diagrams created in a MBSE approach.

In this context, Müller et al. [31] present a solution that enables the analysis of system safety in early design stages by the use of SysML. In other words, the authors aim at combining system design and safety analysis in a single model-based approach. The approach consists of three main phases:

1. Requirements specification and definition of use cases due to specific operational aspects of the system
2. Functional architecture setup
3. Architecture implementation by physical components

The authors propose several SysML diagrams to model the three phases. It is worth underlining that in this approach, Müller et al. modify and adapt the standard SysML diagrams to perform the safety analysis directly from the system models. Nevertheless, hereafter only the standard SysML diagrams used for each phase are described for general purposes. *SysML Requirement Diagrams* model functional, non-functional, and safety requirements addressed in the first phase. *SysML Use Case Diagrams* identify the use cases and misuse. In the second phase, *SysML Activity Diagrams* model the functional architecture, which includes the functions performed by the system. These SysML diagrams indeed represent the main functions that the different components of the system fulfill. Finally, in the third phase, *SysML Block Definition Diagrams* and *SysML Internal Block Diagrams* model the system architecture. In particular, the former diagram lists all the components of the system, while the latter depicts the connections among components. According to the authors, *Activity Diagrams* enable the execution of the modeled FHA. So, focusing on the early design stages, malfunctions are identified through the *Activity Diagrams* by the negation of each function, including the case in which the function is not executed (e.g., "Function 1" of the *Activity Diagram* depicted in Fig. 2) and/or is not executed properly (e.g., "Function 2" in Fig. 2). A malfunction can have different consequences. For example, it can propagate to other functions, therefore degrading them. All these effects can be indeed modeled in the *SysML Activity Diagram*. If instead the malfunction causes a process interruption, it is linked to the *end node*, a specific SysML element that concludes a flow of functions represented in the *SysML Activity Diagram*. The malfunction identification with their causes and effects is so defined, and thus the FHA is performed.

To assess a deeper investigation and complete the trade-off of the system architecture, Brusa et al. [29] suggest to associate a dysfunctional analysis with the functional one. With their test case on the fuel system of a civil aircraft, the authors demonstrate that *SysML Sequence Diagrams* allow identifying all the critical issues for an eventual dysfunction, thus driving the dysfunctional analysis. *Sequence Diagrams* commonly show the interactions between the system and external users and the interactions among system components. In this context, these types of diagrams help in detecting the failure modes and defining the corresponding safety requirements. It is worth noting that *Sequence Diagrams* may be used for both the

Fig. 2 Example of *SysML Activity Diagram* proposed by Müller et al. to support the FHA. (Adapted from [31])

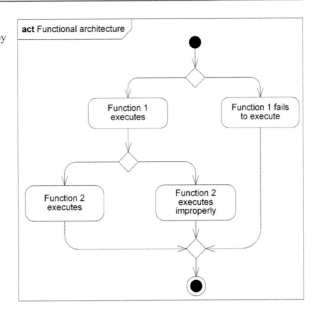

functional and dysfunctional analyses. A difference between a functional and dysfunctional *Sequence Diagram* is that to describe a dysfunction often it is required to resort to several additional details or links and some more functions. As consequence, the use of dysfunctional *Sequence Diagram* increases the nodes of possible failure in the related architecture of the system, which have to be enclosed inside the FHA. The authors propose to start the safety analysis from a system functional model. System use cases and functions are first determined, and then functional *Sequence Diagrams* are drawn. It is sufficient negating the operations defined in the *Sequence Diagrams* to create several dysfunctional behaviors. According to Brusa et al., this process can be a relevant support for the safety analyst in the judgment (e.g., severity, probability) of the different malfunctions. Finally, [29] highlights how *Activity Diagrams* (as in [31]) may complete the description of the system behavior and immediately identify a dysfunction wherever the action flow is stopped.

Fault Tree Analysis (FTA)

Fault Tree Analysis (FTA) is one of the main reliability analysis techniques, widely used in the aerospace and electronics industries as evidenced by the research works that this section describes. FTA is defined as a deductive procedure in which an undesired event at the system level, called a *top event*, is identified on the basis of various combinations of *components failures* [32]. A *Fault Tree* is a graphical way to represent the result of this analysis. Classified as a top-down analytical method, the FTA is usually performed once an architecture of the system is defined.

Consequently, system designers have to update safety analyses according to design changes, but this process is usually performed manually, therefore with the risk of generating inconsistent results. This inconsistency may be avoided by using processes to automatically derive Fault Tree Analysis from SysML diagrams. In this way, safety analyses are automatically generated from the latest system model version. Several studies propose methodologies that aim at generating automatically FTA from SysML system models. This section widely describes these studies.

Xiang et al. [32] implement an approach that extracts logical information needed for reliability analysis from the SysML system models and converts it into Reliability Configuration Model (RCM). The RCM is defined as general logical system independent from specific system modeling language, so that it can be applied to different system models if similar logic structure can be extracted from their modeling languages. The RCM includes different information about the system architecture. Namely, it models the system components (i.e., hardware and software) and the functional dependencies between them. The Fault Tree Analyses are performed using the RCM specifications. The researchers apply the proposed methodology to a relatively complex IT system, namely, a Fault-Tolerant Parallel Processor. In this case, the use of SysML is also advantageous due to the more familiarity of IT engineers with this modeling language. Moreover, another advantage is specifying system architectures intuitively by using SysML extensions. SysML extensions or stereotypes are modifications of the standard SysML elements, which are introduced to address specific contexts (IT system in this case) to perform the safety analysis. The reader may refer to [32] for more details on how to adapt some of the SysML standard elements to model IT systems. Two SysML diagrams, namely, *Internal Block Diagrams* and *Sequence Diagram*, represent system models in [32]. *Internal Block Diagrams* describe the internal system structure, while *Sequence Diagrams* show the system behavior. Other diagrams of SysML, such as *Use Case Diagrams*, are also adopted. Since *Sequence Diagrams* are coupled with use cases by *Use Case Diagrams*, it is possible to assess how the system behaves according to the different use cases. Then, specific algorithms convert these diagrams in RCM specification to perform FTA. In this case, SysML diagrams don't directly generate FTA, but SysML is a mean to automatically generate FTA from RCM.

A step forward is due to Mhenny et al. [33], who propose a methodology that aims at generating semi-automatic safety analyses directly from SysML models. According to them, SysML is the most adapted language for the modeling of complex systems, in their case a mechatronic system. The authors use *Internal Block Diagrams* to automatically generate FTA from SysML models through the use of traversal algorithms and block design patterns. For a given undesired top event, a graph traversal algorithm finds out components related to each other and identifies different patterns based on these components' relationships. *Internal Block Diagrams* are so divided in these patterns to facilitate the Fault Tree generation. In the representative example of Fig. 3a, four patterns are identified and marked by the labels *Entry, Redundant, Feedback loop*, and *Exit*. Each one of these patterns entails a partial Fault Tree, as the ones represented in Fig. 3b, relative to the *Exit* pattern, and in Fig. 3c, focusing on the *Redundant* pattern. Then, the final tree is progressively

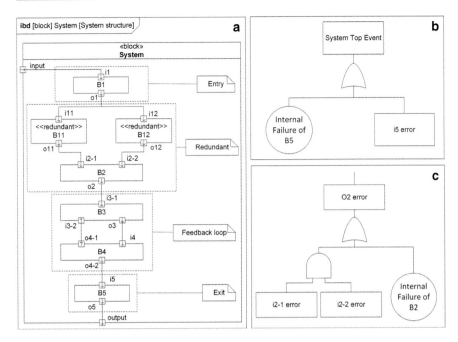

Fig. 3 *SysML Internal Block Diagram* representing the structure of a generic system (**a**) and parts of Fault Trees (**b–c**) generated from the diagram by following the guidelines proposed by Mhenny et al. (Adapted from [33])

assembled considering the partial Fault Tree of each pattern. These Fault Trees can be generated in a desired format to be used in any of the existing Fault Tree Analysis tools to make the quantitative assessment of failure rates. As a test case, the authors apply the proposed methodology for the safety analysis of an electromechanical actuator (EMA) for aircraft ailerons.

Izygon et al. [34] instead propose and describe an approach to automatically generate FTA directly from SysML diagrams. Their study focuses on developing techniques and toolsets to ease the integration of FTA early in the design process. The modeling techniques based on SysML language are applied to the Cascade Distiller System, an advanced Water Recovery System being developed at NASA Johnson Space Center. In this model, *SysML Block Definition Diagram* and *Internal Block Diagram* capture the information about the structural aspects of the physical architecture. *State Machine Diagrams* instead capture the behavior of the components, by modeling operational states, failed states, and the triggers that entail state transitions, including *signal* (e.g., a command from an external actor) and *time events* (e.g., when a given instant of time has been reached). From the *State Machine Diagrams*, a set of tools within the selected SysML modeling tool (i.e., MagicDraw [35]) extracts the Fault Trees.

Finally, it is worth mentioning the most recent study found in literature, attributed to Melani et al. [36]. In this study, not only Fault Trees are automatically obtained by mapping SysML diagrams, but this novel structured procedure also allows the creation of different Fault Trees for each operating mode of a system. In fact, a system can operate in different ways, performing different functions depending on the situation or on decisions made by its operator. This step is important because with this approach, it is possible to perform an FTA for different system operating modes. As the previous approach [34] proposes, also here the *Block Definition Diagram* models the system components in a logical and hierarchical way, while the *State Machine Diagrams* represent the operating modes of the system under study. However, in this approach, an additional *Activity Diagram* considers each operating system mode. As case study, the approach is applied to a bearing lubrification system of a hydrogenator.

Concluding, the introduced research works show an increasing automatic generation of the FTA from SysML models over years. Furthermore, the application of the so-far presented methods to systems belonging to different fields (IT, aeronautical, space) highlights the adaptability of SysML models to such complex systems. This is due to the availability of more SysML diagrams capturing structural, behavioral, and architectural system information.

Failure Model and Effects Analysis (FMEA)

Multiple works address the challenge to bridge the gap between design process and the already well-assessed techniques used in reliability and safety studies. The Failure Modes and Effects Analysis (FMEA) is the standard method for risk analysis usually addressed in the design phase. It is a well-known inductive and qualitative method that proposes to explore the system component by component. For each system component, the analyst searches the relative failure modes and their effects on the system, detailing their severity and occurrence rate. FMEA tables then collect and store these aspects [37].

Three research works are hereafter described to provide the reader with an overview on how to link functional design phase using SysML with FMEA. The aim of these works is to automatically generate FMEA tables starting from SysML diagrams. The elements that the modeling language should be able to represent to then conduct a FMEA are the architecture of the system and the functionality of the system components, the relationships between components, and the data and flow transmission between components.

Pierre David et al. emphasize in [38] that SysML can model the needed information required to generate FMEA tables. This information includes the functional behavior and the structure of the system. *Internal Block Diagrams* can hence represent the system logical and physical architectures, made of various components. *Sequence Diagrams* instead show the interactions between components in a given operative mode. For the component states identification and representation, SysML *State Machine Diagrams* should be used. The various state machines of the system model represent the behavior of components as a state history. *Activity Diagrams* depict the inputs, outputs, sequences, and conditions for coordinating system behaviors. In addition, the authors propose to set up the *Dysfunctional Behaviour Database* (DBD) using SysML. This database would contain the lesson learnt on the failure modes of utilized components, and it would be essential to automatically generate FMEA from SysML diagrams. In other words, the authors employ data analysis and organization techniques using a DBD written in a SysML to automatically exploit document traditionally built in engineering companies as the FMEA table. The DBD collects and maintains data and results of previous studies conducted on similar systems. This means that data presented in a FMEA can either be directly used to construct the final reliability study model or be injected in the DBD in an updating process. In the last case, the process leads to a larger DBD exploitable for future studies. Specific algorithms automatically identify various information for FMEA construction from DBD. Multiple columns of the FMEA table can be automatically filled by using this approach. They encompass the component, the failure modes, the causes, and the effects on the component behavior. In addition, the solution proposed by the authors supports the identification of the effects of component failures on the system and on the requirements. This is possible thanks to the SysML model of the system under development, which represents also system components and their connections, requirements, and relationships between components and requirements. More specifically, the relationship *satisfy* indicates which system components are involved in requirements satisfaction. Therefore, this approach entails to identify which requirements and which system components are affected in case of failures, hence supporting the safety analyst in assessing the effects of malfunctions. Figure 4 schematizes the main elements of the method proposed by the authors. The authors apply the proposed method to a Level Control System (LCS), which is an equipment installed in a tank that insures that the fluid

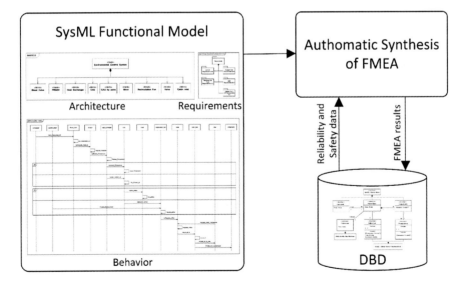

Fig. 4 Schema representing the method proposed by Pierre David et al.

never exceeds the maximum capacity of the container. This example of application illustrates the declaration of safety requirements, the use of SysML to describe the system architecture and behavior, and eventually the creation of FMEA through the proposed algorithms. Unfortunately, the authors don't provide details on constructing and analyzing the diagrams. However, their proposed approach is useful to map the FMEA generation process steps with the SysML models. In particular, this approach allows to determine per each component all the potential failure modes and possible causes and effects.

Mhenni et al. extend this approach in [39] where *Internal Block Diagrams* model not only system main components as traditionally done but also the connections among them. In more details, these connections are represented as *blocks*, like the system main components. In addition, the multi-physical flows transferred from one component to another (flows that highly impact the functioning of mechatronics systems) are depicted as *flow ports*. The choice of including connections between components is crucial since they have an impact on the probability of failure occurrences. As result of this approach, SysML models automatically generate several columns of the FMEA table, particularly providing information regarding the component, the failure mode, the causal factors, the immediate effects, the system effect, the recommended actions, and the severity.

SysML diagrams may automatically generate many more FMEA columns according to the approach introduced by Hecht et al. [40]. Also, analysts may use this approach as a guide to automatically generate FMEA tables from system models represented by SysML diagrams. The approach indeed enables multiple analyses to be performed during each design stage, and it integrates the FMEA into the design process. In particular, the approach tabulates all the components subjected to the most frequent failure modes and the symptoms linked to each failure mode, assessing and listing for each failure mode all the failure propagation paths. The mentioned paper well describes the four major steps characterizing this method, but a synthesis is hereafter reported to give an overview to the reader. In the first step, *Block Definition Diagrams* represent the system components to be analyzed. In the second step, these diagrams model the failure propagations within each component, assigning to the *blocks* representing the components not only design properties (e.g., mass, power) but also the component failure modes. To investigate the failure propagation path among the components as part of the FMEA technique, *Internal Block Diagrams* are used in the third step. Finally, in the last step, the failure propagations and transformations between physical components are performed by using *Internal Block Diagrams*.

Reliability Analysis

Traditionally, the reliability of a system is computed through a Reliability Block Diagram (RBD), which depicts the *reliability values* of the different system components. These values range from 0 to 1, and they represent the probability of each component to operate satisfactorily for a specified period of time in the actual application for which it is intended [6]. The RBD then allows the computation of the total *reliability value* of the entire system.

Modeling activities may indeed enhance reliability studies, but reliability models (e.g., represented by RBD) are often built once the system architecture and performance are baselined. However, this entails that in case of changes to the system design, a lot of time and effort is required to effectively update the performed reliability analyses. Therefore, a reliability modeling method conducted during the development of a system is needed.

Pushed by this motivation, Liu et al. present a solution to automatically create RBDs during the system design phase [41]. This solution is based on the SysML modeling language. *SysML Internal Block Diagrams* indeed describe the internal structure of the system, including the connections between components and the exchanged energy, material, and signal. A *fault* happens when a component is incapable to accomplish an intended function. This fault negatively affects the exchange between components. A matrix named *reachability matrix* then transforms the *Internal Block Diagrams* in RBD, on the basis of the graph theory. In other words, this matrix is a connection between the system model represented in SysML and the RBD. Thus, the system model is first transformed into the *reachability matrix* that derives the reliability diagram. The work in [41] exhaustively describes this process, including the algorithms for the creation of the matrix and the following generation of the RBD. Unfortunately, it does not provide a universal algorithm capable to solve complex system characterized by complex logical relationships among system components.

Summary Table

Table 1 reports all the research works introduced in this section. The objective of this table is to provide the reader with an overview of SysML diagrams used to (automatically) generate reliability and safety analyses starting from SysML models.

A Recommended Approach for Model-Based Reliability and Safety Assessment

The previous section presents several studies for the generation of reliability and safety analyses starting from SysML models. In the majority of cases, the results derived from the application of these approaches are reliability and safety analyses in the form of documents. However, the best practice approach described in the present section aims at producing reliability and safety models instead of documents, due to the multiple advantages introduced in section "From Document-Based to Model-Based Reliability and Safety Assessment" and explained in detail in section "Advantages of the Model-Based Reliability and Safety Assessment." This is the approach proposed and illustrated in [42]. More specifically, the present section illustrates the guidelines for the development of models for the investigation of four reliability and

Table 1 Summary table with the research works introduced in this section dealing with the (automatic) generation from SysML diagrams of the Functional Hazard Analysis, Fault Tree Analysis, Failure Model and Effects Analysis, and Reliability Analysis

Source	Ref.	Analysis	SysML Diagrams	Application
Brusa et al. 2016	[29]	Functional Hazard Analysis	IBD, BDD, SD, AD	Fuel System (Aeronautics)
Müller et al. 2016	[31]	Functional Hazard Analysis	IBD, BDD, AD	Coffee Maker (Industrial Application)
Xiang et al. 2011	[32]	Fault Tree Analysis	IBD, SD, UCD, RD	Fault-Tolerant Parallel Processor (IT)
Mhenni et al. 2014	[33]	Fault Tree Analysis	IBD, SD, UCD, RD	Electromechanical Actuator (Aeronautics)
Izygon et al. 2016	[34]	Fault Tree Analysis	BDD, IBD, SMD	Cascade Distiller System (Space)
Melani et al. 2020	[36]	Fault Tree Analysis	BDD, IBD, SMD, AD	Bearing Lubrification System (Mechanics)
Pierre David et al. 2010	[38]	Failure Model and Effects Analysis	IBD, SD, SMD, AD	Level Control System (Industrial Application)
Mhenni et al. 2014	[39]	Failure Model and Effects Analysis	IBD, SD, SMD	Mechatronic System
Hecht et al. 2019	[40]	Failure Model and Effects Analysis	BDD, IBD	Water System (Industrial Application)
Liu et al. 2013	[41]	Reliability	IBD	-

Acronyms: *AD* Activity Diagram, *BDD* Block Definition Diagram, *IBD* Internal Block Diagram, *RD* Requirement Diagram, *SD* Sequence Diagram, *SMD* State Machine Diagram, *UCD* Use Case Diagram

safety analyses, namely, FHA (section "Model-Based Functional Hazard Analysis"), FTA (section "Model-Based Fault Tree Analysis"), FMEA (section "Model-Based Failure Modes and Effects Analysis"), and RBD (section "Model-Based Reliability Block Diagram"). Standard SysML diagrams and SysML elements are used in this approach.

Model-Based Functional Hazard Analysis

The Functional Hazard Analysis (FHA) is one of the first analyses of the safety assessment process. Traditionally, tables like the one represented in Fig. 5 collect FHA results. This table contains the following information:

- The list of functions performed by the system
- The hypothetical failure modes, for instance, "Loss of a function" or "Function provided when not required"

Function	Failure mode	Effect	Risk

Fig. 5 Typical table collecting FHA results in a traditional document-based approach

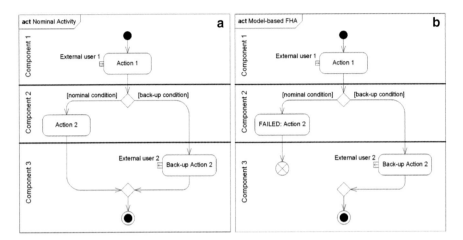

Fig. 6 Model-based FHA represented through SysML Activity Diagram, showing functional failures and failure effects. (Adapted from [42])

- The effects of each hypothetical failure mode, e.g., other functions cannot be performed anymore by the system
- The associated risk factors, i.e., the severity and probability of the failure modes

This traditional approach however implies often several difficulties [30]. The first difficulty is the identification of functions at the right level of abstraction. Functions may be indeed too abstract or detailed for the specific system level being examined. A second difficulty regards the correct determination of the effects associated with functional failures. Finally, the coupling and inter-dependency of functions represent a third difficulty.

Nevertheless, a structured approach based on MBSE may be exploited to overcome all the just mentioned difficulties. The publication [43] collects and describes examples of MBSE structured approaches. In principle, they prescribe how to make use of MBSE technologies to effectively develop systems. Indeed, many of these MBSE approaches focus on the modeling of system functions at different levels.

The best practice approach here described to perform the FHA by employing an MBSE approach is therefore based on the modeling of systems functions. This process employs a *SysML Activity Diagram* showing the functions that should be performed by the system under investigation. In other words, the *SysML Activity Diagram* represented in Fig. 6a represents the functional behavior of a system, i.e., all the actions that the system should perform. Some of these actions require inputs from external users, i.e., "External user 1" represented in the specific case of Fig. 6a. Other actions instead provide outputs to external users, for instance, as represented by "Back-up Action 2." Moreover, the diagram may be divided in *partitions*, in order to allocate functions to system components. In addition, the functions may represent both the nominal and the off-nominal behavior of the system, as some back-up functions might be included and allocated to redundant components. The nominal and off-nominal behaviors are described by different *functional flows* in the *Activity Diagram*, i.e., by different series of actions driven by *decision nodes* and *merge nodes*, which are represented by the two diamonds depicted in Fig. 6a.

The safety process continues with the assessment of each function. Functional failure modes are identified for each function, and consequences are investigated. The *Activity Diagram* also supports this step: each function represented in the diagram might bring to a failure mode, causing an interruption to the sequent part of the functional branch, bringing in some cases to a stop of the entire activity. The *Activity Diagram* of Fig. 6b indeed represents the model-based FHA, where a functional failure entails an interruption to part of the functional flow.

Model-Based Fault Tree Analysis

Figure 7a shows a generic Fault Tree that can be realized during a Fault Tree Analysis (FTA) for the analysis of system faults [44]. A Fault Tree includes Boolean logic gates that identify and assess all the possible causes (e.g., component failures) that originate a top event, which represents an undesired condition, as a system-level fault. These causes are indicated in Fig. 7a as *events*, and they are related to the system components. For example, an *event* can be a component failure. However, a traditional Fault Tree doesn't depict the relation between system components and events.

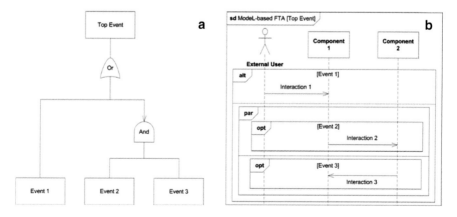

Fig. 7 Traditional FTA (**a**) model-based FTA (**b**). The SysML Sequence Diagram (**b**) represents events and Boolean logic gates as shown in the Fault Tree (**a**)

A Fault Tree shows logically and graphically all the different combinations of possible events that might lead the system to an undesired condition. These combinations of events are represented by logical paths characterized by cause-effect relationships, which are modeled through Boolean logic gates. The logic gate *and* conveys the logical path if all its inputs occur together. In the example Fault Tree of Fig. 7a, the logical path proceeds through the logic gate *and* toward the top event if both *Event 2* and *Event 3* are occurring. The logic gate *or* instead conveys the logical path only if at least one of its inputs occurs.

A *SysML Sequence Diagram* like the one of Fig. 7b is proposed in this chapter as best practice approach to represent the model-based version of the traditional Fault Tree that is peculiar of this analysis. Generally, this type of diagram shows the *behavior* of a system in terms of its interactions with all external users, other systems, and structural elements of the system itself.

The diagram header (e.g., *sd Model-based FTA [Top event]*) in this case reports the top event investigated with the FTA. The *Sequence Diagram* includes the *blocks* representing the system components. *Component 1* and *Component* 2 of Fig. 7b are two examples of *block*. The SysML *actor* element represents external users and other systems interacting with the system under design. Other SysML elements of the *Sequence Diagram* are used to represent all the information collected in the FTA, although this approach is limited to only two types of Boolean logic gate: *and* – *or*. The interaction operator *alt* represents the logic gate *or*, while the logic gate *and* is represented by the interaction operator *par*. The *guard* that generally specifies the condition of the operands defines now all the events of the Fault Tree (see, for instance, the guard *[Event 1]* into the interaction operator *alt* of Fig. 7b). Since the interaction operator *par* is not characterized by guards, a third type of interaction operator (*opt*) may be introduced to specify the events linked to the Boolean logic gate *and*.

It is worth noting that this approach based on the *Sequence Diagram* contains more information than the traditional Fault Tree. In particular, the interactions among the system components and users can be represented inside the three types of interaction operator. Hence, it is clear which interactions cannot happen during the different failure events. This additional information can be helpful for the engineers during the safety assessment. Moreover, the relation between events and component is explicit in the model-based FTA.

Model-Based Failure Modes and Effects Analysis

Another analysis of the safety assessment process is the Failure Modes and Effects Analysis (FMEA). Traditionally, this analysis is document-based since it results in tables. Several examples of FMEA worksheets have been proposed over the years, as reported in [44]. An example of FMEA table is shown in Fig. 8, which includes the following information:

Item	Failure modes	Failure causes	Effects	Failure rates

Fig. 8 Example of table collecting FMEA results in a traditional document-based approach

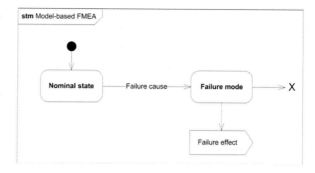

Fig. 9 Model-based FMEA represented through SysML State Machine Diagram, showing failure modes, causes, and effects

- Generic **item** being analyzed, which may be a system component, hardware or software
- All the possible **failure modes** of the item
- All the possible **causes** that may lead to each failure mode
- Possible **effects** of the failure modes, including those most immediate and direct but also those the ultimately affect the entire system
- **Failure rates**, i.e., failure probabilities of each failure mode

A FMEA might include additional information, as severity of the failure modes, explanation for the detection of the failure mode, and additional recommendations.

This chapter suggests a different approach based on models represented by *SysML State Machine Diagrams*. These diagrams represent different *states* of the system or its components. A *state* is defined in SysML as a significant condition in the life of a block (i.e., a system or its component), and it represents some change in how the block responds to events and what behaviors it performs [13]. *States* may correspond to nominal but also off-nominal conditions. Therefore, *states* represent failure modes that are collected in a FMEA. The *SysML State Machine Diagram* of Fig. 9 shows a *nominal* state and a *failure mode* state. In addition, *terminate pseudostate nodes* are connected to failure states to terminate the behavior of the state machine. *Triggers* model the causes that entail the transition from one state to another one. *Triggers are* represented by unidirectional connections between *states*. In case of change from a nominal condition to a failure mode, *triggers* represent failure causes. Eventually, *send signal nodes* are linked to the failure state to identify its effects. It is worth noting that the presence of a *terminate pseudostate node* and *send signal nodes* distinguishes nominal states from failure states. Figure 9 collects all the elements of the *State Machine Diagram* that represent the most relevant information generally included in a FMEA table. It should be noted that the proposed model doesn't represent other details that may be part of a FMEA, for example, the probability of the failure in terms of failure rate. In addition, single-state machines should represent every item being analyzed. This aspect might represent an additional burden entailed by the recommended approach, but several are the advantages over a traditional document-based FMEA. First of all, state machines can be simulated, therefore used to predict component failures and consequences. Moreover, models of FMEA can show more clearly the nominal and faulty behavior of components, while this information might not be highlighted in traditional FMEA tables. Finally, a database containing several state machines can be created once, from where safety analysts can select specific component models that can be reused every time a new model-based FMEA has to be performed.

Model-Based Reliability Block Diagram

The fourth technique investigated in the present chapter is the creation of the Reliability Block Diagram (RBD) for the evaluations of the reliability of a system. Figure 10a depicts an example of RBD. It shows all the components of the system being analyzed. The components are represented by *blocks*, which generally display the *reliability* of the component itself, other than its name.

The *reliability* of the whole system may be calculated through this kind of diagram. The six components of Fig. 10a are characterized by six different values of reliability indicated by the generic parameter R. All the components in the diagram are connected together in order to fulfill the functions expected from the system. Components may be connected in series and parallel configurations. In a series configuration, the functionality that should be provided by the group of components belonging to the configuration can't be realized in case of at least one component is failed. In a parallel configuration instead, the functionality can't be realized if all the components belonging to the configuration are failed. The parallel configuration represents in engineering components with redundancies. Different algorithms compute the reliability or failure rate of the whole system according to the connections of the system components.

The best practice approach illustrated in this chapter for the modeling of RBDs is based on *SysML Internal Block Diagrams*, since this kind of representation is the most appropriate for the evaluation of the reliability of systems and system components, as suggested also by Liu et al. [41]. However differently from [41], the aim in the here described approach is not to extract an RBD from an *Internal Block Diagram*, but to model the system reliability by means the SysML diagram.

The *Internal Block Diagram* depicted in Fig. 10b can generated from the RBD of Fig. 10a by representing all the components as SysML *parts*, and for each of them specifying the *value* "reliability," in the compartment beneath the name of the component (e.g., *:Component 1*). The *Internal Block Diagram* shows also the connections among all the *parts*, defining groups of components in parallel and in series, analogously to the relative RBD. It may be noted that *Component 5* represents a component which is outside the system under design. However, the system

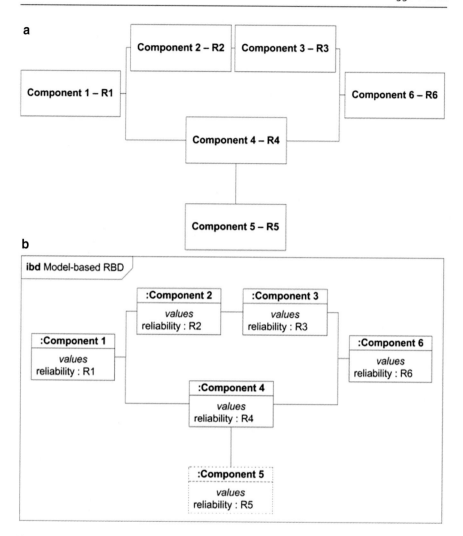

Fig. 10 Traditional RBD (**a**) and model-based RBD (**b**). The SysML blocks in (**b**) report the reliability values of the system components

reliability might be affected by it, and thus it must be included in the diagram. *Component 5* is therefore represented in the *Internal Block Diagram* as a *Reference element* (see the dashed boundary).

Advantages of the Model-Based Reliability and Safety Assessment

The best practice approach described in this chapter based on SysML models to perform reliability and safety analyses has the following advantages over a traditional document-based approach:

- The safety and reliability models may be reused to quickly model and evaluate alternative system solutions, for instance, when multiple system architectures are identified and assessed. This advantage gains importance and relevance for the selection of the system architecture, since the reliability and safety analyses are used in a trade-off study together with other system performance, e.g., masses, efficiencies, and cost.
- The models improve and facilitate the communication and understanding between the teams involved in the development.
- Analysts may quickly query reliability and safety models, to check completeness and consistencies and to verify the system solution against the reliability and safety requirements and constraints.
- In the case of the FHA, since the Activity Diagram represents the functional model of the system, all the functional failures can be considered, without neglecting anyone. Moreover, the diagram supports the identification of effects in case of functional failure.
- The Sequence Diagram of the model-based FTA contains more information of the traditional Fault Tree. In particular, all the actions that can't be performed in case of failure events are included in the model.
- The State Machine Diagram representing the model of the FMEA may be simulated to predict the system behavior in case of malfunctioning.
- The model used to represent the RBD is more intuitive of the traditional technique, and its modeling can be supported by the Internal Block Diagram representing all the system components and their connections.

All the advantages of the proposed approach reflect the benefits stated by several authors and organizations, including INCOSE, about the model-based approach (see section "From Document-Based to Model-Based Reliability and Safety Assessment"). It is worth noting that since the reliability and safety models can be validated, queried, and simulated, development risks and time may be reduced, and the complexity of innovative systems can be managed more efficiently. Moreover, the adoption of a standard modeling language is essential to enhance the comprehension and collaboration between teams performing different tasks, not only about system architecting but including also safety assessment.

Examples of Application of the Recommended Model-Based Reliability and Safety Assessment

In order to give an example of application of the best practice approach previously illustrated, the guidelines described in section "A Recommended Approach for Model-Based Reliability and Safety Assessment" are applied for the safety and reliability assessment of two architectures of an aircraft flight control system (FCS). Having the FCS the function to constantly control the aircraft flight, it could be certainly defined as one of the more safety critical systems of the aircraft. However, considering the reliability of actuators and other FCS equipment and the number of flight hours between detailed checks, a failure of one of these components is a possible event. Therefore, as for the other aircraft on-board systems, the FCS architecture, its number of parallel lanes, and the number of redundancies are directly defined to satisfy the safety requirements. Lately, this subject is even more interesting since the use of new technology that gives to the designer more redundancy options. This example is taken from [42]. Due to the extent and complexity of a reliability and safety evaluation of an entire on-board system, the application case described in this section is simplified with the introduction of some assumptions. The main assumption concerns to the scope of the application case: of the whole control system, only the aileron command is evaluated. Other minor assumptions are provided below.

The aircraft selected as reference for the application is an Airbus A320, since some data is available in literature (e.g., [45]). The first system architecture evaluated in this section is the conventional one. Figure 11a shows the schema of the whole conventional FCS. All the control surfaces are moved by hydraulic actuators, which are supplied by three hydraulic circuits: "Blue" (B), "Green" (G), and "Yellow" (Y). In particular, two hydraulic linear actuators move each one of the two ailerons. One actuator is always active, while the other one is in standby mode, and it operates only in case of failure of the main actuator. The second architecture considered in this study is a "more-electric" one, and it is outlined in Fig. 11b. The proposed solution is similar to the one installed on the Airbus A380, as described in [46]. In this innovative architecture, the "Blue" hydraulic circuit is removed and replaced by an electric line. As assumption, only a single electric line is installed in place of the hydraulic circuit, although the solution adopted on the Airbus A380 is characterized by two electric lines. In the case of the ailerons, each of them is again moved by two actuators, one active and the other one in stand-by mode. The active one is a linear hydraulic actuator, while an Electro-Hydrostatic Actuator (EHA) is selected as redundancy.

The *SysML Internal Block Diagram* depicted in Fig. 12 represents the main components of the conventional architecture of the part of FCS dealing with the movement of the ailerons and the connections between the system components. The two hydraulic actuators connected to two hydraulic circuits provide mechanical force to each aileron, which generates an aerodynamic force with its rotation. Two computers named *ELAC* (Fig. 10 Aileron Computer) electrically control the actuators on the basis of the pilot and co-pilot commands and the aircraft conditions (e.g., speed, altitude) determined by the avionic system. It may be noted that the pilot commands are electrically provided to the ELAC, according to the Fly-By-Wire technology.

Since this section aims at presenting an example of application of the proposed reliability and safety model, the reference system is here simplified. The reason of this choice is to limit the size and complexity of the reliability and safety model; otherwise it might affect the clarity of the proposed example. The following assumptions are identified:

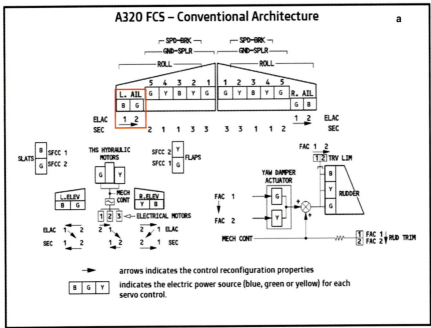

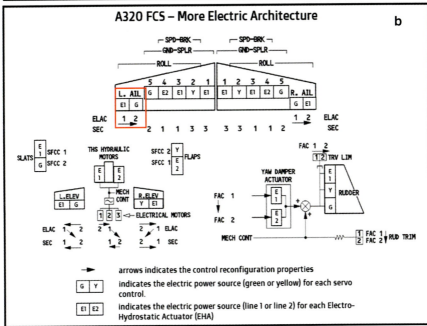

Fig. 11 Schema of A320 FCS: conventional architecture [45] (**a**) and more-electric architecture (**b**). The present application focuses on the left aileron (highlighted in red)

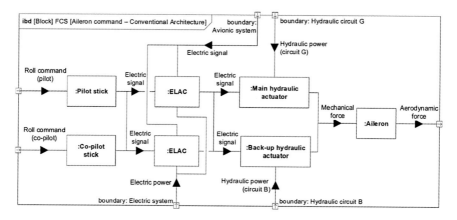

Fig. 12 SysML Internal Block Diagram representing the aileron command: conventional architecture. (Adapted from [42])

- Only the pilot gives the roll command. This application case doesn't include co-pilot or auto-pilot commands.
- Only a single ELAC is considered in the system. Actually, two ELACs are installed for safety purposes. In addition, the ELAC should be connected with the avionic system, since it receives information about the aircraft and flight needed to command the actuators. However, this connection is not considered in the present use case.
- Actuators and aileron position feedback is not considered.

By adopting all these assumptions, the *Internal Block Diagram* of Fig. 13a represents a simplified version of the conventional architecture. A similar diagram is derived to represent the more-electric system architecture, as shown in Fig. 13b. In this case, an innovative EHA supplied by the electric system replaces the back-up hydraulic actuator.

A model may also provide the description about the functioning of the system. The *Sequence Diagram* of Fig. 14 shows the interactions between components and external entities of the conventional system. The five system components represented in the diagrams are the *blocks* of the *SysML Internal Block Diagram* of Fig. 13a. The four external entities instead are users of the system (in this case only the pilot is represented) and other aircraft systems that supply power to this part of the FCS, namely, the two lines B and G of the hydraulic system and the electric system. It is worth noting that the diagram shows *nominal behavior*, i.e., the main actuator moves the aileron, and *off-nominal behavior*, i.e., in case the redundant actuator is utilized.

The system model represents the starting point for the reliability and safety evaluation by means of the proposed model-based approach. The four analyses explained in section "A Recommended Approach for Model-Based Reliability and Safety Assessment" are applied and described in the following sections.

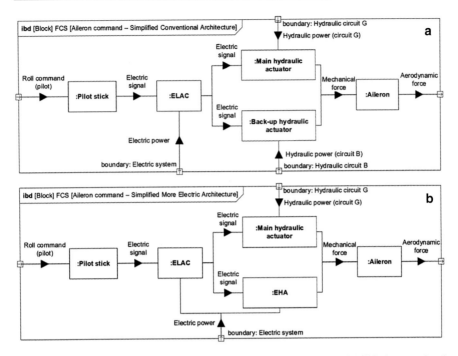

Fig. 13 SysML Internal Block Diagrams representing aileron command: simplified conventional (**a**) and simplified more-electric (**b**) system architectures. (Adapted from [42])

Model-Based Functional Hazard Analysis: Aileron Command

As previously described, *SysML Activity Diagrams* represent the model-based FHA.

The *Activity Diagram* of Fig. 15 collects the functions performed by the conventional system. All the system main functions are allocated to the different components. The reader may note the presence of a conditional element, which directs the flow of functions from the main actuator to its redundancy, in case of failure.

The same *Activity Diagram* may be reused to show what happens in case of a functional failure. Figure 16a depicts the event of functional failure involving the function associated with the main actuator. The corresponding branch of the *Activity Diagram* is interrupted, and as a consequence, the alternative function is performed by the system. The same happens in case of innovative architecture (Fig. 16b), where instead the same back-up function is allocated to a different component.

It may be easily verified that the proposed approach based on models improves the traditional FHA. In fact, an important and challenging activity required to perform this safety analysis is the identification of all the system functions, which are then assessed to determine the effects in case they are not executed or not executed properly. An MBSE approach improves the activity of system functional modeling, therefore enhancing the FHA.

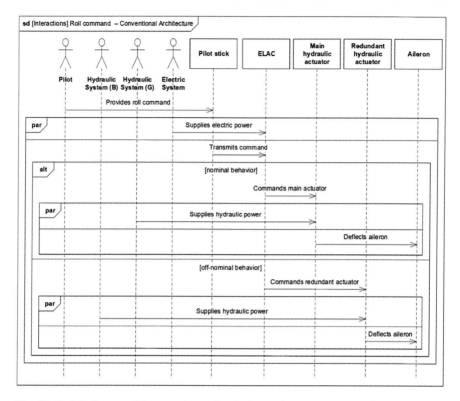

Fig. 14 SysML Sequence Diagram representing the interactions among users and system components of the simplified conventional system architecture focusing on the aileron command. (Adapted from [42])

Model-Based Fault Tree Analysis: Aileron Command

The reliability and safety assessment of the aileron command proceeds with the FTA. Figure 17 depicts the Fault Tree of the conventional system. The identified top event is the not deflection of the aileron. This failure happens if at least one of the following conditions occurs:

- The pilot stick cannot transmit the electric signal.
- Issues affect the ELAC, i.e., it doesn't receive power from the electric system or it results in failure mode.
- Both the actuators have a failure, which may be a generic issue (in this case the actuator is broken) or a more specific low-pressure problem affecting the hydraulic circuit.

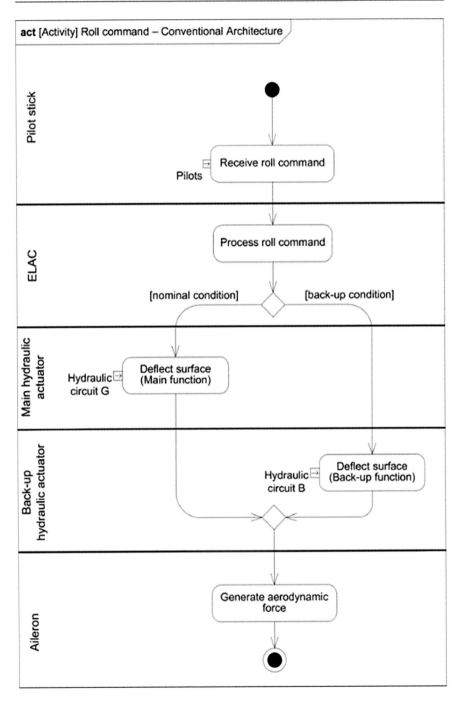

Fig. 15 SysML Activity Diagram representing nominal functions of a conventional system architecture focusing on the aileron command. (Adapted from [42])

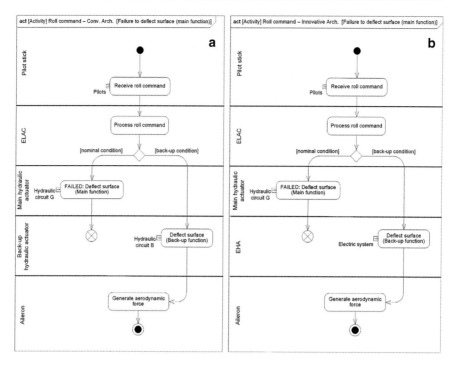

Fig. 16 SysML Activity Diagrams representing model-based FHA applied to a simplified conventional (**a**) and more-electric system architecture (**b**), both focusing of the aileron command. (Adapted from [42])

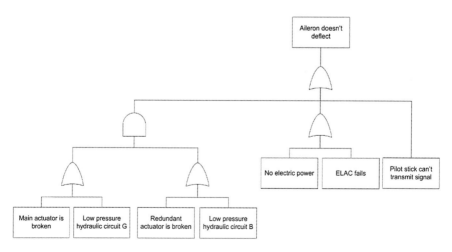

Fig. 17 Traditional FTA applied to a simplified conventional system architecture focusing on the aileron command. (Adapted from [42])

The same Fault Tree of Fig. 17 may be modeled and represented through a *Sequence Diagram*, where the Boolean logic gate *or* is represented by the interaction operator *alt* and the Boolean logic gate *and* is represented by the interaction operator *par*. Figure 18 shows this *Sequence Diagram*. The same diagram shows the interactions of Fig. 14, but in this diagram they are all considered as failed interactions.

The *Sequence Diagram* may be reused to realize the model-based FTA of the more-electric architecture, as Fig. 19 shows. The actor *Hydraulics System (B)* is removed, and the block *EHA* replaces the *Redundant hydraulic actuator*. The final part of the diagram is slightly modified, since a change in failure event is introduced: in the case of innovative architecture, the events *EHA is broken* and *No electric power to EHA* are considered. The modifications of the FTA of the more-electric architecture are highlighted in red. The possibility of reusing models to evaluate the safety of multiple alternatives is one of the major advantages of the proposed model-based FTA. A lot of time and effort may be therefore saved with respect to a traditional document-based analysis.

Model-Based Failure Modes and Effects Analysis: Aileron Command

The following reliability and safety assessment technique is the FMEA. In this application case, two components are evaluated. The former is the hydraulic actuator, which is installed in both the conventional and more-electric architectures. The latter is the EHA, which is peculiar of only the innovative solution. The two *State Machine Diagrams* of Fig. 20 show the failure modes, their causes, and their effects. Both the actuators may fail and change their status to *damping mode*. As a consequence, they wouldn't be able to actuate the ailerons. However, two different causes may bring to this failure condition. In the case of hydraulic actuator, a low pressure in the hydraulic circuit might cause the failure, while in case of the EHA, a shortage of electric power would entail the fail state.

Although not demonstrated in the present chapter, it should be noted that possibility of simulating the models representing the FMEA may actively assist the safety analyst in the identification of system failure modes and their effects.

Model-Based Reliability Block Diagram: Aileron Command

The reliability and safety analysis of the proposed use case ends with the reliability evaluation. The guidelines proposed in section "A Recommended Approach for Model-Based Reliability and Safety Assessment" are applied for the study of the present application case. Figure 21 shows the model-based RBD of the conventional and more-electric system architectures. The two diagrams are similar to the ones of Fig. 13, with the addition of the value *reliability*. Moreover, the components belonging to interface systems – namely, the hydraulic and electric systems – are represented as *Reference elements*, since characterized by a reliability value.

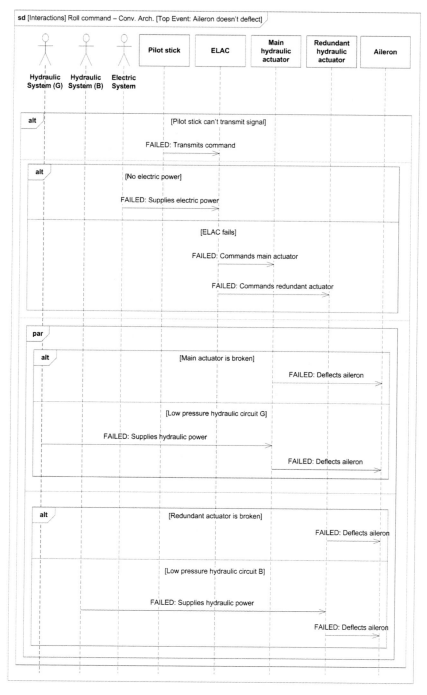

Fig. 18 SysML Sequence Diagram representing the model-based FTA applied to a simplified conventional system architecture focusing on the aileron command. (Adapted from [42])

18 Model-Based Mission Assurance/Model-Based Reliability, Availability,... 581

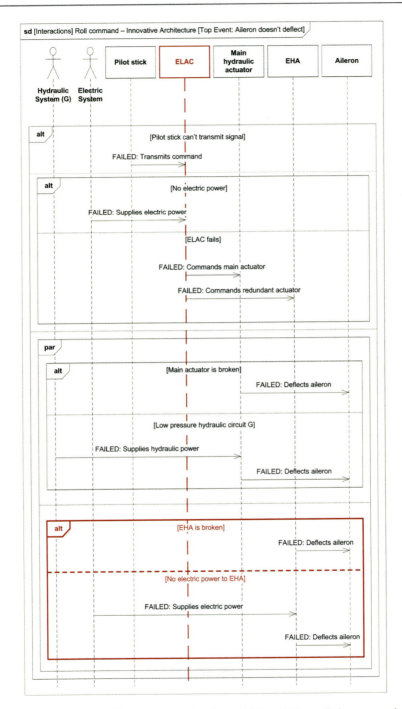

Fig. 19 SysML Sequence Diagram representing the model-based FTA applied to a more-electric system architecture focusing on the aileron command. (Adapted from [42])

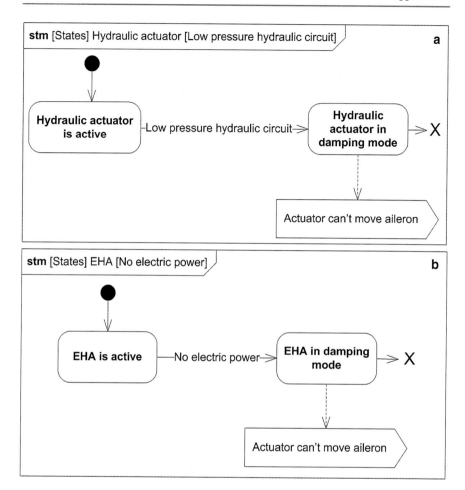

Fig. 20 SysML State Machine Diagram representing the model-based FMEA applied to a simplified conventional (**a**) and more-electric system architecture (**b**), focusing on the aileron command. (Adapted from [42])

The similarity between the diagrams of Fig. 13 representing the systems architecture and those depicted in Fig. 21 showing the model-based RBD is another advantage of the proposed approach. RBDs may be generated very easily and quickly updated in case some changes of the system architecture are made.

The redundant actuator represents the main difference between the two diagrams. Different values of reliability characterize the hydraulic actuator installed in the conventional solution and the EHA of the more electric architecture. Furthermore, the removal of the hydraulic line B changes the system reliability of the innovative alternative.

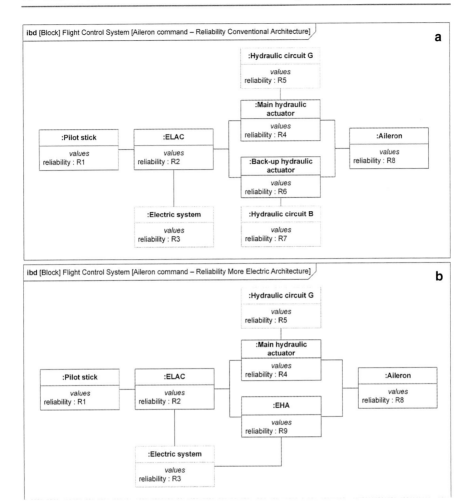

Fig. 21 SysML Block Definition Diagrams representing the model-based RBD applied to a simplified conventional (**a**) and more-electric system architecture (**b**), both focusing of the aileron command. (Adapted from [42])

Chapter Summary

The transition from a document-based to a model-based approach is one of the most relevant enablers to develop the next generation of aerospace systems. Several studies and research projects focus on the creation and usage of system models due to all the advantages of this approach over a traditional document-based approach. Due to all the claimed advantages, these studies are proposing the adoption of a model-based approach in the context of reliability and safety evaluations.

The present chapter collects the most relevant research studies addressing the integration of safety and reliability analyses in a MBSE context. More specifically, the majority of these selected studies proposes methods to conduct these analyses from a system model represented by multiple SysML diagrams. SysML models built according to the guidelines proposed in literature may generate tables collecting results from *Functional Hazard Analyses* and *Failure Modes and Effects Analyses*, *Fault Trees*, and *Reliability Block Diagrams*.

However, almost of the studies described in the chapter that aim at conduct reliability and safety techniques from system models still collect the obtained results in documents, e.g., tables and reports. Therefore, it is unfortunately not possible to exploit all the advantages of MBSE. Undoubtedly, more consistent and complete safety and reliability results may be generated by implementing these proposed approaches, but the entire information is still spread among multiple sources, which hampers communication and understanding between development teams and makes difficult the synchronization of design results and safety and reliability results in case design changes are implemented.

This chapter therefore recommends and describes an additional MBSA approach for the development of reliability and safety models. The models and their application proposed below demonstrate all the advantages previously mentioned. Moreover, these models may be always updated in case of variations of design, for instance, those regarding the system architecture, hence facilitating trade-off studies including reliability and safety results.

Concluding, it can be positively affirmed that the transition from a document to a model-based approach is effectively and successfully addressing also reliability and safety domains since the latest years, although additional research should be made for improvements and to tackle challenges that are still unsolved. First, the methods should be extended to model all the other reliability and safety techniques other than those addressed in this chapter. The whole process of reliability and safety assessment supported by a model-based approach will be indeed furtherly enhanced when models will be created to perform other analyses, e.g., Zonal Safety Analysis and Common-Cause Analysis. Secondly, the development of new modeling tools and plugins can foster all the benefits of a model-based reliability and safety assessment. The transition from a document to a model-based approach might in fact require investments of time and resource (for instance, for the training of the analysts with the new methods), but eventually it will bring to all its advantages.

Cross-References

▶ MBSE in Architecture Design Space Exploration
▶ MBSE Mission Assurance

Acknowledgment The research presented in this chapter has been performed in the framework of the AGILE 4.0 project (toward cyber-physical collaborative aircraft development) and has received funding from the European Union Horizon 2020 Programme under grant agreement n° 815122.

References

1. U.S. DoD, "Guide for achieving reliability, availability, and maintainability," 2005.
2. American Society for Quality (ASQ), 2011. [Online]. Available: https://asq.org/quality-resources/quality-glossary/. [Accessed 3 December 2020].
3. Society of Automotive Engineers (SAE), "ARP4761 – guidelines and methods for conducting the safety assessment process on civil airborne systems and equipment," 1996.
4. IEC 60812 Technical Committee. "Analysis techniques for system reliability-procedure for failure mode and effects analysis (FMEA)," 2006.
5. Society of Automotive Engineers (SAE), "ARP4754A – guidelines for development of civil aircraft and systems," 2010.
6. M. Rausand and A. Høyland, System reliability theory: models, statistical methods, and applications (Vol. 396), Wiley, 2003.
7. E. Zio, F. A. Mengfei, Z. E. Zhiguo and K. A. Rui, "Application of reliability technologies in civil aviation: lessons learnt and perspectives," Chinese Journal of Aeronautics, vol. 32, no. 1, pp. 143-158, 2019.
8. U.S. Department of Defense, "MIL-HDBK-217F," Washington, DC, 1991.
9. G. P. Pandian, D. A. Diganta, L. I. Chuan, E. Zio and M. Pecht, "A critique to reliability prediction techniques for avionics applications," Chinese Journal of Aeronautics, vol. 31, no. 1, pp. 10-20, 2018.
10. Technical Operations – INCOSE. "Systems Engineering Vision 2020 – INCOSE-TP-2004-004-02," 2007.
11. International Organization for Standardization, "ISO/IEC 15288 – systems and software engineering – software life cycle processes," 2002.
12. A. L. Ramos, J. V. Ferreira and J. Barceló, "Model-based systems engineering: an emerging approach for modern systems," IEEE Transactions on Systems, Man, and Cybernetics, Part C (Applications and Reviews), vol. 42, no. 1, pp. 101-111, 2011.
13. S. Friedenthal, A. Moore and R. Steiner, A practical guide to SysML – The systems modeling language, Waltham: Elsevier, 2012.
14. A. Joshi and M. Heimdahl, "Model-based safety analysis of simulink models using SCADE design verifier," *Computer Safety, Reliability, and Security. SAFECOMP 2005. Lecture Notes in Computer Science*, vol. 3688, 2005.
15. A. Joshi, M. Whalen and M. Heimdahl. "Model-based safety analysis final report," NASA Techreport, 2005.
16. J. B. Dabney and T. L. Harman, Mastering simulink, Pearson, 2004.
17. "Scade suite product description," Esterel Technologies, [Online]. Available: https://www.ansys.com/products/embedded-software/ansys-scade-suite. [Accessed 3rd May 2020].
18. M. W. Whalen, "A formal semantics for RSML-e," Master's thesis, University of Minnesota, 2000.
19. N. Halbwachs, P. Caspi, P. Raymond and D. Pilaud, "The synchronous data flow programming language LUSTRE," Proceedings of the IEEE, vol. 79, no. 9, pp. 1305-1320, 1991.
20. J. H. Bussemaker, P. D. Ciampa and B. Nagel, "System architecture design space exploration: An approach to modeling and optimization," in *AIAA AVIATION 2020 FORUM*, Virtual Event, 2020.
21. J. H. Bussemaker and P. D. Ciampa, "MBSE in architecture design space exploration," in *Handbook of model-based systems engineering*, Springer, To be accepted.
22. C. Leangsuksun, H. Song and L. Shen, "Reliability Modeling Using UML," *Software Engineering Research and Practice*, pp. 259-262, 2003.
23. Z. Pap, I. Majzik, A. Pataricza and A. Szegi, "Methods of checking general safety criteria in UML statechart specifications," Reliability Engineering & System Safety, vol. 87, no. 1, pp. 89-107, 2005.

24. F. Iwu, A. Galloway, J. McDermid and I. Toyn, "Integrating safety and formal analyses using UML and PFS," Reliability Engineering & System Safety, vol. 92, no. 2, pp. 156-170, 2007.
25. Object Management Group (OMG), "Unified Modeling Language (UML)," [Online]. Available: https://www.omg.org/spec/UML/About-UML/.
26. A. Joshi, S. Vestal and P. Binns, "Automatic generation of static fault trees from AADL models," in DSN workshop on architecting dependable systems, vol. 10, Berlin (DE), Springer, 2007.
27. INCOSE, Systems Engineering Handbook v.3, 2006.
28. Object Management Group (OMG). "System Modeling Language (SysML)," [Online]. Available: https://www.omg.org/spec/SysML/About-SysML/.
29. E. Brusa, D. Ferretto, C. Stigliani and C. Pessa, "A model based approach to design for reliability and safety of critical aeronautic systems," in *Proceedings of INCOSE Conference on System Engineering*, Turin (IT), 2016.
30. P. J. Wilkinson and T. P. Kelly, "Functional hazard analysis for highly integrated aerospace systems," 1998.
31. M. Müller, M. Roth and U. Lindemann. "The hazard analysis profile: linking safety analysis and SysML," in *2016 annual IEEE Systems Conference (SysCon)*, 2016.
32. J. Xiang, K. Yanoo, Y. Maeno and K. Tadano, "Automatic synthesis of static fault trees from system models," in *Fifth International Conference on Secure Software Integration and Reliability Improvement. IEEE,* pp. 127–136. 2011.
33. F. Mhenni, N. Nguyen and J. Choley, "Automatic Fault Tree Generation From SysML System Models," in *IEEE/ASME International Conference on Advanced Intelligent Mechatronics (AIM)*, Besançon (FR), 2014.
34. M. Izygon, H. Wagner, S. Okon, L. Wang, M. Sargusingh, and J. Evans. Facilitating R&M in spaceflight systems with MBSE. *Annual Reliability and Maintainability Symposium (RAMS)*, pp. 1–6. 2016.
35. Dassault Systems CATIA/No Magic, [Online]. Available: https://www.nomagic.com/products/magicdraw. [Accessed 10 12 2020].
36. A. H. Melani and G. F. Souza. Obtaining fault trees through SysML diagrams: A MBSE approach for reliability analysis. In *Annual Reliability and Maintainability Symposium (RAMS)*. 2020.
37. U.S. Department of Defense, "MIL-STD-1629A," Washington (DC), 1980.
38. P. David, V. Idasiak and F. Kratz, "Reliability study of complex physical systems using SysML," Reliability Engineering & System Safety, vol. 95, no. 4, pp. 431-450, 2010.
39. F. Mhenni, J. Y. Choley and N. Nguyen. Extended mechatronic systems architecture modeling with SysML for enhanced safety analysis. In *IEEE International Systems Conference Proceedings*, 2014.
40. M. Hecht and D. Baum, "Use of SysML for the creation of FMEAs for reliability, safety, and cybersecurity for critical infrastructure," INCOSE International Symposium, vol. 29, no. 1, pp. 145-158, 2019.
41. X. Liu, Z. Wang, Y. Ren and L. Liu, "Modeling method of SysML-based reliability block diagram," *Proceedings 2013 International Conference on Mechatronic Sciences, Electric Engineering and Computer (MEC) – IEEE,* 206–209, 2013.
42. F. Bruno, M. Fioriti, G. Donelli, L. Boggero, P. D. Ciampa and B. Nagel, "Methodology for innovative aircraft on-board systems developed in a MDO environment," in *AIAA Aviation Forum 2020*, Virtual event, 2020.
43. J. A. Estefan, „Survey of Model-Based Systems Engineering (MBSE). Methodologies," 2008.
44. C. A. Ericson, Hazard analysis techniques for system safety, John Wiley & Sons, 2015.
45. "Airbus A319-320-321 [Flight Controls]." [Online]. Available: http://www.smartcockpit.com/aircraft-ressources/A319-320-321-Flight_Controls.html. [Accessed 1st April 2020].
46. D. van den Bossche. "The A380 flight control electro-hydrostatic actuators, achievements and lesson learnt," in *25th International Congress of the Aeronautical Sciences (ICAS)*, Hamburg (DE), 2006.

Luca Boggero works as a Research Scientist at the German Aerospace Center (DLR) in Hamburg since August 2018. Before joining DLR, he has obtained a PhD in Aerospace Engineering at Polytechnic of Turin with a dissertation on preliminary design of aircraft subsystems. He currently takes part in German and EU H2020 funded research projects. His research activities focus on Multidisciplinary Design and Optimization, Systems Engineering and Model-Based Systems Engineering.

Marco Fioriti is Assistant Professor of Aerospace Systems at Politecnico di Torino. He is a member of the AeroSpace Systems Engineering Team at the Mechanical and Aerospace Engineering department from 2006. He spent his Ph.D. at the Preliminary Aircraft Design office of LEONARDO Aircraft Division developing new aircraft design and cost estimation models. His expertise ranges from aircraft on-board systems design and integration, aircraft conceptual design, hybrid-electric propulsion, multidisciplinary design optimization, MBSE, life cycle cost, and RAMS estimation. He has been involved in several national and international research projects in the framework of Aerospace National Cluster, Clean Sky, Clean Sky 2, and Horizon2020. He is a lecturer of B.Sc. and M.Sc. academic courses on avionic system design, on-board system design, RAMS, and cost disciplines. He is author of several scientific articles, congress proceedings, and book chapters.

Giuseppa (Pina) Donelli works as a Research Scientist at the German Aerospace Center (DLR) in Hamburg since September 2019. She graduated in Aerospace Engineering at the University of Naples with a Master Thesis in the frame of the European project "IRON" targeting a campaign of CFD analysis to evaluate the HTP/nacelle interference effects of an innovative turboprop configuration. She is currently involved in German and international research projects dealing with MDO and Model-Based Systems Engineering.

Pier Davide Ciampa is Head of the System Integration & Multidisciplinary Design Optimization research group at DLR Institute of Systems Architecture in Aeronautics, in Hamburg. His research activities focus on digital engineering for development and optimization of complex aeronautical systems, novel methodologies for MDO and MBSE. He is member of the AIAA MDO Technical Committee and INCOSE and guest lecturer for MBSE and MDO at several organizations and universities. From 2015 to 2018, he coordinated the large-scale EU-funded research project AGILE, receiving the "ICAS Award for Innovation in Aeronautics" in 2018. From the 2019 he is coordinating the EU-funded research project AGILE4.0, in a consortium of 16 international organizations developing the next generation of MBSE systems with applications in aeronautics.

MBSE in Architecture Design Space Exploration

19

J. H. Bussemaker and Pier Davide Ciampa

Contents

Introduction	590
The Need for Quantitative Evaluation in the Early Design Stage	590
System Architecting in the Systems Engineering Process	592
Challenges	593
Developments at the DLR Institute of System Architectures in Aeronautics	594
State of the Art	595
Architecture Modeling and Tightly Coupled Evaluation	595
Generating Architectures by Morphological Matrix Enumeration and Reasoning	599
Domain-Specific Systematic Architecting Methods	602
Dynamic Mapping of Function to Form	604
Summary Table	605
Proposed Best Practice Approach for Systematic Architecture Design Space Exploration	606
Modeling the Architecture Design Space	607
Formalizing the Architecture Optimization Problem	613
Evaluating the Performance of Generated Architectures	616
Illustrative Examples	618
NASA Apollo Mission Design	618
Hybrid-Electric Propulsion System Architecture	622
Chapter Summary	624
Cross-References	626
References	626

Abstract

Architecting decisions play a large role in the final system performance. However, the architecting process suffers from a combinatorial explosion of alternatives, often requiring judgment based on expert experience to reduce the design space size. Systematic design of complex system architectures enables a more

---

J. H. Bussemaker (✉) · P. D. Ciampa
Institute of System Architectures in Aeronautics, MDO Group, German Aerospace Center (DLR), Hamburg, Germany
e-mail: jasper.bussemaker@dlr.de; pier-davide.ciampa@dlr.de

© Springer Nature Switzerland AG 2023
A. M. Madni et al. (eds.), *Handbook of Model-Based Systems Engineering*,
https://doi.org/10.1007/978-3-030-93582-5_36

thorough exploration of the complete design space and less exposure to expert bias, subjectivity, and conservatism. Two aspects of system architecture design are treated in this chapter: the automated generation of architecture candidates by capturing architecting decisions and quantitative evaluation strategies. An overview of different design space modeling techniques and their applicability to design space exploration is given at first. A graph-based method that models architecture decisions related to function assignment, component characterization, and component structure is presented in more depth, along with how a formalized optimization problem can be constructed from this design space model. Thereafter, the main features of architecture evaluation are discussed. Applications of the introduced method to the Apollo mission architecting case and a hybrid-electric propulsion system architecture problem are presented.

Keywords

System architecture · Architecture optimization · Architecture evaluation · Architecture design space · ADSG · MBSE · MDO

Introduction

A system architecture is the description of what components a system consists of, what functions they perform (i.e., why they are there), and how they are connected and related to each other [14]. It provides a bridge between upstream systems engineering product development activities, like identifying stakeholders, needs, and requirements, and downstream phases, like detailed design and operation of the system. However, the design of system architectures is anything but trivial, having to take many stakeholders and potentially conflicting needs into account and dealing with an extremely large design space due to the combinatorial explosion of alternatives. Structured approaches for solving these issues, like systems engineering, are available; however, often the capability to quantitatively evaluate the performance of system architectures is missing.

This chapter provides an overview of how the design of complex system architectures can be approached in a systematic way, such that the whole design space that governs the choices available in designing the system architecture can be explored satisfactorily. The objective of such a systematic approach to system architecture design space exploration is to get more confidence in the final design by attaining more knowledge of the behavior of the design space while reducing the influence of expert bias, subjectivity, and conservatism.

The Need for Quantitative Evaluation in the Early Design Stage

Decisions taken when designing the system architecture have a large impact on the final system design, especially on the performance and how well the system meets its goals. At the same time, normally little is known about the behavior of the system during the design stage, making it more difficult to take informed decisions. As the design process progresses, more is known about the behavior of the system and the impact of decisions, but it is more difficult to retroactively change decisions taken earlier on in the design process. This phenomenon has been called the knowledge paradox [31].

Additionally and especially for complex systems, the architecture design space, the space spanning all possible architectures as made up by the combination of all architectural choices, can be extremely large. This is because during the architecting phase, decisions are often of a discrete nature: for example, a choice between several technology alternatives. An example of such a choice would be between wing-mounted or fuselage-mounted aircraft engines. Already for a low number of discrete decisions, a design space can suffer from a **combinatorial explosion of alternatives** [26]. For such large design spaces, it is infeasible to exhaustively search all possible combinations of options.

Expert judgment and experience can address the problems of taking decision in uncertain and large design spaces. These approaches, however, may suffer from expert bias, subjectivity, conservatism, or overconfidence [40], and are not applicable to novel systems that bear little resemblance to existing systems [35].

Systematic design space exploration couples the use of **automatic exploration and exploitation** (e.g., optimization) algorithms to the **quantitative evaluation** of candidate solutions. Systematic thus means that the design (or at least the searching of the design space) is done by the computer rather than the human engineer, although the engineer is still in charge of defining the design space and evaluation metrics. This enables the thorough but non-exhaustive search of the design space in a timely manner while considering all relevant design decisions. The systematic exploration concept is visualized in Fig. 1.

In order to prepare the architecture design space for systematic exploration, a mechanism for generating candidate architectures must be defined, and it must be possible to quantitatively evaluate these candidate architectures. This chapter mainly deals with the modeling and definition of the generic architecture design space, such that the design space can be formalized and existing optimization algorithms might be used for efficiently generating candidate architectures. Nevertheless, modeling and formalizing architecture design spaces are only possible if the development process itself is based on models, such as by leveraging a model-based systems engineering (MBSE) context.

Several approaches for quantitative architecture evaluation can be distinguished; however, implementation specifics vary greatly between different system types. It is important to note that such an evaluation procedure should be fully automated (i.e., require no human interaction) to enable design space exploration within a feasible timeframe (e.g., by not being bound to office hours) and performance metrics should meaningfully capture important (physical) phenomena in sufficient detail with respect to the considered architecting decisions [46].

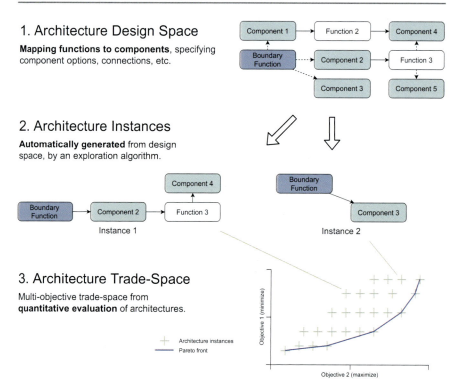

Fig. 1 Systematic architecture design space exploration concept. First, the architecture design space is modeled: functions are mapped to components, component options and connections are defined, and architectural choices are identified. Then, architecture instances are automatically generated from architecture decisions using an exploration algorithm (e.g., optimization algorithm). Finally, architecture instances are quantitatively evaluated, and results can be used to inspect and visualize the trade space to select the best architecture

System Architecting in the Systems Engineering Process

System architecting as one of the phases within a wider systems engineering development process. Systems engineering in general is a structured approach for designing products by treating them as a system, where interconnected components work together to perform functions that they would not be able to provide by themselves [14].

Therefore, a way of decomposing a system is needed for studying its individual components and their interactions and take decisions on which components to include in the system and how to connect them. There are many options for decomposing a system, ranging from more generic to more domain-specific. In the context of aircraft design, systems may, for example, be decomposed based on physical location (e.g., wing, fuselage) or engineering discipline (e.g., aerodynamics, structures), or the decomposition may be maintenance-oriented like the ATA (Air Transport Association) chapters [34]. Domain-specific methods of decomposition, however, always include a certain level of solution and experience bias. The most generic way to decompose system architectures is **functional decomposition**: it provides a generic breakdown of functions to be satisfied by the system and its components which are independent of the architecture alternative, and it is less subject to bias toward a specific solution or technology. Additionally, functions explicitly specify how system requirements are fulfilled.

With a function-based decomposition technique, care is taken to define *what* the system does (its functions) before *how* the system does it (its form). Form is the thing that will be implemented in the physical system. Form exists and is also referred to as the components of the system. Function describes what should be able to do in order to meet stakeholder needs and requirements and ultimately provides the reason the system exists. Function is implemented by form.

The system architecture then defines how functions are fulfilled, by assigning components to functions, based on available knowledge about which components can fulfill which functions. Additionally, it is possible that components themselves need additional functions to be performed within the system, thereby requiring the inclusion of additional components. This leads to an iterative zigzagging procedure, which is completed once all functions have at least one component assigned to them. The finalized system architecture describes this function-form allocation, together with relationships among the components [14]: it is a function-form-structure mapping.

Challenges

The preceding sections show that there is a need for quantitative exploration and evaluation of the architecture design space during the MBSE product development process, in which architectures are defined by a function-form-structure mapping. To enable this, candidate architectures must be generated, and their performance should be evaluated. This chapter therefore aims to answer the following questions:

- What types of architecture **decisions** and accompanying architecture **elements** might be present in an architecture design space?
- How can architecture decisions be identified?
- How can architecture decisions be captured and formalized such that existing **exploration and optimization algorithms** might be used?
- What are the characteristics of a formalized architecture design space?
- What types of optimization algorithms are capable of dealing with these characteristics?
- How are **architecture candidates generated** by taking decisions?
- How does the architecture generation step fit in the design space exploration process?
- What generic approaches to **architecture evaluation** can be distinguished?

Developments at the DLR Institute of System Architectures in Aeronautics

At the Institute of System Architectures in Aeronautics of the German Aerospace Center (DLR), these aforementioned questions are addressed as part of a larger effort to improve the way complex aeronautical systems are designed. This started from the observation that the development of modern complex systems needs to account for an ever-increasing number of capabilities to be delivered and deal with organizational boundaries, integration and communication challenges, and constraints stemming from all stages of the product life cycle. Such complexity impacts not only the product itself (e.g., the aircraft) but also the so-called development systems deployed to support the development of a product and requires a shift to a novel development paradigm.

The DLR institute is developing a "model-based conceptual framework" for streamlining the architecting, designing, and optimizing complex aeronautical systems. It is expected that development time and costs are reduced through modeling and increased transparency, efficiency, and traceability of the design and decision-making process. The conceptual framework is introduced by [13], as bridging the downstream product design and optimization phases of complex system development (e.g., a multidisciplinary design and optimization process) to the upstream architecting phases (i.e., a systems engineering approach).

The conceptual framework scope is shown in Fig. 2. On one side, efforts focus on the acceleration of upstream architecting phases, including identification and trade-off of goals and capabilities, the specification of scenarios and requirements accounting for all the stakeholders involved, and the design and optimization of the architecture of the system of interest (e.g., an aircraft), or system of systems, under development. On another side, the framework aims to accelerate the downstream product design phases, including the selection of the capabilities needed for every design stage (e.g., conceptual, preliminary, detailed), the integration of capabilities into a design processes (e.g., a MDO), and the deployment and operation of design

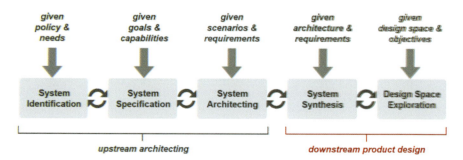

Fig. 2 Model-based conceptual framework in development at DLR Institute of System Architectures in Aeronautics, bridging upstream architecting and downstream product design. (Figure adapted from [13])

systems (e.g., computational environments), needed for the exploration of the design space and the selection of the optimal solution.

The implementation of the concept is supported by the development of novel design methods and approaches, leveraging digital design engineering and modeling technologies.

State of the Art

This section presents the state of the art in systematic architecture design space exploration techniques. Several types of approaches are distinguished based on the degree to which they model all possible decisions and how tightly integrated with evaluation methods they are.

In section "Architecture Modeling and Tightly Coupled Evaluation," architecture modeling methods that tightly integrate with evaluation of architectures are discussed. Section "Generating Architectures by Morphological Matrix Enumeration and Reasoning" presents architecture generation methods based on combinatorial enumeration of morphological matrices. Then, in section "Domain-Specific Systematic Architecting Methods," domain-specific systematic architecting methods that integrate architecture evaluation are discussed. Section "Dynamic Mapping of Function to Form" presents dynamic function-form-structure mapping approaches, which offer the most flexible and complete architecture generation capabilities while being solution-agnostic as far as architecture evaluation is concerned. Finally, section "Summary Table" presents a **summary table** giving an overview of the literature discussed in the preceding sections.

Architecture Modeling and Tightly Coupled Evaluation

Part of model-based systems engineering (MBSE) methods is the modeling of system architectures. That by itself is an essential step in the direction of enabling systematic design system architectures due to the semantic element connections realized in such models. However, no architecture decisions are modeled, and only architecture instances can be described [11]. Architecture decisions describe the architecture design space by specifying the different options for fulfilling the system requirements; an architecture instance is a specific realization (or design point) within the architecture design space, implicitly or explicitly constructed by taking architecture decisions. Directly modeling architecture instances can be useful for semantically describing architecture candidates, but it quickly becomes cumbersome to do this for more than a dozen alternatives. Combined with the combinatorial explosion of alternatives often encountered in architecture design spaces, this approach quickly becomes infeasible.

However, another aspect of systematic design space exploration is the evaluation of candidate architectures. MBSE languages such as SysML do not have integrated methods for quantitatively evaluating system performance. SysML [24] does include parametric modeling capabilities, but these are normally used for specifying custom requirements verification behavior, rather than simulation of the system itself. Behavior can be simulated using state and activity diagrams; however, this will not give any quantitative measures of performance. One way of quantitatively evaluating candidate architectures then is to directly translate the architecture model into a simulation model: all the knowledge for setting up a simulation model is already contained in the system architecture model, and only a one-to-one translation is needed to make the model executable. Such an executable model can then be used to evaluate the performance of the architecture. These approaches work well if the system describes a mechanical or physical system and if a single simulation model can indeed capture all relevant physical phenomena sufficiently to compare different architecture alternatives.

Several such conversion methods exist for different simulation frameworks. One example is SysML4Modelica [39], a SysML profile that allows the annotation of SysML models with simulation-specific information, which can then be used to transform the SysML model into an executable Modelica model. SysML parametric and block diagrams are extended with specification of, among others, flow types and Modelica standard library parts. An example of a SysML and an equivalent Modelica model is shown in Fig. 3.

An application of SysML4Modelica has been presented by [21]. Their software package ArCADia Architect offers an interactive way of defining system architectures. Architectures are defined by assigning components from a component database to system functions and inducing additional functions from selected components. A logical flow view is then constructed that determines the flow of quantities (e.g., mass, information, data) between components. This is then converted to a SysML model, at which point SysML4Modelica is used to create an executable simulation model, as shown in Fig. 4. Application to the design of an Environmental Control System (ECS) is demonstrated.

For system architectures with fewer components or that take fewer physical phenomena into account, for example, during early conceptual design, a method by [8] allows the direct translation from the logical flow field into a computational workflow. This method also allows the inclusion of black box elements in the workflow, as long as their source-sink behavior and connection to other elements is correctly defined in the system architecture model. This does assume that all system elements can be mapped to computational blocks. They demonstrate the application of this method to aircraft conceptual design.

For simulations involving more system components or if more types of simulations are needed to correctly predict the performance of an architecture, it might be necessary to translate the architecture into a more sophisticated multidisciplinary design optimization (MDO) workflow in a Process Integration and Design Optimization (PIDO) environment, enabling the coupling of multiple engineering analyses and convergence toward a design consistent across all involved disciplines. [5] present a method for translating SysML models into OpenMDAO workflows, and the same principle applies to other PIDO environments. [36] note that for some simulation tools, it might require significant effort to map their interfaces to SysML,

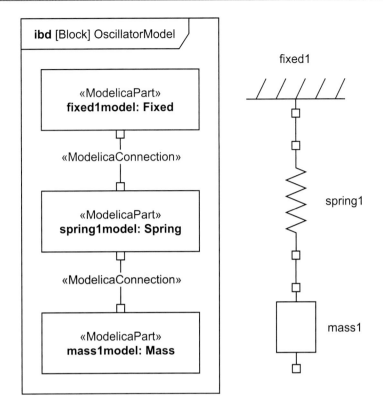

Fig. 3 Example of a mass-damper system in SysML (left) and its equivalent Modelica representation using SysML4Modelica (right), showing how a Modelica model can automatically be generated from the systems model, by including additional simulation parameter directly in the SysML model. (Figure reproduced from [39])

prohibiting the use of direct translation methods described before. However, To automatically convert a SysML model to a PIDO workflow, the computational elements themselves, including data interfaces, need to be modeled. It might be helpful to have information on the composition of the system (i.e., its functions and components) together with the description of the computational workflow for gaining insight into design decisions.

A method presented by [43] integrates function-based architecture definition with behavioral modeling and parametric models to evaluate the architecture performance, implemented in ParADISE tool, using a custom modeling language. Simple equations are modeled directly in the tool, but for more sophisticated computations, external interfaces to various computational environments are available. This method has later been extended to also include decision-making and trade-off analysis of different architecture candidates [44]. Different architecture candidates, however, are not automatically generated, and mainly vary in their design and sizing parameters rather than function-form assignments.

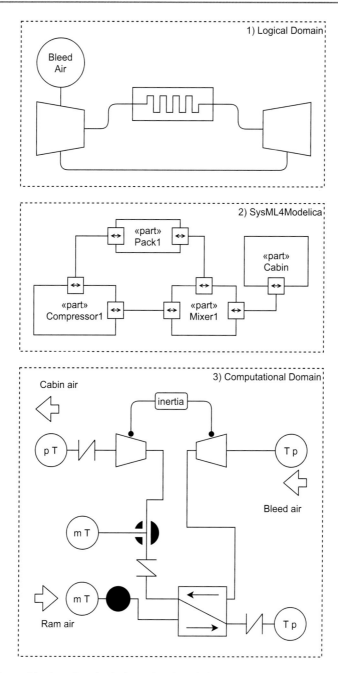

Fig. 4 Views of the three domains in the ArCADia tool: first, a logical model of the architecture is defined, which is then converted into SysML, and then SysML4Modelica is used to create a computational model in Modelica to simulate the architecture and evaluate its performance. (Figure reproduced from [21])

Beernaert and Etman [7] present a method for recursively modeling a system architecture and its components that can then be directly converted into an analytic target cascading (ATC) MDO formulation in order to size the specified architecture. ATC is a formulation for coordinating multilevel MDO models, which works well for representing system architectures specified in terms of decomposition of components: high-level targets are propagated down the component hierarchy chain. The system architecture is specified using the Elephant Specification Language (ESL), originally developed by [50]. ESL defines components and their design variables, design goals, function transformations, sub-components, and connections between components at different system levels. From this, the ATC formulation with all its intermediate design variables, objectives, and constraints is automatically defined. Solving the MDO problem then depends on implementing analysis models describing each of the ATC levels. [7] demonstrate the method using a car powertrain architecture; however, they do not discuss the implementation of the analysis models.

Recently, a graph-based approach was presented by [41] that allows representing engineering systems in terms of components, behaviors, and flows. Analyses are specified in terms of node patterns that effectively detect the presence of certain technologies. A computation graph is constructed that defines the computational dependencies between analyses and can be used for simulating the behavior of the system. Pattern matching is used to replace technologies (a set of components, flows, and behaviors) by other technologies to evaluate the impact of the change of technology on the performance of the system. This graph-based method has been specifically designed for the analysis of mechanical systems that include continuum mechanics, for example, in aircraft conceptual design. The method has been used for automatically detecting compatibilities and synergies between technologies [42], thereby creating a set of feasible technology portfolios.

The different approaches discussed in this section highlight the importance of the evaluation kind for different architecture candidates is an important part of the systems engineering process. It also shows, however, that translation between a systems modeling language (like SysML) and a simulation environment is far from trivial. This is even more so when dealing with complicated multidisciplinary black box simulation tools, which might require the architect to model the computation process in addition to the system architecture itself.

Generating Architectures by Morphological Matrix Enumeration and Reasoning

The morphological matrix method is a popular method for conceptualizing design solutions [48]. It presents an organized way of identifying different possible solutions and combining them to synthesize system designs. At first, independent system functions are identified: these are the functions that the system should perform in order to meet its design goals. Then, for each function, solutions for performing these functions are identified. Solutions are mutually exclusive, meaning that only one solution can be included in an architecture at a time. [46] note that if combinations are possible, these should be added as an option on its own. This way of proceeding through the conceptualization process enables a structured approach to innovation and makes sure all possible system alternatives are considered. Optimally, functions and solutions are independent [47], but in real-life system designs, this is almost never the case. The morphological matrix methodology has been extended many times to include ways of identifying incompatibility, compatibility, and/or synergism: for example, using a compatibility matrix. An example of a morphological matrix is presented in Fig. 5. It shows the large number of architecture alternatives making up the design space (1.16×10^{19}) by combining 37 system functions with up to 6 options each.

[34] present the Adaptive Reconfigurable Matrix of Alternative (ARMA) that is based on the morphological matrix. Compared to the original method, it adds support for function induction: the derivation of new functions due to the selection of a solution for a higher-level function. [3] applied the methodology to the design of a propulsion system architecture. With the help of the function mapping matrix (FMM),

	Functions			Solutions			
	Configuration Type	Tube and Wing	BWB				
	Fuselage Cross-section	Elliptical	Circular	Rectangular	Diamond	Triangular	
	Wing Shape	Elliptical	Circular	Rectangular	Diamond	Triangular	
	Wing Sweep	Forward	Backward	Variable	Straight	Switch	
	Wing Structure	Internal	Truss	Strut	Cable	Other	
AIRFRAME	Number of Wings	1	2	Other			
STRUCTURE	Horizontal Empennage	Tail	Canard	None			
	# of Hor. Emp.	0	1	2	3		
	Vertical Empennage	Tail	Canard	None			
	# of Vert. Emp.	0	1	2	3		
	Materials	Aluminum	Steel	Composite			
	Wing Morphing	Yes	No				
	Loading Door	Front	Mid	Rear	Two Sided	Other	
	High Lift Devices	Slats	Flaps	LFC	Cooling Wings	Plasma	Vortex Gen
	Number of Engines	1	2	3	4		
AERO/PROPULATIO	Engine Type	Turboprop	Turbofan	Turbojet	Propfan	Piston	
N INTEGRATED	Engine Position	Over Wing	Under Wing	Fuselage	Tail	Embedded	Behind wings
SYSTEM	Moving Engines	Thrust Vectoring	Tilt-rotor	Variable Placement	None		
	Hybrid Engines	Yes	No				
	Fuel Type	Conventional	Biofuels	Synthetic	Hydrogen	Natural Gas	Fuel Cells
	Takeoff	Traditional	Floating	Assisted	Vertical		
	Assisted Takeoff Types	JATO	Vert. Catapult	Hor. Catapult	Mag. Catapult	None	
	Landing	Traditional	Assisted	Vertical	Spiral	Steep	
OPERATIONS	Design Range	1000	1500	2000			
	Cruise Speed	0.7	0.74	0.78	0.8	0.82	
	Cruise Altitude [k ft]	15	20	25	30	35	40
	In Air Refuelable	Yes	No				
	Machining	Conventional	Non-conventional				
MANUFACTURING	Paint	Single Coating	Multi-coating				
	Part Transportation	Ground	Air				
	Passengers View	Digital	Real	None			
OTHER SYSTEMS	Pilots View	Digital	Real				
	Piloting	Manned	Autonomous	Ground Pilots			
	Safety Features	Fire Suppression	Crash Safety	Alternative Exits	Fuselage Chute	Fuel Dump	
OTHER	APU	Conventional	Electric	Fuel Cells	Piezo-electric	Batteries	
CONSIDERATIONS	Bleed Air	Yes	No				
	On Board Security	Metal Detector	Explosive Detector	Combination	None		

Fig. 5 An example of a morphological matrix of an aircraft. The left columns show the different functions to be fulfilled, grouped by subsystem. The right columns show the different alternatives (e.g., technologies, numbers, configuration) that can fulfill the corresponding functions. It also shows the combinatorial explosion of alternatives. (Figure reproduced from [29])

connections between architecture elements are established, for example, for transmitting power between components. Furthermore, each function is coupled to analysis code to estimate the performance impact for different solution alternatives, enabling enumeration and evaluation of different architecture candidates.

ARMA has been combined with coordinated optimization approaches in order to compare locally optimized architecture concepts, to fairly compare innovative architectures with established architectures [15]. Using Overall Evaluation Criteria (OEC), multiple conflicting design objectives are combined into a system-level value function.

Another aspect of system architecting is the decision-making process, which can be captured by the architecture decision graph (ADG) [46]. This graph explicitly identifies architecting decisions and their interrelations, allowing reasoning about and effectively communicating the structure of the decision problem (see Fig. 6). The ADG is automatically constructed from the combination of a morphological and compatibility matrix. This approach allows identifying different impact levels of decisions, and it allows for a systematic enumeration of all possible architectures, ensuring solution feasibility out of the box.

A method of constructing a formalized optimization problem from a morphological matrix is presented by [37]. Each function is mapped to a discrete design variable that enables an optimization algorithm to search through all architecture options, and solution incompatibilities are represented by constraint functions. Solution themselves can also have sizing parameters; these are mapped to continuous design variables. A Tabu search algorithm is used to search the design space.

A similar approach is taken by [18], in which architectures are generated by interpreting the morphological matrix and the compatibility matrix. Functions are assigned to design variables, and similar combinations of functions are considered to be architectures. Different selections of design variable values (i.e., function solutions) are then called alternatives, and since these are spanned by the same design variables, they define a formalized design space and can be optimized using existing optimization algorithms. Multi-objective genetic algorithms (e.g., NSGA-II) were applied that account for the mixed-integer nature of the optimization problems (i.e., the presence of both integer design variables coming from architectural choices and continuous design variables coming from sizing parameters), and to account for the fact that system stakeholders often have conflicting needs. Additionally, Frank recognized the benefit of decoupling the generation of architectures (i.e., identifying decisions and ensuring feasibility) with the search of the design space (i.e., by an optimization algorithm), which enables detailed investigations of both steps and in the end facilitates a more effective way of searching the design space.

It is shown in this section that the morphological matrix can be a powerful tool for identifying different solutions for the system functions. However, additional steps are needed for specifying incompatibility or synergy between solution alternatives. Although the system functions are represented, the solutions often represent very different types of elements, like components, multiplicity selectors, or subsystem selections, and component connections are not explicitly represented. This might

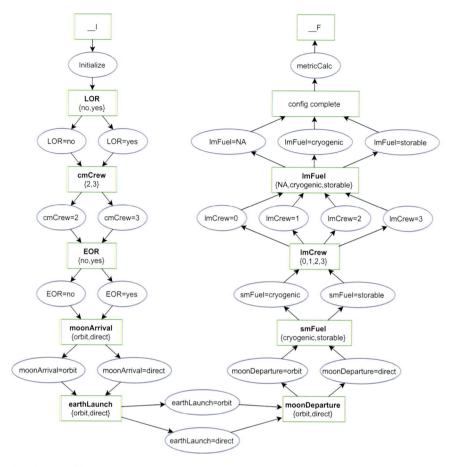

Fig. 6 An architecture decision graph (ADG), showing the different decisions (green boxes) and corresponding options (blue ovals) for each decision. The ADG defines a specific order for taking the decisions, and more complicated decision paths are possible than the one shown here. (Figure reproduced from [46])

make it difficult to translate to a function-form-structure mapping as is normally used in system architecting in an MBSE context.

Domain-Specific Systematic Architecting Methods

System architecting methods can also be more systematic by looking at the design process and involved decisions for a specific subdomain of engineering. This way, it is ensured that all relevant decisions can be modeled and architecture alternatives are efficiently enumerated and evaluated. Since developing a domain-specific approach to system architecting only seems natural looking from a process optimization perspective, it can be helpful to review several methodologies here. It must be noted, however, that this overview in no way claims to be exhaustive.

Liscouët-Hanke et al. [32] developed a method for sizing and evaluating the performance of aircraft power systems. These power systems are defined using a function-based approach, which makes it convenient for modeling a wide range of power system architectures: from conventional architectures that provide electric, pneumatic, and hydraulic power to novel ones such as bleedless and more electric architectures. The method enables quick evaluation of system-level and aircraft-level impacts in the conceptual design phase, including mass, drag, and fuel consumption estimates. Analysis is performed by automatically constructing a Simulink model that simulates the operation of the system under various conditions over the aircraft mission.

One aircraft design subdomain that enjoyed early adaptation of systematic architecting methods is that of flight control system (FCS) design. The FCS enables the safe control of air vehicles and usually consists of a set of control surfaces, actuators, power sources, and management computers. There are many available technology alternatives that can be applied for all of these categories. Additionally, safety considerations make for redundancy in elements and therefore greatly increase the problem of assigning actuators to control surfaces, actuators to power sources, and so on. Genesys is a methodology for design and evaluation of aircraft onboard systems [9]. First, aircraft-level functional requirements are determined. These are then allocated to system components, which are then allocated to control and monitor sources, on the one hand, and power sources, on the other. Then, a choice is made for the power system concept, and finally different architectures are sized to estimate their power consumption, safety parameters, and other relevant metrics. Sizing and evaluation of the generated architectures is built-in, since the architecture models closely represent simulation models by design.

More recently, a method for modeling and visualizing flight control systems using Capella has been developed by [27]. A catalog of FCS and power system components based on composition from generic functions is defined, which is then used to model different flight control systems and their impact on consumption requirements to the power generation system.

A method aimed at the generation and evaluation of aircraft onboard systems is presented by [23]. Their method is applied for the design of thermal management systems (TMS). A graph is constructed that contains various system components that are connected based on known possible connections (see Fig. 7 for a visual representation). The graph represents the cold air and hot air channels and their interactions and can be directly translated to a Modelica simulation model for evaluation. Several graph transformations and structure constraints are defined to enable the automated enumeration of all possible system architectures.

A method presented by [19] enables the representation of system-of-systems (SoS) scenarios that can then be used to automatically derive SoS solutions that fulfill needed capabilities. Generated solutions then can be interpreted and fed into an agent-based simulation (ABS) program to evaluate the performance of the solutions.

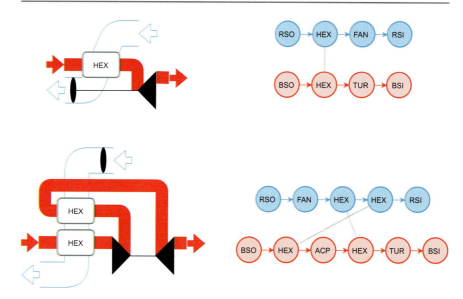

Fig. 7 Schematics of two common thermal management system (TMS) architectures: a simple cycle (top) and a two-wheel bootstrap cycle (bottom). On the right side, the equivalent graph-based representations of the architectures as used by [23] are displayed. The graph-based representation allows the architectures to be systematically manipulated to discover new efficient architectures. (Figure reproduced from [23])

Dynamic Mapping of Function to Form

The most generically applicable systematic system architecting methods allow the dynamic and iterative mapping of form to function and the representation of component connection choices. Such a method would then offer good integration with upstream MBSE steps, sufficiently represent reason for the inclusion of different system components, be free of solution-bias, and be applicable to any kind of system architecting problem.

A formal architecture synthesis and optimization method that models functions and functional flows is presented by [1]. Function flows determine the functional flow path of the system and used for determining which functions are needed for meeting the main system requirements. Functions define required capabilities, which are used to search for components fulfilling these capabilities. Component instances are created that fulfill the capabilities. Components may define ports with flow types and directions that may be connected to other components. Components can induce additional functions that are needed for defining a consistent architecture. Architectures are compared using criteria that are interpreted as objectives and constraints. The corresponding values are calculated by component attributes (e.g., mass) or models (e.g., analysis code). This method allows the automated synthesis of consistent architectures using (multi-objective) evolutionary optimization algorithms such

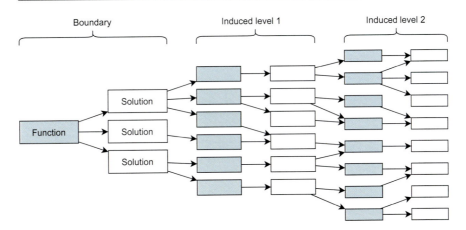

Fig. 8 Function tree showing possible solutions for each function for multiple function-induction levels. Each level has its own associated morphological matrix representation. (Figure reproduced from [28])

as NSGA-II. The method is demonstrated using a cockpit design problem that includes five optimization objectives and shows the feasibility of treating the architecture design process as a multi-objective optimization problem. No architecture structural constraints such as flow connections between components were considered, meaning that structural constraints are only treated at the evaluation phase, leading to many infeasible architectures.

Judt and Lawson [28] present a method for architecture synthesis based on a function tree that is constructed from an iterative procedure that trades finding solutions that fulfill a given function (starting with the boundary function) and inducing functions needed for implementing the chosen solution (see Fig. 8). Architectures may be generated automatically from a definition of all possible function-solution pairs. Filters may be defined for excluding solution combinations (representing incompatible technology combinations). The authors apply this method for the design of aircraft subsystems and therefore couple the architecture generation procedure with an analysis code that automatically sizes architecture component for a predefined aircraft mission. An ant colony optimization (ACO) algorithm [17] is used for searching through the architecture design space, and a genetic algorithm (GA) searches through the design parameters (continuous design variables) of the architecture components for a given architecture.

Summary Table

The state of the art presented in the preceding sections is summarized in Table 1.

Table 1 Summary of the presented literature review, comparing approaches to architecture design space exploration

Class	Definition	Exploration	Evaluation	Reference
Coupled architecture evaluation	SysML	N/A	Modelica	Paredis, et al., [39]
				Guenov, et al., [21]
			Integrated	Bile, et al., [8]
			OpenMDAO	Balestrini-Robinson, et al., [5]
	Custom		Integrated	Schumann, et al., [43]
				Roelofs, et al., [42]
			MDO	Beernaert & Etman, [7]
Morphological matrix methods	Morphological matrix	Enumeration	N/A	Mavris, et al., [34]
				Simmons & Crawley, [46]
		Optimization		de Tenorio, et al., [15]
		Tabu search		Ölvander, et al., [37]
		NSGA-II		Frank, [18]
Domain-specific methods	Custom	N/A	Integrated	Liscouët-Hanke, et al., [32]
	Capella		N/A	Jeyaraj & Liscouet-Hanke, [27]
	Components, ports	Enumeration	Integrated	Bornholdt, et al., [9]
	Graph-based		Modelica	Herber, et al., [23]
	Semantic web		ABS	Franzén, et al., [19]
Dynamic mapping	Func-comp, ports	NSGA-II		Albarello, et al., [1]
	Function tree	ACO/GA		Judt & Lawson, [28]

Abbreviations: *Func-comp* function-component mapping, *MDO* multidisciplinary design optimization, *NSGA-II* nondominated sorting genetic algorithm 2, *ABS* agent-based simulation, *ACO* ant colony optimization, *GA* genetic algorithm

Proposed Best Practice Approach for Systematic Architecture Design Space Exploration

The system architecture design space exploration method developed by the authors of this chapter is based on two pillars [10]:

1. *Architecture Generation:* It should be possible to automatically generate architectures in the relevant design space, thereby yielding architecture representations compatible with MBSE concepts like requirements, functions, and components.

2. *Architecture Evaluation:* It should be possible to quantitatively evaluate every generated architecture, thereby capturing all relevant phenomena to enable the comparison of architectures and identification of the optimal architecture(s) for the design problem.

The separation of these pillars allows a flexible approach to setting up a design space exploration toolchain. When it comes to architecture evaluation, there might be different computational environment, data formats, fidelity levels, and expertise involved for different systems of interest. Coupling such a diverse set of evaluation methods to one architecture generation method then allows the consistent coupling with both the MBSE and the optimization processes.

This section focuses on how to **model architecture design spaces** to yield both architectures in a formulation that integrates well with upstream MBSE processes and allows the definition of a formal optimization problem to enable systematic exploration of the design space. Additionally, compared to previously discussed system architecting methods, the presented best practice approach yields a **semantic representation** of the architecture design space, and next to mapping form to function also takes component structure decisions into account. The method is evaluation agnostic on purpose, to allow the method to be applied to the design of a wide range of system of interests. However, to guide the reader when it comes to enabling quantitative architecture evaluation, this section ends with a discussion on possible architecture evaluation approaches.

Modeling the Architecture Design Space

The proposed architecture design space modeling method arises from the observation that design choices in system architecting are mainly represented by decisions related to [14, 45]:

1. *Function fulfillment*: How are the boundary functions of the system fulfilled? Do the selected solutions require additional functions to be fulfilled?
2. *Component characterization*: How are components modified to be better suitable for the given architecture? How many component instances are included in the architecture?
3. *Component connections*: Which component (instance) is connected to which component (instance)? How many connections are needed?

The architecture design space model therefore needs to be able to represent functions, components, component characteristics, component connections (ports), and their related architecture decisions. Additionally, to enable optimization of architecture, it also should be able to define performance metrics to be used as objectives or constraints.

The elements and their connections are modeled using the **Architecture Design Space Graph (ADSG)**. The ADSG is a directional graph that contains architecture elements and their derivation relationship: if one element points to another, this means that the latter is included in an architecture if the former is also included. Some exceptions include decision nodes, incompatibility edges, and connection edges, all of which are explained in the subsequent subsections. The ADSG is used for modeling the architecture design space and extracting architecture decisions, generating architecture alternatives, and formulating optimization problems.

The following sections describe the elements that may be represented in the ADSG (section "Design Space Elements"), how the ADSG is constructed (section "The Architecture Design Space Graph (ADSG)"), and how architecture decisions are represented (section "Generating Architecture Instances").

Design Space Elements

The architecture design space model contains the information needed for constructing the Architecture Design Space Graph (ADSG). A description of possible elements in this model follows. For a more complete discussion and overview of possible ADSG nodes, see [10].

Functions specify what the system should do. The functions that deliver the value of the system to the system context (all that with which the system interacts but is not part of the system itself) are defined as boundary functions [34]. The designer chooses the boundary functions when defining the design space based on identified stakeholder needs and requirements. These functions are always included in an architecture instance: the architectural choices then define how these functions are fulfilled.

Components fulfill functions and are only included in an architecture if they are needed for fulfilling a function. Each component described in the design space defines the functions it *fulfills* and the functions it *needs* (also known as function induction [34]) that fulfill its functions. Figure 9 shows an example architecture with components (COMP) fulfilling and deriving functions (FUN) for a centrifugal pump architecture: the casing diffuses water, the impeller accelerates water, and the electric motor rotates the impeller.

Only one component can fulfill a function at a time: they are mutually exclusive. This means that there are architectural choices if multiple components can fulfill a

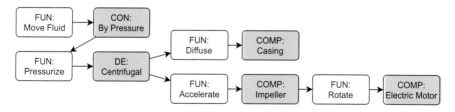

Fig. 9 Mapping components to functions: concepts (CON), function decomposition (DE), and component (COMP) nodes that fulfill (incoming connections) and derive (outgoing connections) functions (FUN). Shown is part of centrifugal pump architecture, adapted from [14]. (Figure from [10])

19 MBSE in Architecture Design Space Exploration

function. It should be noted that it is possible to model non-mutually exclusive options as additional options themselves [46].

A **concept** is a notional mapping between solution-neutral and solution-specific form and function [14]. Figure 9 shows a concept (CON) mapping the solution-neutral function "move fluid" to the solution-specific function "pressurize" (another solution-specific functions could, for example, be "displaced").

Function decompositions simply map one function to one or more other functions and can be convenient for managing complexity. Figure 9 shows the decomposition of "pressurize" into "diffuse" and "accelerate." A decomposition means that the originating function performs the same as the combination of the target functions: the originating function *emerges* from the target functions [14]. Moving in the opposite direction can be seen as function *zooming*, since the emerged function is represented by more detailed functions.

Two types of component characterizations are distinguished (see Fig. 10):

1. *Characterization by number of instances*: Components have component **instances**, multiple instances of components can exist, and the number of instances can be an architecture decision. For example, this can be used for modeling component distribution or redundancy.
2. *Characterization by attributes*: Components or component instances can have **attributes**, and the selection of values for the attributes is an architecture decision.

Connections between components are modeled using **ports**. Connections can represent anything from spatial relationships to mass/energy/information flows and intangible relationships like membership and sequence [14]. No restrictions are made on the type of connection; this will be entirely up to the interpretation of the component connections when evaluating the performance of the generated architectures. Connections are made from a *source* to a *target* and for ports; these are output and input ports, respectively. See Fig. 11 for an example of components with ports and their connections. Ports are characterized by:

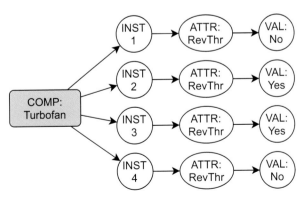

Fig. 10 Component (COMP) characterization: by number of instances (INST) and by attributes. Attributes are represented by attribute nodes (ATTR) pointing to value nodes (VAL). Shown is a turbofan with four instances, where two have reverse thrust capabilities and two have not (e.g., Airbus A380). (Figure from [10])

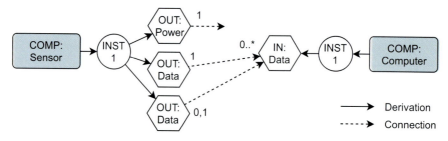

Fig. 11 Component connection: components (COMP) and their instances (INST) are connected through ports, from output ports (OUT) to input ports (IN). Ports can only connect to ports with the same identifier, and their connection degrees are specified near to the connection arrows: "1" for exactly one, "0,1" for zero or one connections, "0..*" for any number, "1..*" for one or more, etc. Note that the dashed arrows represent a connection relationship, not a derivation relationship. (Figure from [10])

1. *Identifier*: Connections are only possible between ports with the same identifier.
2. *Number of instances*: The number of port instances per component instance can be an architecture decision.
3. *Allowing self-connections*: Whether connections between ports of the same component are allowed or not.
4. *Allowing repeated connections*: Whether repeated connections between two port instances are allowed or not.
5. *Connection degree specification*: Specifies the number of connections per port instances.
6. *Relevant of connection sequence*: If relevant, the connection problem turns into a permutation problem; otherwise, it is a selection problem.

These types of characterizations make it possible to specify complex connection problems, with many possible connection sequences possible, and can be a major contribution to the explosion of alternatives often seen in architecture design problems [26]. If for a given port connection problem it is not possible to make the connection (e.g., if there are two output ports with one connection each trying to connect to only one input port), the architecture is not feasible.

The mechanism of fulfilling functions and deriving additional functions can be used to implicitly establish component compatibility. To enable a more explicit (in) compatibility relationships between elements, **incompatibility constraints** can be used. Incompatibility constraints can be specified between any function, component, concept, or decomposition and are bidirectional: if any of the connected elements exists in an architecture instance, the other element cannot. They are non-transitive: if there exists an incompatibility constraint between elements A and B and between B and C, elements A and C can still exist together in one architecture instance.

Performance metrics record quantifiable performances of the architecture as a whole or of individual architecture elements and specify how different architecture instances can be compared to each other. They can be interpreted as objectives or constraints in an optimization problem. Performance metrics can be associated with functions, components, or component instances. It is also possible to define additional **design variables** for these elements, to be able to model the complete optimization problem in the architecture design space.

The Architecture Design Space Graph (ADSG)

The architecture design space is represented by the Architecture Design Space Graph (ADSG). The ADSG is a directed graph, where nodes represent some element of the architecture, like functions (FUN), components (COMP), and ports (IN/OUT). An overview of all different node types can be found in [10]. Edges can mean a derivation or a connection relationship. An example of ADSG is shown in Fig. 12: the boundary node (BND) derives a neutral function (FUN), which is mapped to a specific function by a concept (CON). This function is further decomposed by a function decomposition (DE) and several components (COMP). Components have instances (INST), and instances can have attributes (ATTR) and ports (IN/OUT). The ADSG also contains decision nodes that represent the architecture decisions to be taken. Decisions come in three forms:

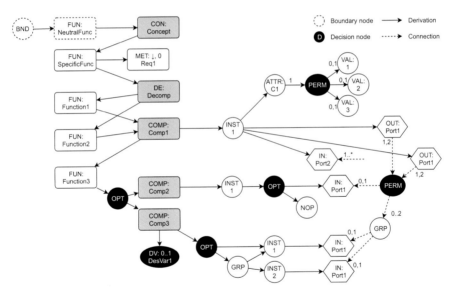

Fig. 12 An example of Architecture Design Space Graph (ADSG), showing a boundary (BND) solution-neutral function (FUN) being fulfilled using concepts (CON), function decompositions (DE), and components (COMP). Nodes external to the system are shown with dashed borders. A solid arrow denotes a node derivation, a dashed arrow a node connection. Decision nodes, option decision (OPT) and permutation decision (PERM), are shown in black, along with design variable (DV) nodes. A performance metric (MET) is associated with one of the functions (FUN). Components are characterized by component instances (INST), attributes (ATTR), and associated attribute values (VAL). Connections between ports are made from output ports (OUT) to input ports (IN), and their connection degree is specified. Grouping (GRP) and no operation (NOP) nodes are used for modifying behavior of decision. (Figure from [10])

1. *Option decisions*: decisions with mutually exclusive options. For example, thrust is provided by either a turbofan or a turboprop. In Fig. 12, these are shown as OPT nodes: the out edges point toward the different options.
2. *Permutation-decisions*: decisions regarding the connection of multiple sources to multiple targets. For example, specifying the connections between energy sources and users. In Fig. 12, these are shown as PERM nodes: the source nodes are connected through the decision node to the target nodes.
3. *Additional design variables*: additional design variables that are relevant to the design problem, but not to the system architecture. For example, the bypass ratio of a turbofan. Shown in Fig. 12 as DV nodes.

Generating Architecture Instances

An architecture instance is created from the ADSG by selecting options for all architecture decisions. Option decisions are assigned first, because these determine which nodes are present in the graph. Once all option decisions have been assigned, the permutation decisions are assigned, creating connections between their corresponding source and target nodes. The reasoning for selecting options for the decisions comes from an external source like an optimization algorithm. As also discussed in a later section, quantitative performance evaluation is needed to guide the optimization algorithm toward the most optimal set of options.

Option decisions are decisions with mutually exclusive options. When creating an architecture instance from the ADSG, the assignment of an option leads to the removal of the decision node from the graph, the connection of the originating node to the option node, and the removal of the remaining options and their derived nodes from the graph. Figure 13 shows an example of a graph with an option decision node. Any nodes that derive directly from the boundary functions (i.e., not through a

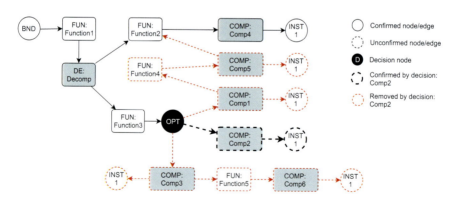

Fig. 13 Option decision (OPT) behavior: edges and nodes are confirmed or removed based on which option is assigned for fulfilling function (FUN) "Function3." Confirmed edges and nodes are shown with solid borders, unconfirmed with dashed borders. Edge confirming and removing is shown for the case where component (COMP) "Comp2" is chosen to fulfill the function: its derived edges are confirmed (thick, black dashes), whereas edges derived from other options are removed (red dashes). (Figure from [10])

decision node) are considered to exist unconditionally. All other nodes exist conditionally and can either become confirmed after being assigned in an option decision or removed from the graph.

Option decisions are not defined explicitly in the design space definition. Rather, an option decision is automatically inserted for different graph patterns, like function fulfillment, component instantiation, and port instantiation. For more details on this behavior, refer to [10]. Just like any other node, decision nodes can exist conditionally, meaning they can be removed based on some other decision. This hierarchical dependence between decisions is considered by the order in which decisions are taken: only decisions coming from confirmed decision nodes are taken.

Permutation decisions determine the connections between a set of source nodes and a set of target nodes. The architecting sub-problem roughly translates to an assigning problem as defined by [45]; however, additional behavioral rules are present. Each connection slot (a source or target) is characterized by whether repeated connections are allowed and the number of connections coming from or going to the slot. All possible connection sets are then determined by the sets of source and target nodes and an optional list of excluded connections. Permutation decisions are applied for port connections and attribute value assignments.

Formalizing the Architecture Optimization Problem

As discussed before, finding the best architecture in a design space plagued by a combinatorial explosion of alternatives is enabled by using systematic design space exploration techniques, like optimization. To enable this, the design space has to be well-defined in terms of a set of design variables, objectives, and constraints. A design space exploration algorithm, like an optimizer or design of experiments, can then be used to generate points in the design space and asking an analysis model to evaluate the performance of these design points. The results of the evaluations are expressed quantitatively in terms of objectives and constraint and are used by the exploration algorithm to generate new design points for the next iteration, until some stopping criterion has been met.

The rest of this section discusses how design variables, objectives, and constraints are defined from the ADSG, how the interaction between the optimizer and the analysis model is structured, and what the main characteristics of architecture optimization problems are.

Decision Hierarchy: Architecture Generation In the Loop
Design variables are derived from the architecting decisions and additional design variables in the ADSG. Option decisions are encoded as discrete design variables. Permutation decisions are also encoded as simple discrete design variables in order to maintain compatibility with optimization frameworks that can only handle continuous and integer design variables. Decisions are ordered based on their type and hierarchy, ensuring that option decisions come before permutation decisions, and conditionally existing decisions come after their confirming decisions. Additional

design variables as defined in the design space appear at the end of the design vector. This ensures that from the beginning to the end, the potential impact of the design variables on the architectures decreases.

One special behavior of architecture optimization problems is that design variables can be **conditionally active** based on other design variables: they are of **hierarchical nature**. Zaefferer et al. [52] mention three ways to deal with this: ignoring the effects, imputation, or explicit consideration. Ignoring the effects can confuse the optimization algorithm, because it can be possible that multiple different design vectors might map to the same architecture instances. Explicitly considering the hierarchy effects needs modification of the optimization algorithm. Therefore, as the default way to handle the hierarchy effect, imputation is chosen. With **imputation**, inactive design variables are set to a predefined value: zero for discrete design variables and domain center for continuous design variables, for example. Due to this effect, there is a difference between the apparent design space and the feasible design space. The **apparent** design space spans all combinations of all design variables, whereas the **feasible** design space takes out combinations of design variables either leading to infeasible architectures or including inactive design variables not on their imputation values.

During a design space exploration loop, design vectors that specify a design point outside the feasible design space are automatically moved to the nearest design point in the feasible design space. This logic is implemented in the **architecture generator step**, and therefore a feedback mechanism is needed for notifying the optimization algorithm of the modified design vector. Figure 14 shows the coupling of the architecture generation step with the design space exploration algorithm and analysis model and shows the feedback of the updated (imputed) design vector to the exploration algorithm.

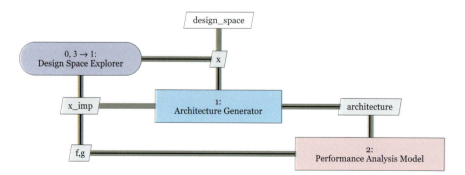

Fig. 14 Structure of the coupling between the design space explorer and the analysis model for an architecture optimization problem. An architecture generator step is inserted in between, which uses the design space to interpret the design vector x into an architecture instance, which is then used by the analysis model to determine the architecture performance. Due to design variable hierarchy and the possibility of infeasible architectures, the design vector can be modified at this stage, which needs to be communicated back to the design space exploration algorithm. (Figure from [10])

19 MBSE in Architecture Design Space Exploration

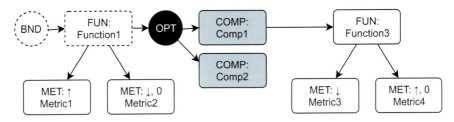

Fig. 15 Possible uses of performance metrics (MET). Metrics 1 and 2 exist unconditionally (i.e., in all architecture instances) and can therefore be used as objectives. Metrics 2 and 4 have a reference value and can therefore be used as constraints. Metric 3 can neither be used as an objective nor as a constraint. The up- and down-pointing arrows denote the direction of the performance metric: whether a higher or lower value is considered "better." (Figure from [10])

Performance Metrics as Objectives and Constraints

Performance metrics are interpreted as objectives or constraints in the optimization problem. A performance metric is defined by two properties: a direction and optionally a reference value. The direction specifies what is preferred: a lower or a higher value. How a metric can be interpreted then depends on some additional factors:

- A performance metric may be used as an **objective** if it exists unconditionally, because an objective serves as a way to compare different architectures, so it should exist in all possible architectures.
- A performance metric may be used as a **constraint** if it has a reference value, which is needed to determine whether the constraint is violated or not.

Note that it is possible to define more than one objective, as generally there are conflicting trade-offs to be made when dealing with multiple stakeholders in the design of complex systems. Figure 15 shows the four possible options for possible usage of performance metrics: metrics without a reference value can only be used as objective if they exist unconditionally, and metrics with a reference value can always be used as constraints.

Properties of System Architecture Optimization Problems

Problems with conditional design variables are hierarchical optimization problems [52], it is possible to define more than objective, variables can be either discrete or continuous, and nothing is known about the performance analysis model in advance: in general, the architecture optimization problem is a **black box, hierarchical, mixed-integer, multi-objective** optimization problem. The optimization problem can be expressed as:

$$\begin{aligned} \text{minimize} \quad & f_m(\boldsymbol{x}, \boldsymbol{y}) & m = 1, 2, \ldots, \mathcal{M} \\ \text{where} \quad & x_i \in \mathbb{R} & i = 1, 2, \ldots, n \\ & y_j \in \mathbb{Z} & j = 1, 2, \ldots, \mathcal{J} \\ \text{w.r.t.} \quad & g_k(\boldsymbol{x}, \boldsymbol{y}) & k = 1, 2, \ldots, \mathcal{K} \\ & x_i^{(L)} \leq x_i \leq x_i^{(U)} \end{aligned}$$

where x and y are the vectors of continuous and discrete design variables and f and g are the objective and constraint functions. The optimization problem has $\mathcal{M}$ objectives, n continuous design variables, $\mathcal{J}$ discrete design variables, and $\mathcal{K}$ constraints. The continuous variables have bounds, described by $x^{(L)}$ and $x^{(U)}$ for the lower and upper bounds, respectively.

The main implication of the nature of the optimization problem is that it is not possible to use gradient-based methods for solving it and the optimization algorithms should be able to solve multi-objective optimization problems. Additionally, the optimization algorithms should be able to deal with the implications of hierarchical design variables or accept the architecture generation engine to modify the design vector based on this hierarchy. Algorithms that can handle such problems include multi-objective evolutionary algorithms (MOEAs) like NSGA-II [16].

Evaluating the Performance of Generated Architectures

There are different approaches to evaluating the performance of generated architecture instances. For flexibility and reuse, it makes sense to develop a method that is as generally applicable as possible for the types of system that are being designed. Because the method presented in the preceding sections represents architectures as mappings between function and components, normally some kind of additional input is needed to create models that can be executed and/or simulated, especially if high-fidelity and/or multidisciplinary analysis is needed. It is important to note that any architecture evaluation method should be free of user interaction, to enable the optimization algorithm to autonomously search the design space.

Architecture proxy metrics attempt to quantify some of the characteristics of the architecture itself, without needing to convert to an actual simulation model for evaluation. These kinds of metrics are relatively simple and are often based on empirical data and/or experience that only applies for a relatively narrow set of architectures. This approach could reduce computational cost by filtering out architectures that have a low chance of being effective, before analyzing them with a higher-fidelity analysis. Examples of proxy metrics are architecture complexity to represent design, manufacturing and operating costs [30, 51], and component technology readiness levels to quantify risk [4, 20]. Proxy metrics can also be used to implement common design heuristics that are not possible to represent using the ADSG, such as [6] did for flight control system architectures.

When defining system architectures, there is always some sort of decomposition or modularity of elements involved; otherwise, it would not be able to dynamically generate architecture by (re)combining elements and connections. Therefore, when it comes to evaluating the architecture by performing (physics-based) simulation, it makes sense that the evaluation environment would maintain a similar level of flexibility as the architecture generator, because the overall flexibility of the systematic architecture approach is determined by the *least* flexible element in the chain. To ensure a high level of flexibility, the evaluation environment should therefore expose a **modular interface**. This would mean that an almost one-to-one translation from

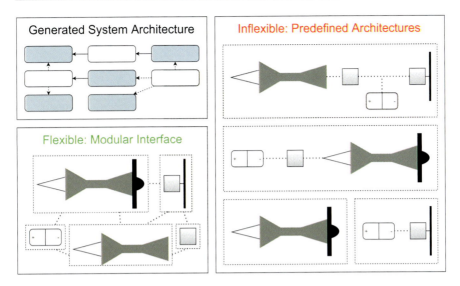

Fig. 16 Difference between types of architecture evaluation environments: with predefined system architectures or with a modular interface. The latter approach enables higher flexibility when combined with the architecture generator

the architecture elements and connections to the analysis environment becomes possible. The only additional input needed would then be boundary and operating conditions to evaluate the architecture for. The concept of interface modularity is visualized in Fig. 16, comparing between inflexible predefined (or explicitly defined) architectures and a flexible interface that allows the implicit definition of architectures.

Examples of analysis environments that expose modular interfaces are Modelica [39], SUAVE [33], and pyCycle [22]. All of these tools specialize in a certain type of engineering analysis and apply these analyses to a wide variety of elements that all share some kind of interface that makes this specialization possible. For example, Modelica is a general purpose continuous and discrete time differential equation solver but comes with a large library of elements that can be used to simulate among others mechanical, electrical, hydraulic, and control systems. All of the available elements are implemented such that their inputs and outputs are well-defined and each block defines the representation of its physics as differential equations. Another example is pyCycle, a thermodynamic cycle analysis framework. It comes bundled with a library of elements that can be used to model aircraft engine architectures (e.g., compressors, combustors, turbines), but at its basic level, each element defines a transformation of airflow in terms of thermodynamic quantities such as pressure, entropy, and temperature.

The defining feature of tools with modular interfaces is that in order to build their analysis models, they rely on the object-oriented programming (OOP) concepts of *abstraction* and *inheritance*. Abstraction and inheritance allow concepts to be represented as hierarchies, which all share a common interface but progressively refine the internal workings of the concepts. This allows on the one hand the development of specialized solvers (e.g., for differential equations or thermodynamic cycles) and on the other hand locating the element-specific logic within the elements themselves. This makes it straightforward both to implement new elements without needing to change the solver and to build complex and novel architectures by (re)combining elements. There is even evidence that thinking in terms of abstraction and concept hierarchies is precisely how we as humans reason about the world, and it is therefore easier for us to reason about complex systems if they are defined using OOP concepts [31].

The concept of a modular interface also applies to multidisciplinary design optimization (MDO) processes. Dealing with the challenge of accelerating the development of complex products, the integration of heterogeneous engineering disciplines into collaborative MDO processes should follow structured approaches promoting agility and modularity [12]. An enabling element is the development of a common data language for data exchange between the disciplines. Such a common data language can then be used to automatically determine data connections between engineering disciplines [38]. However, only if such a data exchange format supports the flexible definition of architectures, and all tools correctly implement the data format, this flexibility precipitates into the whole MDO toolchain. In the field of aircraft design, an example of such a common data language is CPACS [2]. Currently, CPACS, for example, supports the modular definition of lifting surfaces by defining them from more abstract concepts such as segments and guide curves. Effort is underway to extend the data format to also support a flexible definition of onboard systems, in terms of elements and connections.

To conclude, architecture evaluation environments should be built using OOP principles and expose their elements in a modular interface so that one-to-one translation from the system architecture to the simulation model is possible. As a last note, it is important to realize that physical limits on the modularity of systems exist, especially for high-power mechanical systems like aircraft [49]. However, also in the field of aircraft design (a discipline known for its low modularity of parts due to aerodynamic efficiency considerations), it is possible to define aircraft configurations by (re)combining a limited set of "primitive" elements, as is shown by [31].

Illustrative Examples

In this section, the previously discussed systematic architecting method is demonstrated using two design problems: the NASA Apollo mission architecture and a hybrid-electric aircraft propulsion system.

NASA Apollo Mission Design

Simmons et al. [46] present a mission architecting problem based on the Apollo moon program. Decision variables include several mission phase-related decisions

19 MBSE in Architecture Design Space Exploration

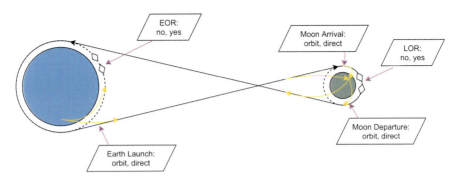

Fig. 17 Mission-related decisions in the Apollo mission architecting problem from [46]. The decisions relate to the launch from Earth, the arrival at the moon, and the departure from the moon. The launch from Earth can be either direct or first to a low Earth orbit (LEO). LEO is required if there is to be an Earth orbit rendezvous (EOR): two separate spacecraft that join in orbit before continuing to the moon. At the moon, the arrival and departure can either be direct or via lunar orbit, except if a lunar orbit rendezvous (LOR) is needed, in which case arrival and departure orbits are required. LOR is needed if a spacecraft (lunar module) separate from the main spacecraft (command module) makes the actual moon landing. (Figure reproduced from [46])

(see Fig. 17), crew assignment decisions, and fuel-type selections. Two performance metrics are used: initial mass to low Earth orbit (as a proxy for cost, to be minimized) and probability of mission success (to be maximized).

Simmons et al. present the problem using a morphological matrix and a set of logical constraints that rule out infeasible combinations of choices. The apparent design space consists of 1536 architectures, of which 108 are feasible [10].

Figure 18 shows the Architecture Design Space Graph (ADSG) modeling the architecture of the Apollo mission design problem, up to the mission phase functions. The decisions related to the mission phases and crew assignments are shown in Figs. 19, 20, 21 and 22. The main mission phases are represented by four functions to be fulfilled: launch (from Earth), land on moon, arrive (at moon), and return (from moon). Among these four, the "fly to moon" function emerges (coupled by a function decomposition node), to which both performance metrics are assigned: mass, to be minimized, and success probability, to be maximized. The choices of launch mode (Fig. 19), whether there is a lunar module or not, and the type of moon arrival and departure (Fig. 20) are modeled by function fulfillment. The constraints related to the lunar orbit rendezvous (LOR) are modeled using incompatibility edges (Fig. 21). The choices for the amount of crew members and their assignment to the lunar module if applicable are modeled by component instances and port connections (Fig. 20): the fuel-type choices by component attributes (Fig. 22). This problem serves as a clear example that component nodes do not necessarily have to represent physical components (in this problem they, for example, can represent maneuvers), and that port connections do not have to represent actual ports (in this problem used to represent crew assignments).

[10] show that modeling the architecture problem using the ADSG reduces the apparent design space to 576 architecture, which is a 63% reduction from the

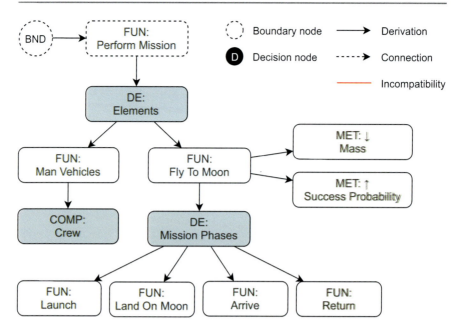

Fig. 18 ADSG showing the main elements of the Apollo mission design problem. The boundary function is "Perform Mission," which is decomposed into manning the vehicles and flying to the moon. The former is done by the crew; the latter is decomposed into four further functions related to the mission phases: launch, landing on moon, arrive (at moon), and return (from moon). Both performance metrics are associated with the "fly to moon" function: the mass (to be minimized) and the probability of success (to be maximized). (Figure from [10])

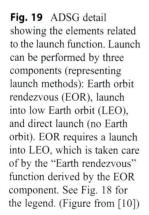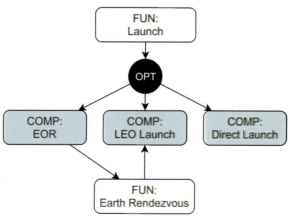

Fig. 19 ADSG detail showing the elements related to the launch function. Launch can be performed by three components (representing launch methods): Earth orbit rendezvous (EOR), launch into low Earth orbit (LEO), and direct launch (no Earth orbit). EOR requires a launch into LEO, which is taken care of by the "Earth rendezvous" function derived by the EOR component. See Fig. 18 for the legend. (Figure from [10])

morphological matrix representation. This is mainly due to including the choice for Earth orbit rendezvous (EOR) in the launch-mode choice (LEO or direct) and by eliminating the choice for lunar orbit rendezvous (LOR), as it is implied base on the

19 MBSE in Architecture Design Space Exploration

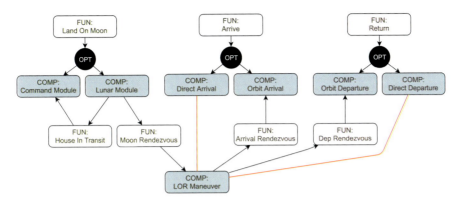

Fig. 20 ADSG detail showing the elements related to the moon landing, arrival, and return functions. Landing on moon is performed by either the command module or the lunar module. If the lunar module is chosen, a lunar orbit rendezvous (LOR) is needed. The command module will always be included in any architecture, either through a direct choice from "land on moon" or by the induced "house in transit" function. The LOR maneuver requires an orbit when arriving and departing from the moon, which is modeled by incompatibility constraints between the LOR and the direct arrival and departure components and derived functions to the orbit arrival and departure components. See Fig. 18 for the legend. (Figure from [10])

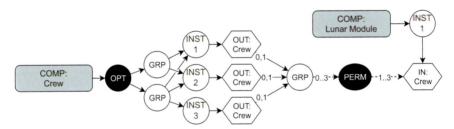

Fig. 21 ADSG detail showing how the crew size and assignment selection is modeled. The number of crew members is represented by the number of instances of the crew component, which by the grouping nodes is restricted to two or three. Each crew member (component instance) has an output "crew" port, each of which need to be connected either zero or one time. The connections of these three ports are grouped, because the sequence of the connections is not important, resulting in a "total" connection degree of between zero and three (automatically adjusted to two if there are only two crew members). The lunar module has an input "crew" port, which accepts between one and three connections. This permutation decision results in two main possibilities: if the lunar module component exists, there will be one, two, or three (if possible) crew members assigned to the lunar module; if the lunar module does not exist, there will be no crew members assigned to the lunar module. See Fig. 18 for the legend. (Figure from [10])

choice of whether there is a lunar module or not. Nevertheless, the 108 feasible architectures only represent 19% of the apparent design space, showing that the impact of decision hierarchy cannot be ignored.

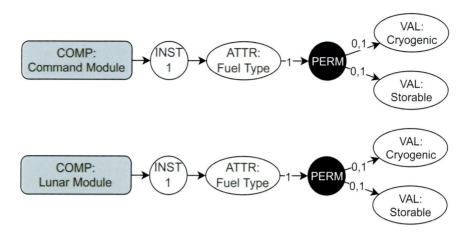

Fig. 22 ADSG detail showing the selection of the fuel types for the service and lunar modules. Since the command module will always be there, the service module fuel type is modeled as an attribute of the command module. Note that if the lunar module does not exist, the fuel type will not have to be selected. See Fig. 18 for the legend. (Figure from [10])

Hybrid-Electric Propulsion System Architecture

An important topic in the electrification of aviation is the study of hybrid-electric propulsion systems for aircraft. Such systems use both electrical energy and chemical energy (i.e., fossil fuels) to power the propellers in order to increase efficiency and reduce emissions. Different hybrid-electric architectures are possible, of which the most popular include the parallel and series hybrid architectures [25]. In parallel architectures, electric motors are used together with combustion engines to mechanically power the propellers, possibly using a gear box. In series architectures, the combustion engine is used to generate electrical power using electric generators, which is then used by electric motors to power the propellers, optionally with additional electric energy coming from batteries.

Figure 23 shows the architecture design space model for a hybrid-electric propulsion system. The main function is to generate traction by powering the propeller. This can then be done either by a turboprop engine directly, by an electric motor, or by a combination of the two. Additionally, air is pressurized either by the turboprop (i.e., bleed air) or by a separate compressor. Electric energy is provided to the battery either by a generator connected to the turboprop engine or by a battery. The bleed air tap and compressor cannot exist together in an architecture instance, thereby ruling out some infeasible architectures.

When considering the design of hybrid-electric propulsion systems, usually also fully electric and conventional (i.e., using only chemical energy) are included in the design space for comparison. The architecture design space model is able to implicitly model all four different architecture based on two decisions:

19 MBSE in Architecture Design Space Exploration

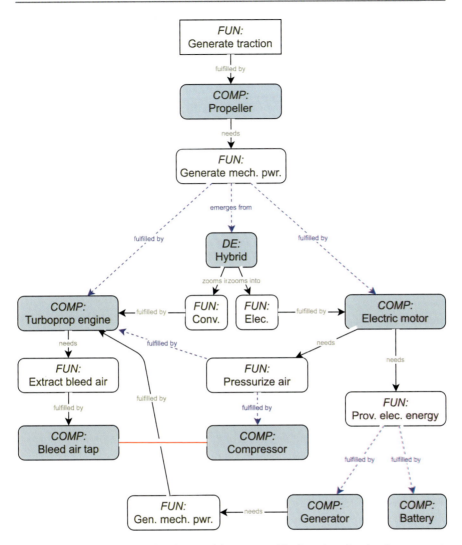

Fig. 23 ADSG of a hybrid-electric propulsion system. The boundary function is to generate traction, which is done by a propeller, which gets its mechanical power either from a turboprop, an electric engine, or both. Additionally, air is pressurized either by the turboprop (i.e., bleed air) or by a separate compressor. Electric energy is provided to the battery either by a generator connected to the turboprop engine or by a battery. The bleed air tap and compressor cannot exist together in an architecture instance, thereby ruling out some infeasible architectures. Decisions are shown as blue dashed arrows and derivations are annotated

1. How to generate mechanical power: using the turboprop, electric motor, or both (hybrid)
2. How to provide electric energy: using the generator or batteries

The second choice is only relevant if the electric motor is used to generate mechanical power. This serves as a good example of a hierarchical architecture

Table 2 Showing the relation between the two main decisions in the architecture design space and how they relate to the available explicitly defined architectures. Note that the combination of hybrid and battery is not possible, because the turboprop engine and the generator always have to exist together

Architecture decisions		Included components		Architecture
Generate mechanical power	Provide electrical energy	Turboprop	Electric motor	
Turboprop	N/A	Yes	No	Conventional
Hybrid	Generator	Yes	Yes	Parallel hybrid
Electric motor	Battery	No	Yes	Fully electric
Electric motor	Generator	Yes	Yes	Series hybrid

decision. Table 2 shows the relation between these two decisions and how these lead to implicitly define the architectures discussed in literature.

By implicitly defining the architectures, more flexibility is introduced in the design process, and there will be a better chance of finding novel or unconventional architectures. This aspect also relates to the discussion on the flexibility of the architecture evaluation environment in section "Evaluating the Performance of Generated Architectures," where it is determined that in order to keep this flexibility in the architecting process the architecture evaluator needs to be able to handle implicitly defined architectures as well.

Chapter Summary

System architecting decisions have a large impact on the performance of the designed system. To solve the conflict between design freedom and design knowledge, to deal with large design spaces as is common in system architecting, and to prevent expert bias, systematic design space exploration techniques should be applied to the design of system architectures. In order to enable the automated exploration of the design space, a mechanism for generating candidate architectures must be defined, and it must be possible to quantitatively evaluate these candidate architectures.

Functional decomposition enables the architecture to be specified in terms of what to do (the functions) and how to do it (form) and thereby enables a generic and solution-bias-free approach to system architecting. Currently, several different approaches to systematic architecting can be distinguished: for example, methods for connecting SysML and simulation environments, approaches based on the morphological matrix, domain-specific architecting processes, and dynamic function-to-form mapping approaches.

The best practice approach explained in more depth in this chapter is based on the Architecture Design Space Graph (ADSG). The ADSG is used to model the architecture design space in terms of function assignment, component characterization, and component connection decisions. This model can then be used to enumerate the design space, or construct a formal optimization problem from it to enable optimization, and at the same time offers an interface to upstream MBSE steps free of solution-bias. Next to the discrete nature of architecture decisions and the possibility of multiple conflicting objectives, architecture optimization problems also suffer from decision hierarchy: decisions that themselves may be active or not based on some other decision. Several strategies for dealing with decision hierarchy are discussed, the main being design vector imputation by an additional architecture generator step between the optimizer and evaluation model. In general, the system architecture optimization problem is of *black box, hierarchical, mixed-integer, multi-objective* nature.

Quantitative architecture evaluation can be assisted by architecture proxy metrics: some metrics that attempt to quantify characteristics of the architecture based on the architecture itself, rather than an engineering model of the architecture. Examples are architecture complexity to represent costs and component readiness levels to represent risk. These proxy metrics can be used to filter out likely unfeasible architecture.

Architecture evaluation tools or toolchains should be implemented with a modular interface, which allows a high level of flexibility when it comes to defining the system architecture. Then when combined with operating and boundary conditions, the translation from architecture definition to executable model is almost one to one. The design of a tool with a modular interface is achieved by using abstraction and inheritance: engineering concepts can be represented as hierarchies where all share a common interface but progressively refine their implementation details. A practical example is pyCycle, an engine thermodynamic cycle analysis framework: components like compressors, turbines, combustors, and ducts can be connected together to represent an engine architecture, whereas at their highest abstract level these elements simply represent some thermodynamic transformation of the airflow.

The ADSG method is demonstrated using two examples: the design of the NASA Apollo mission and a hybrid-electric aircraft propulsion system. It is shown\demonstrated that the ADSG method can be successfully applied to model different types of realistic architecture design spaces and create formalized design (optimization) problems from it so that a design space exploration algorithm can be used to perform the exploration of the design space. The design space model explicitly represents the architecture decisions and offers detailed insight into their interdependent behavior. In the future, this method will be applied to a wider variety of system architecting problems, and a practical integration of the method within an MBSE process will be demonstrated. This also includes the development of an intuitive and practical graphical user interface for creating, inspecting, and modifying the ADSG and for running design space exploration studies.

Additionally, more research should be performed on how to best approach architecture evaluation and formulate more detailed recommendations for tool or toolchain creators. The ADSG method should be coupled with heterogeneous MDO toolchains that use common data formats to communicate data to enable architecting of systems taking several engineering disciplines into account during evaluation.

It should be made sure that these common data formats and multidisciplinary tools can handle the flexibility required for modeling architectures using the ADSG.

Cross-References

▶ MBSE Methodologies
▶ Role of Decision Analysis in MBSE

Acknowledgments This work has been performed in the work of the AGILE 4.0 project (Towards Cyber-Physical Collaborative Aircraft Development) and has received funding from the European Union Horizon 2020 Programme under grant agreement nr 815122.

References

1. Albarello, N., Welcomme, J. B. & Reyterou, C., 2012. *A formal design synthesis and optimization method for systems architectures.* Bordeaux, s.n.
2. Alder, M., Moerland, E., Jepsen, J. & Nagel, B., 2020. *Recent Advances in Establishing a Common Language for Aircraft Design with CPACS.* s.l., s.n.
3. Armstrong, M., de Tenorio, C., Garcia, E. & Mavris, D., 2008. *Function Based Architecture Design Space Definition and Exploration.* Anchorage, s.n.
4. Austin, M. F. & York, D. M., 2015. System Readiness Assessment (SRA) an Illustrative Example. *Procedia Computer Science* 44, p. 486–496.
5. Balestrini-Robinson, S., Freeman, D. F. & Browne, D. C., 2015. An Object-oriented and Executable SysML Framework for Rapid Model Development. *Procedia Computer Science*, Volume 44, p. 423–432.
6. Bauer, C., Lagadec, K., Bès, C. & Mongeau, M., 2007. Flight Control System Architecture Optimization for Fly-By-Wire Airliners. *Journal of Guidance, Control, and Dynamics,* 7, 30, p. 1023–1029.
7. Beernaert, T. F. & Etman, L. F. P., 2019. Multi-level Decomposed Systems Design: Converting a Requirement Specification into an Optimization Problem. *Proceedings of the Design Society: International Conference on Engineering Design,* 7, Volume 1, p. 3691–3700.
8. Bile, Y., Riaz, A., Guenov, M. D. & Molina-Cristobal, A., 2018. Towards Automating the Sizing Process in Conceptual (Airframe) Systems Architecting. *2018 AIAA/ASCE/AHS/ASC Structures, Structural Dynamics, and Materials Conference,* p. 1–24.
9. Bornholdt, R., Kreitz, T. & Thielecke, F., 2015. *Function-Driven Design and Evaluation of Innovative Flight Controls and Power System Architectures.* s.l., SAE International.
10. Bussemaker, J. H., Ciampa, P. D. & Nagel, B., 2020. *System Architecture Design Space Exploration: An Approach to Modeling and Optimization.* s.l., American Institute of Aeronautics and Astronautics.
11. Chang, G., Perng, H. & Juang, J., 2008. A review of systems engineering standards and processes. *Journal of Biomechatronics Engineering.*
12. Ciampa, P. D. & Nagel, B., 2020. AGILE Paradigm: The next generation of collaborative MDO for the development of aeronautical systems. *Progress in Aerospace Sciences,* 11.Volume 119.
13. Ciampa, P. D., La Rocca, G. & Nagel, B., 2020. *A MBSE Approach to MDAO Systems for the Development of Complex Products.* Reno, American Institute of Aeronautics and Astronautics.
14. Crawley, E., Cameron, B. & Selva, D., 2015. *System architecture: strategy and product development for complex systems.* s.l.: Pearson Education.

15. de Tenorio, C., Armstrong, M. & Mavris, D., 2009. *Architecture Subsystem Sizing and Coordinated Optimization Methods.* Reston, American Institute of Aeronautics and Astronautics, p. 1–11.
16. Deb, K., Pratap, A., Agarwal, S. & Meyarivan, T., 2002. A fast and elitist multiobjective genetic algorithm: NSGA-II. *IEEE Transactions on Evolutionary Computation*, Volume 6, p. 182–197.
17. Dorigo, M. & Socha, K., 2018. An Introduction to Ant Colony Optimization. In: *Handbook of Approximation Algorithms and Metaheuristics, Second Edition.* s.l.: Chapman and Hall/CRC, p. 395–408.
18. Frank, C. P., 2016. *A Design Space Exploration Methodology to Support Decisions under Evolving Requirements Uncertainty and its Application to Suborbital Vehicles,* s.l.: s.n.
19. Franzén, L. K., Staack, I., Jouannet, C. & Krus, P., 2019. *An Ontological Approach to System of Systems Engineering in Product Development.* s.l., Linköping University Electronic Press.
20. Garg, T., Eppinger, S., Joglekar, N. & Olechowski, A., 2017. *Using TRLs and System Architecture to Estimate Technology Integration Risk.* Vancouver, s.n.
21. Guenov, M. et al., 2018. Aircraft Systems Architecting – Logical-Computational Domains Interface. *31st Congress of the International Council of the Aeronautical Sciences*, p. 1–12.
22. Hendricks, E. S. & Gray, J. S., 2019. pyCycle: A Tool for Efficient Optimization of Gas Turbine Engine Cycles. *Aerospace,* 8, Volume 6, p. 87.
23. Herber, D. R. et al., 2020. *Architecture Generation and Performance Evaluation of Aircraft Thermal Management Systems Through Graph-based Techniques.* s.l., American Institute of Aeronautics and Astronautics.
24. Holt, J. & Perry, S., 2013. *SysML for systems engineering: 2nd edition: A model-based approach.* s.l.: Institution of Engineering and Technology.
25. Hung, J. Y. & Gonzalez, L. F., 2012. On parallel hybrid-electric propulsion system for unmanned aerial vehicles. *Progress in Aerospace Sciences,* 5, Volume 51, p. 1–17.
26. Iacobucci, J. V., 2012. *Rapid Architecture Alternative Modeling (Raam): a Framework for Capability-Based Analysis of System of Systems Architectures,* s.l.: s.n.
27. Jeyaraj, A. & Liscouet-Hanke, S., 2018. *A model-based systems engineering approach for efficient flight control system architecture variants modelling in conceptual design.* Toulouse, s.n.
28. Judt, D. M. & Lawson, C. P., 2016. Development of an automated aircraft subsystem architecture generation and analysis tool. *Engineering Computations,* 7, Volume 33, p. 1327–1352.
29. Kernstine, K. et al., 2010. Designing for a Green Future: A Unified Aircraft Design Methodology. *Journal of Aircraft,* 9, Volume 47, p. 1789–1797.
30. Kinnunen, M. J., 2006. *Complexity Measures for System Architecture Models,* s.l.: s.n.
31. La Rocca, G., 2011. *Knowledge Based Engineering Techniques to Support Aircraft Design and Optimization,* s.l.: s.n.
32. Liscouët-Hanke, S., Maré, J.-C. & Pufe, S., 2009. Simulation Framework for Aircraft Power System Architecting. *Journal of Aircraft,* 7, Volume 46, p. 1375–1380.
33. Lukaczyk, T. W. et al., 2015. *SUAVE: An Open-Source Environment for Multi-Fidelity Conceptual Vehicle Design.* s.l., American Institute of Aeronautics and Astronautics.
34. Mavris, D., de Tenorio, C. & Armstrong, M., 2008. *Methodology for Aircraft System Architecture Definition.* Reston, American Institute of Aeronautics and Astronautics, p. 1–14.
35. McDermott, T. A., Folds, D. J. & Hallo, L., 2020. *Addressing Cognitive Bias in Systems Engineering Teams.* Virtual Event, s.n.
36. Min, B. I., Kerzhner, A. A. & Paredis, C. J. J., 2011. *Process Integration and Design Optimization for Model-Based Systems Engineering With SysML.* s.l., ASME, p. 1361–1369.
37. Ölvander, J., Lundén, B. & Gavel, H., 2009. A computerized optimization framework for the morphological matrix applied to aircraft conceptual design. *Computer-Aided Design,* 3, Volume 41, p. 187–196.
38. Page-Risueño, A., Bussemaker, J., Ciampa, P. D. & Nagel, B., 2020. *MDAx: Agile Generation of Collaborative MDAO Workflows for Complex Systems.* s.l., American Institute of Aeronautics and Astronautics.

39. Paredis, C. J. J. et al., 2010. An overview of the SysML-Modelica transformation specification. *INCOSE International Symposium,* Volume 2, p. 1–14.
40. Roelofs, M. & Vos, R., 2018. *Correction: Uncertainty-Based Design Optimization and Technology Evaluation: A Review.* Reston, American Institute of Aeronautics and Astronautics, p. 1–21.
41. Roelofs, M. N. & Vos, R., 2019. *Formalizing Technology Descriptions for Selection During Conceptual Design.* Reston, American Institute of Aeronautics and Astronautics, p. 1–14.
42. Roelofs, M. N., Vos, R., Amadori, K. & Jouannet, C., 2019. *Automatically Generating the Technology Compatibility Matrix Using Graph Transformations.* Reston, American Institute of Aeronautics and Astronautics, p. 1–14.
43. Schumann, H. et al., 2013. *ParADISE for Pre-Designing Aircraft and Systems.* s.l., s.n.
44. Schumann, H., Berres, A., Escher, S. & Stehr, T., 2015. PARADIGMshift: A Method for Feasibility Studies of New Systems. *Procedia Computer Science,* Volume 44, p. 578–587.
45. Selva, D., Cameron, B. & Crawley, E. F., 2016. Patterns in System Architecture Decisions. *Systems Engineering,* 11, Volume 19, p. 477–497.
46. Simmons, W. L. & Crawley, E. F., 2008. *A Framework for Decision Support in Systems Architecting,* s.l.: s.n.
47. Suh, N., 2001. *Axiomatic design : advances and applications.* New York: Oxford University Press.
48. Weber, R. G. & Condoor, S. S., 1998. *Conceptual design using a synergistically compatible morphological matrix.* s.l., IEEE, p. 171–176.
49. Whitney, D. E., 2004. Physical limits to modularity. *Engineering Systems Symposium 2004,* p. 17.
50. Wilschut, T., 2018. *System specification and design structuring methods for a lock product platform,* s.l.: s.n.
51. Wyatt, D. F., Wynn, D. C., Jarrett, J. P. & Clarkson, P. J., 2012. Supporting product architecture design using computational design synthesis with network structure constraints. *Research in Engineering Design,* 1, Volume 23, p. 17–52.
52. Zaefferer, M. & Horn, D., 2018. A First Analysis of Kernels for Kriging-Based Optimization in Hierarchical Search Spaces. In: R. Schaefer, C. Cotta, J. Kołodziej & G. Rudolph, eds. *Parallel Problem Solving from Nature, PPSN XI.* Berlin(Heidelberg): Springer Berlin Heidelberg, p. 399–410.

Jasper Bussemaker, Researcher, MDO Group, Aircraft Design and System Integration, Institute of System Architectures in Aeronautics, German Aerospace Center (DLR), Hamburg, Germany.

Jasper Bussemaker received his MSc in Aerospace Engineering with a focus on aircraft design and multidisciplinary design optimization (MDO) from Delft University of Technology. Since 2018, he is working at the German Aerospace Center (DLR) Institute of System Architectures in Aeronautics (Hamburg, Germany) in the MDO group. There, he is working on bridging the gap between systems engineering and MDO by developing system architecture optimization capabilities. He is developing a method for modeling system architecture design spaces, applying multi-objective mixed-integer optimization algorithms to explore the design space, integrating with a collaborative MDO framework to leverage diverse disciplinary expertise, and testing the new method on relevant aeronautical design problems.

Pier Davide Ciampa, Head of MDO Group, Aircraft Design and System Integration, Institute of System Architectures in Aeronautics, German Aerospace Center (DLR), Hamburg, Germany.

Pier Davide Ciampa holds his MSc degree in Aerospace Engineering from Delft University of Technology, Netherlands. In 2008, he joined the German Aerospace Center, DLR, as researcher with focus on unconventional aircraft configurations. From 2015, he is leading the MDO and MBSE activities at the Aircraft Design and Systems Integration Department in Hamburg. His research activities focus on digital engineering for development and optimization of complex aeronautical systems and novel methodologies for MDO and MBSE. He is active as principal investigator and coordinator of projects supported by national and international research frameworks. He is member of the AIAA MDO Technical Committee and INCOSE and guest lecturer for MBSE and MDO at several organizations and universities. From 2015 to 2018, he coordinated the large-scale EU-funded research project AGILE, receiving the "ICAS Award for Innovation in Aeronautics" in 2018. From 2019, he is coordinating the EU-funded research project AGILE4.0, in a consortium of 16 international organizations developing the next-generation of MBSE systems with applications in aeronautics.

Part V

Digital Engineering and MBSE

Digital Twin: Key Enabler and Complement to Model-Based Systems Engineering

20

Azad M. Madni and C. C. Madni

Contents

Introduction	634
Digital Twin Rationale and Potential Uses	636
Exploiting Synergy Between Digital Twins and MBSE	639
Level 1: Pre-digital Twin	642
Level 2: Digital Twin	643
Level 3: Adaptive Digital Twin	644
Level 4: Intelligent Digital Twin	644
Integration of Digital Twins with Related Technologies	646
Digital Twin and Simulation	646
Digital Twin and Machine Learning	647
Digital Twin and Internet of Things (IoT)	648
Digital Twin and Cost	648
Chapter Summary and Future Prospects	649
Cross-References	651
References	651

Abstract

Digital twin, a concept introduced in 2002 and popularized in the last several years, has become increasingly more relevant to model-based system engineering (MBSE) with the growing convergence between systems engineering and digital engineering. A digital twin is a dynamic virtual representation of a physical system. However, unlike a virtual prototype, it is a virtual instance that reflects the state and status of the physical twin in the real world. A digital twin is

---

A. M. Madni (✉)
Systems Architecting and Engineering, Astronautical Engineering Department, University of Southern California, Los Angeles, CA, USA
e-mail: azad.madni@usc.edu

C. C. Madni
Intelligent Systems Technology, Inc, Los Angeles, CA, USA
e-mail: cmadni@intelsystech.com

© Springer Nature Switzerland AG 2023
A. M. Madni et al. (eds.), *Handbook of Model-Based Systems Engineering*,
https://doi.org/10.1007/978-3-030-93582-5_37

continually updated with the state and status data acquired from the physical twin throughout its life cycle. It is enabled by data and test-driven through simulation. It has several uses that make it invaluable for MBSE. These include real-time monitoring, optimization, control, and decision-making of the physical entity as well as prediction of future states that require intervention. This chapter presents an overall vision and rationale for incorporating digital twin technology into MBSE. The chapter discusses the different levels of sophistication of digital twins, and their integration with system simulation and Internet of Things (IoT) in support of MBSE. It provides examples of how digital twin technology is benefiting different industries. It recommends that digital twin technology be made an integral part of MBSE.

Keywords

Model-based systems engineering · Digital twin · Digital thread · Internet of Things · Simulation · Virtual prototype

Introduction

Model-Based Systems Engineering (MBSE) is increasingly being adopted in various industries because it is beginning to be viewed not just as a technical capability but a source of competitive advantage [1]. At the same time, there is a growing convergence between digital engineering and MBSE that opens up new opportunities to advance MBSE. In particular, digital twin technology from digital engineering is beginning to be leveraged within MBSE [2].

The term concept of "digital twin" was first introduced in 2002 in a white paper by Michael Grieves of the University of Michigan [3]. In recent years, with the advent of Digital Engineering, Internet of Things (IoT), and Industry 4.0, the digital twin concept has evolved. Today a digital twin is defined as a dynamic virtual model of a system, process, or service. Thus, a product, factory, or a business service can all have digital twins [4, 5]. The digital twin concept today calls for incorporating both contextual (including business) and sensor data from a physical system (or process) into the virtual system model of the digital twin to facilitate analysis, circumvent problems, and make informed decisions about, for example, physical twin operation and maintenance. By integrating the virtual and physical worlds, a digital twin enables real-time monitoring of systems and processes and timely analysis of data to head off problems before they arise, schedule preventive maintenance to reduce/prevent downtimes, uncover new business opportunities, and plan for future upgrades and new developments. In short, a digital twin is enabled by data, and its behavior is explored and analyzed through simulation. It can then be used for real-time monitoring, optimization, control, decision-making, and prediction. Madni et al. [2] distinguish between a virtual prototype and a digital twin by defining a virtual prototype as a generic representation of a system, component or process, and a digital twin as a representation of an instance (i.e., a particular system or process). Hicks [6] describes a digital twin as an appropriately synchronized body of useful information (structure, function, behavior) of a physical entity in virtual space, with flows of information that enable convergence between the physical and virtual states.

Therefore, a digital twin, by definition, requires a physical twin for data acquisition and context-driven interaction. The virtual system model in the digital twin can change in real time as the state and health status of the physical system changes during operation. Today, a digital twin consists of connected products, typically utilizing the IoT, and a digital thread. The digital thread provides connectivity throughout the system's lifecycle and collects data from the physical twin to update the models in the digital twin.

Today we are witnessing the convergence of digital engineering and systems engineering, and more specifically, MBSE. The digital twin captures up-to-date data on the physical twin while the digital thread provides the requisite connectivity within the system life cycle processes and between the real world and the virtual environment. In this chapter, we describe how digital twin technology can be exploited within MBSE. Figure 1 presents the concept of a digital twin-enabled MBSE framework for complex systems engineering.

As shown in Fig. 1, the digital twin serves to link the virtual and physical environments. The physical environment includes the physical system, onboard and external sensors, communication interfaces, and possibly other vehicles operating in an open environment with access to GPS data. Both operational and maintenance data associated with the physical system are supplied to the virtual

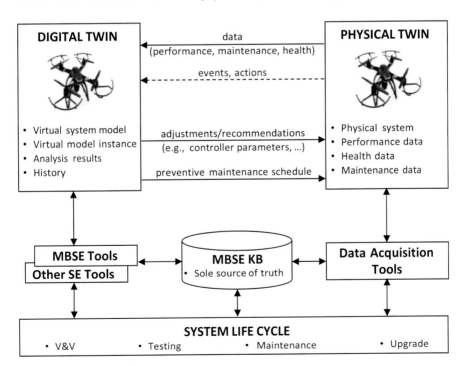

Fig. 1 Digital twin enabled MBSE concept

environment to update the virtual model in the digital twin. Thus, the digital twin becomes a precise and up-to-date representation of a physical system that also reflects the operational context of the physical twin. Importantly, the relationship with the physical twin can continue even after the sale of the physical twin, thereby making it possible to track the performance and maintenance history of each physical twin over time, detect and report anomalous behavior, and recommend/ schedule maintenance.

Figure 1 provides a functional description of an MBSE testbed that exploits digital twins in a variety of ways. The MBSE testbed includes system modeling methods (e.g., SysML models, Design Structure Matrix [7], process dependency structure matrix, probabilistic models such as POMDP [8], discrete event simulation [9], agent-based simulation, model-based storytelling [10, 11]), an MBSE knowledge base (that constitutes the authoritative sources of truth), and systems engineering life cycle process models. The virtual system model can range from lightweight models to full-up models. The lightweight models reflect simplified structure (e.g., simplified geometry) and simplified physics (e.g., reduced order models) to reduce computational load, especially in upfront engineering activities. These lightweight models allow simulations of complex systems and system-of-systems (SoS) with fidelity in the appropriate dimensions to answer questions with minimal computation costs. These models can be shared within the organization and with the supplier network thereby increasing their understanding of the system being engineered. The digital twin concept also subsumes automated and manual processes in manufacturing environments. From the physical twin, performance, maintenance, and health data can be collected and supplied to the digital twin. This data includes operational environment characteristics, engine and battery status, and other such factors. The digital twin and the physical twin can both be supported by a shared MBSE repository, which also supports SE and data collection tools. The MBSE models constitute the authoritative sources of truth. This configuration also ensures bi-directional communication between the digital twin and the physical twin.

This chapter is organized as follows. Section "Digital Twin Rationale and Potential Uses" presents the digital twin rationale and potential uses. Section "Exploiting Synergy Between Digital Twins and MBSE" presents how the synergy between digital twins and MBSE can be exploited and describes the levels of sophistication of a digital twin. Section "Integration of Digital Twins with Related Technologies" presents the integration of digital twins with other MBSE-related technologies such as simulation, machine learning, Internet of Things (IoT), and cost. Section "Chapter Summary and Future Prospects" summarizes the chapter and presents the future prospects of digital twin technology in advancing MBSE.

Digital Twin Rationale and Potential Uses

Digital twin technology has the potential to reduce the cost of system verification and testing while providing early insights into system behavior. Today digital twin technology is an integral part of the DoD's Digital Engineering initiative that seeks to connect systems and operations in a world of increasingly interconnected and smart physical devices [12–14]. The key uses of a digital twin include real-time monitoring and control of remote systems, predictive and condition-based maintenance, what-if rapid and cost-effective exploration of physical twin behavior leading to informed decision-making on the part of the physical twin, and support for up-to-date documentation and reporting.

With the advent of the Internet of Things (IoT), digital twin technology has become cost-effective to implement and is gaining increasing acceptance in the Industrial Internet of Things (IIoT) community [15] which tends to focus on large, complex, capital-intensive equipment. At the same time, the aerospace and defense industry, which continues to invest in Industry 4.0, has started to invest in digital twin technology. According to BCC Research, the global market for digital twins is estimated to increase from $4.9 billion in 2021 to about $50.2 billion by 2026, at a compound annual growth rate (CAGR) of 59.0% during the forecast period of 2021–2026 [16].

A digital twin is different from the traditional CAD/CAE model in the following important ways:

a) It is a specific instance that reflects the structure, performance, health status, and mission-specific characteristics such as miles flown, malfunctions experienced, and maintenance and repair history of the physical twin.
b) It helps determine when to schedule preventive maintenance based on knowledge of the system's maintenance history and observed system behavior.
c) It helps in understanding how the physical twin is performing in the real world, and how it can be expected to perform with timely maintenance in the future.
d) It allows developers to observe system performance to understand, for example, how modifications are performing, and to get a better understanding of the operational environment.
e) It promotes traceability between life cycle phases through connectivity provided by the digital thread.
f) It facilitates refinement of assumptions with predictive analytics – data collected from the physical system and incorporated in the digital twin can be analyzed along with other information sources to make predictions about future system performance.
g) It enables maintainers to troubleshoot malfunctioning remote equipment and perform remote maintenance.
h) It combines data from the IoT with data from the physical system to, for example, optimize service and manufacturing processes and identify needed design improvements (e.g., improved logistics support, improved mission performance).
i) It reflects the age of the physical system by incorporating operational and maintenance data from the physical system into its models and simulations.

NASA and USAF researchers describe a digital twin as an integrated multiphysics, multiscale, probabilistic simulation of an as-built vehicle or system that uses the best available physical models, sensor updates, fleet history, etc., to mirror the life of its corresponding flying twin [17]. The digital twin is ultra-realistic and may consider one or more important and interdependent vehicle systems such as airframe, propulsion and energy storage, life support, avionics, and thermal protection. The extreme requirements of the digital twin necessitate the integration of the design of materials and innovative material processing approaches. Manufacturing anomalies, that may affect the vehicle, are also "explicitly considered, evaluated and monitored." In addition to the high-fidelity physical models of the as-built structure, the digital twin integrates sensor data from the vehicle's on-board integrated vehicle health management (IVHM) system, maintenance history, and all available historical and fleet data obtained through techniques such as data mining and text mining [17]. By combining data from these different sources of information, the digital twin can continually forecast vehicle health status, the remaining useful vehicle life, and the probability of mission success. The digital twin can also be used to predict system response to safety-critical events and uncover new issues before they become acute by comparing predicted and actual responses. Finally, the systems on board the digital twin are capable of mitigating damage or degradation by activating self-healing mechanisms, or by recommending changes in mission profile to decrease loads, thereby increasing both the life span and the probability of mission success [17].

Digital twin technology is both an enabler and complement to MBSE. It enables superior verification and validation of system models. It helps MBSE span the system life cycle by providing valuable information for maintenance and manufacturing. Digital twins can be exploited within an instrumented MBSE testbed in MBSE tools (e.g., system modeling and verification tools) and operational scenario simulations (e.g., discrete event simulations and agent-based simulations) can be used to explore the behavior of virtual prototypes in a what-if simulation mode under the control of the experimenter. Insights from the operational environment can be used to modify the system models used in the virtual prototype. Data supplied by the physical system can be used by the virtual prototype to instantiate a digital twin. Subsequently, the digital twin is updated on an ongoing basis, so it faithfully mirrors the characteristics and history of the physical twin. Importantly, systems that collect and supply data from the operational environment also contribute to insights needed for business intelligence (BI). For example, the DoD is presently altering its systems to comprehensively record all electronics warfare (EW) signals (data) from operational missions to analyze, better understand the operational environment, and provide focused updates to their systems.

A digital twin, being a virtual representation of a physical system, is easier to *manipulate* and *study* in a controlled testbed environment than its physical counterpart in the operational environment. This flexibility enables cost-effective exploration of system behaviors and sensitivities to various types of system malfunctions and external disruptions. Second, data produced by the digital twin under various what-if conditions can be used to: improve future system designs; optimize maintenance cycles; surface ideas for new system applications; validate preliminary design decisions; and predict system response to different types of disruptions in the field. Specifically, the digital twin can be used to:

- *Validate system model with real-world data*
  Operational environment data and the interactions of the system with that environment can be incorporated into the digital twin to validate its models and to make assessments and predictions.
- *Provide decision support and alerts to users*
  After incorporating operational, maintenance, and health data, the digital twin can be employed in a what-if analysis mode to produce tailored decision-support information and alerts to operators/users of the physical system.
- *Predict changes in the physical system over time*
  Simulation-based analysis of operational, maintenance, and health data from the physical twin can facilitate optimization of operations (including satisfying requirements and identifying root causes), enhance contingency planning, and predicting system performance. The digital twin can also be embedded in the control loop to predict changes in the physical system and adjust/modify physical system parameters to deal with contingencies.
- *Discover new application opportunities and revenue streams*
  With a digital twin, different versions of systems can be evaluated to determine which features provide the "biggest bang for the buck." Machine learning and other data science technologies can facilitate the timely analysis of significant volumes of data generated, thereby providing insights into potential new uses and revenue streams.

Exploiting Synergy Between Digital Twins and MBSE

From the preceding sections, it becomes evident that there is unmistakable synergy between MBSE and digital twin technology. Digital twins can potentially extend MBSE coverage of the system life cycle by contributing to verification and validation testing. MBSE provides a convenient framework for evaluating digital twins using simulation within an instrumented testbed. Digital twins have been employed in a variety of industries including petrochemical [18], aircraft, health monitoring [19], industrial production [20], and education [21].

A digital twin is first created when the physical system can start providing data to the virtual system model to create a model instance that reflects the operational state and health status of the physical system [2]. The physical system can be said to be "fit for purpose" if the digital twin's behavior is analyzed and can be appropriately adjusted for a variety of contingency situations. For example, computer simulations of the braking system of a car can be run to understand how the vehicle (model) would perform in different real-world scenarios. This approach is faster and cheaper than building multiple physical vehicles to test. However, computer simulations tend to be confined to currently known events and operational environments. Therefore, they do not have the ability to predict system (e.g., vehicle) responses to future/envisioned scenarios. Also, braking systems are software-intensive systems, not merely a combination of mechanical and electronic subsystems. This is where digital twin and IoT can play important roles within the MBSE rubric. A digital twin uses data collected from connected sensors onboard the physical vehicle to support virtual testing. Also, with IoT-provided data (e.g., temperature and humidity), it becomes possible to incorporate and reflect system performance and health status data in the digital twin and use that information to make predictions about the physical twin. Equipped with this information, the digital twin can provide the operational story (i.e., events, experiences, and history) of its physical twin over the physical twin's life cycle.

Thus far, the initial series of digital twin initiatives have focused on complex, high-cost systems. However, today with advances in cost models for sensors, communication networks, analytics, and simulation, it is possible to develop a digital twin for almost any entity. The specific industries that are exploiting digital twin technology include healthcare, manufacturing, education, energy, construction, real estate, transportation, and meteorology. By scrutinizing the use of digital twins in these different industry sectors, it becomes possible to gauge the state-of-the-art, common challenges, and promising directions for the use of digital twins.

There are two key modeling technologies that can be exploited in the development of digital twins. These include physics-based modeling and data-driven modeling [22].

Physics-based modeling is typically used in the design phase to create a digital twin. It involves observing and understanding a physical system or process of interest and translating the limited understanding into an appropriate mathematical representation (e.g., equations) with appropriate assumptions, and solving the equations. Due to a limited understanding of the system or process, it will necessarily be the case that certain aspects of physics will go unaddressed at this stage. The physics-based approach includes: (a) *experimental modeling* to understand a system or process and developing models or correlations for variables that either do not lend themselves to direct measurement or are too expensive to measure; (b) *3D Modeling*, the starting point for a digital twin and the first step in numerical modeling, is concerned with developing a mathematical representation of any surface of an object; and (c) *high fidelity numerical methods* (e.g., Finite Difference Method, Finite Element Method, Discrete Element Method), to solve equations derived through physical modeling.

Data-driven Modeling is becoming increasingly popular with the wide availability of data today. The data-driven approach assumes that data is a manifestation of both known and unknown physics, and therefore data-based modeling can potentially account for the full physics of the physical system or process of interest. The key enablers of this approach are open-source libraries (TensorFlow, PyTorch, and OpenAI), inexpensive computational infrastructure (CPU, GPU, and TPU), and training resources. The key functions associated with this approach are data generation, data preprocessing and management, data ownership, ethics and data privacy, and machine learning and AI. Data generation is facilitated by the availability of inexpensive, miniaturized sensors (e.g., vehicle-mounted and building-mounted) that can record all types of data (e.g., audio, video, text, RGB images, and hyperspectral images). Data preprocessing and management is concerned with the management of high data volume, data ownership, detection of outliers, and filling missing data using, for example, Generative Adversarial Networks (GANs). Data ownership is complicated by the fact that there are multiple stakeholders with conflicting objectives who need to compromise and cooperate in digital twin generation, operation, and testing.

Ethics and data privacy are key to ensuring trust when digital twins are used in safety-critical operations such as self-driving vehicles. Technologies such as blockchain, which provides distributed, secure, verifiable records of data and transactions, hold great promise in this regard. This technology is already being pursued in health record-keeping and data sharing. Today organizations are working to integrate blockchain with IoT to overcome challenges associated with scalability, privacy, security, and ownership.

Machine learning and AI are two critical technologies when there is incomplete knowledge of the system, and the environment is partially known or unknown. Machine learning enables learning exclusively from data and without explicit programming. The three popular types of machine learning techniques that are applicable in the digital twin context are supervised learning, unsupervised learning, and reinforcement learning. In *supervised learning*, the objective is to learn a function that can map inputs (independent variables) to outputs (dependent variables) using methods such as regression or classification on a labeled set of training data. A limitation of this approach is that it needs labeled data (i.e., dependent variables) which is not always available. In *unsupervised learning*, learning occurs when through algorithms that learn patterns from unlabeled data through exploration of datasets without pre-existing labels. Unsupervised learning is well-suited to anomaly detection in, for example, online health monitoring. It is ideally suited when labeled data is not available. In *reinforcement learning*, the goal is to maximize the cumulative reward by taking actions in an environment, while balancing exploration and exploitation. Learning occurs as the learning agent interacts within an environment and uses feedback from its own actions to adjust its behavior. Reinforcement learning uses the concept of rewards (or punishment) to steer its behavior toward its goal. Reinforcement learning is well suited to problems with sparse data. Instead of learning directly from data, the agent determines an optimal policy to maximize cumulative long-term gain. This type of learning is well-suited to the operation and maintenance of power grid applications, intelligent planning and decision-making in autonomous vehicles, and smart building energy management.

Today digital twin is defined as any digital representation of a system, component, process, or person. With this broader interpretation, several questions arise including:

- Does a physical system have to exist before a digital twin is created?
- What are the key challenges in building digital twins and what methods can be used to build them?
- Does the physical system with onboard sensors and processors need to report operational and health status data to the virtual system model before the latter can be called a digital twin?
- Does the definition of a digital twin have to expand because any physical asset today can be made smart with the advent of the Industrial Internet of Things (IIoT)?

Table 1 Digital twin levels [2]

Level	Model sophistication	Physical twin	Data acquisition from physical twin	Machine learning (operator preferences)	Machine learning (system/environment)
1 (Pre-digital twin)	Virtual system model with emphasis on technology/technical-risk mitigation	Does not exist	Not applicable	No	No
2 (Digital twin)	Virtual system model of the physical twin	Exists	Performance, health status, maintenance; batch updates	No	No
3 (Adaptive digital twin)	Virtual system model of physical twin with adaptive user interface (UI)	Exists	Performance, health status, maintenance; real-time updates	Yes	No
4 (Intelligent digital twin)	Virtual system model of physical twin with adaptive UI and reinforcement learning	Exists	Performance, health status, maintenance, environment; both batch/real-time updates	Yes	Yes

- What outputs are typically expected from a digital twin and in what way do those outputs help?
- What information needs to be collected from the stakeholders to derive maximum benefit from the use of digital twins in MBSE?

These and many more such questions need to be answered about digital twins and their various uses. To this end, we define four levels of virtual representation. Each level has a specific purpose and scope and helps with decision-making and answering questions throughout the system's lifecycle. Table 1 presents these different levels and their characteristics.

Level 1: Pre-digital Twin

Level 1 is the traditional virtual prototype created during upfront engineering. It supports decision-making during the concept design and preliminary design. The virtual prototype is a virtual generic executable system model of the envisioned system that is typically created before the physical prototype is built. Its primary purpose is to mitigate technical risks and uncover issues in upfront engineering. We call this virtual prototype – Pre-digital Twin. Like most model-driven approaches,

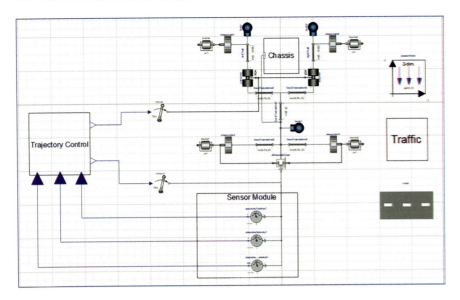

Fig. 2 Virtual vehicle model built with planar mechanics open-source library

virtual prototyping involves a model of the system early in the design process. However, a virtual prototype is not usually used to derive the final system. This is because a virtual prototype can be a "throwaway" prototype or a "reusable" prototype. The latter can be used to derive the final system. A virtual prototype is mostly used to validate certain key decisions about the system and mitigate specific technical risks early in the design process. Figure 2 presents an example of a Level 1 virtual representation of a car using the Planar Mechanics library provided by Modelica. This model consists of ideal wheels with dry friction contact patch rolling on a surface. It employs a simple (i.e., low fidelity) model of differential gear to distribute torque equally to the wheels, and reflects properties such as inertia, mass, fixed translation, and torque to realize a basic structure of vehicle with mass properties. A PlanarWorld component is used to define gravity and the global coordinates system. The sensors measure absolute position, velocity, and acceleration. The trajectory control module provides torque values to the steering and differential gear mechanism. Such low fidelity models can be used, for example, in testing, planning, and decision-making algorithms related to, for example, trajectory control in autonomous vehicles performing lane changes.

Level 2: Digital Twin

Level 2 is a digital twin in which the virtual system model is capable of incorporating operational data from the physical twin. The virtual representation, an instantiation of the generic system model, receives batch updates from the physical system that it uses to support high-level decision making in conceptual design, technology specification, preliminary design, and development. Data collection from the physical sensors and computational elements in the physical twin includes both health status data (e.g., battery level) and mission performance data (e.g., flight hours). The data is reported back to the digital twin which updates its model and reflects the impact of the update on, for example, the maintenance schedule for the physical system. Since interaction with the physical system is bidirectional, there is ample opportunity for the physical twin to use the knowledge acquired from one or more digital twins to improve its performance during real-time operation. The digital twin at this level is used to explore the behavior of the physical twin under various what-if scenarios. Being an executable digital representation, it is easy to manipulate when exploring the behavior of the system in the controlled simulation environment of the testbed. Any deficiencies discovered are used to modify the physical twin with the changes reflected in the digital twin. Figure 3 presents an example of a Level 2 system model of a passenger car. This model is built with Vehicle Interface Library of the Modelica tool. The model has a passenger car with a power split hybrid power train. The chassis model has single degree of freedom with mass-and speed-dependent drag properties. The braking subsystem uses the brake pedal position resulting from driver action to calculate brake torque. This affects the driveline. The driveline model consists of four wheels with frontwheel drive and ideal differential. The power split device consists of ideal epicyclic gear without losses. An ideal battery with constant voltage source powers the DC motor model with inductor, resistor, and emf component connected to the shaft hub. The engine model with a flywheel consists of drive-by-wire accelerator, where the accelerator inputs are converted to output torque. Models of Road, environment, and WorldFrame are used to define the inertial frame, gravity, air temperature, wind velocity, gas constant for air, and air pressure. The data used in this model is from the physical twin.

Level 3: Adaptive Digital Twin

Level 3 is the Adaptive Digital Twin. It offers an adaptive user interface (in the spirit of a smart product model) to the physical and digital twins. The adaptive user interface is sensitive to the preferences and priorities of the user/operator. A key capability at this level is the ability to learn the preferences and priorities of human operators in different contexts [23]. These characteristics are captured using a neural network-based supervised machine learning algorithm. The models employed within this digital twin are continually updated based on the data being "pulled" from the physical twin in real-time. It can also accept information in batches after system use. This digital twin can support real-time planning and decision-making during operations, maintenance, and support.

Level 4: Intelligent Digital Twin

Level 4 is the Intelligent Digital Twin. It has all the capabilities of a level 3 digital twin (including supervised machine learning). In addition, it has unsupervised

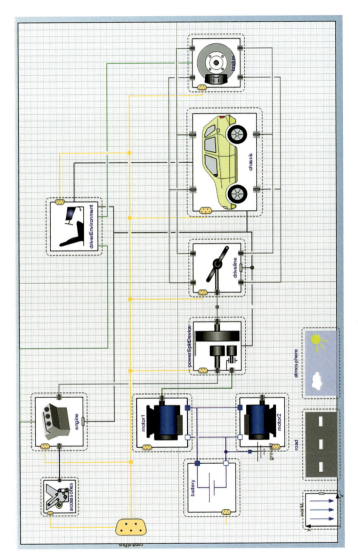

Fig. 3 Digital twin model built with vehicle interfaces open-source library

machine learning capability to discern objects and patterns encountered in the operational environment, and reinforcement learning of system and environment states in uncertain, partially observable environment. This digital twin at this level has a high degree of autonomy. At this level, the digital twin can analyze more granular performance, maintenance, and health data from the real-world counterpart.

The above-described levels of the digital twin can be used at different stages in the system life cycle within the MBSE rubric. In the initial stages, levels 1 and 2 can be employed to support tradeoffs in upfront engineering (e.g., concept development, preliminary design). Levels 3 and 4 come into play in detailed design and in making decisions such as when to do maintenance next.

Integration of Digital Twins with Related Technologies

Digital twins can be integrated through the digital thread to MBSE tool suites, and potentially become a core element of Model-Based Systems Engineering (MBSE). In fact, MBSE can serve as the starting point for the digital thread [24]. Using data gathered from IoT, system simulations can be run to explore failure modes, leading to progressive design improvements over time. For example, a manufacturer can link a digital twin to its service history, manufacturing process, design history, real-time IoT data, configuration-specific simulation models, and expected failure modes using MBSE tools such as Design Failure Mode and Effects Analysis (DFMEA). The ability to compare simulation outputs with actual results can provide valuable insights about the physical twin. Using an appropriate MBSE tool, engineers can generate event-driven or agent-based simulations to explore the behavior and interactions of the digital twin, respectively. The digital twin can incorporate 3D data and simulations, along with their characterizations using methods such as response surface models. To gauge customer experience and the impact of innovation on that experience, the digital twin can be employed to simulate a plant, product, or service at the right level of abstraction. For example, the DoD has a policy of barcoding line replaceable units for analysis in logistics support. The benefits of this policy would apply to digital twins.

Digital Twin and Simulation

Engineers can potentially use simulations linked to the digital twin to predict how the physical twin can be expected to perform in the real world [25–27]. Contrast this with having to rely on the ideal and perceived worst-case conditions typically employed in the design process. Actual system performance data can be compared to data from the digital twin, prompting adjustment decisions that can contribute to successful mission outcomes [22]. Also, by incorporating data from the physical twin into the digital twin, engineers can improve system models, and subsequently use the results of the analysis with the digital twin to improve the operation of the physical system in the real world.

The value added by a digital twin to its physical twin stems from its ability to optimize both physical twin operation and maintenance schedule. Simulating digital twin behavior enables the determination and adjustment of real-world system behavior. Specifically, insights gained through simulation can guide changes needed in system design and manufacturing, with the digital thread providing the necessary connectivity across the system life cycle. The fidelity of the simulation will typically vary with the purpose of the simulation and the stage in the system life cycle [2]. A potential benefit would be to rehearse missions in specific operational conditions (terrain, weather, etc.) using the capabilities of the digital twin. For example, in the system design phase, a relatively slow, non-real-time simulation may suffice so long as it enables exploration and investigation of multiple different use cases under real-world conditions. With access to IoT data, actual operational conditions can be simulated (with high confidence) to yield insights into expected outcomes.

An important use of simulation is in the assessment of the expected operational life of the system (i.e., how long the system can be expected to be operational). In this regard, a digital twin can keep track of its mortality based on the wear and tear experienced by the physical twin. By employing simulation, the digital twin can estimate the remaining working life of the physical twin and proactively schedule maintenance. In other words, predictive maintenance can be used to estimate how long the physical system can be expected to operate normally and use that knowledge to proactively schedule and perform system shut down rather than wait for the breakdown of the physical twin which can be both expensive and potentially catastrophic.

Digital Twin and Machine Learning

For simple applications, digital twin technology offers value without having to employ machine learning. Simple applications are characterized by a limited number of variables and an easily discoverable linear relationship between inputs and outputs. However, most real-world systems which have to contend with multiple data streams stand to benefit from machine learning and analytics to make sense of the data. Machine learning, in this context, implies any algorithm that is applied to a data stream to uncover/discover patterns that can be subsequently exploited in a variety of ways. For example, machine learning can automate complex analytical tasks. It can evaluate data in real time, adjust behavior with minimal need for supervision, and increase the likelihood of desired outcomes. Machine learning can also contribute to producing actionable insights that can lead to cost savings. Smart buildings are an excellent example of applications that stand to benefit from machine learning capabilities in the digital twin [2]. Machine learning uses within a digital twin include supervised learning (e.g., using a neural network) of operator/user preferences and priorities in a simulation-based experimentation testbed [28], unsupervised learning of objects and patterns using, for example, clustering techniques in virtual and real-world environments [29, 30], and reinforcement learning of system and environment states in uncertain, partially observable operational environments [31].

Digital Twin and Internet of Things (IoT)

Linking digital twins to the IoT brings the data needed to understand how the physical twin (e.g., manufacturing assembly line, autonomous vehicles network) behaves and performs in the operational environment [3]. Furthermore, the combination of IoT and digital twins can enhance preventive maintenance and analytics/AI-based optimization of the physical system and operational processes. Acting as a bridge between the physical and virtual worlds, the IoT can deliver performance, maintenance, and health data from the physical twin to the digital twin. Combining insights from the real-world data with predictive modeling can enhance the ability to make informed decisions that can potentially lead to the creation of effective systems, optimized production operations, and new business models. Multi-source/multi-sensor information (e.g., outside temperature, moisture content, and production status of current batch) can be delivered to the digital twin along with information from traditional sensors (e.g., SCADA) to facilitate predictive modeling. Furthermore, the IoT enables much-needed flexibility when it comes to system mobility, location, and monetization options. Such flexibility contributes to the creation of business options such as selling a capability (i.e., product as a service) versus selling the product itself. For example, Caterpillar sells the capability to move dirt (i.e., a service) versus selling the equipment (i.e., a product).

Importantly, the combination of digital twin and IoT allows an organization to gain insights into how a system/product is being used by customers. Such insights can enable customers to optimize maintenance schedules and resource utilization, proactively predict potential product failures, and avoid/reduce system downtimes. Ultimately, the digital twin is a key enabler for improving system maintenance over time based on system operation and maintenance history. The benefits of the IoT can be further amplified through the integration of the IoT with multiple digital twins, each monitored from a central location that oversees maintenance schedules and cycles. In this latter case, data ownership becomes somewhat complicated especially when equipment is rented.

While access to execution data can contribute to improving manufacturing operations, the reality is that most manufacturers tend to have limited equipment data for optimization. In contrast, suppliers typically have a multitude of interconnected equipment and devices supplying data that can be exploited. Importantly, vendors indicate that customers would be willing to share data with suppliers. If this is the case, then timely optimization is achievable. Perhaps the greatest potential benefit of the IoT is in the service area. For example, a service that is continually informed about the operational state and health status of the system can be effective in ensuring cost savings and high availability. In a similar vein, predictive analytics can be employed, for example, to pre-fetch and rapidly deliver a required part to a maintenance crew.

Digital Twin and Cost

Cost is always going to be an important consideration when deciding to integrate digital twins into MBSE and systems engineering processes [32]. Clearly defined scope and purpose are prerequisites for estimating the cost of implementing digital twins. While a digital twin requires a greater upfront investment, the inclusion of digital twin(s) can be expected to provide a significant return on investment down the line in the systems life cycle. Based on the sophistication level, time, and effort required for generating a virtual system representation, costs can vary. However, in today's world, most organizations are pursuing the creation of virtual system models anyway because of their value in reducing verification and test durations and costs. The digital twin is a logical next step in this progression. The cost of the digital twin is a function of the number of components in the system, the interfaces and dependencies among the components, the complexity of the algorithms employed to implement specific functions, and the knowledge and know-how required to build the digital twin. Importantly, architecting the digital twin for reuse can further reduce costs.

Chapter Summary and Future Prospects

This chapter has presented the potential synergy between MBSE and digital twin technology and discussed how the latter has become a key enabler and complement to MBSE. Digital twin technology has the potential of becoming a central capability in MBSE because of its ability to help MBSE span the full system life cycle [2]. The growing acceptance of digital twins can help MBSE penetrate new markets such as manufacturing [33], city planning and design [34], education [2], energy, medical devices, and healthcare [35, 36]. Digital twins can be expected to enhance upfront engineering (e.g., system conceptualization and model verification), testing (e.g., model-based system validation), system maintenance (e.g., condition-based maintenance), and smart manufacturing (e.g., co-evolving product and manufacturing processes). MBSE can contribute different system modeling constructs and languages that digital twins can employ in their virtual representation of the physical system.

Digital twin technology is seeing increasing use in manufacturing to enhance early identification and correction of design and process problems [33, 37, 38]. For manufacturers, digital twin technology can provide a window into system performance. For example, a digital twin can potentially help identify equipment faults and troubleshoot equipment remotely thereby alleviating a key customer concern [39].

Digital twin technology is also becoming a key enabler of the shift from schedule-driven maintenance to condition-based/predictive maintenance thereby increasing system availability while substantially reducing system maintenance costs. For example, digital twins are beginning to be exploited in aircraft engine maintenance. Today, aircraft engines are routinely taken apart and rebuilt based on the number of hours flown, regardless of whether most of those hours are simply cruising at altitude or performing high-G maneuvers. With the introduction of digital twin technology, the maintenance activity will be able to better understand maintenance needs and schedule maintenance accordingly. For example, the Nunn-McCurdy breach on the

C-130 AMP, which was primarily due to "tail-number diversity" would be better understood with digital twins.

The construction and real estate industries are exploiting digital twins in the different phases of construction (e.g., design, construction management, operation, and demolition). The ability to monitor and track system data in real time is essential for smart functions such as adaptive energy management which can reduce energy costs through smart usage. The increasing use of digital twins in the construction industry has been largely enabled by the availability of sophisticated dashboard software which allows users to view the state and status of their projects and properties in intuitively appealing form and format. However, as the real estate universe continues to digitize, and building and property data becomes increasingly more complex and abundant, there has been an exponential growth in the number of tools, data formats, and data services. As a result, generating actionable information from diverse data supplied by different tools is becoming increasingly difficult. Also, dashboard technology today is largely disconnected from the tools and data services that building owners rely on today. A next-generation dashboard can potentially remedy this situation. Specifically, it can tie into one or more digital twins and access information in near real-time thereby enhancing decision-making and plan execution on property construction and maintenance projects. In these industries, digital twins will continue to increase return-on-investment (ROI) from multiple projects and properties over their life cycle. In the initial stages, architects, planners, and property managers will be able to realize gains in the planning and design stages. In the later stages, they will derive benefits from the digital twin-provided ability to optimize energy management and maintenance functions.

For architects, engineers, and planners, digital twin technology can become a source of sustainable competitive advantage by linking projects and properties to real-time data with user-customizable smart dashboards. Zoning, permitting, traffic, and air quality data can be linked to the digital twin, and used when making siting and structural component placement decisions. For example, pedestrian traffic information can be used to determine where to place an entrance to a building. Similarly, property managers interested in reducing energy consumption will be able to exploit the digital twin to rapidly determine the energy use and usage profile of each tenant and use that information to optimize and dynamically adjust energy usage in the building. Real-time sensor data from throughout the building such as those connected to HVAC and lighting can make this possible [40, 41]. Timely maintenance can also be expected to contribute to tenant satisfaction, while faster time-to-market will contribute to increase in sales.

In addition to improving virtual system models, a digital twin can potentially improve physical system operations and sustainment. A digital twin can also help with product differentiation, product quality, and add-on services. Knowing how customers are using the product post-purchase can provide useful insights including identifying and eliminating unwanted product functionality and features, as well as unwanted components thereby saving both time and money. A digital twin can enable visualization of remote physical systems (e.g., a system engineer in Washington D.C. can use the digital twin to troubleshoot a landing gear problem of an airplane parked at a gate of Los Angeles airport). Multimodality sensors (e.g., sight, sound, vibration, and altitude) can serve to deliver data from physical systems to digital twins anywhere in the world. These flexible capabilities can potentially lead to a clear understanding of the state of remote systems through multi-perspective visualization [10].

The digital twin story and its impact on MBSE is just beginning to be written with leadership provided by the U.S. Department of Defense (through its Digital Engineering initiative), the IIoT community, and Industry 4.0. Digital twin technology will become key to acquiring early insights into system performance and technical risks [14]. The digital thread will be able to facilitate the upstream and downstream transfer of knowledge from one program to another. Within the MBSE rubric, digital twins will contribute to the authoritative sources of truth about the system over its life cycle. By exploiting simulation-based modeling and data analytics, digital twins can help quantify margins and uncertainties in cost and performance. The combination of digital connectivity provided by the digital thread and the trusted data and knowledge supplied by the digital twin can accelerate the transformation of systems engineering processes employed in MBSE [1, 42].

Despite the promise of digital twin technology for MBSE, there are a few concerns that need to be resolved before wide-scale adoption of digital twin technology can be achieved. For example, being a relatively new concept, the digital twin raises concern about privacy and ownership. To many, the extensive sharing of data with suppliers and potential customers is disconcerting. There are also concerns about intellectual property (IP), and legal considerations. For example, will operators want to report operational data to the manufacturer? And who owns the data provided by the digital twin? Ultimately, who reaps the benefit? And does the operator want to share manufacturing execution data with the device manufacturer? Does the operator want to share such data with competing device manufacturers? If the operator decides not to share such data with everyone, then who owns that data? The digital engineering community is working diligently on answering these questions and more. In the meantime, digital twin technology is continuing to make impressive inroads in several major industries including aerospace and defense, manufacturing, building construction and real estate, and medical devices and healthcare.

Cross-References

▶ Model-Based System Architecting and Decision-Making
▶ Overarching Process for Systems Engineering and Design

References

1. Madni AM, Sievers M (2018) Model Based Systems Engineering: Motivation, Current Status, and Research Opportunities. Systems Engineering 20th Anniversary Special Issue, vol. 21, issue 3, pp 172–190

2. Madni AM, Madni CC, Lucero DS (2019) Leveraging Digital Twin Technology in Model-Based Systems Engineering. MDPI Systems, special issue on Model-Based Systems Engineering, Feb
3. Grieves M (2014) Digital Twin: Manufacturing Excellence through Virtual Factory Replication. A White Paper, Michael Grieves, LLC
4. Matthews S (2018) Designing Better Machines: The Evolution of the Digital Twin explained. May 21, 2018, Keynote delivered at Hannover Messe.
5. Marr B (2017) What is Digital Twin Technology – And why is it so Important? Forbes, Mar 6
6. Hicks B (2019) Industry 4.0 and Digital Twins: Key lessons from NASA. https://www.thefuturefactory.com/blog/24
7. Eppinger SD, Browning TR (2012) Design Structure Matrix Methods and Applications. The MIT Press
8. Hollinger G (2007) Partially Observable Markov Decision Processes (POMDP). Graduate Artificial Intelligence, Fall
9. Zeigler B, Muzy A, Kofman E (2018) Theory of Modeling and Simulation: Discrete Event and Iterative System Computational Foundations. 3rd Edition, Academic Press, August
10. Madni AM, Spraragen M, Madni CC (2014) Exploring and Assessing Complex System Behavior through Model-Driven Storytelling. IEEE Systems, Man and Cybernetics International Conference, invited special session Frontiers of Model Based Systems Engineering, San Diego, CA, Oct 5–8
11. Madni AM, Nance M, Richey M, Hubbard W, Hanneman L (2014) Toward an Experiential Design Language: Augmenting Model-Based Systems Engineering with Technical Storytelling in Virtual Worlds. 2014 CSER, Eds.: Madni AM et al., Redondo Beach, CA, Mar 21–22
12. Madni AM (2017) Transdisciplinary Systems Engineering: Exploiting Convergence in a Hyperconnected World. (foreword by Augustine N), Springer, September
13. West TD, Pyster A (2015) Untangling the Digital Thread: The Challenge and Promise of Model-Based Engineering in Defense Acquisition. INSIGHT, 18(2), pp 45–55
14. Kraft E (2017) Challenges and Innovations in Digital Systems Engineering. NDIA 20th Annual Systems Engineering Conference, Springfield, VA, October 25, 2017.
15. O'Connor C (2017) IBM IoT Platform, July 11
16. The Business Research Company (2022) Digital Twin Global Market Report 2022, Jun 2022, https://www.bccresearch.com/partners/tbrc-market-briefs/digital-twin-global-market-report.html
17. Glaessgen EH, Stargel DS (2012) The Digital Twin Paradigm for Future NASA and U.S. Air Force Vehicles. AIAA 53rd Structures, Structural Dynamics, and Materials Conference: Digital Twin Special Session
18. Brandtstaedter H, Ludwig C, Hubner L, Tsouchnika E, Jungiewicz A, Wever U (2018) Digital twins for large electric drive trains. in Proc Petroleum Chem Ind Conf Eur (PCIC Eur), June, pp 1–5
19. Li C, Mahadevan S, Ling Y, Choze S, Wang L (2017) Dynamic Bayesian network for aircraft wing health monitoring digital twin. AIAA J, vol 55, no 3, pp 930–941, Mar
20. Lou X, Guo Y, Gao Y, Waedt K, Parekh M (2019) An idea of using digital twin to perform the functional safety and cybersecurity analysis. in Proc INFORMATIK, pp 283–294
21. Madni AM, Erwin D, Madni A (2019) Exploiting Digital Twin Technology to Teach Engineering Fundamentals and Afford Real-World Learning Opportunities. 2019 ASEE 126th Annual Conference and Exposition, Tampa, FL, June 15–19
22. Rasheed A, San O, Kvamsdal T (2020) Digital Twin: Values, Challenges and Enablers from a Modeling Perspective. IEEE Access, Volume 8, pp 2198022012
23. Madni AM, Samet MG, Freedy A (1982) A Trainable On-Line Model of the Human Operator in Information Acquisition Tasks. IEEE Transactions of Systems, Man, and Cybernetics, Special

issue on Human Factors in Computer Management of Information for Decision Making, Vol SMC-12, No 4, July/August, pp 504–511
24. Schluse M, Atorf L, Rossmann J (2017) Experimentental Digital Twins for Model-Based Systems Engineering and Simulation-Based Development, 2017 Annual IEEE International Systems Conference (SysCon), Montreal, QC, Canada, April 24–27
25. Schluse M, Rossman J (2016) From simulation to experimentental digital twins: Simulation-based development and operation of complex technical systems. 2016 IEEE International Symposium on Systems Engineering (ISSE), Edinburgh, UK, October 3–5
26. Madni AM, Purohit S (2021) Augmenting MBSE with Digital Twin Technology: Implementation, Analysis, Preliminary Results, and Findings. submitted to 2021 IEEE Systems, Man, and Cybernetics International Conference, Melbourne, Australia
27. Dahmen U, Rossmann J (2018) Simulation-based Verification with Experimentental Digital Twins in Virtual testbeds. in Sxhuppstuhl T, Tracht K, Franke J (eds) Tagungsband des 3. Kongresses Montage Hand habung Industrierboter, Springer Vieweg, Berlin, Heidelberg
28. Madni AM (2021) MBSE Testbed for Rapid, Cost-Effective Prototyping and Evaluation of System Modeling Approaches, Applied Sciences Journal, 11(5), 2321, https://doi.org/10.3390/app11052321
29. Madni AM, Sievers M, Erwin D, Madni A, Ordoukhanian E, Pouya P (2019) Formal Modeling of Complex Resilient Networked Systems, AIAA Science and Technology Forum, San Diego, California, January 7–11
30. Madni AM, Sievers, M, Ordoukhanian E, Pouya P, Madni A (2018) Extending Formal Modeling for Resilient Systems. 2018 INCOSE International Symposium, July 7–12
31. Pouya P, Madni AM (2020) Expandable POMDP Framework for Modeling and Analysis of Autonomous Vehicle Behavior. IEEE Systems Journal, July
32. West TD, Blackburn M (2017) Is Digital Thread/Digital Twin Affordable? A Systemic Assessment of the Cost of DoD's Latest Manhattan Project. Procedia Computer Science, Vol 114, 47–56
33. Kritzinger W, Karner M, Traar G, Henjis J, Sihn W (2018) Digital Twin in Manufacturing: A categorical literature review and classification. IFAC-PapersOnLine, Vol 51, Issue 11, pp 1016–1022
34. Kent L, Snider C, Hicks B (2019) Early stage digital-physical twinning to engage citizens with city planning and design. in Proc. IEEE Conf Virtual Reality 3D User Inter. (VR), Mar, pp. 1014–1015
35. Bruynseels K, de Sio FS, van den Hoven J (2018) Digital twins in health care: Ethical implications of an emerging engineering paradigm. Frontiers Genet, vol 9, p 31, Feb
36. Jimenez JI, Jahankhani H, Kendzierskyj S (2020) Health care in the cyberspace: Medical cyber-physical system and digital twin challenges. in Digital Twin Technologies and Smart Cities, Cham, Switzerland, Springer, pp 79–92
37. Haag S Anderl R (2018) Digital Twin: Proof of Concept. Manufacturing letters, Vol 15, Part B, pp 64–66
38. Uhlemann TH-J, Lehmann C, Steinhilper R (2017) The Digital Twin: Realizing the Cyber-Physical Production System for Industry 4.0. Procedia CIRP, Vol 61, pp 335–340
39. Laaki H, Miche Y, Tammi, K (2019) Prototyping a digital twin for real time remote control over mobile networks: Application of remote surgery. IEEE Access, vol 7, pp 20325–20336
40. Cityzenith (2018) What are Digital Twins? What the Building and Real Estate Industries Need to Know. October 9
41. Scholten A (2017) Smart Buildings and Their Digital Twins. Realcomm20 Advisory Newsletters, Vol 17, No 31, August 2
42. Madni AM, Sievers M (2017) Model-Based Systems Engineering: Motivation, Current Status, and Needed Advances. Conference on Systems Engineering Research, March 23–25, Redondo Beach, CA

Azad M. Madni, University Professor of Astronautical Engineering and Northrop Grumman Foundation Fred O'Green Chair in Engineering, University of Southern California. Dr. Azad Madni is a member of the National Academy of Engineering and University Professor of Astronautical Engineering and holder of the Northrop Grumman Foundation Fred O'Green Chair in Engineering at the University of Southern California. He holds a joint appointment in Aerospace and Mechanical Engineering and Civil and Environmental Engineering Departments. He is the Executive Director of USC's Systems Architecting and Engineering Program and the founding director of the Distributed Autonomy and Intelligent Systems Laboratory. He is the founder and CEO of Intelligent Systems Technology, Inc., a hi-tech company that offers educational courses in intelligent systems for education and training. He received his Ph.D., M.S., and B.S. degrees in Engineering from UCLA. His recent awards include *2021 Joint INCOSE/ASEE Outstanding Educator Award*, *2021 INCOSE Benefactor Award*, *2021 IEEE AESS Judith A. Resnik Space Award*, *2020 IEEE SMC Norbert Wiener Award*, *2020 NDIA's Ferguson Award* for Excellence in Systems Engineering, *2020 IEEE-USA Entrepreneur Achievement Award*, *2019 IEEE AESS Pioneer Award*, *2019 INCOSE Founders Award*, *2019 AIAA/ASEE Leland Atwood Award*, *2019 ASME CIE Leadership Award*, *2019 Society for Modeling and Simulation International Presidential Award*, and *2011 INCOSE Pioneer Award*. He is a Life Fellow/Fellow of IEEE, INCOSE, AIAA, AAAS, SDPS, IETE, AAIA, and WAS. He is the co-founder and current chair of IEEE SMC Technical Committee for Model Based Systems Engineering. He is the author of *Transdisciplinary Systems Engineering: Exploiting Convergence in a Hyper-Connected World* (Springer, 2018). He is the co-author of *Tradeoff Decisions in System Design* (Springer, 2016).

Carla C. Madni, EVP for R&D and COO, Intelligent Systems Technology, Inc. Carla is a systems engineer specializing in the engineering of complex and intelligent human-machine systems for human performance enhancement. She has served as Program Manager and Co-Principal Investigator on numerous R&D projects sponsored by DARPA, AFRL, ONR, ARL, NIST, DOE, and DHS S&T. Her specific areas of interest are simulation-based training systems, intelligent decision aiding systems, digital engineering and uses of digital twin technology in Model Based Systems Engineering and education. She received her B.S. degree from Tulane University in Biomedical Engineering, and her M.S. degree from University of California, Los Angeles in Engineering Systems with specialization in Distributed Problem Solving and Artificial Intelligence. She is a member IEEE and Human Factors and Ergonomics Society. She has supported the Conference on Systems Engineering Research as a sponsor. She has led several successful Small Business Innovation Research projects sponsored by various government agencies. She is currently involved in supporting General Motors and Cruise LLC in conducting studies in the safety of autonomous vehicles.

Developing Industry 4 Systems with OPM ISO 19450 Augmented with MAXIM

21

D. Dori

Contents

Introduction	656
Model-Based Systems Engineering	656
The Digital Transformation	656
Digital Engineering: The Ultimate Blend of Hardware and Software	657
High-Level DoD Acquisition Community Goals for the DE Transformation	658
INCOSE Model-Based Capabilities Matrix	659
The System-Software Engineering Gap	659
Why Is MBSE Not Picking Up Momentum in the Expected Pace?	660
The UML/SysML Dominance of the MBSE Landscape	661
The UML/SysML Software/System Focus Difference	662
OPM ISO 19450 and Its MAXIM Extension	663
OPM ISO 19450	663
MAXIM: Methodical Approach to Executable Integrative Modeling	667
OPM Cyber-physical System Applications	668
Chapter Summary	670
Cross-References	672
References	672

Abstract

Industry 4.0 and the transition to digital engineering go hand-in-hand, mandating a paradigm shift for model-based systems engineering. In this chapter, we discuss what this new environment requires in terms of modeling capabilities, and why current languages are less than adequate for the task at hand. We then describe the basics of OPM ISO 19450 and its MAXIM extension. We explain why this combination of high-level conceptual modeling and detailed computational modeling, where hardware and software components are described consistently

---

D. Dori (✉)
Technion, Israel Institute of Technology, Haifa, Israel
e-mail: dori@technion.ac.il

© Springer Nature Switzerland AG 2023
A. M. Madni et al. (eds.), *Handbook of Model-Based Systems Engineering*,
https://doi.org/10.1007/978-3-030-93582-5_38

and seamlessly, is a solution to the problem that model-based systems engineering is facing as it is trying to adapt to the Industry 4.0 environment.

Keywords

Model-based systems engineering · Industry 4.0 · Object-Process Methodology · OPM ISO 19450 · Conceptual modeling · Computational modeling

Introduction

Over the past decade, the digital revolution has been making a significant impact on our lives. We have entered the era of the Fourth Industrial Revolution, Industry 4.0, and the Internet of Things (IoT) is quickly becoming a prevalent reality. Given these trends, there is an urgent and growing need for a new paradigm that integrates systems and software engineering.

Model-Based Systems Engineering

Model-based systems engineering (MBSE) is a systems engineering approach that views modeling and models as a centerpiece source of truth for a system throughout its lifecycle. Adoption and utilization of MBSE have been growing [19]. The basic tenet of MBSE is that the conceptual model of the system-to-be is expressed as early as possible, ideally from the requirements stage, at which a solution-neutral model of the problem at hand can be constructed. MBSE is often defined as the formalized application of modeling to support system requirements, design, analysis, verification, and validation activities. MBSE advocates that formal models are the authoritative set of artifacts that anchor product and system development processes and reflect the evolution of the design [27].

The formal model evolves along the system lifecycle stages, and each related engineering discipline continues the system design using its domain-specific language and toolset, while the systems engineer keeps the holistic system model updated, integrating the details provided by the vertical disciplines and ensuring that they are coordinated and aligned. James [15] summarized the challenges that MBSE is facing in Fig. 1.

The Digital Transformation

The digital transformation is progressing from being initially involved mainly in advanced manufacturing with robotics up the value chain through product design to systems engineering and business management [38]. According to a market research by Builta et al. [2], 87% of the various industries are adopting one or more of the transformative technologies that include IoT, artificial intelligence (AI), blockchain, augmented and virtual reality smart manufacturing, software continuous integration

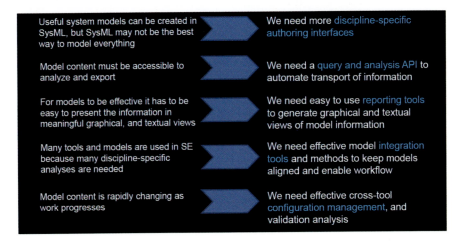

Fig. 1 The challenges of MBSE [15]

and delivery (CI-CD), cloud computing, and 5G – the recent generation of cellular communication. Among these, the top three transformative technologies most ready for adoption by industry are AI, cloud and virtualization, and IoT.

Digital Engineering: The Ultimate Blend of Hardware and Software

The leading technologies which Industry 4.0 features are represented in Fig. 2. Each technology name is color-coded, with words that pertain to software colored purple, while those that relate to hardware are light blue. Examining this list, it becomes crystal clear that without any exception, each technology has a major software component and a major hardware one. For example, in **Smart Manufacturing**, **Manufacturing** is obviously hardware, but the **Smart** component is achieved through massive investment in software.

The digital revolution is transforming the traditional engineering into the new digital engineering (DE) paradigm. DE has been defined as *an integrated digital approach that uses authoritative sources of systems' data and models as a continuum across disciplines to support lifecycle activities from concept through disposal* [29]. This definition includes both data and models, so it is in line with the spirit of MBSE with emphasis on the cross-disciplinary continuum throughout the lifecycle of the system. The DE ecosystem features an interconnected infrastructure, along with a methodology for exchanging data, information, and knowledge in a digital form from an authoritative source of truth.

Yet, while modeling of software systems and hardware systems have been evolving in parallel, little effort to integrate software engineering with systems engineering has been made. Despite the growing adoption of MBSE and its image as a critical enabler of the digital revolution – the transformation into the DE

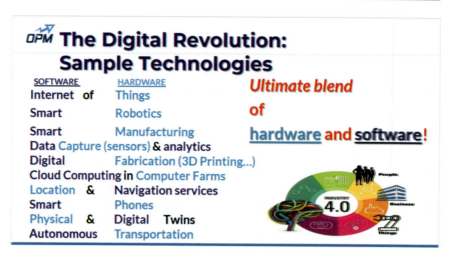

Fig. 2 Sample technologies that Industry 4.0 features

paradigm – a recent cross-industry survey of MBSE maturity and adoption [24] has shown that MBSE is still perceived as immature.

High-Level DoD Acquisition Community Goals for the DE Transformation

Based on the US Department of Defense (DoD) DE Strategy, which considers digital transformation an ongoing change process across the enterprise that is linked to enterprise value, McDermott and Van Aken [23] identified the following five high-level transformation goals for the DoD acquisition community and associated industries, described here as requirements:

1. **Use of Models** – the enterprise shall have a comprehensive strategy for using models, which shall be used consistently across all programs. Models shall constitute the basis for all business practices and guide program decisions. Consistent metrics must show that model-based practices provide measurable value.
2. **Authoritative Data** – digital artifacts shall provide a basis for enterprise decisions, and programs shall use an established authoritative source of truth (ASOT). Data and information shall be accessible and discoverable for decision-making related to systems' lifecycles. Processes shall be established for curating and managing the ASOT across program lifecycles and program supply chain.
3. **Technical Innovation** – the enterprise shall establish mature digital-based approaches to program planning, adoption, and implementation across the enterprise. The enterprise shall have processes to examine and anticipate how new

technologies can bring value, and it shall be able to measure and assess return on investment of new technologies.
4. **Supporting Infrastructure** – based on policies, guidance, and planning, a digital ecosystem shall be established to collaborate digitally across organizations, disciplines, and lifecycle phases. Programs shall apply common practices to protect intellectual property and critical information across enterprises. Engineering and program management activities shall be able to rapidly discover, manage, and exchange models and data. Information technologies (IT), software, and tools shall support model and data exchange, visualization, collaboration, and decision processes. Infrastructure changes shall provide measurable improvement over existing enterprise practices.
5. **Culture and Workforce** – the enterprise shall have a clear DE vision, strategy, culture, and processes, with experts and champions leading effective change and transformation processes. Enterprise leadership shall understand DE and be committed to introducing DE at all levels. DE transformation shall be linked to enterprise strategy with clearly defined outcomes and a path to communicate the benefits of DE based on success stories. The enterprise shall have training programs to establish and staff roles for professionals with appropriate knowledge, skills, and abilities for DE. Systems engineers shall be recognized and rewarded for using DE processes and tools.

INCOSE Model-Based Capabilities Matrix

The International Council on Systems Engineering (INCOSE) Model-Based Capabilities (MBCA) Matrix [13] aims to help organizations determine their modeling capability for the enterprise, system, and program at both the current and needed levels. The matrix rows are 42 organization modeling capabilities, and the columns are 5 increasing stages of capability:

Stage 0: No MBSE capability or MBSE applied ad hoc to gain experience.
Stage 1: Modeling efforts are used to address specific objectives and questions.
Stage 2: Modeling standards are applied, including ontology, languages, tools.
Stage 3: Program- or project-wide capabilities and models are integrated with other functional disciplines; digital threads and digital twins are defined.
Stage 4: Enterprise-wide capabilities contribute to the entire enterprise, as its programs and projects use enterprise-defined ontologies libraries, and standards.

The System-Software Engineering Gap

Despite the fact that systems engineering and software engineering are complementary, sister disciplines, their evolution paths have been largely disconnected, preventing integration between the system aspects of hardware from the software ones. Accurate modeling and design of the system-to-be is essential for streamlining the development process and smoothing the interfaces between the two [36]. While operating the system and executing the software program that controls it using real-life data, new insights are gained [4, 9], and problems in industrial product development are discovered [1, 10, 11]. These insights often reveal that important issues are modeled incorrectly or not modeled at all, requiring significant rework, which delays the project's successful completion.

The major problem that arises from the present development process is that the transition from a system-level model to software and other disciplinary models creates a "Grand Canyon" [8, 12]. This gap causes engineers in the various disciplines to lose the common big picture that the original conceptual model expresses and the critical information on design considerations that are associated with this model. Misalignments between early and late design start to emerge, causing functionality and quality problems, cost overruns, and project delays. Because of the difference between the physical system parts and the informatical, intangible software that is supposed to control it, the system-software transition is where these phenomena are most acute.

Why Is MBSE Not Picking Up Momentum in the Expected Pace?

The number of MBSE supporters is growing, but it is still much smaller than the number of those who continue to use traditional tools like word processors, electronic worksheets, or text-intensive requirement management tools [3]. There is a gap between the growing recognition of the criticality of MBSE and its rate of adoption in industry, and this is hindering smooth integration of disparate digital and non-digital components of the expected end-to-end DE paradigm.

If systems engineering in general and MBSE in particular are to remain relevant and useful in the era of this rapidly changing ecosystem, it must adopt and adapt the evolving technologies that the digital transformation is promoting, primarily modeling that is based on cloud computing as a common ubiquitous infrastructure. Yet, as late as June 2020, McDermott and Van Aken [23] found that "best practices do not yet exist in the DE and MBSE community, and the transformation process is not yet mature enough across the community to standardize best practices and success metrics."

Lack of a common executable modeling framework that integrates systems engineering and software engineering, as well as other engineering domains, is a major cause of problems and impediments in product development processes. Moreover, current MBSE techniques apply a variety of model kinds, each with its own fidelity and exactness level. The gap between the modeling level and the software level leads to a great fidelity gap; it is often the case that the software that is supposed to control a cyber-physical system does not quite fit to the model of the physical system it is supposed to control, opening the door to an often hazardous or even fatal cyber-physical gap [26].

What are the root causes for this sluggish rate of adoption? There are two main reasons: (1) the UML/SysML dominance of the MBSE landscape and (2) the need to model cyber-physical systems with both its physical and cybernetic components and how they affect each other.

The UML/SysML Dominance of the MBSE Landscape

When using the term MBSE, most people imply either implicitly or explicitly the use of the OMG family of modeling language – UML for software-intensive systems and SysML for hardware intensive systems. There is a whole set of derivatives in this family: OMG has tried to integrate conceptual modeling with computational and execution capabilities by developing xUML [22] and fUML [35], but none succeeded to include the entire system without dividing it into several subsystems that have little or no interaction between them.

Researchers have found major drawbacks of using SysML. For example, Hampson [12] has identified many gaps in SysML, a partial list of which is listed in Table 1. Three of the six gaps pertain to model execution. The SysML executable diagrams are activity and parametric diagrams. The former has complex syntax and requires too much detail for practical applications. The latter is too low fidelity, and there exist dedicated languages and programs, such as *Modelica* and *Matlab/Simulink* that do a better job at engineering computations. Other gaps identified include lack of clear relations between requirements and behavior and between structural and behavioral aspects.

Indeed, the major deficiency with the UML/SysML family of languages is rooted deeply in the fact that they use multiple kinds of diagrams (14 in UML and 9 in SysML) to represent a system. This model multiplicity problem [33] has adverse implications, including lack of clear and unambiguous semantics and the excessive

Table 1 A partial list of high severity gaps in SysML identified by Hampson [12]

Relative Severity [Low-Med-High]	Category	Gaps Identified
High	Model Execution	For executable activity diagrams, syntax becomes extremely complex and places a huge burden on the modeler
High	Model Execution	Executable activity diagrams require a significant amount of detailed modeling to make even a simple model simulation. Level of detailed modeling needed is impractical
High	Model Execution	Parametric diagrams built for parametric analysis are generally low fidelity compared to external system analysis
High	Specification	It is not clear how requirements and behavioral aspects of a model relate to one another
Medium	Specification	Structural/Behavioral relations are only laid out at a top level and could be specified further
High	Graphical	Majority of elements have similar symbol representation which can confuse reviewers

cognitive load on the shoulders of both modelers and the audience to which the models are to be communicated – see last gap in Table 1. Weak semantics of SysML and UML may be considered beneficial in that it enables broad usage, but it hinders formal reasoning and interoperability. Moreover, SysML cannot represent complex timing and synchronization that systems of systems often require.

The UML/SysML Software/System Focus Difference

Given the clear coalescence between hardware and software in the DE era, there is a clear need for a modeling language that does not favor either hardware or software. This need highlights a major problematic difference in the focus between UML and SysML: While the former is software-engineering-focused, the latter has a systems engineering focus. This is evident from the definitions of the scopes of UML and SysML as they appear in the official OMG publications, copied in Fig. 3. Software and hardware are underlined in purple (mostly in UML) and light blue (in SysML), respectively, clearly showing the focus differences.

UML and SysML have many diagram options to choose from, but most applications and most modelers use only a handful, and for new users, understanding which ones to use is a challenge. People who use UML or SysML often choose a subset of diagrams that have a small overlap (usually class diagrams) or no overlap, yet they claim to use the same language. In fact, every UML or SysML diagram kind is a language in its own right, with its own syntax and semantics and with many symbols that look the same but mean different things in different diagram kinds. This diagram multiplicity makes it difficult both to agree on a common semantics across the diagram kinds and to cognitively obtain a holistic view of the system with its function,

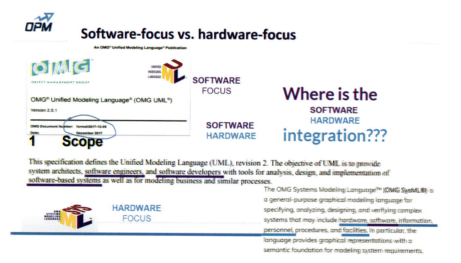

Fig. 3 The software focus of UML vs. the systems focus of SysML

structure, and behavior aspects, as each aspect is spread across at least two diagram kinds. For example, to understand the behavior, dynamic aspect, one must mentally integrate activity diagrams with sequence diagrams. To understand the structure, static aspect, class diagrams and package diagrams must be mentally integrated. Things get worse when interactions between structure and behavior need to be explored, as this requires concurrent examination and comprehension of the four diagram kinds listed above, but there may be more, and we have not yet talked about the functional aspect, which requires as a minimum the use case diagram.

The transition from UML 1.X to 2.X has not improved the situation in a significant way [5]. Similarly, there is no basis to assume that the transition from SysML 1.X to v2 will contribute to alleviating the problematic aspects, such as model multiplicity and the lack of concise semantics. Specifically, SysML v2 still does not focus on hardware-software integration, a mandatory requirement for Industry 4.0. Rather, it is expected in vague general terms that "The capabilities provided by SysML v2 should enable improved effectiveness and broader adoption of MBSE" [30]. As noted in OMG SysML v2 [31], "...the emphasis for SysML v2 is to improve the precision, expressiveness, interoperability, and the consistency and integration of the language concepts relative to SysML v1." Indeed, it looks like SysML v2 may well be able to achieve these goals [32], but it still does not address the hardware-software divide, and until this divide is bridged, the struggle to have a complete, end-to-end modeling environment will continue, with the requirements of Industry 4.0 and the DE severely exacerbating the situation.

With this in mind, OPM ISO 19450 with its MAXIM extension offer a fresh, formal yet intuitive paradigm that is elaborated briefly next.

OPM ISO 19450 and Its MAXIM Extension

OPM [1–3, 6, 7] ISO 19450 [14] is a systems modeling language and methodology that represents the function, structure, and behavior of any system using a minimal universal (upper) ontology with only two kinds of things, shown in Fig. 4: stateful objects, which are things that exist, possibly with states, and processes, things that transform objects by creating or consuming them or by changing their state.

OPM ISO 19450

Any OPM thing has a property (metamodel attribute) called **Essence**, which can assume one of two values: **informational** (designated graphically as not shaded) or **physical** (shaded), as shown in Fig. 5. This property is key to OPM's ability to model

Fig. 4 Example of an OPM object (left) and process (right)

Fig. 5 Example of OPM objects and processes with different Essence values

Fig. 6 Examples of OPM objects and processes with different Affiliation values

physical systems, which are characterized first and foremost by having a cybernetic (informatical) aspect alongside the obvious physical aspect. The same model can naturally contain these two kinds of things (objects and processes), and they can interact with each other. Another property of an OPM thing is **Affiliation**, which can be **systemic** (designated graphically by a solid contour) or **environmental** (dashed contour), as shown in Fig. 6.

OPM is both a language and a methodology. OPM's methodological part includes guidelines for limiting the size of a single OPD and refining the model in lower-level OPDs, catering to the second multimedia assumption of Mayer [20, 21] of humans' limited channel capacity. OPM's language part is implemented in OPCloud (https://www.opcloud.tech/) [9, 10], ensuring correct-by-construction OPM models while the diagram is constructed, as it constantly validates the OPM language syntax. It does so, for example, by offering the modeler who seeks to connect two things in the model only the subset of links that produce syntactically correct constructs and therefore might make sense semantically, preventing the modeler, whether junior or senior, from introducing wrong constructs into the OPM model. Moreover, the ability to execute the model at any stage increases the model correctness, since execution is the highest level in the model fidelity hierarchy [17].

OPM things are connected by links, which graphically express relations. There are two kinds of links: procedural (Table 2) and structural (Table 3). Any OPM model consists of two parts, which express the same set of model facts in two modalities: (1) the graphical part, the OPD set, a hierarchically organized set of one or more Object Process Diagrams (OPDs), and (2) the textual part, the OPL spec, a collection of sentences in a subset of English called Object Process Language (OPL). The OPD set is a set of OPDs related to each other in a tree structure, such that each diagram lower in the hierarchy refines its ancestor, usually by zooming into one of the processes in the ancestor OPD. OPD is the only kind of diagram of an OPM model. It can contain things (objects and processes), with links connecting them to express structural and procedural relations. OPL is the counterpart textual representation of the OPD set. Each OPD construct – two or more things connected by one or

Table 2 OPM procedural links

Link name	Example	OPL
Result	Creating → File	**Creating** yields **File**
Consumption	Eating ← Food	**Eating** consumes **Food**
Effect	Pedal Pressing ↔ Speed	**Pedal Pressing** affects **Speed**
Agent	Eating —● Person	**Person** handles **Eating**
Instrument	Eating —○ Fork	**Eating** requires **Fork**
Invocation	Test Finishing ⤳ Test Submitting	**Test Finishing** invokes **Test Submitting**
Instrument Condition	Buying ○c, Store {open, close}	**Buying** occurs if **Store** is at state **open,** otherwise **Buying** is skipped
Instrument Event	Running ○e, Whistle	**Whistle** initiates **Running,** which requires whistle

more links – is reflected textually in one or more OPL sentences. This bimodal representation caters to the second dual channel multimedia assumption [20, 21].

An OPM model can be presented at various levels of detail in different, interconnected views, each being an OPD. The top-level OPD is called system diagram (SD). SD usually consists of one systemic process and its *operand* – the object which that process transforms. Together, the process and the operand are the *function* of the system. SD also includes the *beneficiary group*, i.e., the person or people benefiting from the system's function, and *enablers*: *agents* (humans) and *instruments* (non-humans). Enablers are linked to the main process and enable it, but unlike *transformees* (input and output objects), they are not transformed by it.

Each OPD can be refined to expose deeper levels of detail by one of the following three refinement-abstraction complexity management mechanisms:

1. In-zooming-out-zooming: In-zooming refines a thing, usually a process, to show subprocesses, of which the main process consists and their temporal (sequential

Table 3 OPM structural links

Link name	Example	OPL
Aggregation-Participation	Animal → Head, Tail	**Animal** consists of **Head** and **Tail**.
Generalization-Specialization	Animal → Cat	**Cat** is an **Animal**.
Classification-Instantiation	Cat → My Cat	**My Cat** is an instance of **Cat**.
Exhibition-Participation	Cat → Color	**Cat** exhibits **Color**.
Unidirectional Tagged	People —4 live in→ House	4 **People** live in **House**.
Bidirectional Tagged	Exam —contains→ Question, ←appears in	**Exam** contains **Question**. **Question** appears in **Exam**.

or parallel) execution order. Out-zooming is the inverse of in-zooming – it reduces a set of things into a more abstract thing.
2. Unfolding-folding: Unfolding refines a thing, usually an object, to show parts (sub-objects), of which the main object consists, or features (attributes and operations), or specializations of that thing. Folding is the inverse – it abstracts an unfolded thing by hiding its parts, features, or specializations.
3. State expression-suppression: Each object can have one or more states, and each state can be expressed (shown) or suppressed (hidden), based on what the modeler would like to emphasize in the diagram.

The entire OPM model of a complex system is expressed in a tree of OPDs, each created by zooming into or unfolding some process or object in its ancestor OPD. The first OPD, SD, is the root of the OPD tree, and it is the only OPD at detail level 0. This is the bird-eye's view of the system, which provides a quick overview of the system's function and benefit. Lower-level OPDs are denoted SD$n1.n2....nm$, where $nj, j = 1, 2...m$, is the number of the detail levels (layer number in the OPD tree) and m is the number of the refined (in-zoomed or unfolded) subprocess in the ancestor OPD.

OPM provides strong semantics for representing non-digital components and the relationships of those components to each other and to digital components. For example, OPM might be used to model cyber-physical systems [26], where the fusion between physical and informatical things (objects and processes) is ultimate. OPM simplifies some of the complexities inherent in modeling complicated interactions in SoS using SysML activity or sequence diagrams. A major development in OPM is its ability to seamlessly fuse in the same modeling framework both conceptual-qualitative and computational-quantitative aspects due to its MAXIM extension, described next.

MAXIM: Methodical Approach to Executable Integrative Modeling

OPM overcomes the hardware-software modeling gap [1, 34] by bringing systems engineering and software engineering closer together in a stepwise fashion. To this end, OPM has been extended with MAXIM – a Methodical Approach to Executable Integrative Modeling. MAXIM enables integrating computational, software engineering capabilities into OPM ISO 19450 [14], extending OPM's power as a model-based systems engineering methodology. The implemented integration of MAXIM is part of OPCloud (https://www.opcloud.tech/) [9, 10]. In MAXIM, the system model, which comprises both physical (human and hardware) and informatical (software) components, represents not only the structure and behavior of the system, which enables it to perform its intended function, but also its quantitative requirements and computational, software-driven aspects. This combined humans-hardware-software model constitutes a complete and accurate executable specification of the system from its top-level abstract view, which expresses the system's main function, beneficiaries, and value, all the way to a set of the most detailed views, which may specify how leaf-level, mostly computational processes transform the system's atomic parts. This detailed information provides a solid basis for model simulation and execution. The domain-neutral nature of OPM does not favor systems engineering, software engineering, or any other engineering domain, such as mechanical, aerospace, or electrical engineering. It is therefore conducive to cross-domain collaboration not just among engineers with various educational backgrounds but also among system architects and designers on one hand and other system stakeholders, such as beneficiaries and users, on the other hand. From a software engineering viewpoint, MAXIM is a bridge between standard, traditional textual programming and visual programming. MAXIM offers a smooth, seamless transition that ideally blends these two approaches. There is always a certain point along the detail-level spectrum, beyond which it is no longer desirable or "cost-effective" to stick to visualization. At that point, the transition to text-based program specification makes more sense, as demonstrated in Dori et al. [9]. MAXIM enables the visual and textual approaches to be complementary rather than competing, so systems and software engineering benefit from both: Visualization is good at expressing high-level and abstract aspects of the system, while text is good for concrete, low-level computations.

The MAXIM-enhanced OPM unifying framework facilitates integrative modeling of different domains governed by several engineering disciplines. Being OPM-based, MAXIM is not limited by the kinds of domains that comprise the underlying system, because OPM is based on a minimal universal ontology, which has indeed been applied in multiple domains, from civil aviation [28] to molecular biology [37]. MAXIM has enabled building a new kind of model – a diagnostic model, applied in the medical domain to diagnose the potential for pediatric FTT during the perinatal and postnatal stages [18].

One of MAXIM's design criteria has been to extend OPM with computational capabilities without augmenting its minimal yet universal ontology. Therefore, MAXIM is built of computational constructs comprising objects, processes, states (which can have numeric or symbolic values), and relations between them that convey meanings that are analogous to their non-computational counterparts. While the conceptual model is represented at increasing levels of detail and refinement, computations are performed only at the most detailed, atomic, leaf-level processes. Similarly, the inputs and outputs of these computational processes come from and are stored in computational objects.

OPM Cyber-physical System Applications

In this section, we provide a couple of examples that use MAXIM to take OPM beyond ISO 19450:2015, extending it from a conceptual modeling language and methodology to one that enables seamless fusion of qualitative and quantitative modeling aspects and their simulation. The first example is a model of a braking system of a large civil aircraft. Figure 7 shows part of an OPM model in OPCloud of an **aircraft braking system** [17].

21 Developing Industry 4 Systems with OPM ISO 19450 Augmented with MAXIM 669

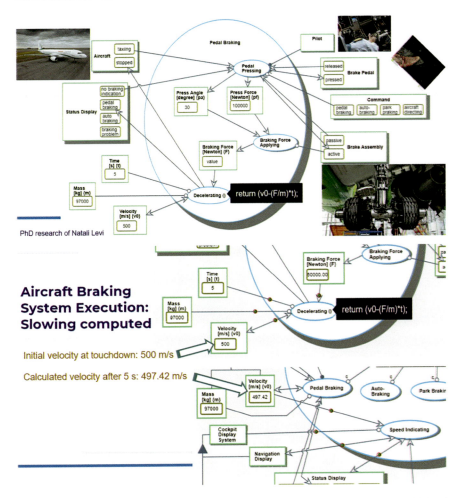

Fig. 7 The **aircraft braking system** example. Top, **Pedal Braking** in-zoomed; Bottom, Initial and calculated **Velocity**

At the top part, the **Pedal Braking** process is in-zoomed, showing its three subprocesses – **Pedal Pressing**, **Braking Force Applying**, and **Decelerating**. The latter is a computational process, as expressed by the () symbol at the end of its name. Hovering over a computational process shows a tooltip with the TypeScript code inside the process, in this case Newton's second law. At the bottom part of Fig. 7 are the initial velocity at touchdown, **500 m/s**, and the velocity after a **5 s** simulation period, which is calculated to be **497.42 m/s**.

In Levi Soskin et al. [17] we show how an OPM model is used to define parameters of an aircraft landing gear, such as the static load for the nose (front) and main (rear) landing gears, as shown in Fig. 8.

Fig. 8 Calculating **Load Rating** of **Nose Tire** and of **Main Tire**

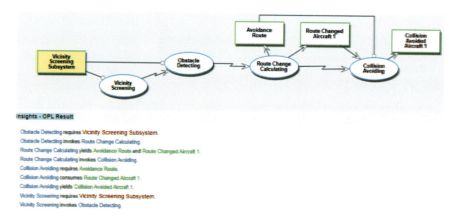

Fig. 9 An **Airplane Collision Avoidance** OPM model query

In Medvedev et al. [25], we apply a graph-theoretic approach to query OPM models. Figure 9 is an example of a query that shows how a collision is avoided thanks to a **Vicinity Screening Subsystem**.

Finally, Kohen and Dori [16] have shown how OPM stereotypes are used to enhance modeling of IoT systems, as exemplified in Fig. 10.

Chapter Summary

Industry 4.0 and the transition to digital engineering are progressing rapidly toward an economy that is wholly based on digital systems lifecycles. For model-based systems engineering (MBSE) to remain relevant and cope with this pace of change, a new modeling paradigm that combines conceptual and computational modeling capabilities in the same framework must be adopted. OPM with its MAXIM extension caters to this very need and provides an environment that extends the traditional role of MBSE to detailed, quantitative design, enabled by seamlessly and gradually transitioning from abstract, high-level system architecting to handling the fine details of complex systems without losing the big picture: The MAXIM framework enables concurrent modeling of the hardware and software system aspects, avoiding the need to make the painful and information-leaking transition

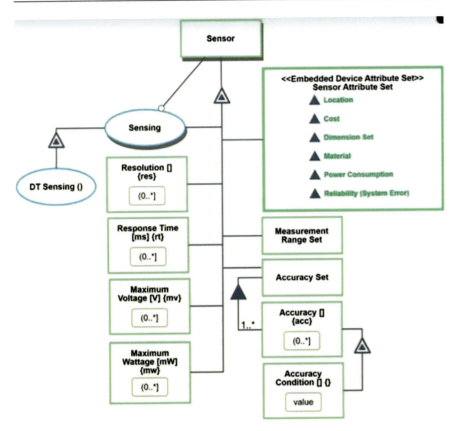

Fig. 10 An **Embedded Device Attribute Set** as an OPM stereotype anchored to **Sensor Attribute Set**

from the abstract, qualitative conceptual system architecting stage to the concrete, detailed, quantitative design stage. Implementing the "holy grail" of MBSE, MAXIM provides for involving systems engineering and software engineering professionals, as well as mechanical, electric, electronic, aerospace, and human factors engineers, all relating to the same conceptual-computational executable model as the most reliable, dynamic, up-to-date, and evolving authoritative source of truth. OPCloud is the implementation platform for OPM and the MAXIM paradigm. It is a cloud-based collaborative environment that uses OPM ISO 19450 to integrate, execute, and provide animated simulation of the combined hardware-software system model. The model is developed and presented from its top, abstract system level via increasingly operational and quantitative details, all the way to the nuts and bolts of the system's hardware and to the bits and bytes and the basic arithmetic operations of the system's software. This environment is especially suitable to cope with the new challenges that Industry 4.0 presents, with focus on the design of cyber-physical IoT-intensive systems and products.

Cross-References

▶ Digital Twin: Key Enabler and Complement to Model-Based Systems Engineering
▶ Exploiting Digital Twins in MBSE to Enhance System Modeling and Life Cycle Coverage
▶ MBSE Methodologies
▶ Model Interoperability
▶ Model-Based Hardware-Software Integration
▶ Modeling Hardware and Software Integration by an Advanced Digital Twin for Cyber-physical Systems: Applied to the Automotive Domain
▶ Toward an Engineering 3.0

References

1. Böhm W, Henkler S, Houdek F, Vogelsang A, Weyer T (2014) Bridging the Gap between Systems and Software Engineering by Using the SPES Modeling Framework as a General Systems Engineering Philosophy. *Procedia Comput. Sci.*, vol. 28, pp. 187–194.
2. Builta J, Howell J, De Ambroggi L, Short M, Grossner C, Morelli B, Tait D, Hall T (2019) Digital Orbit – Tracking the development, impact, and disruption caused by transformative technologies across key industries. Available: https://cdn.ihs.com/www/pdf/0419/ihs-markit-digital-orbit-brochure.pdf. Accessed 14 Dec 2020.
3. Cameron B, Adsit DM (2018) Model-Based Systems Engineering Uptake in Engineering Practice. IEEE Trans. Eng. Manag. 67, 152–162. https://doi.org/10.1109/TEM.2018.2863041.
4. Christie AM (1999) Simulation: An Enabling Technology in Software Engineering CrossTalk. Available: https://resources.sei.cmu.edu/library/asset-view.cfm?assetid=29627. Accessed: Dec. 24, 2020.
5. Dori D (2002) Why Significant Change in UML is Unlikely. *Communications of the ACM*, 11, pp. 82–85.
6. Dori D (2002) Object-Process Methodology – A Holistic Systems Paradigm, Springer Verlag, Berlin, Heidelberg, New York, 2002. eBook version: http://link.springer.com/book/10.1007/978-3-642-56209-9/page/1
7. Dori D (2016) Model-Based Systems Engineering with OPM and SysML, Springer, New York, 2016. http://www.springer.com/gp/book/9781493932948
8. Dori D and Goodman M (1996) On bridging the analysis-design and structure-behavior grand canyons with object paradigms. *Rep. Object Anal. Des.*, vol. 2, no. 5, pp. 25–35.
9. Dori D, Jbara A, Levi N, and Wengrowicz N (2018) Object-Process Methodology, OPM ISO 19450 – OPCloud and the Evolution of OPM Modeling Tools. Syst. Eng. Newsl. (PPI SyEN) 61, 6–17.
10. Dori D, Kohen H, Jbara A, Wengrowicz N, Lavi R, Levi Soskin N, Bernstein K, and Shani U (2020) OPCloud: An OPM Integrated Conceptual-Executable Modeling Environment for Industry 4.0. In Kenneth R, Zonnenshain A, and Swarz RS (eds.) *Systems Engineering in the Fourth Industrial Revolution: How Big Data and Novel Technologies Affect Modern Systems Engineering*. Wiley.
11. Dori D, Renick A, and Wengrowicz N (2016) When quantitative meets qualitative: Enhancing OPM conceptual systems modeling with MATLAB computational capabilities. Res. Eng. Des. 27(2) 141–164.
12. Hampson K (2015) Technical evaluation of the Systems Modeling Language (SysML). Conference on Systems Engineering Research, Procedia Computer Science 44, 403–412.

13. Hoheb A, Hale J (2019) Leading the Transformation of Model-Based Engineering: The Model-Based Capability Matrix. Available: https://www.incose.org/docs/default-source/default-document-library/leading-mbse-transformation_v5.pdf?sfvrsn=48e59bc6_0. Accessed: Dec. 14, 2020.
14. ISO 19450 (2015) Automation systems and integration – Object-Process Methodology. Available: https://www.iso.org/obp/ui/#iso:std:iso:pas:19450:ed-1:v1:en. Accessed: Dec. 26, 2020.
15. James L (2020) Model-based Systems Engineering at JPL. Presentation at CSER 2020.
16. Kohen H, Dori D (2021) Improving Conceptual Modeling with Object-Process Methodology Stereotypes. 2021, 11, 2301. https://doi.org/10.3390/app11052301.
17. Levi Soskin N, Jbara A and Dori D (2020) The Model Fidelity Hierarchy: From Text to Conceptual, Computational, and Executable Model. IEEE Systems Journal, 2020. Early Access. https://doi.org/10.1109/JSYST.2020.3008857.
18. Levi-Soskin N, Shaoul R, Kohen H, Jbara A, and Dori D (2019) Model-Based Diagnosis with FTTell: Assessing the Potential for Pediatric Failure to Thrive (FTT) During the Perinatal Stage. SIGSAND/PLAIS: 2019 EuroSymposium on Systems Analysis and Design – Information Systems: Research, Development, Applications, Education, Proc. 12th SIGSAND/PLAIS, Gdansk, Poland, September 19, 2019. Lecture Notes in Business Information Processing book series (LNBIP), volume 359, pp. 37–47.
19. Madni AM, Sievers M (2018) Model-based systems engineering: Motivation, current status, and research opportunities. Syst. Eng. 21, 172–190. https://doi.org/10.1002/sys.21438.
20. Mayer RE (2003) The promise of multimedia learning: using the same instructional design methods across different media. Learn. Instr. 13(2) 125–139.
21. Mayer RE and Moreno R (2003) Nine ways to reduce cognitive load in multimedia learning. Educ. Psychol. 38(1) 43–52.
22. Mellor SJ and Balcer MJ (2003) Executable and Translatable UML. Embed. Syst. Program., vol. 16, no. 2, pp. 25–30.
23. McDermott T, Eileen Van Aken E (2020) Task Order WRT-1001: Digital Engineering Metrics. Technical Report SERC-2020-TR-002. Available: https://apps.dtic.mil/sti/citations/AD1104591. Accessed: Dec. 14, 2020.
24. McDermott TA, Hutchinson N, Clifford M, Van Aken E, Slado A, Henderson K (2020) Benchmarking the Benefits and Current Maturity of Model-Based Systems Engineering across the Enterprise.
25. Medvedev D, Shani U, Dori D (2021) Gaining Insights into Conceptual Models: A Graph-Theoretic Querying Approach. Applied Science 11 (766).
26. Mordecai Y, Dori D (2017) Minding the Cyber-Physical Gap: Model-Based Analysis and Mitigation of Systemic Perception-Induced Failure, Sensors, 17(1644), 1644. https://doi.org/10.3390/s17071644.
27. Mordecai Y, de Weck LO, Crawley EF (2020) Towards an Enterprise Architecture for a Digital Systems Engineering Ecosystem. CSER 2020.
28. Mordecai Y, Orhof O, and Dori D (2016) Model-Based Interoperability Engineering in Systems-of-Systems and Civil Aviation. *IEEE Trans. Syst. Man. Cybern.* 1. http://ieeexplore.ieee.org/stamp/stamp.jsp?arnumber=7571127
29. ODASDE (2017) Office of the Deputy Assistant Secretary of Defense (Systems Engineering) [ODASD (SE)], "DAU Glossary: Digital Engineering," Defense Acquisition University (DAU), 2017. Available: https://www.dau.edu/glossary/Pages/GlossaryContent.aspx?itemid=27345. Accessed: Dec. 14, 2020.
30. OMG SysML v2 (2017) OMG SysML v2: The next-generation systems modeling language. Available: https://www.omgsysml.org/SysML-2.htm. Accessed: Dec. 26, 2020.
31. OMG SysML v2 (2017) Document – ad/17-12-02 (Systems Modeling Language (SysML) v2 RFP). Available: https://www.omg.org/cgi-bin/doc?ad/17-12-02.pdf. Accessed.
32. OMG SysML v2 SST (2020) Introduction to the SysML v2 Language Textual Notation. Available: https://drive.google.com/file/d/1iW2pVnsCxuhSadQhT0ut9yMJRDLvz29v/view?usp=sharing. Accessed: Dec. 26, 2020.

33. Peleg M and Dori D (2000) The Model Multiplicity Problem: Experimenting with Real-Time Specification Methods. IEEE Transaction on Software Engineering, 26, 8, pp. 742–759.
34. Pyster A et al. (2015) "Exploring the Relationship between Systems Engineering and Software Engineering," Procedia Comput. Sci. 44 708–717.
35. Seidewitz E (2014) UML with Meaning: Executable Modeling in Foundational UML and the Alf Action Language. Proc. 2014 ACM SIGAda Annual Conference on High Integrity Language Technology, New York: ACM, pp. 61–68.
36. Selic B (2003) The pragmatics of model-driven development, IEEE Softw., vol. 20, no. 5, pp. 19–25.
37. Somekh J, Haimovich G, Guterman A, Dori D, and Choder M (2014) Conceptual Modeling of mRNA Decay Provokes New Hypotheses. PLoS ONE 9(9): e107085. https://doi.org/10.1371/journal.pone.0107085.
38. Ustundag A, Cevikcan E (2018). Industry 4.0: Managing the Digital Transformation, Springer Series in Advanced Manufacturing. https://doi.org/10.1007/978-3-319-57870-5.

Dov Dori is Fellow of IEEE, INCOSE, and IAPR, a Professor of Systems Engineering and Head of the Enterprise Systems Modeling Laboratory at the Technion, Israel Institute of Technology. During 2020 he was Visiting Professor at the Aeronautics and Astronautics Department at MIT – Massachusetts Institute of Technology, Cambridge, MA, USA. He has intermittently been Visiting Professor at MIT since 1999. In 1993 Dr. Dori invented Object-Process Methodology (OPM ISO 19450:2015) and has been central to the field of model-based systems engineering (MBSE). Prof. Dori has authored about 400 publications and supervised 60 graduate students. He chaired nine international conferences and was Associate Editor of IEEE T-PAMI and Systems Engineering and Founding Co-Chair of the IEEE Society of Systems, Men, and Cybernetics Technical Committee on MBSE. His 2002 and 2016 books are the basis for MBSE edX certificate program and MOOC series. He has received various research and innovation awards and is a member of Omega Alpha Association-International Honor Society for Systems Engineering.

MBSE Testbed for Unmanned Vehicles

System Concept and Prototype Implementation

A. M. Madni and D. Erwin

22

Contents

Introduction	676
MBSE State-of-the-Art	677
MBSE Testbed for Unmanned Vehicles: A Best Practice Approach	678
Testbed Ontology	678
MBSE Testbed Concept	680
Key Features	681
Predefined Scenarios	681
Dashboard Tool	681
System Modeling	682
Scenario Elements	682
Human-Computer Cooperation	682
Extensible Architecture	682
Current Testbed Components	683
Software Programming Environment	683
Optimization, Control, and Learning Algorithms	683
Graphical User Interface	683
Exemplar Repositories, Packages, Libraries (Not Fixed)	684
Simulation Platforms	684
Hardware and Connectors	684
Logical Architecture	684
Testbed Benefits	685
Distributed Hardware Environment	685
Prototype Testbed Implementation	687
System Modeling and Verification	687
Model and Scenario Refinement	687
Testbed Repository	688
Experimentation Support	688
Rapid Scenario Authoring	689
Multi-perspective Visualization	691
Smart Dashboard	692

A. M. Madni (✉) · D. Erwin (✉)
University of Southern California, Los Angeles, CA, USA
e-mail: azad.madni@usc.edu; erwin@usc.edu

© Springer Nature Switzerland AG 2023
A. M. Madni et al. (eds.), *Handbook of Model-Based Systems Engineering*,
https://doi.org/10.1007/978-3-030-93582-5_39

Preliminary Experiments 698
Lessons Learned 700
Summary 702
Cross-References 703
References 703

Abstract

The number of digital artifacts created by Model Based Systems Engineering (MBSE) continues to increase with greater collaboration among teams and increasing coverage of the system life cycle. Also, with the growing convergence between MBSE and Digital Engineering (DE), digital twin technology is being exploited within MBSE to improve system model accuracy and facilitate system verification and validation. With these advances, the MBSE community to needs a flexible framework to efficiently create, organize, access, manipulate, and manage MBSE artifacts and simplify experimentation. The latter is key to evaluating modeling approaches, facilitating analysis, and enabling comparative evaluation of models and algorithms. This chapter presents the concept implementation and capabilities of a prototype MBSE innovation testbed developed for addressing these needs. The capabilities of the testbed are presented using an illustrative real-world problem scenario (i.e., aircraft perimeter security). The chapter concludes with key findings, and lessons learned, potential uses, and future extensions.

Keywords

Model based systems engineering · MBSE · Digital engineering · Testbed · Digital twin · Simulation · Ontologies

Introduction

Model Based Systems Engineering (MBSE) continues to make significant strides in increasing systems life cycle coverage and modeling increasingly more complex systems [1]. MBSE has also begun to leverage digital twin concepts to enhance system verification and validation capabilities, thereby increasing system model accuracy and life cycle coverage [2]. As a result of these advances, MBSE produces increasingly more digital artifacts that need organization, categorization, and meta-tagging for rapid retrieval and reuse during model development and experimentation [3, 4]. Also, there are limitations in MBSE practice that need to be overcome.

For example, MBSE practitioners to tend to work with specific models and simulations that address particular problems. This practice does capture helpful knowledge and lessons learned for future MBSE projects. In some industries, lessons learned are documented but not useful. As a result, lessons learned are often ignored in the "next" MBSE project. Furthermore, MBSE teams collaborating on projects do not have a convenient means to share MBSE artifacts. The recognition of these deficiencies led to creating a *MBSE innovation testbed* for a particular domain.

This chapter introduces the concept, usage, and value proposition of a MBSE innovation testbed for unmanned vehicles. This testbed organizes and manages MBSE artifacts (including lessons learned) and supports system modeling and experimentation with new modeling concepts and innovative algorithms. This chapter presents a prototype implementation of the testbed using a real-world operational scenario to convey the value proposition of the testbed. The chapter concludes with a discussion of findings, lessons learned, and implications for future enhancements.

MBSE State-of-the-Art

MBSE state-of-the-art is primarily reflected in the MBSE tools available to. While MBSE tools have advanced significantly, their integration with simulation that support simulation-based experimentation is expensive, complicated, and subject to tool constraints. To, several simulation tools (e.g., Simulink, ModelCenter, Mirabilis Architect, Open MDAO) work with MBSE tools. However, these simulations have limitations. For example, Ptolemy can contribute significantly with respect to heterogeneous systems, multiple time domains, and more. However, Ptolemy is difficult to use. Architecture frameworks such as the Department of Defense Architecture Framework (DODAF) and Unified Architecture Framework (UAF), which are now widely available, only minimally contribute to the development of a flexible simulation and experimentation infrastructure [5].

In recent years, the use of simulation within MBSE has increased for system model verification and testing and for acquiring new insights [6, 7]. This advance departs from traditional brute-force standalone simulations that tend to be limited in the range of operational conditions (e.g., contingencies, threats) they can test. More recently, the MBSE community has turned to formal representation of semantically rich ontologies and the use of metamodels to facilitate assessment of model completeness, syntactic correctness, semantic consistency, and requirements traceability. The use of metamodels and ontologies in MBSE is not new. SysML-based models are metamodel-based and support requirements traceability with normative relationships between requirements and other elements in a model. For example, it is possible to generate maps and tables that show, for example, requirement-to-requirement, requirement-to-use case, requirement-to-V&V, requirement-to-logical and physical entities, requirements-to-activities relationships. However, the system representation needs to be flexible and adaptable to learn from new information for some problems. Such problems require a flexible representation and a semantically richer language than SysML.

And finally, systems are becoming increasingly more complex due to the increase in interconnectedness and interdependencies. These complex systems are typically a combination of legacy and third-party components from previously deployed systems. Therefore, some components tend to be fully verified under certain operating conditions that may or may not apply when considering their reuse. Furthermore, complex systems are subject to unreliable interactions (e.g., sporadic/incorrect sensor inputs, control commands that are not always precisely followed) because they frequently interact with the physical world. Finally, complex systems are increasingly vulnerable to security threats.

MBSE Testbed for Unmanned Vehicles: A Best Practice Approach

Given the preceding discussion, this chapter presents the overall concept, architecture, implementation, and value proposition of a MBSE testbed for exploring innovative concepts for planning and decision-making in unmanned vehicles. The MBSE testbed concept is broader than conventional hardware-in-the-loop (HWIL) testbeds. HWIL testbeds are used for early integration and testing of physical systems and formal verification and validation (V&V). Typical HWIL testbeds consist of hardware, software modules, and simulations in which system components are either physically present or simulated. Physical components progressively replace simulated components as they become available [6, 7]. HWIL testbeds often include software-based physics models. For example, an abstract control system model can be created using a MATLAB Kalman filter that represents the approximate behavior of a complicated controller yet to be implemented. The MBSE testbed construct can potentially extend HWIL capabilities by including the means for developing and exercising *abstract models*. These models can be exercised independently or made to interoperate with HWIL. The modeling, simulation, and integration environment of the MBSE testbed can also be used to develop and evaluate digital twins [6, 7].

The MBSE innovation testbed is a best practice approach for conducting MBSE projects and managing MBSE artifacts. Central to this approach are facilities that manage core MBSE activities, i.e., organize and manage MBSE artifacts and experiment with new modeling constructs, model optimization algorithms, model refinement techniques, and model reuse options in a variety of scenarios. Other testbed activities include artifact configuration management, model reviews, discrepancy tracking, and open issues tracking at this time. Future versions of the testbed will address these latter capabilities. Users of the testbed are system modelers, test engineers, and project managers. The target application selected for prototype capabilities demonstration is unmanned aerial vehicle (UAV) team operations [6, 7]. The testbed enables the creation of MBSE projects associated with different experiments and enables the capture of project particulars and use cases. A formal ontology that describes experimentation entities is employed that assures completeness of the model with respect to questions that need to be answered through experimentation. The following subsections present the key elements of the testbed.

Testbed Ontology

The testbed ontology comprises the key elements needed in MBSE projects for developing and evaluating collaborative unmanned vehicle operations [6, 7]. The ontology is sufficiently general to support various scenarios in the unmanned air and ground vehicle problem domains. Sufficient generality implies that the ontology can support multiple instantiations in the unmanned air and ground vehicle application domains. The ontology supports both question-answering and reasoning in these domains. The key elements of the ontology are:

- *Mission:* the goal(s) to be achieved in the scenario; the missions may be time-independent monitoring (e.g., high-value asset-perimeter surveillance) or time-dependent (e.g., time-specific waypoints). The missions may have a discrete goal (e.g., destroy bridge) or a quantitative objective function (e.g., maximize fitness function).
- *Scenario:* specification of a specific instance of a mission that provides the contextual backdrop for experiments.
- *World (or Terrain):* the environment (e.g., geospatial region) that supports experiments. World with respect to the selected application domains pertains to the physical area or volume in which the scenario unfolds. For UAV scenarios, the terrain is 3D, with the horizontal extent being typically much larger than the vertical. The terrain can be fictitious, in which case it is defined by a set of graphics-rendered models, or actual, meaning that the terrain is a region of Earth's surface with an altitude bound. The terrain is simply connected, and may have one or more regions of interest (ROIs). For instance, in a search and rescue mission, the staging area and the primary search area may be two regions of interest (ROIs).
- *System of Interest (SoI):* an envisioned or existing physical system (e.g., UAV). In MBSE parlance, it is called the "domain."
- *Agent:* an active, persistent element.
- *Event:* a change in condition or a disruption that may occur during the conduct of the scenario. For example, UAVs may be subject to random wind gusts, occasional engine malfunction, and other failures. An event need not be adverse (e.g., landing of C-130 aircraft).
- *Evaluation:* a systematic procedure to test models and algorithms or demonstrate a capability.
- *Laboratory:* elements participating in an evaluation communicating with each other. Communication could be over Wi-Fi, through Bluetooth, IR links, or direct-wired connection. Communication can also happen within a simulation.
- *Project:* consists of all elements associated with a particular experiment; it should be properly scoped to achieve desired results (e.g., system performance evaluation).
- *Scenario Builder:* this software facilitates the creation and modification of scenarios; it also supports the import of scenarios from third-party applications; scenario definition includes entry and exit criteria, agents, events, activities, etc.
- *Dashboard:* a customizable and smart visualization, monitoring and control software that allows the user to import scenarios from the scenario builder.
- *Predefined Scenarios:* part of a "starter kit" for systems engineers to quickly start using the testbed; created using a scripting or graphical modeling language.
- *Data Collector:* software that collects data from simulation execution.
- *Digital Twin:* a software replica of a physical system that reflects both operational and maintenance history of the physical system.

The current implementation of the testbed has the following capabilities to enable modeling, simulation, and analysis of the system of interest.

MBSE Testbed Concept

Figure 1 presents the MBSE testbed concept. The testbed comprises: (a) a user interface that supports scenario authoring, dashboard capabilities for scenario execution monitoring visualization, and control, and report generation; (b) a suite of modeling and analysis tools including system modelers, machine learning and data analytics algorithms; (c) simulation engines for discrete event simulation, hybrid simulation, and component simulation; and repositories of operational scenario vignettes, system models, component libraries, and lessons learned.

The prototype testbed offers the ability to:

- Conceptualize models using modeling approaches using testbed components.
- "Test-drive" concept of operations (CONOPS) for system of interest (e.g., multi-UAV teams)
- Verify system models of cyber-physical-human systems using testbed.
- Experiment with formal and probabilistic models in the safety of a simulation environment.
- Develop a deeper understanding of state-space modeling, self-learning, and adaptive control concepts.
- Generate additional data for use in learning algorithms that define the probabilities as well as to update utility functions.
- Evaluate system safety and resilience properties using state-space analysis.

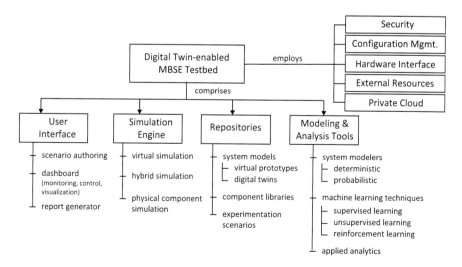

Fig. 1 MBSE testbed concept [6, 7]

Key Features

The core features characteristics of the testbed enable users to "hit the ground running" in MBSE projects. The key features are presented next.

Predefined Scenarios

A key testbed concept is a "starter kit" that allows users to start using the testbed on MBSE projects quickly. The starter kit comprises predefined unmanned vehicle scenarios that can be used "out of the box" in a simulation environment and later in the physical laboratory. The illustrative scenario provided as part of the starter kit is concerned with perimeter security of a stationary C-130 aircraft using a team of quadcopters with downward facing cameras and building-mounted video cameras.

Dashboard Tool

A customizable dashboard program allows importing scenarios, provides default views such as a plan view (or "mission view") of the World showing all systems of interest (e.g., vehicles). For example, for an aircraft perimeter security scenario in which a team of quadcopters with downward-facing video cameras, the dashboard provides views from one or more vehicle cameras, state indicators for all vehicles of interest, and controls for starting and stopping an experiment. The dashboard can also be configured from an externally defined scenario description.

Figure 2 shows an example of a dashboard tool. In this example, the dashboard is used to monitor and control UAVs assigned to maintain perimeter security of a

Fig. 2 Exemplar dashboard

parked C-130 troop transport aircraft. The UAVs are equipped with downward-facing cameras to monitor the aircraft perimeter. Buildings in the vicinity are equipped with video cameras that monitor the airstrip on which the aircraft is parked.

System Modeling

In the exemplar problem, the systems being modeled are quadcopters with downward facing video cameras. The quadcopter model is an instantiation of the system class. It is an engineering model at a level that enables ease of assembly and assures crashworthiness (ability to crash without damage). On the software side, a discoverability component is provided that automates vehicle network communication by providing a "plug and play" capability, so that manual setup of the vehicle IP addresses is not needed.

Scenario Elements

The scenario elements are presented through the dashboard display interface (Fig. 2). A plan view of the terrain (mission view) is typical; hi-resolution views of important regions of interest (ROIs) can also be provided.

Human-Computer Cooperation

Human-computer cooperation comes into play in decision-making and control, for example, quadcopter teams. In general, human-computer cooperation is needed when the machine cannot respond autonomously, and human involvement becomes essential. In this case, mission execution can be partly automated and partly interactive. The dashboard allows selection of automated execution versus human-in-the loop execution by checking a box. Also, the concept of augmented intelligence [8] can be introduced and evaluated. Augmented intelligence involves human-AI partnership in which AI enhances human cognitive performance including learning, decision-making, and responding to new experiences.

Extensible Architecture

The testbed allows expansion of its capabilities in support of digital threads, enabling distributed collaborative development. The testbed architecture allows the addition of future capabilities such as utilities, APIs, analysis and measurement tools, and system modeling tools. Extensibility is achieved through an architecture that enables "plug-and-play" of various vehicles (e.g., quadcopter, 1/16 scale robot vehicle) using available APIs.

Current Testbed Components

The core elements of the testbed include the programming environment; optimization, control, and learning algorithms; repositories, software packages, and libraries; simulation platforms; and hardware and connectors. The specific elements under each category is presented below.

Software Programming Environment

- Operating system(s): Linux, Windows.
- Programming environment: Python 3.x programming language. Python is used in the quadcopter testbed instantiation. It is not a testbed constraint.
- Other language: C#.

Modeling Methods (Descriptive, Analytic; Deterministic, Probabilistic)

- Systems Modeling Language (SysML).
- Decision trees.
- Probabilistic methods: Hidden Markov Model (HMM), Partially Observable Markov Decision Process (POMDP) Model.

These modeling methods are both custom software and Python packages. For example, SysML can be used to model the structure of the digital twin for a quadcopter. In the future, the testbed could use pomdp py for POMDPs, scikit for HMMs, gym, PyGad.

Optimization, Control, and Learning Algorithms

- Optimization using fitness functions.
- Utility maximization using utility functions.
- N-Step Look-Ahead decision-processing algorithm.
- Traditional deterministic control algorithm (e.g., PID algorithm).
- Q-learning algorithm.
- These applications integrate with the model using middleware, e.g., Cameo's OpenAPI integrates with Jython.

Graphical User Interface

- Dashboard GUI is written in Python using the wxWidgets library.
- Panda3D game engine GUI provides display graphics.

Exemplar Repositories, Packages, Libraries (Not Fixed)

- NumPy – a Python library for manipulating large, multidimensional arrays and matrices.
- Pandas – a Python library for data manipulation and analysis, specifically numerical tables, and time series.
- Scikit-learn – a machine learning library for Python; built on top of NumPy, SciPy, and matplotlib.
- OpenCV – a computer vision library with Python bindings.
- Note: any Python package can be installed using PIP.

Simulation Platforms

- Hardware-Software Integration Infrastructure: used for visualization, evaluation, and data collection from scenario simulations for ground and airborne systems. This is a custom infrastructure created for the application domains.
- DroneKit Platform: an open-source platform used to create models, and algorithms that run on onboard computers installed on quadcopters; provides Python APIs for experimenting with simulated quadcopters and drones. The code is available on GitHub [5].
- Unity 3D Game Engine: with facilities for Python scripting, machine learning, and analytics.

Hardware and Connectors

- Raspberry Pi (onboard computer, e.g., computers on a quadcopter).
- Quadcopters (very small UAVs used, for example, in surveillance).
- Video cameras (typically PiCams): mounted on quadcopters; PiCams are camera modules that connect to Raspberry Pi (plug into connector directly on Raspberry Pi).
- Socket communication: to communicate between processes on different computers.

Logical Architecture

Figure 3 presents the logical configuration of the testbed prototype. Incorporating the various components is performed in stages.

As shown in Fig. 3, the testbed prototype comprises: (a) a user interface for scenario definition and system modeling as well as for the dashboard used for monitoring, visualizing, and controlling scenario execution; (b) models created by the systems engineer or researcher that reflect an existing or envisioned system and stored in the MBSE repository; (c) a multi-scenario capable simulation engine, that dynamically responds to "injects" from the user interface and collects experiment results that area sent to the repository and user interface; (d) experiment scenarios,

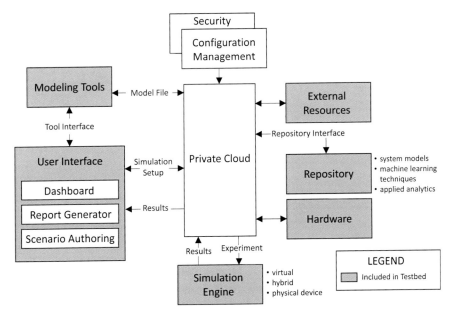

Fig. 3 Prototype MBSE testbed logical architecture

stored in the repository or entered from the GUI; and (e) a Private Cloud, that provides testbed connectivity to external resources, repository, and hardware (e.g., quadcopters), and protects MBSE assets.

Testbed Benefits

The key benefits of the testbed approach are:

- *Easy entry into the field:* teams will be able to start quickly with zero-dollar investment by installing the software and running pre-built simulations.
- Reduced risk in experimentation: experimental work can be initiated using known, proven designs without the need to design new vehicles.
- Easy comparison: it will be possible to compare research results from different research teams in a straightforward way.
- Best practices: rapid dissemination of extensions (new and modified scenarios or vehicles) to distributed collaborative teams.

Distributed Hardware Environment

Typically, tests that run on the testbed require additional assets besides the primary testbed computer for the following reasons:

- Simulations may involve humans or actual hardware such as vehicles.
- An agent or algorithm may run only on specific hardware (e.g., a legacy processor) or on a machine licensed to run proprietary software.
- The total simulation load exceeds the capacity of the primary computer. (In the future, the testbed could display CPU, memory, disk, and I/O usage.)

In such cases, the communication between the distributed assets and the primary computer is paramount. To this end, standard protocols such as TCP socket communications, both software and hardware agnostic, are used where possible. As in all parallel computations, communication bandwidth needs to be sufficiently low so as to not burden the distributed computers. Online multiplayer games provide an apt example of this concept. In such games, the players' machines communicate the actions of their avatars using only summary data such as location, orientation, and joint angles to reduce communication overhead, rather than deliver video rendering information which would require orders of magnitude higher bandwidth.

Clock synchronization and communication latency are important issues. Clock synchronization using network time protocol (NTP) is sufficient for most cases, since if the hardware resides on a local area network, NTP accuracy is around one millisecond. In the context of the kind of work discussed here, this accuracy would be insufficient only if a high-speed control loop (such as needed to run a quadcopter doing racing or aerobatic maneuvers) were distributed across the network, an unlikely scenario.

Communication with legacy programs can potentially require special attention. For instance, standard Fortran IV programs can only read and write the filesystem, so embedding such a program in a distributed test will require file-based communications such as named pipes. In cases where specific hardware is required (second bulleted reason above), the legacy or proprietary software must be wrapped by a program that runs the software and communicates with the primary computer typically via socket communication. How would this work if the user only had compiled code? If the code reads and writes standard input and output (units 5 and 6 in Fortran), there is no issue at all. However, in many cases, the inputs and outputs are hardcoded file names. This would require the following approach: a request for computation would result in creation of the input file(s) (deleting the old versions if present); running the code until the process completes; and reading of the generated output files.

Hardware test articles such as vehicles are often set up to allow remote operation over a wireless link. While the hardware may record its sensor information, it is often desirable to obtain independent "ground truth," for instance, time histories of vehicle performance concerning the terrain; this is easier done in an indoor laboratory than in the field. However, indoor environments have constraints that prevent understanding realistic conditions such as up/down drafts, glint, and terrain irregularities. For the latter, outdoor tests are necessary.

Human involvement is straightforward if the humans use standard tools such as keyboard-mouse-monitor for interacting with the test computer or with a distributed asset such as a laptop or tablet, or even a smartphone, which communicates with the testbed computer. Alternatively, in a large-scale field test where humans operate vehicles, the vehicles themselves provide the human-machine interface. Fortunately, these cases cover a significant fraction of the cyber-human systems cases of interest. It should be noted that the constraints of indoor operation prevent understanding of system operation in realistic conditions such as up/down drafts, glint, and terrain.

The prototype testbed implementation supports virtual, physical, and hybrid simulations. It supports virtual system modeling and interoperability with the physical system. It is able to access data (initially manually and eventually autonomously) from the physical system to update the virtual system model thereby making it into a digital twin of the physical system. The testbed supports proper switching from the physical system to the digital twin and vice versa using the same control software.

Prototype Testbed Implementation

The testbed currently offers the following capabilities in support of system modeling and experimentation:

System Modeling and Verification

Both deterministic and probabilistic modeling capabilities are available in the testbed. For example, the testbed offers SysML modeling capability for descriptive system modeling and Partially Observable Markov Decision Process (POMDP) modeling for probabilistic system modeling. Exemplar models of both types are provided in the online "starter kit" allow users to copy before commencing system modeling. The idea is that a user copies something from the starter kit and modifies it for their project. Verification in this context pertains to ascertaining model correctness (i.e., model completeness with a request to questions that need to be answered, syntactic correctness, semantic and syntactic consistency, and model traceability to specified system requirements).

Model and Scenario Refinement

The requisite scripting and graphical modeling capabilities for scenario definition and model refinement are available in the testbed. The testbed also enables straightforward substitution of coarse models with refined models. Also, virtual models can be replaced with hardware components. For example, a user of the testbed can start the analysis by graphically depicting the scenario in SysML diagrams. Users can identify and list all the scenario elements in the SysML block definition diagram. As a next step, the testbed can map the scenario elements in Unity 3D and rapidly develop a 3D scenario. Behaviors can be assigned to model elements in the 3D scenario, and simulations can be conducted. In Unity 3D, it is possible to extract relevant properties (e.g., velocities, locations, states) of scenario objects. Entity behaviors are assigned to objects using Unity 3D scripts written in C#. This capability affords greater flexibility in experimentation and added capabilities. A Python interface is used to test various machine learning (ML) algorithms. We use the Unity 3D ML Agents package to test various reinforcement learning models such as Proximal Policy Optimization (PPO), Soft Actor-Critic (SAC), PPO with Generative Adversarial Imitation Learning (GAIL), PPO with Behavioral Cloning (BC), and PPO combined with GAIL and BC [19–23]. Having flexibility and extensibility embedded in the architecture of the testbed allows greater experimentation capabilities. Finally, we provide the capability to use real-world data to input the model elements to execute digital twin behavior. For example, the battery level of real-world UAVs can be communicated to the Digital Twin model view in Unity 3D simulation, ultimately providing more accuracy and fidelity for experimentation.

Testbed Repository

The testbed repository comprises scenario libraries, scenario objects, 3D objects (e.g., virtual system models), object behaviors, and digital twins. For example, 3D objects such as UAV models and camera models are part of the scenario object repository. The repository consists of a hierarchical folder structure where various repository elements can be categorized. Testbed repository consists of model elements that can be reused. Repositories are used for rapid scenario authoring. The 3D objects repository consists of 3D models of UAVs, ground vehicles, air vehicles, buildings, and road segments. Users can model a scenario using these model elements. The repository also contains a library that associates specific behaviors with the objects in the scenario. Testbed repository consists of behavior elements such as line following behavior, keyboard control behavior, predefined trajectory behavior, custom Python behavior, pathfinding behavior, etc. Users can store behavior scripts in the library and assign behaviors to model elements in the scenario. Similarly, the testbed repository consists of SysML diagram templates that users can use as a starting point for modeling. Instead of starting from scratch, having a diagram template allows modelers to streamline the modeling process.

Experimentation Support

The testbed's instrumented environment supports collecting and storing data from experiments for subsequent analysis. Also, machine learning algorithms can use data collected from experimentation to train models. The MBSE testbed provides access to the properties of scenario objects (e.g., velocity, size, shape, and location of static objects and auxiliary agents) are directly extracted from the virtual environment. This capability enables the definition of abstract perception systems. Abstract perception system allows creation of virtual perception system in the Unity 3D virtual environment. Behavior scripts are used to extract, process, and transfer data to other components of the dashboard. Users can customize behavior scripts based on the experimentation requirements. The virtual environment supports manual control of objects, thereby allowing experimentation and testing flexibility. Users are able to interact with virtual objects thereby realizing complex behaviors during experimentation.

Rapid Scenario Authoring

The exemplar scenario used to illustrate the prototype testbed's capabilities is concerned with maintaining perimeter security of a stationary aircraft (e.g., C-130) using a team of small UAVs called quadcopters. Buildings in the immediate vicinity are equipped with video cameras that provide complimentary perimeter coverage of the stationary aircraft.

Eclipse Papyrus, in conjunction with Unity 3D virtual environment, is used for scenario authoring and definition of behaviors for entities in the scenarios. We are using Papyrus because it is open source and free. In other words, the source code is available to public for customization. Additionally, support for new languages can be developed too. We built a translation tool based on XML to extract data elements from the SysML model. The extracted elements are then mapped in the Unity 3D environment. We use the Asset Library (comprising 3D models, algorithms, behaviors) to populate the Unity 3D environment with model elements. If no matching elements are found in the Asset Library, empty objects are created in the Unity 3D environment. These empty objects then can be modified by modelers in Unity 3D. Modelers can add more 3D objects, modify shapes, add behaviors, define scenarios, etc. The testbed offers multi-perspective visualization. For example, if one SysML diagram uses "battery" and another uses "power supply," these can be related in Unity 3D with "battery" being a type of "power supply." The modeler has discretion in how to model, for example, an x-cell Li-ion battery and a y-cell Li-ion battery. The modeler can model them individually or instantiate them from a "battery" object. Figures 4 and 5

Fig. 4 Scenario with UAV and surveillance cameras

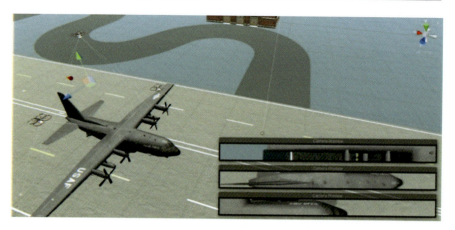

Fig. 5 Scenario with multiple surveillance UAVs

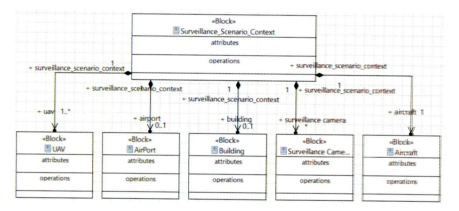

Fig. 6 Scenario context definition

show exemplar visualizations for the aircraft perimeter security problem. In Fig. 4, one UAV and one static camera are watching the aircraft. Figure 5 shows three UAVs with downward cameras watching the aircraft. In the bottom right corner of the figures, camera views are shown. The exemplar scenario is used to experiment with different models (e.g., planning and decision-making) and algorithms (e.g., optimal resource allocation). The initial scenario context is represented in SysML, add behaviors, define scenarios, with Python script used to extract data from the SysML model stored in XMI and populates the Unity 3D virtual environment. In other words, the Python script reads the XMI and creates 3D assets in Unity 3D.

Figure 6 is a SysML representation of the scenario domain for the aircraft perimeter security exemplar problem. In standard MBSE terminology, context refers to the environment that the system of interest (i.e., domain) interacts with. Thus, context includes the UAV, adjacent buildings, building-mounted surveillance cameras, and the aircraft. In addition, Fig. 6 lists the different entities that participate in the surveillance scenarios. A block definition diagram is used to identify different entities and their multiplicities. The scenario can be defined using Sequence Diagram or Activity Diagram.

Multi-perspective Visualization

The virtual environment offers insightful visualization of system behaviors during simulation execution. Exemplar visualizations for multi-UAV operations are presented in Figs. 7, 8, and 9.

Figures 7 and 8 show the change in the state of a UAV from "safe-state" (blue cloud surrounding the UAV) to "unsafe-state" (red cloud surrounding the UAV). The proximity from the nearing object is used to define "safe-state." If the UAVs are less

Fig. 7 All UAVs in "safe states"

Fig. 8 One UAV in "unsafe state"

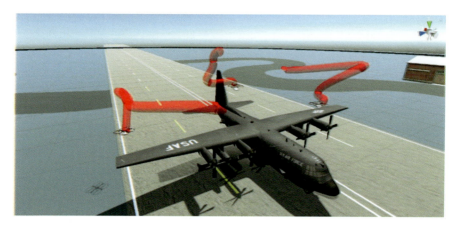

Fig. 9 Trajectory visualizations of UAVs

than 1 m from the nearest objects, the state turns to an "unsafe state." Figure 9 shows a different perspective for visualizing UAV trajectories during experimentation. Visualization assets and corresponding scripts are stored in the repository. The experimenter can "drag and drop" an asset on the scenario object and then incorporate the object within the experiment. Asset visualization parameters can be customized through a user interface.

Smart Dashboard

The smart dashboard is scenario-configurable, supports users in various functions (e.g., monitoring, visualization, control, optimization), analysis, and integration. (The "Preliminary Experiments" section provides additional details about the application of the dashboard.) Specifically, the dashboard allows users to:

- *Analyze* an arbitrary scenario represented by one or more descriptive files.
- Employ and *compare* the performance of automated control algorithms.
- *Execute* scenario simulations in batch-mode ensembles of experiments.
- *Incorporate* physical and virtual vehicles (digital twins).

Graphical User Interface. The dashboard graphical user interface (GUI) is written in Python using the wxWidgets library. The Panda3D game engine provides display graphics and performs computations involving collisions, distance from terrain, etc. Since both wxWidgets and Panda3D are platform-independent, the resulting dashboard can run on Windows, MacOS, and Linux.

The dashboard's specific design goal is to allow existing algorithms (e.g., intelligent UAV flight sequencing code) to be run as is, without the need for translation to Python code to be incorporated into the dashboard. This capability results from the ability of the dashboard to communicate with external programs using network socket communications. We have demonstrated this capability with algorithms running on machines different from the dashboard.

Critical to the analysis of a scenario is the ability to run experiments, test different algorithms, conditions, and probabilistic outcomes (e.g., various failure modes which occur with varying probabilities), or a combination. The dashboard is designed with a batch capability to run ensembles of experiments and maintain detailed logs to facilitate results post-processing. The dashboard implementation employs:

- A Panda3D graphics engine and code to invoke/terminate engine use.
- A scenario specification language and terrain description files.
- A library of standard views and panels used to configure the GUI.
- A mechanism through which users may supply external code (e.g., for terrain and model definition) and invoke it using the Python import mechanism. For example, the airport from our earlier C-130 perimeter surveillance scenario is now defined in a text file with the line Airport: python airport CreateAirport which imports the module defined in airport.py containing the function CreateAirport, which returns a set of graphical models representing the airport.

Exemplar Views and Panels. Examples of standard panels are shown in Figs. 10 and 11.

It should be noted that the views can include not only the actual objects in the experiment (terrain and actors) but computed artifacts which can dramatically increase the ability to interpret the scene. Figure 12 shows the layout of the smart dashboard that provides visualization, monitoring, and control of the scenario simulator. The problem being addressed is perimeter security of a C-130 aircraft

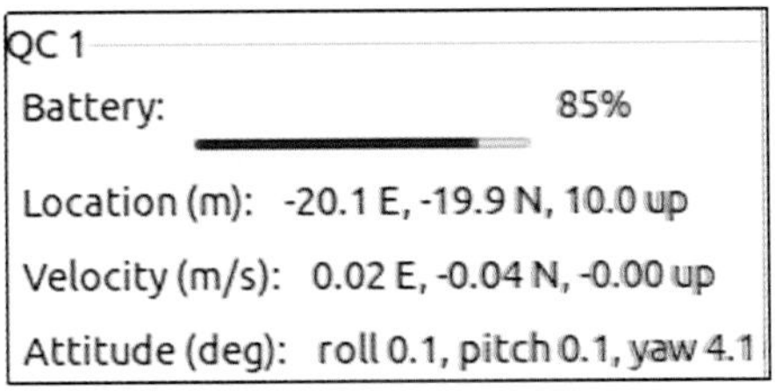

Fig. 10 Quadcopter indicator panel

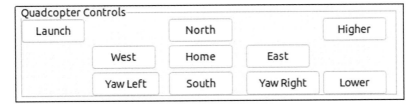

Fig. 11 Quadcopter control panel

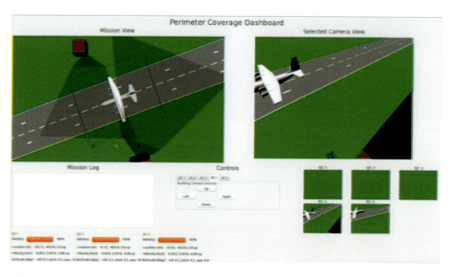

Fig. 12 Smart dashboard prototype

pursuant to its landing. This particular scenario was chosen because it affords the opportunity to demonstrate three resilience aspects of the solution: adaptive coverage; human in the loop decision-making; and collaboration among multiple agents. The problem is to control the collection assets (UAVs and fixed cameras) to optimize multi-sensor coverage of the aircraft perimeter.

The top left window presents the Mission View. The Mission View is a plan view of the aircraft perimeter and surroundings. Two buildings are visible in this view. Video cameras are mounted on each building with views of the aircraft from different directions. The shadows on the ground indicate the intersection of the viewing volume of each camera with the ground.

The top right window presents the selected camera views. Three quadcopters, assigned to this surveillance mission, are ready for launch. These quadcopters can be seen on the ground at the bottom center of the mission view. There are five cameras in all (three quadcopters QC 1–QC 3 and two building-mounted cameras BC 1 and BC 2). The views from each of the five cameras are shown at the lower right. The quadcopter cameras do not show anything because the quadcopters are still on the ground.

Below the mission view to the left is the mission log that presents all state changes for all the vehicles involved in the mission. The location, velocity, and altitude of each vehicle is shown right below the battery indicators.

The controls section in the bottom center of the simulator dashboard allows manual (human) control of the quadcopters and of the azimuth and elevation of the building cameras [9]. The selected camera view shows the field of view for the camera corresponding to the currently selected control Table (BC 1). The controls are associated with human-in-the-loop control or automated control algorithm. Once simulation begins, the perimeter coverage achieved appears below the control window. To the right of controls window are the camera views from each of the three quadcopters and the building-mounted video cameras.

Figure 13 shows the dashboard view during scenario simulation. In this view, one quadcopter is flying. Views from the building-mounted cameras is shown on the right of the dashboard. Right above that is the view from the single quadcopter that is airborne. The goal is to maximize coverage provided by this quadcopter in conjunction with the coverage provided by the building-mounted cameras. Here there is an opportunity to maximize perimeter coverage using different optimization algorithms.

This visualization shows the results of optimization with one quadcopter. In this view, the control "Automatic fitness optimization" is checked. A fitness function is a particular type of objective function that is used to summarize, as a single figure of merit, how close a particular solution is to achieving the set of aims. Fitness functions are used in genetic programming and genetic algorithms to guide simulations towards optimal solutions.

The quadcopter moves in a manner to maximize this fitness function. The fitness function is a sum of coverage functions from each of five cameras (three on the quadcopters and two on the building cameras) and each tile corner. The coverage of a tile corner from a camera is defined as the inverse of the distance from the tile corner to the camera, or zero if the corner is not in the camera's field of view. The total coverage of a single tile corner is clipped to an empirically chosen value, which represents the fact that additional coverage past a certain point is of no value.

It should be noted that this fitness function is chosen only to demonstrate reasonable collective behavior of the quadcopters in maximizing coverage. There is no claim that the chosen function is in any way the best one, and it leaves out realistic effects, for instance difference in vantage points or different spectral regions.

The coverage area here shows the coverage due to the quadcopter. The messages in the mission log show the quadcopter's autonomous search for the optimal location. In this case, the optimization method being used is "fitness function optimization."

Fig. 13 Dashboard with one quadcopter during optimization of fitness function

Figure 14 shows the optimal location for a single quadcopter achieved by maximizing the fitness function. This shows the single flying quadcopter has found the optimal position. Note that the quadcopter has climbed to 60 meters and yawed to −20 degrees to fit its field of view with the aircraft perimeter.

Figure 15 shows the optimal location for three quadcopters that maximize aircraft perimeter coverage. Optimality is considered to be when the quadcopters stop

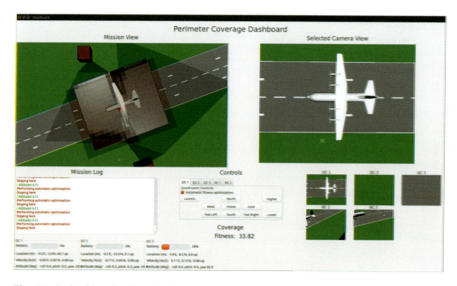

Fig. 14 Optimal location for a single quadcopter

Fig. 15 Optimal location for three quadcopters

moving, which occurs when no quadcopter finds no way to move that increases the fitness function. There is no guarantee that the maximum so found is global.

In this instance, the views from the three quadcopters, which are airborne can be seen in the three panes on the right-hand side of the dashboard. Each view shows the aircraft from the perspective of the corresponding quadcopter. They have deliberately separated to increase the quality of coverage for the entire perimeter. Note that the selected quadcopter has rotated its field of view to concentrate on the east end of the perimeter.

Figures 16 and 17 show the terrain for the perimeter surveillance scenario and the computed artifacts including the viewing volumes of the cameras.

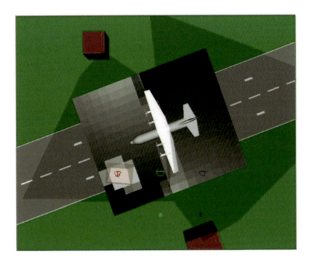

Fig. 16 Plan view of terrain. This is a nadir-pointed view with perspective

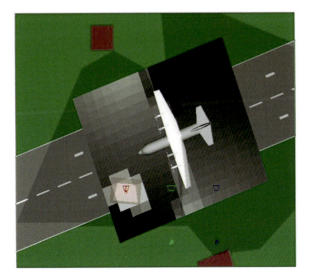

Fig. 17 Plan view of terrain – a nadir-pointed view orthographic projection

The views can include not only the actual objects in the experiment (terrain and actors) but computed artifacts which can dramatically increase the ability to interpret the scene. For instance, Fig. 16 shows the terrain for the perimeter surveillance scenario, the computed artifacts include the viewing volumes of the cameras. Nadir is the local vertical direction pointing in the direction of the force of gravity at that location.

Figure 17 is an orthographic projection of a nadir-pointed view. Orthographic projection is a means of representing 3D objects in two dimensions. It is a form of parallel projection, in which all the projection lines are orthogonal to the projection plane, resulting in every plane of the scene appearing in affine transformation on the viewing surface.

Preliminary Experiments

The initial set of experiments with the testbed explored the creation and update of a UAV digital twin using data from the operation of the real-world counterpart (Fig. 18). In the experiment, the contextual model of the system is used to demonstrate the value of digital twins (i.e., modifying the model during system operation in the real world). The contextual model captures the state of the quadcopters and external entities present in the system's environment as well as the physical properties such as size, shape, location, and velocity. Properties can be owned by entities; the instantiation of properties can be considered in the rest of the model views.

The experimentation setup employed two UAVs operating within an indoor maze environment. Each UAV system consists of a sensing, communication, battery, propulsion, and control subsystem. The structured indoor maze environment,

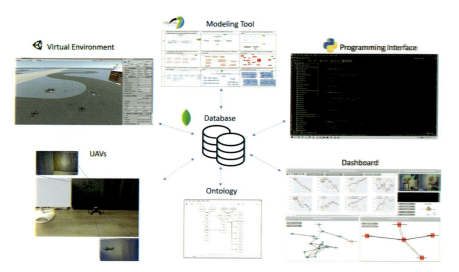

Fig. 18 Experimentation setup

comprising static and dynamic entities, is subject to external forces such as wind. The mission of the UAV system is to search for a predefined object in the environment. The UAVs communicate the external visual and acoustic data, internal subsystem information related to battery subsystem health, internal subsystem temperature, external temperature altitude, location, communication bandwidth, and speeds. The data are transferred to the data storage system through a communication network for preprocessing. The inference/analytics server performs data collection, integration, filtering, feature extraction, transformation, and analysis. Figure 19 presents the process of clustering static and dynamic entities in the environment. The model is updated using the clustering analysis. The new element is updated in the context model for a dynamic entity in the environment.

Each UAV has an onboard camera. The video stream from each UAV is converted into frames of fixed size, color threshold, and adjustable contrast values. Video frames are calibrated for distortion and then passed through a trained Recurrent Convolutional Neural Network (RCNN) model, where the objects in the video feed are identified and tagged. The RCNNs are commonly used machine learning models for object detection. Velocity vectors of objects are evaluated based on the movement of objects in the frames and UAV velocity vectors.

Objects with similar characteristics are clustered into the same group. The contextual model of the system is continuously updated based on results from the analysis.

The contextual model is mapped to SysML model which is integrated to the 3D virtual environment. To map data to the SysML model and map data from the SysML model, we used the Python XMI translation tool discussed earlier.

Objects in virtual environment model are stored in an asset container. The structured packaging of SysML model entities facilitates the extraction of model elements. Figure 20 presents the portion of the model update architecture. When a new entity is updated in the context model, the consequent Unity 3D model is also updated. As shown in Fig. 20, objects from the environment are extracted from the contextual model. These block names are then matched with existing 3D objects in the repository. When a match is found, the object is duplicated from the repository to the asset container of the 3D virtual environment. When a matching 3D object is not present in the repository, the translation program creates an empty object in the asset container.

Data is collected continuously, and the model once constructed is updated continually. Dealing with a massive influx of data requires prior data filtering and processing. Our approach to containing the combinatorial explosion in the state space is to employ heuristics to filtering during simulated or real-world system operations. Making model updates is challenging. Specifically, system representation needs to be sufficiently flexible (and extensible) to incorporate new information and changes. Observations require the capability to capture and store data in digital repositories to support decision-making. This requirement, in turn, creates a need for the system architecture to have a rigorous perception subsystem. We found that to make modeling "closed-loop," two aspects are essential (a) physical or virtual sensors during system prototyping, development, integration, testing, operation, and maintenance; and (b) rudimentary testbed instrumentation for data collection.

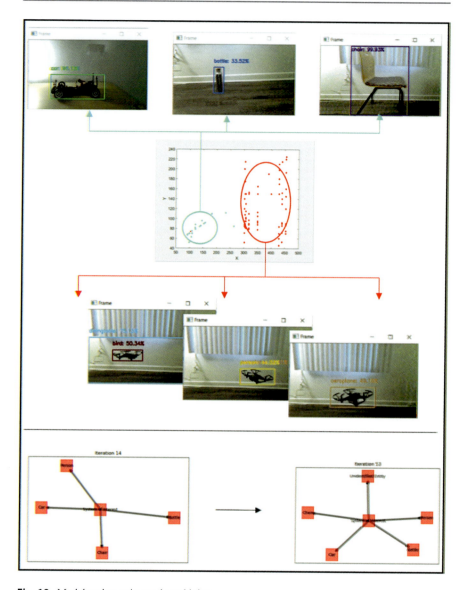

Fig. 19 Model update using real-world data

Lessons Learned

Several important lessons were learned from the implementation of the testbed prototype. First, a minimal testbed ontology [9] with essential capabilities is quite useful to begin initial experimentation with models and algorithms. Second, it is possible to adapt system model complexity to scenario complexity and minimize computation load. For example, for simple navigation scenarios, a finite state

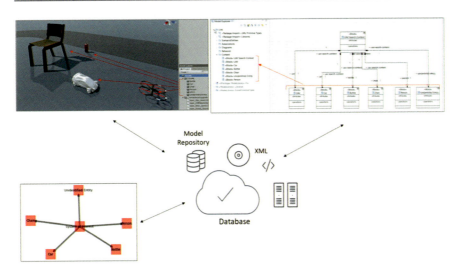

Fig. 20 Mapping of data within digital twin

machine (FSM) can suffice for vehicle navigation. However, as the navigation scenario gets more complicated (e.g., poor observability, uncertainties in the environment), more complex probabilistic system models such as POMDP can be employed to cope with environment uncertainty and partial observability. The value of probabilistic modeling becomes evident when operating in complex uncertain environments. Third, when it comes to system development, it pays to start off with simple scenarios and ensure that basic navigation requirements are met. Then, progressively complicate the scenarios and employ more complex models while ensuring satisfaction of safety and resilience requirements. The most straightforward way to implement this strategy is to control the complexity of the operational environment by imposing constraints and then systematically relaxing them. Therefore, after the simple model has been shown to satisfy safety requirements, constraints can be systematically relaxed (e.g., no obstacles, fixed obstacles, moving obstacles, partial observability) to create more complex navigation scenarios. In the latter case, more complex system models can be employed, verified, and validated with respect to safety and resilience requirements. Fourth, system and environment models can be reused, thereby reducing development time. To facilitate model reuse, the models can be metadata-tagged with usage context. Then contextual similarity between a problem situation and system models can be employed to determine suitability of a particular system model for reuse in a specific context. This reuse feature can accelerate experimentation and development. Fifth, a smart, context-sensitive, scenario-driven dashboard can be used to dynamically adapt monitoring capability and dashboard display to maximize situation awareness with a manageable cognitive load. To this end, the military's familiar mission-enemy-troops-terrain-time available-civilian (METT-TC) construct was used as the underlying context ontology [10]. Based on the prevailing context, either the entire set or a subset of variables in METT-TC are needed to characterize context. The flexible architecture of the scenario-driven dashboard can support both human-in-the-loop and autonomous operations. Importantly, it provides a convenient and cost-effective environment to try out different augmented intelligence concepts [8]. Sixth, a hybrid, distributed simulation capability enables integration of virtual simulation with real-world systems (e.g., unmanned aerial vehicles or UAVs). This integration, in turn, enables the creation of digital twins which can enhance system verification and validation while helping MBSE extend system life cycle coverage [4]. Furthermore, by assuring individual control of simulations, the testbed offers requisite flexibility in experimentation with system/system-of-system simulation. Also, by allowing different models (e.g., scenario model, threat model) to run on different computers, simulation performance can be significantly increased.

Summary

This chapter presented the system concept, architecture, prototype implementation, and capabilities of a MBSE innovation testbed. The testbed allows innovators to experiment with different models, algorithms, and operational scenarios, including edge cases.

The testbed supports enables exploration and exploitation of different modeling and simulation capabilities during system development and experimentation. Also, it allows experimentation with different human-machine teaming constructs and machine learning concepts. The chapter introduced the state-of-the-art in MBSE and needed capabilities, followed by an overview of the MBSE innovation testbed. Then, the enabling technologies were presented, including the testbed ontology, key elements of the software environment, including models, algorithms, repositories, simulation platforms, and hardware and connectors. Thereafter, the MBSE testbed concept is presented in detail, starting with the capabilities of the testbed, the logical architecture of the testbed prototype, the distributed computing environment of the testbed, and its support for virtual and hybrid simulations, and the key benefits of the testbed. The key capabilities of the testbed include system modeling and verification, rapid scenario authoring, model and scenario refinement, repository, multi-perspective visualization, and experimentation support. An illustrative example (i.e., aircraft perimeter security) is then described. The example is followed by a detailed description of scenario characterization, dashboard implementation, and exemplar views and panels within the context of the aircraft perimeter security problem. Thereafter, a series of dashboard capabilities are presented, including context-specific visualization, automated and human-in-the-loop optimization of perimeter coverage using multi-UAV teams. The chapter concludes with lessons learned.

Future advances in the testbed include more extensively leveraging formal ontologies and metamodel [6, 7] to guide systems integration and human-systems integration [11], formalizing the modeling of digital twins for both products and processes [2, 12], incorporating ontology-enabled artifact reuse [13, 14], expanding model-based verification [15], integrating reinforcement learning into system modeling to cope with incompleteness and uncertainties in knowledge about the system and the environment [4], increasing support for interoperability [16], distributed simulation standards (i.e., IEEE 1278.2-2015) [17], and collaborative workflow management [18].

Cross-References

▶ Model-Based System Architecting and Decision-Making
▶ Overarching Process for Systems Engineering and Design
▶ Problem Framing: Identifying the Right Models for the Job

References

1. Madni AM, Sievers M (2018) Model-Based Systems Engineering: Motivation, Current Status, and Research Opportunities. Systems Engineering, special 20th anniversary issue 21(3)
2. Madni AM, Madni CC, Lucero D (2019) Leveraging Digital Twin Technology in Model-Based Systems Engineering. MDPI Systems, special issue on Model-Based Systems Engineering
3. Siaterlis C, Genge B (2014) Cyber-Physical Testbeds. Communications of the ACM, 57(6): 64-73
4. Madni AM (2021) MBSE Testbed for Rapid, Cost-Effective Prototyping and Evaluation of System Modeling Approaches. Applied Sciences Journal
5. Zeigler BP, Mittal S, Traore MK (2018) MBSE with/out Simulation: State of the Art and Way Forward. Systems, 6(4):40:1–18
6. Madni AM, Sievers M, Purohit S, Madni C (2020) Toward a MBSE Research Testbed: Prototype Implementation and Lessons Learned. IEEE SMC International Conference, Toronto, Canada, Oct 11–14
7. Madni AM, Sievers M (2014) SoS Integration: Key Considerations and Challenges. Systems Engineering 17(3):330-347
8. Madni AM (2020) Exploiting Augmented Intelligence in Systems Engineering and Engineered Systems. INSIGHT Special Issue, Systems Engineering and AI, March
9. Madni AM, Erwin D, Sievers M (2020) Architecting for Systems Resilience: Challenges, Concepts, Formal Methods, and Illustrative Examples. MDPI System, January 19
10. Madni AM, Madni CC (2004) Context-driven Collaboration During Mobile C^2 Operations. Proceedings of The Society for Modeling and Simulation International 2004 Western Simulation Multiconference, January 18–22, San Diego, CA
11. Madni AM (2010) Integrating Humans with Software and Systems: Technical Challenges and a Research Agenda. Systems Engineering, 13(3):232–245, Autumn
12. Kapteyn MG, Pretorius VR, Wilcox KE (2021) A Probabilistic graphical model for enabling predictive digital twins at scale. Nature Computational Science, 1:337–347 www.nature.com/natcomputsci
13. Trujillo A, Madni AM (2020) An MBSE Approach Supporting Technical Inheritance and Design Reuse Decisions. AIAA ASCEND Conference, Designed to Accelerate Our Off-World Future, Nov 16–18
14. Trujillo A, Madni AM (2020) Exploration of MBSE Methods for Inheritance and Design Reuse in Space Missions, 2020 Conference on Systems Engineering Research, Redondo Beach, CA, Mar 19–21
15. Engels G, Kuster JM, Heckel R, Lohmann M (2003) Model-Based Verification and Validation of Properties, Elsevier Science B. V.
16. Chapurlat V, Daclin N (2012) System interoperability: definition and proposition of interface model in MBSE context. IFAC Proceedings 45(6); 1523-1528
17. Bridges S, Zeigler BP, Nutaro J, Hall D (2012) Evolving Enterprise Infrastructure for Model and Simulation-Based Testing of Net-Centric Systems. ACIMS, Arizona State University

18. Bretz L, Tschirner C, Dumitrescu R (2016) A concept for managing information in early stages of product engineering by integrating MBSE and workflow management systems. 2016 IEEE International Symposium on Systems Engineering (ISSE), pp. 1–8.
19. Schulman, J., Wolski, F., Dhariwal, P., Radford, A., & Klimov, O. (2017). Proximal policy optimization algorithms. arXiv preprint arXiv:1707.06347.
20. Haarnoja, T., Zhou, A., Hartikainen, K., Tucker, G., Ha, S., Tan, J., ... & Levine, S. (2018). Soft actor-critic algorithms and applications. arXiv preprint arXiv:1812.05905.
21. Torabi, F., Warnell, G., & Stone, P. (2018). Behavioral cloning from observation. arXiv preprint arXiv:1805.01954.
22. Ho, J., & Ermon, S. (2016). Generative adversarial imitation learning. arXiv preprint arXiv:1606.03476.
23. Juliani, A., Berges, V., Teng, E., Cohen, A., Harper, J., Elion, C., Goy, C., Gao, Y., Henry, H., Mattar, M., Lange, D. (2020). Unity: A General Platform for Intelligent Agents. arXiv preprint arXiv:1809.02627. https://github.com/Unity-Technologies/ml-agents.

Dr. Azad Madni is a member of the National Academy of Engineering, a University Professor, and the Northrop Grumman Foundation Fred O'Green Chair in Engineering Professor of Astronautics and of Aerospace and Mechanical Engineering in the University of Southern California. He also has a joint appointment in the Sonny Astani Department of Civil and Environmental Engineering. He is the Executive Director of USC's Systems Architecting and Engineering Program and the founding director of the Distributed Autonomy and Intelligent Systems Laboratory. He is the founder and CEO of Intelligent Systems Technology, Inc., a hi-tech company that conducts research and offers educational courses in intelligent systems for education and training. He is the Principal Systems Engineering Advisor to The Aerospace Corporation. He received his Ph.D., M.S., and B.S. degrees in Engineering from UCLA. His recent awards include *2021 INCOSE Benefactor Award, 2021 IEEE AESS Judith A. Resnik Space Award, 2020 IEEE SMC Norbert Wiener Award, 2020 NDIA's Ferguson Award* for Excellence in Systems Engineering, *2020 IEEE-USA Entrepreneur Achievement Award, 2019 IEEE AESS Pioneer Award, 2019 INCOSE Founders Award, 2019 AIAA/ASEE Leland Atwood Award, 2019 ASME CIE Leadership Award, 2019 Society for Modeling and Simulation International Presidential Award, and 2011 INCOSE Pioneer Award.* He is a Life Fellow/Fellow of IEEE, INCOSE, AIAA, AAAS, SDPS, IETE, and WAS. He is the co-founder and current chair of IEEE SMC Technical Committee for Model Based Systems Engineering. He is the author of *Transdisciplinary Systems Engineering: Exploiting Convergence in a Hyper-Connected World* (Springer 2018). He is the co-author of *Tradeoff Decisions in System Design* (Springer, 2016).

Dr. Dan Erwin is a Professor and Chair of Astronautical Engineering, and an Associate Fellow of AIAA. He is also the faculty member who oversaw the all-student research team that set the student altitude record for rockets to pass the Karman line into outer space. He is the Co-Director of USC's Distributed Autonomy and Intelligent Systems Laboratory. His awards include the 2017 Engineers' Council Distinguished Engineering Project Achievement Award, 2016 Engineers' Council Distinguished Engineering Educator Award, 2006 USC-LDS Student Association Outstanding Teaching Award, 1995 USC School of Engineering Outstanding Teaching Award, and 1993 TRW, Inc. TRW Excellence in Teaching Award. He received his B.S. in Applied Physics from California Institute of Technology and his M.S. and Ph.D. in Electrical Engineering from the University of Southern California.

Transitioning from Observation to Patterns: A Real-World Example

23

S. Russell, B. Kruse, R. Cloutier, and D. Verma

Contents

Introduction and Context	706
State of the Art	706
Background	706
Best Practice Approach	708
Energy Storage System	708
Pattern Mining	709
Using Patterns	719
Chapter Summary	720
Cross-References	720
References	721

Abstract

The systems engineering community of practitioners, researchers, and educators are in the process of transitioning to model-based systems engineering within a broader context of digital engineering. Industry is investing significantly to support this transition to ensure an appropriate and responsive digital engineering

---

S. Russell
Johnson Space Center, NASA, Houston, TX, USA
e-mail: samuel.p.russell@nasa.gov

B. Kruse
e:fs TechHub GmbH, Gaimersheim, Germany
e-mail: benjamin.kruse@efs-auto.com

R. Cloutier
University of South Alabama, Mobile, AL, USA
e-mail: rcloutier@southalabama.edu

D. Verma (✉)
Stevens Institute of Technology/SERC, Hoboken, NJ, USA
e-mail: dverma@stevens.edu

© This is a U.S. Government work and not under copyright protection in the U.S.; foreign copyright protection may apply 2023
A. M. Madni et al. (eds.), *Handbook of Model-Based Systems Engineering*, https://doi.org/10.1007/978-3-030-93582-5_76

infrastructure to support research, engineering, and development teams. While this infrastructure must contend with the requisite computational resources, sufficient attention must also be given to tools, methods, and the underlying systems engineering and architecting methodology. An element of this methodology is a well-curated library of architectural and design patterns available to engineers. These patterns represent deep domain knowledge and experience within an application or technology domain if done right. Therein lies the focus of this chapter: How to best leverage patterns and a pattern language as a methodology for capturing and documenting tacit domain knowledge inherent in any high-performance engineering organization? This chapter outlines an approach to develop design patterns in a technology domain.

Keywords

Patterns · Pattern mining · Pattern language · Energy storage · Systems engineering

Introduction and Context

As part of an overall strategy and infrastructure for model-based systems engineering and digital engineering, the use and management of architectural and design patterns are emerging as a critical enabler. When addressing systems problems of increasing complexity, this approach offers a means of reducing the cognitive load associated with describing complex systems [1–3]. A pattern is an individual construct that represents an observed and proven regularity in a system for consumption and application by the pattern user [3–5]. By identifying patterns, and the complementary relationships between patterns within an observed system, the Systems Engineer gains insight into the architectural context and implicit knowledge used in developing the system [1, 2, 4, 5]. Pattern-Based Systems Engineering (PBSE) is the application of Model-Based Systems Engineering (MBSE) techniques in the management and configuration of system patterns [4]. However, the question remains, "how does a Systems Engineer identify and document a design pattern?" This chapter summarizes applying a pattern identification and refinement framework in a technical domain using an actual example.

State of the Art

Background

A fundamental tenant of patterns is that patterns are identified or mined and not created [1–3]. A pattern is a type of heuristic, an efficient mental process, or shortcut that reduces complex problem solving to a simpler judgmental operation [6–9] by encapsulating and hiding the unnecessary details [10]. Patterns differ from engineering drawings. Like a heuristic, a pattern considers regularities in the system, such as proximity between objects, as part of the engineering solution [5, 8]. Created using deep domain and technical knowledge [11, 12], patterns describe observed solutions as a balance of interacting forces (or interests) operating within the system [1, 3, 11]. Patterns are documented textually using the Pattern Form (a template) [1, 13], or graphically using MBSE (SysML or IDEF0) tools [4, 14, 15]. A fully developed pattern describes the context, solution, attributes, and applying patterns may provide potential solutions to an engineering problem [16]. The holistic nature of patterns affords the user insight into the expert knowledge used in developing the practical design solution and provides a basis for an architecture description of the observed system [1, 2, 5, 17].

Design patterns describe and relate reusable objects of design experience [14, 17] as "an arrangement of parts or elements that together exhibit behavior or meaning that the individual constituents do not" [18]. The design pattern is constrained by a boundary or an observable field of force that isolates the points of energy use in the observed design from the surroundings [1, 2, 19–21]. A design pattern captures a more abstract view of the interacting elements contained within the system boundary in a usable form [5, 22, 23] for configuration and management in Pattern-Based Systems Engineering (PBSE) [4]. The pattern-mining process offers the systems engineer insight into the architectural themes balanced by design and how the designer achieved the observed balance.

Pattern mining is the process of identifying expert knowledge used in problem-solving. Initially proposed to describe the design of buildings and towns [3], the technique has proven useful in such fields as software [14], human behavior [12], and architecture [1, 2]. Leitner [11] and Iba [12] proposed frameworks for the pattern-mining process. Both frameworks use a three-step candidate identification process, candidate consolidation and refinement into patterns, and pattern documentation. The principal difference between the two methods is preexisting knowledge of the system. Iba's methodology uses an interview process for pattern refinement [12], while Leitner's framework requires prerequisite knowledge of the system of study [11]. Both methods identify tacit knowledge by studying a solution to a problem in a context. Tacit knowledge is expert knowledge that is known but difficult to describe by specification or standard [9, 12]. The interview method assumes the interviewer has no prior knowledge of the solution and must deduce the problem-solving logic through a series of conversational interviews [24]. Conversely, the prerequisite knowledge method assumes that the solution designer performs candidate identification [11]. Once pattern candidates are identified, the consolidation and refinement process and the pattern documentation method are common between the two frameworks. The example in this chapter relies on preexisting knowledge and uses the approach from Leitner [11].

The following example represents the development of three electrical energy storage devices designed to tolerate catastrophic electrochemical cell failure by preventing the loss of a single cell from propagating throughout the design. Each of the three batteries is energy, mass, and volume constrained and supported human spaceflight in low earth orbit. Early prototypes designed according to heritage methods [25] demonstrated a tendency toward thermal runaway failure propagation [26], necessitating the development of a new approach. Preventing cell failure propagation required developing an understanding of battery failure modes, the energy transfer mechanisms responsible for failure propagation, and how new components or packaging considerations can prevent thermal runaway. Approached serially with increasing design complexity, each battery presented unique design challenges. Device performance and the development time line improved with each development activity due to growing domain knowledge of construction materials, manufacturing methods, energy management techniques, and understanding the relationship between components, component arrangement, and the observed failure response.

The following sections demonstrate the utility of patterns in capturing domain knowledge by applying a pattern identification and refinement framework to the results of the multiyear, sequential design project and capturing the resulting pattern in a digital engineering format for further exploration, reuse, and archival. The pattern-mining process encourages the expression of architectural and design concepts. SysML use case diagrams, block definition diagrams, internal block diagrams, and activity diagrams provide model-based descriptions of the resulting concepts. Once developed, patterns can be an essential part of a growing body of reusable domain knowledge within an organization, as a critical element of a digital engineering/model-based systems engineering environment.

Best Practice Approach

Energy Storage System

The three batteries of this study replace four existing batteries with a modern lithium ion technology while remaining compatible with a proven battery management schema. Each energy storage system converts power received from the battery management system to chemical energy for storage and, when installed in the application, provides stored energy as electrical power. Each battery is designed to tolerate handling, storage, transport, and usage environments. The operational model is the same for the three batteries. A battery management system located in the habitable environment of a spacecraft provides storage, charge, discharge, and maintenance when the battery is not in use. When needed, each is charged, removed from the battery management system, and installed in the intended application, or load. The load provides the thermal environment during use, and should inadvertent contact occur between the load and structure, the load transmits live loads to the battery. When usage is no longer required, each battery is returned to the battery management system for storage and maintenance. During storage and maintenance, the battery management system provides the thermal environment for the energy storage system. To maintain operational reliability of the heritage systems, each battery provides the telemetry necessary for power management and contains no internal power management functionality beyond that required for safe handling and transport. As energy storage capability diminishes with long-term use, each battery

23 Transitioning from Observation to Patterns: A Real-World Example

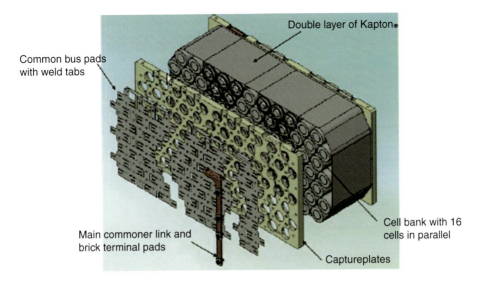

Fig. 1 Heritage cell module

is designed to tolerate multiple refurbishments should the asset be returned for postmission processing.

The energy storage system reflects the heritage approach of a cylindrical lithium ion cell packaged into a cell module within a reusable housing. The module includes electrochemical cells suspended between two structural and nonconductive capture plates. Bonded to the terminal end of each cell, the capture plates secure the bus material necessary for connecting cells in parallel. Energized surfaces in the module are covered in insulating materials to prevent inadvertent contact with the structural shroud of the electrically floating power supply. Paralleled cells are connected to the housing-mounted electrical interface by a wiring harness. The heritage approach to module design is shown in Fig. 1 [26].

Mitigating the risk of single cell thermal runaway required interrupting the unabated flow of thermal energy between failed and exemplar cells and preventing the release of combustion materials to the external system [26]. The batteries of this study employed a common design strategy to develop three unique solutions to achieve this result. Solution selection was based on external constraints including stored energy, volume, mass, and heat transfer. Each design considered the interstial volume between separated cells, unoccupied volume between module and housing, orientation of the cell vent mechanism, and filtration of materials released from the cell vent [26].

Pattern Mining

The pattern mining process begins by identifying a list of pattern candidates, progresses through candidate refinement, and ends with a holistic description of the observed design. Refinement is the process of renaming, dividing, consolidating, adding, or removing candidates. A list of pattern candidates is prepared based on intrinsic knowledge of the design, and each candidate is assigned to one of three hierarchical categories, architecture, design, or solution, in accordance with the stakeholder perspectives of the Enterprise Architecture ontology [27]. The initial pattern candidates and hierarchial categorization are shown in Table 1.

The pattern candidates of Table 1 reflect a strong bias toward physical structure as is expected of the expert knowledge methodology. In this example, physically replaceable items such as conductor, insulator, and cell are categorized as solutions, elements requiring adaptation to chosen items such as module and housing are assigned design, and candidates required by external systems are assigned architecture. For example, the battery designer selects a cell and heat transfer method, adapts the module and housing to accept these solutions, and then incorporates the current interruption device required by the external system. Once candidates are identified and categorized, the refinement process begins.

Using a graphical representation, candidates of Table 1 are arranged into a hierarchy with architectural candidates on the top layer, design candidates in the middle, and solution candidates on the bottom. Beginning with solutions, pattern candidates are rearranged, renamed, created, or deleted to ensure the nondirectional relationship lines between solution and design candidates are straight and do not cross other relationship lines. An example of the resulting network diagram is shown in Fig. 2.

Table 1 Initial pattern candidate list

Pattern	Architecture	Design	Solution
Battery	X		
Prevent failure propagation	X		
Circuit protection	X		
Current collection		X	
Circulating current	X		
Cell		X	
Conduction			X
Insulation			X
Structure	X		
Housing		X	
Module		X	
Insulator			X
Heat spreader			X
Conductor			X
Ventilation	X		
Filter			X
Flow restriction		X	
Thermal protection		X	

23 Transitioning from Observation to Patterns: A Real-World Example

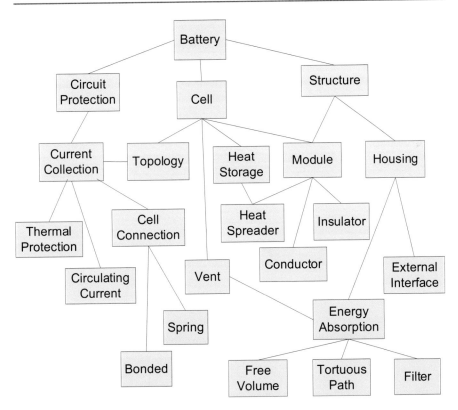

Fig. 2 Result of the initial candidate refinement cycle

Although the pattern is incomplete, Fig. 2 illustrates the refinement process as new candidates are identified and initial candidates renamed. During this phase, the pattern mining practitioner will begin to identify logical aspects of the design. For example, a safe, nonpropagating battery design requires managing the flow of combustion products and materials relieved from the failed cell, and encouraging energy transfer between this flow stream and the physical objects acting to impede ejecta flow. While the physical features responsible for this behavior are identified in the candidates' ventilation, filter, and flow restriction of Table 1, inspection of Fig. 2 shows increasingly abstract notions such as energy absorption and tortuous path are beginning to emerge. In this way, pattern mining encourages the identification of expert knowledge.

Correcting the deficiencies of Fig. 2 begins with an inspection of highly related design candidates. By considering the functional performance achieved by the candidate to candidate relationships, and describing this functionality using action words, architectural and design candidates are refined. At this point, the pattern mining practitioner will have developed a new conceptual abstraction of the design that is readily resolvable into a diagram such as that of Fig. 3.

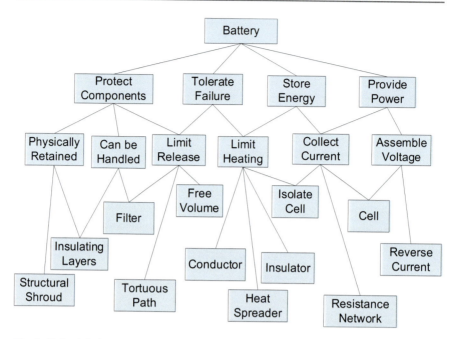

Fig. 3 Refined design pattern

The design pattern of Fig. 3 complies with the pattern mining guidelines and completes the pattern mining process. The network diagram communicates battery design heuristics, demonstrated solutions, and illustrates how the relationships between these heuristics and solutions achieve the observed architecture. For example, the battery tolerates failure by limiting heat transferred between cells and released from the housing. Heat transfer is managed by insulator, heat spreader, and conductor solutions and by sharing the collect current design of the energy storage function to electrically isolate the failed cell. The battery limits the release of cell failure to the external system through a combination of free volume, tortuous flow path, and a filter. The filter acts, along with electrically bonded structural materials, to ensure the battery can be handled without harming the user or external system. This example highlights the utility of patterns in expressing the functional capability of the observed system and deducing tacit knowledge gained by the designer during a development activity.

The usability of the mined pattern is improved with the aid of Model-Based Systems Engineering. Modeled using SysML [28], the interactions between the external system and the energy storage system are shown in the use case diagram of Fig. 4.

In this figure, external system includes the loads, charging system, user, and the environment. The diagram shows that catastrophic failure is an extension of the energy storage activity and that this activity is resolved by the energy storage system without significant interaction with the external system. The use case refines the

23 Transitioning from Observation to Patterns: A Real-World Example

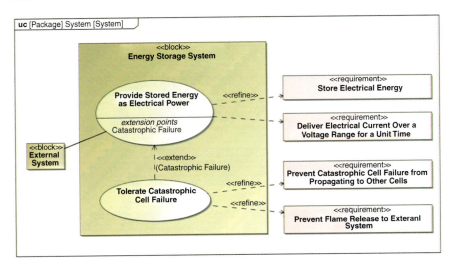

Fig. 4 Use case diagram for the system of interest

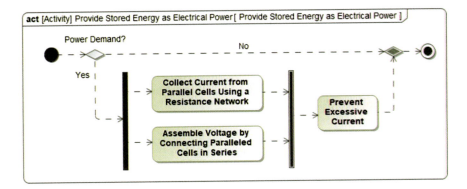

Fig. 5 Provide stored energy as electrical power

functional requirements representing the mined architectural patterns of Fig. 3 by owned behaviors for each activity as indicated by the rake. The behaviors are described in the following figures.

The activity diagram of Fig. 5 describes how the battery behaves when providing stored energy as electrical power. During periods of power demand, the current collection and voltage assembly actions operate in parallel to generate available power, and a single current interruption device serves to isolate the power source if demand exceeds battery capability. The figure also shows that no current flows in periods of no current demand. This behavior reflects nominal battery performance and provides both rationale and context for the resistance network, voltage assembly, and current collection elements of the mined design pattern.

The tolerate catastrophic cell failure activity of Fig. 6 explains how the design responds to a single cell-catastrophic failure by preventing unmitigated cell to cell heating, interrupting locations of high current flow within the resistance network, and reducing the energy of vented materials. Initiated only by cell failure, this behavior can occur anytime after battery assembly.

The functional performance of the battery is achieved by design. Transitioning the functional description of Fig. 3 to a logical description results in Fig. 7.

The block definition diagram of Fig. 7 includes the six designs of Fig. 3 renamed using descriptive titles. For example, Assemble Voltage is retitled as Voltage Assembly, Limit Heating became Thermal Management, Limit Release is Ejecta Management, and Can be Handled is Component Protection. The solution candidates of the

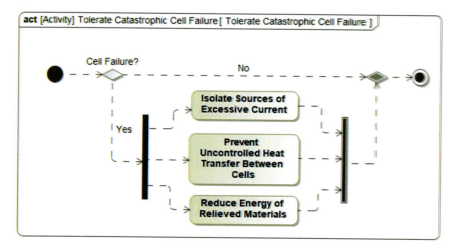

Fig. 6 Tolerate catastrophic cell failure

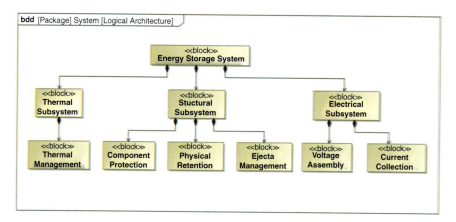

Fig. 7 Logical architecture of the energy storage system

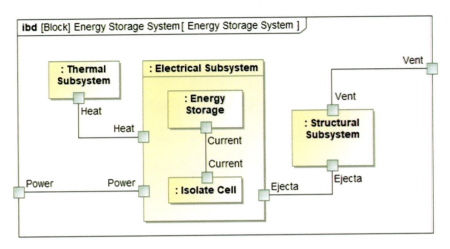

Fig. 8 Subsystem role in tolerating catastrophic cell failure

design pattern are included but not shown in this diagram. Adding flows of energy and materials between the subsystems of Fig. 7 results in an internal structure for the Energy Storage System of Fig. 8.

The internal block diagram of Fig. 8 shows how the subsystems contribute to the functional performance of the battery. The electrical subsystem shares power with the system interface and contains the energy storage and isolate cell properties of the voltage assembly and current collection design. During catastrophic cell failure, the electrical subsystem and the thermal and structural subsystems share heat and ejecta, respectively. The thermal subsystem resolves heat without effecting the system boundary, while the structural subsystem converts ejecta into lower-energy materials for venting to the external system.

The thermal and structural subsystems passively perform the required function using physical features of the system as reflected in the following activity diagrams.

The activity diagram of Fig. 9 describes how the insulation and heat spreader solutions of the mined design pattern transfer heat during cell failure. As this solution allocates heat across the other cells in the battery, a lower limit of applicability exists based on the amount of thermal capacity available, the environmental temperature, and the heat released by the failed cell. As these limits are approached, or for batteries capable of exchanging heat with the external system, a pure insulation or conduction option is more suitable.

The activity diagram of Fig. 10 explains the sequential performance of the structural subassembly in reducing ejecta energy during catastrophic cell failure. As gravimetric energy density improves in battery design, free volume is necessarily reduced, and path tortuosity and filter performance must increase to maintain desired performance. The activity sequence also shows the importance of the housing vent not only in preventing the release of energetic material, but also in preventing the intrusion of conductive particulate or liquid materials. The previous two examples

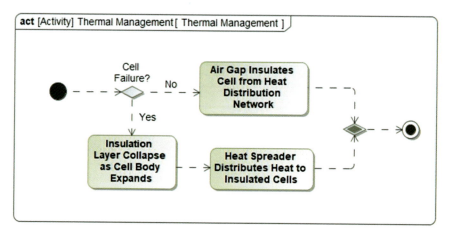

Fig. 9 Heat management during cell failure

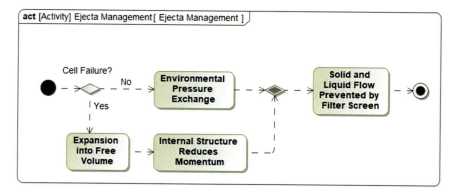

Fig. 10 Ejecta management during cell failure

illustrate the functional relationships between solution candidates of the design pattern in Fig. 3.

The design logic explored thus far is reflected in the physical system in terms of spatial arrangement and geometric dimensions of internal components and the properties of selected materials. The block diagram of Fig. 11 illustrates the physical architecture for the system of interest.

The physical system is organized into three subassemblies consisting of a protective structure, a replaceable battery module, and a harness. Only the module is replaced during refurbishment. The system structure is reflected in the following internal block diagrams.

The internal block diagram of Fig. 12 describes subsystem role in delivering stored energy as electrical power. The housing absorbs impact energy and provide vibration loads to the module while providing structural support for the harness. The harness serializes module current, to create power for delivery to the external system,

23 Transitioning from Observation to Patterns: A Real-World Example

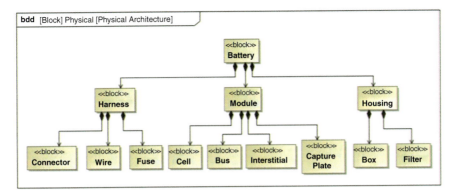

Fig. 11 Physical architecture of the system of interest

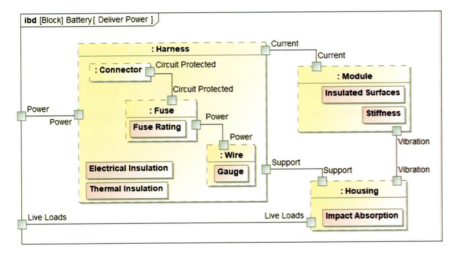

Fig. 12 Achieving power delivery through battery design

and incorporates both a fuse to protect the battery from external fault and an electrical connector for connecting to an external system. Select value properties are identified for each subsystem including wiring harness electrical and thermal insulation, fuse rating and wire gauge, structure impact absorption, module stiffness, and energized surface insulation.

The internal block diagram of Fig. 13 describes the role of the module elements in the deliver power use case of Fig. 4. Again, value properties are shown for parameters critical to achieving the desired functionality, including cell voltage and electrical and thermal capacity, interstitial mass and thermal diffusivity, capture plate electrical and structural properties, and resistance of the current collecting electrical bus. The current, load, heat, and dimensional relationships define the module value relationships of Fig. 12.

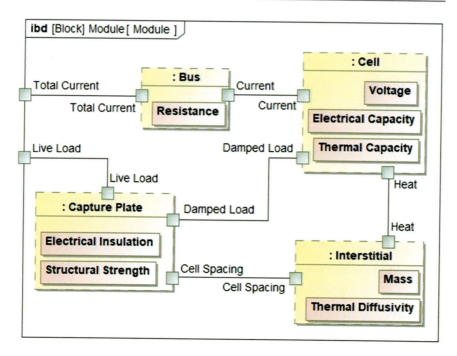

Fig. 13 Achieving power delivery through module design

The internal block diagram of Fig. 14 describes the relationships and critical parameters necessary for achieving the tolerate failure use case of Fig. 4. During failure, the module relieves ejecta to the housing and the housing subsystem tolerates the impulse and uses the physical obstruction of the module to reduce ejecta energy before filtering for the external system. Isolating the failure to a single cell requires failing the interconnecting electrical linkage with the parallel cell network, managing cell body heat using the interstitial, and redirecting ejected materials by the capture plate. As a subsystem, the module manages the location and direction of the ejecta provided to the housing for routing and filtering prior to release to the external system.

The resulting system model incorporates the design logic, functionality, and solution candidates expressed during the pattern mining activity. Since pattern mining was performed on a suite of batteries developments informed by heritage design principles, the resulting pattern represents a family or class of design pattern specific to a multicell, refurbishable battery that relies on heritage principles and customized external battery and environment management systems. Formalizing the relationships between the heritage battery design, external systems, and the environment is necessary for compliance with the pattern mining process [5]. However, the stand-alone model remains useful for exploring aspects unaffected by external relationship, such as material selection trade studies, tacit design

23 Transitioning from Observation to Patterns: A Real-World Example

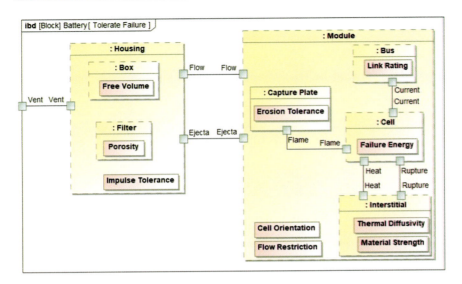

Fig. 14 Enabling the tolerate catastrophic cell failure functional requirement

knowledge communication, performance simulation, and verification and validation activity support.

Using Patterns

The patterns presented in this chapter are derived from a physical system and explained using network diagrams and SysML use case diagrams, activity diagrams, block definition diagrams, and internal block diagrams. The pattern mining activity used expert knowledge and demonstrated solutions to abstract architectural and logical design descriptions. The architectural view refines the specification for the system of interest and enables architectural modeling techniques such as the Functional Architecture for Systems [27]. The logical description provides a conceptual understanding of expert knowledge by describing how the solution manages the flow of information, material, and energy. Sharing this understanding can lead to new solution development or the identification of technological needs. For example, the conceptual understanding provided by this activity leads to performance improvements beyond this experiment. By allocating a portion of the particulate filtration functionality to the tortuous flow path, a solution with less internal volume and a lighter weight and simpler filter was achieved by incorporating mass retention features in physical elements located closer to the cell vent. The pattern mining and modeling activity can also benefit practical users by using parametric analysis to enable solution selection as a form of decision tree [29].

The key to using patterns is to recognize that the patterns are not completed designs. Instead, they represent the most common or most useful aspects of a design or operation and must then be adapted and refined for specific applications. Although the pattern in this chapter contains no relationships to external systems and is functionally incomplete [1, 5], the pattern mining activity and resulting model capture and describe tacit knowledge and provide a basis for knowledge sharing, architectural modeling, and practical decision-making.

It was stated in section "Background" that well-curated patterns become part of the reusable domain knowledge within the organization. Therefore, patterns should be readily available to the systems and design engineers as part of the digital engineering environment. They should include documentation that recommends usage and how they can and should be adapted and refined. Patterns are not cast in stone. Starting a design with proven approaches will always take less time than starting from scratch.

Patterns should be stored in a repository along with other SysML model fragments. When one identifies a pattern that can be used, the model file can be imported into the working model much like software engineers import libraries to use functions in those libraries.

Chapter Summary

This chapter introduces the concept of patterns and describes the process of applying patterns to the results of a sequential design development activity using an MBSE tool. The pattern mining process resulted in incorporating aspects of the design that were missing from the drawing tree but critical to design performance. Identifying and including the "missing" elements of the physical design provides insight into the expert knowledge used to solve the technical challenge of balancing the discrete energy flows resolved by the design.

Patterns offer the systems engineer a means of deriving architectural themes operating within a system, consolidating a family of related products into a singular descriptive diagram, and gaining a more holistic understanding of the observed system. The practitioner of patterns will grow familiar with recognizing the "missing" aspects of a design, understanding when a design deviates from a precedent design, and benefit from the expert knowledge gained by the product designer.

Cross-References

- ▶ Conceptual Design Support by MBSE: Established Best Practices
- ▶ MBSE in Architecture Design Space Exploration
- ▶ Pattern-Based Methods and MBSE
- ▶ Semantics, Metamodels, and Ontologies

References

1. R. J. Cloutier and D. Verma, "Applying the Concept of Patterns to Systems Architecture," *Systems Engineering*, vol. 10, no. 2, pp. 138-154, 2007.
2. J. O. Coplien, "Idioms and Patterns as Architectural Language," *IEEE Software*, vol. 14, no. 1, pp. 36-42, 1997.
3. C. Alexander, Notes on the Synthesis of Form, Cambridge MA: Harvard University Press, 1964.
4. W. D. Schindel and T. Peterson, "An Overview of Pattern-Based Systems Engineering (PBSE): Leveraging MBSE Techniques," INCOSE IS2013 Tutorial, Philadelphia, 2013.
5. C. Alexander, A Pattern Language, New York: Oxford University Press, 1977.
6. P. Todd, "Heuristics for Decision and Choice," in *International Encyclopedia of the Social & Behavioral Sciences*, Elsevier Ltd, 2001, pp. 6676-6679.
7. A. Tversky and D. Kahneman, "Judgment under Uncertainty: Heuristics and Biases," *Science*, vol. 185, no. 4197, pp. 1124-1131, 1974.
8. R. Hertwig and T. Pachur, "Heuristics, History of," *Psychology*, 2015.
9. Random House Living Dictionary Project, Webster's College Dictionary, New York: Random House, 1991.
10. G. Gigerenzer and W. Gaissmaier, "Heuristic Decision Making," *Annual Review of Psychology*, vol. 62, pp. 451-482, 2011.
11. H. Leitner, "Working with Patterns: An Introduction," in *Patterns of Commoning*, Amherst, Levellers Press, 2015.
12. T. Iba, "Pattern Language 3.0 and Fundamental Behavioral Properties," in *In Pursuit of Pattern Languages for Societal Change (PURPLSOC)*, Krems an der Donau, Austria, 2016.
13. H. Leitner, "A Bird's-Eye View on Pattern Research," in *Pursuit of Pattern Languages for Societal Change*, Berlin, 2015.
14. E. Gamma, R. Helm, R. Johnson and J. Vlissides, Design Patterns, Elements of Reusable Object-Oriented Software, Boston: Addison-Wesley, 1995.
15. R. Helm, "Patterns in Practice," New York, 1995.
16. R. Cloutier, G. Muller, D. Verma, R. Nilchiani, E. Hole and M. Bone, "The Concept of Reference Architectures," *Systems Engineering*, vol. 13, no. 1, pp. 14-27, 2010.
17. N. L. Kerth and W. Cunningham, "Using Patterns to Improve our Architectural Vision," *IEEE Software*, vol. 14, no. 1, pp. 53-59, Jan/Feb 1997.
18. H. Sillitto, J. Martin, D. McKinney, R. Griego, D. Dori, D. Krob, P. Godfrey, E. Arnold and S. Jackson, "Systems Engineering and System Definitions," Open Access, San Diego, CA, 2019.
19. H. Leitner, Pattern Theory, Introduction and Perspectives on the Tracks of Christopher Alexander, Helmut Leitner, 2007/2015.
20. C. Alexander, The Nature of Order: An Essay on the Art of Building and the Nature of the Universe, Book 1 - The Phenomenon of Life, vol. Edition 1, New York: Oxford University Press, 2002.
21. J. L. Henshaw, "Guiding Patterns of Naturally Occuring Designs," in *Pursuit of Pattern Languages for Societal Change*, Krems, 2016.
22. C. Alexander, S. Ishikawa and M. Silverstein, A Pattern Language, New York: Oxford University Press, 1977.
23. A. Mouasher and J. Lodge, "The Search for Pedagogical Dynamism – Design Patterns and the Unselfconscious Process," *Educational Technology & Society*, vol. 19, no. 2, p. 274–285, 2016.
24. T. Iba and T. Isaku, "A Pattern Language for Creating Pattern Languages: 364 Patterns for Pattern Mining, Writing, and Symbolizing," in *Proceedings of the 2016 Conference on Pattern Languages of Programs*, October 2016.
25. S. P. Russell, M. A. Elder, A. G. Williams and J. Dembeck, "The Extravehicular Maneuvering Unit's New Long Life Lithium Ion Battery and Lithium Ion Battery Charger," Anaheim, 2010.

26. C. J. Iannello, R. M. Button, E. C. Darcy, J. R. Graika and S. L. Rickman, "Assessment of International Space Station (ISS)/Extravehicular Activity (EVA) Lithium Ion Battery Thermal Runaway (TR) Severity Reduction Measures," NASA Engineering Safety Center, Hampton, VA, 2017.
27. J. A. Zachman, "A Framework for Information Systems Architecture," *IBM Systems Journal,* vol. 26, no. 3, pp. 276-292, 1987.
28. International Organization for Standardization, Information Technology - Object Management Group Systems Modeling Language (OMG SysML), Geneva, 2017.
29. A. J. Myles, R. N. Feudale, Y. Liu, N. A. Woody and S. D. Brown, "An introduction to decision tree modeling," *Journal of Chemometrics,* vol. 18, pp. 2785-285, 2004.

Samuel Russell is a Doctoral Candidate at Stevens Institute of Technology and a Project Manager and Systems Engineer at NASA Johnson Space Center responsible for the design, development, and sustaining of energy storage systems for human spaceflight. With more than 20 years of experience in propulsion and power systems, he managed the first deployment of lithium ion technology in critical human space application, served an active role advancing lithium ion battery design to reduce the risk of catastrophic system failure due to thermal runaway, and managed the development of the first propagation-resistant lithium ion batteries in human spaceflight. Sam holds a BS in Environmental Engineering and an MS in Materials Engineering from New Mexico Institute of Mining and Technology.

Dr. Rob Cloutier is a Professor, Systems Engineering Program Chair, and Graduate Director for the College of Engineering at the University of South Alabama (USA). He also maintains a healthy association with Stevens Institute of Technology. Rob currently has research projects with the US Navy, and with NASA Advanced Concepts Office. Dr. Cloutier is the editor in chief for the System Engineering Body of Knowledge (https://www.sebokwiki.org) which receives >35 k unique visitors and > 85 k page views per month. His interests include systems architecting, model-based systems engineering, and digital transformation. Before joining the USA, Dr. Cloutier was an Associate Professor and Director of Systems & Software Programs at Stevens Institute of Technology in Hoboken NJ. Prior to Stevens, he spent over 20 years at Lockheed Martin and The Boeing Company (where he was an Associate Technical Fellow). Professional roles included system architect, enterprise architect, and principal systems engineer. Dr. Cloutier served 8 years in the US Navy & Navy Reserve. He received his BS from the US Naval Academy, his MBA from Eastern University, and his PhD from Stevens Institute of Technology.

Dr. Dinesh Verma received PhD (1994) and MS (1991) in Industrial and Systems Engineering from Virginia Tech. He served as the Founding Dean of the School of Systems and Enterprises at Stevens Institute of Technology from 2007 through 2016. He currently serves as the Executive Director of the Systems Engineering Research Center/Acquisition Innovation Research Center (SERC/AIRC), a US Department of Defense-sponsored University Affiliated Research Center (UARC) focused on systems engineering research. During his 15 years at Stevens, he has successfully proposed research and academic programs exceeding $150 m in value. Prior to this role, he served as Technical Director at Lockheed Martin Undersea Systems, in Manassas, Virginia, in the area of adapted systems and supportability-engineering processes, methods, and tools for complex system development. His professional and research activities emphasize systems engineering and design with a focus on conceptual design evaluation, preliminary design and system architecture, design decision-making, life cycle costing, and supportability engineering. In addition to his publications, Verma has received three patents in the areas of life-cycle costing and fuzzy logic techniques for evaluating design concepts.